D1328199

Proceedings in Life Sciences

Comparative Studies
of Hearing in Vertebrates

Edited by
Arthur N. Popper and Richard R. Fay

With 223 Figures

Springer-Verlag
New York   Heidelberg   Berlin

Arthur N. Popper
Department of Anatomy
Schools of Medicine and Dentistry
Georgetown University
3900 Reservoir Road, N. W.
Washington, D. C. 20007, USA

Richard R. Fay
Department of Psychology and Parmly Hearing Institute
Loyola University of Chicago
6525 North Sheridan Road
Chicago, Illinois 60626, USA

Library of Congress Cataloging in Publication Data
Main entry under title:
Comparative studies of hearing in vertebrates
    (Proceedings in life sciences)
    Papers based on a workshop given at the joint
meeting of the Acoustical Societies of America and
Japan, held Nov. 28-Dec. 2, 1978 in Honolulu.
    1. Hearing—Congresses. 2. Vertebrates—
Physiology—Congresses. 3. Physiology, Comparative—
Congresses. I. Popper, Arthur N. II. Fay, Richard R.
III. Acoustical Society of America. IV. Nihon
Onkyō Gakkai.
QP460.C65   596'.01825   80-10320

9 8 7 6 5 4 3 2 1

ISBN 0-387-90460-3 Springer-Verlag New York Heidelberg Berlin
ISBN 3-540-90460-3 Springer-Verlag Berlin Heidelberg New York

This volume is dedicated to our wives,
Helen Popper and Catherine Fay,
in appreciation of their support, and to our teachers,
Professor William N. Tavolga and Professor Ernest G. Wever,
in appreciation of their contributions to the field of hearing
and of their guidance during the development of our careers.

Preface

The past two decades have seen an extraordinary growth of interest in the auditory mechanisms of a wide range of vertebrates and invertebrates. Investigations have ranged from auditory mechanisms in relatively simple animals where just a few cells are employed for detection of sound, to the highly complex detection and processing systems of man and the other mammals. Of particular significance to us has been the growing interest in general principles of vertebrate auditory system organization, as opposed to a specific and limited concern for the mammalian or even human systems. Some of the interest in nonmammalian systems has risen from the desire to find simpler experimental models for both the essential components (e.g., the hair cell receptor) and the more complex functions (e.g., frequency analysis) of all vertebrate auditory systems. Interest has also risen from questions about the evolution of hearing and the covariation (or lack of it) in structure and function in a wide variety of biological solutions to the problems of acoustic mechanoreception. Of course, the desire to find simpler experimental models and the need to answer questions about the evolution of hearing are not unrelated. In fact, detailed analyses of a variety of systems have led several times to the realization that some of the "simple systems" are more complex than initially thought. We have felt for some time that an exchange of ideas among workers investigating different vertebrate groups and workers taking different approaches to a single group would be of substantial mutual benefit to all involved. The product of such an interaction would permit a greater appreciation of diversity among the vertebrates, bring together in one place much of the data of comparative hearing, and, at the same time, encourage the growth of general ideas and underlying "themes" of vertebrate auditory mechanisms.

In 1978 in Honolulu, Hawaii, a joint meeting was held of the Acoustical Societies of American and Japan. This meeting brought together a group of investigators representing a range of disciplines and a range of vertebrate classes to discuss their own work and to hear and talk with people interested in parallel problems with different vertebrate groups. Since it is impossible to cover all aspects of vertebrate audition in one workshop or volume, we limited the scope of most of the contributions in this book, an outgrowth of that meeting, to critical reviews of the recent literature on peripheral and central anatomy, peripheral and central neurophysiology, and psycho-physics. In addition, we have emphasized the theme of sound localization in the dis-cussions of the fishes, birds, and mammals. While attempts were made to cover these topics evenly across the vertebrate classes, it was not always possible. Generally speak-ing, we have emphasized peripheral anatomy and physiology and have not dealt in as great detail with the central auditory systems (except for the neuroanatomy of the anamniotic vertebrates). In the case of both reptiles and amphibians, no proper psycho-physical data exist. The coverage of the mammals was most difficult since it was not possible to even begin to review the existing data in a satisfactory way in only a few chapters. In this case, therefore, specific topics of particular comparative interest were chosen.

The joint meeting of the Acoustical Societies of America and Japan was held Novem-ber 28 to December 2, 1978. We would like to thank Dr. John Burgess, chairman of the Honolulu meeting, and Dr. Joseph Hall, chairman of the Physiological and Psycho-logical Acoustics section of the Acoustical Society of America, for their help in organ-izing the workshop. We also thank all the participants in the workshop for their ef-forts. The success of the workshop and the timely completion of this volume is due to them. Finally, we thank Dr. Mark Licker, our editor at Springer-Verlag, for his guidance and assistance in the preparation of this volume.

Arthur N. Popper
Richard R. Fay

Contents

Part I
Fishes 1

Chapter 1
Structure and Function in Teleost Auditory Systems
R. R. Fay and *A. N. Popper* (With 14 Figures) 3

Chapter 2
Underwater Localization—A Major Problem in Fish Acoustics
A. Schuijf and *R. J. A. Buwalda* (With 8 Figures) 43

Chapter 3
Central Auditory Pathways in Anamniotic Vertebrates
R. G. Northcutt (With 12 Figures) 79

Part II
Amphibians 119

Chapter 4
The Structure of the Amphibian Auditory Periphery: A Unique
Experiment in Terrestrial Hearing
R. E. Lombard (With 10 Figures). 121

Chapter 5
Nonlinear Properties of the Peripheral Auditory System of Anurans
R. R. Capranica and *A. J. M. Moffat* (With 15 Figures) 139

Part III
Reptiles 167

Chapter 6
The Reptilian Cochlear Duct
M. R. Miller (With 46 Figures). 169

Chapter 7
Physiology and Bioacoustics in Reptiles
R. G. Turner (With 19 Figures) 205

Part IV
Birds 239

Chapter 8
Structure and Function of the Avian Ear
N. Saito (With 11 Figures) 241

Chapter 9
Behavior and Psychophysics of Hearing in Birds
R. J. Dooling (With 13 Figures). 261

Chapter 10
Sound Localization in Birds
E. I. Knudsen (With 15 Figures) 289

Chapter 11
Response Properties of Neurons in the Avian Auditory System: Comparisons
with Mammalian Homologues and Consideration of the Neural Encoding of
Complex Stimuli
M. B. Sachs, N. K. Woolf and *J. M. Sinnott* (With 21 Figures) 323

Part V
Mammals 355

Chapter 12
Directional Hearing in Terrestrial Mammals
G. Gourevitch (With 7 Figures) 357

Chapter 13
Comparative Organization of Mammalian Auditory Cortex
M. H. Goldstein, Jr. and *P. L. Knight* (With 13 Figures) 375

Chapter 14
Man as Mammal: Psychoacoustics
W. A. Yost (With 14 Figures) 399

Chapter 15
The Evolution of Hearing in the Mammals
W. C. Stebbins (With 5 Figures). 421

Part VI
Future View 437

Chapter 16
Comparative Audition: Where Do We Go from Here?
T. H. Bullock 439

Index 453

List of Contributors

THEODORE H. BULLOCK Neurobiology Unit, Scripps Institution of Oceanography and Department of Neurosciences, School of Medicine, University of California, San Diego, La Jolla, CA 92093, U.S.A.

ROBERT J. A. BUWALDA Laboratory of Comparative Physiology, State University of Utrecht, Jan van Galenstraat 40, Utrecht, The Netherlands

ROBERT R. CAPRANICA Section of Neurobiology and Behavior, Langmuir Laboratory, Cornell University, Ithaca, NY 14853, U.S.A.

ROBERT J. DOOLING The Rockefeller University, Field Research Station, Tyrrel Road, Millbrook, NY 12545, U.S.A.

RICHARD R. FAY Department of Psychology and Parmly Hearing Institute, Loyola University of Chicago, 6525 North Sheridan Road, Chicago, IL 60626, U.S.A.

MOISE H. GOLDSTEIN, JR. Department of Electrical Engineering and Department of Biomedical Engineering, Johns Hopkins University, 720 Rutland Ave., Baltimore, MD 21205, U.S.A.

GEORGE GOUREVITCH Department of Psychology, Hunter College of the City University of New York, 695 Park Ave., New York, NY 10021, U.S.A.

PAUL L. KNIGHT Health Care Technology Center, University of Missouri, Columbia, MO 65201, U.S.A.

ERIC I. KNUDSEN Department of Neurobiology, Stanford University School of Medicine, Stanford, CA 94305, U.S.A.

R. ERIC LOMBARD Department of Anatomy, The University of Chicago, 1025 East 57th St., Chicago, IL 60637, U.S.A.

MALCOLM R. MILLER Department of Anatomy, University of California, San Francisco, CA 94143, U.S.A.

ANNE J. M. MOFFAT Section of Neurobiology and Behavior, Langmuir Laboratory, Cornell University, Ithaca, NY 14853, U.S.A.

R. GLENN NORTHCUTT Division of Biological Sciences, University of Michigan, Ann Arbor, MI 48109, U.S.A.

ARTHUR N. POPPER Department of Anatomy, Schools of Medicine and Dentistry, Georgetown University, 3900 Reservoir Rd., N.W., Washington, D.C. 20007, U.S.A.

MURRAY B. SACHS Department of Biomedical Engineering, Johns Hopkins University School of Medicine, 506 Traylor Research Building, 720 Rutland Ave., Baltimore, MD 21205, U.S.A.

NOZOMU SAITO Department of Physiology, Dokkyo University School of Medicine, Mibu, Tochigi 321-02, Japan

ARIE SCHUIJF Laboratory of Comparative Physiology, State University of Utrecht, Jan van Galenstraat 40, Utrecht, The Netherlands

JOAN M. SINNOTT Department of Biomedical Engineering, Johns Hopkins University School of Medicine, 506 Traylor Research Building, 720 Rutland Ave., Baltimore, MD 21205, U.S.A.

WILLIAM C. STEBBINS Kresge Hearing Research Institute, University of Michigan, Ann Arbor, MI 48109, U.S.A.

ROBERT G. TURNER Otological Research Laboratory, Department of Otolaryngology, Henry Ford Hospital, 2799 West Grand Blvd., Detroit, MI 48202, U.S.A.

NIGEL K. WOOLF Department of Biomedical Engineering, Johns Hopkins University School of Medicine, 506 Traylor Research Building, 720 Rutland Ave., Baltimore, MD 21205, U.S.A.

WILLIAM A. YOST Parmly Hearing Institute, Loyola University of Chicago, 6525 North Sheridan Road, Chicago, IL 60626, U.S.A.

PART ONE

Fishes

Fishes represent, by far, the largest of the vertebrate classes. The more than 25,000 extant species show an extraordinary diversity in their ways of making a living. Over the past 15 years, we have seen a new interest in the teleost auditory system and the development of new data and ideas on the function of all aspects of teleost hearing. We are now learning, for example, that while there are some significant structural and functional similarities between the auditory system of fishes and terrestrial vertebrates (Fay and Popper, Chapter 1), there are a number of other apsects of audition, such as sound localization (Schuijf and Buwalda, Chapter 2), which must be considered in totally new ways. One area in which there has been a significant lack of data has been the central auditory neuroanatomy. Recently, however, new techniques and results have increased our understanding of anamniote neuroanatomy, and these data (North-cutt, Chapter 3) promise to be of substantial value for future studies of the auditory systems of fishes.

Chapter 1

Structure and Function in Teleost Auditory Systems

RICHARD R. FAY[*] and ARTHUR N. POPPER[**]

1 Introduction

In this chapter much of the recent literature on hearing in fishes has been brought together. First, the gross morphological and ultrastructural bases of sensitivity to the pressure and the motional components of underwater sound will be considered. This will be followed by a discussion of the behavioral and physiological literature on signal processing, particularly as it relates to the structure and function of the inner ear. The goal is to contribute to a greater understanding of the organizing principles of auditory processing by fishes, and by vertebrates in general, through emphasis on comparative issues and data. However, the central auditory system or the mechanism of localization, including the possible relationships between labyrinthine and lateral line function will not be considered since they are considered in other chapters (see Schuijf and Buwalda, Chapter 2; Bullock, Chapter 16; and Northcutt, Chapter 3).

Three general themes have emerged from the recent literature on fish hearing. The first is that considerable interspecific variation exists in both the gross and ultrastructural anatomy of the auditory periphery. While species-specific patterns of overall sensitivity and hearing bandwidth may be at least partially understood in these terms, the implications of this structural variation for more complex aspects of auditory processing, such as time and frequency analysis, are not yet clear. Second, the processing of various aspects of the acoustic waveform in time appears to be relatively more important for the teleost auditory system than analysis in the frequency domain. The preoccupation with peripheral frequency analysis in mammalian auditory theory has led to a number of conceptions (e.g., "the auditory filter"), which should only be applied to fish systems with the greatest caution and critical evaluation. Third, it is now clear that the otolithic ears of fishes are inherently directionally sensitive by virtue of rather complex hair cell orientation patterns. The potential value of these

*Department of Psychology and Parmly Hearing Institute, Loyola University of Chicago, 6525 North Sheridan Road, Chicago, Illinois 60626.
**Department of Anatomy, Schools of Medicine and Dentistry, Georgetown University, 3900 Reservoir Road, N. W., Washington, D. C. 20007.

directional patterns in extracting high quality information from the acoustic waveform is great, and they may play a role in signal detection in noise, sound source localization, and other aspects of complex information processing (Schuijf and Buwalda treat this in more detail in Chapter 2).

2 Anatomy of the Auditory Periphery

Sound detection by fishes involves an inner ear and, in many cases, peripheral structures that respond to sound and carry the acoustic energy to the inner ear. In the following sections the anatomy of these systems will be considered as a prelude to discussing functional significance in the system.

2.1 Inner Ear Morphology

The morphology of the fish inner ear is, in a number of ways, similar to that in other vertebrates (Fig. 1-1). The *pars superior* of the labyrinth consists of three semicircular canals and their associated ampullary regions as well as one of the otolithic organs, the utriculus. These structures are, in most species, involved with detection of the animal's orientation with respect to gravity and angular and linear accelerations (see Lowenstein 1971 for review). There is also some evidence that the utriculus may be an auditory structure in the clupeids, or herring-like fishes (O'Connell 1955, Denton and Blaxter 1976, Platt and Popper in press). The *pars inferior* includes two other otolithic organs, the sacculus and lagena, and is generally considered to be involved in audition. Each of the otolithic organs consists of a membranous wall, a sensory epithelium (or macula) that contains a large number of sensory and supporting cells, and a single, dense, calcareous otolith that lies in close contact with the sensory epithelium. The macula and otolith appear to be connected by a thin otolithic membrane.

One of the most striking aspects of the gross morphology of the teleost ear is the marked interspecific variability of the *pars inferior* (Retzius 1881). This variation is especially clear when comparing the ostariophysans (fishes having a series of bones, the Weberian ossicles, connecting the swimbladder to the inner ear), where the sacculus is elongate and the lagena large and round (Fig. 1-1A), with non-ostariophysans (Figs. 1-1B and C) where, in general, the lagena is small compared with the sacculus.

Figure 1-1. (Opposite) Gross morphology of the ears in several species of fishes. (A) An ostariophysan, *Cyprinus idus* (redrawn from Retzius 1881). Note the relatively large lagena and smaller sacculus. (B) *Zebrasoma veliferum*, an Acanthurid from Hawaiian reefs. (C) *Salmo salar*, the Atlantic salmon (redrawn from Retzius 1881). (D) A presumably primitive chondrostean fish, *Polypterus bichir* (from Popper 1978b). Letter designations within the figures are as follows: A-anterior semi-circular canal, CC-crus commune; H-horizontal semi-circular canal; L-lagena; LM-macula; LO-lagenar otolith; S-sacculus; SI-sinus impar; SM-saccular macula; SO-saccular otolith; M-macula; O-otolith; U-utriculus; UO-utricular otolith.

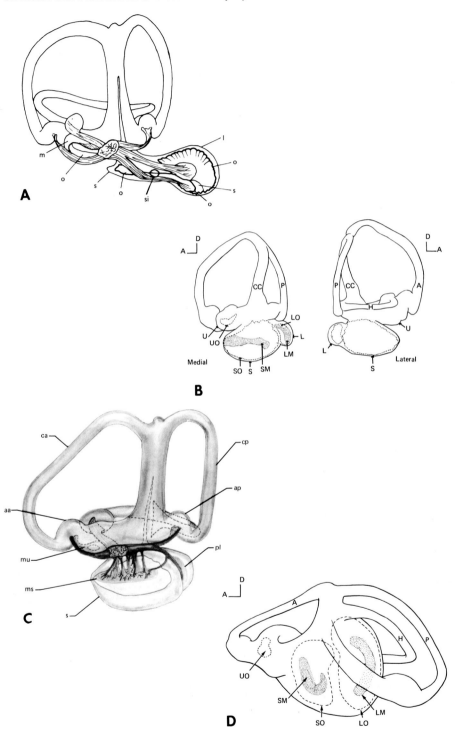

This pattern is further contrasted with that in several primitive fishes, such as the bichir (*Polypterus bichir*) (Fig. 1-1D) and shovel-nosed sturgeon (*Scaphirhynuchus platorynchus*), where both the sacculus and lagena are large and in a single chamber.

In addition to the variation in general shape of the regions of the *pars inferior,* there is also substantial interspecific variation in specific features of the sacculus and lagena that may play an as yet undefined role in signal detection and processing. This variation includes the connections between the sacculus and lagena, the general shapes of the two organs, and the size and shape of the otolith.

The interconnections between the sacculus and lagena are of three primary types. The ostariophysans have an opening in the medial wall of the saccular chamber that leads directly into the lagena. In many non-ostariophysans, such as *Zebrasoma* (Fig. 1-1B), the lagena lies on the dorsal-posterior margin of the saccular chamber with there being only a small opening between the two chambers. In still other species, such as the salmonids (Fig. 1-1C) and several chondrosteans (Fig. 1-1D), the sacculus and lagena lie in a single chamber (Popper 1976, 1977, 1978a).

The otolithic material in fishes consists of a single structure (see Fig. 1-2) that has a density about three times that of water. The presence of a single otolith is unique to teleost fishes; other vertebrates generally have a large number of small crystals, otoconia or statoconia, embedded in a gelatinous matrix (Carlstrom 1963). The saccular otoliths in fishes have distinct and complex species-specific shapes (Fig. 1-2), while there is considerably less complexity, and less interspecific variation, in the lagenar otoliths (Popper 1977).

2.2 Sensory Epithelia

The otolithic organs contain, in addition to the otoliths, a sensory epithelium, or macula. A groove, or sulcus, on one side of the otolith lies close to the sensory epithelium (Fig. 1-2) but is separated from it by a thin otolith membrane (Werner 1960). Several recent scanning electron microscopic studies have shown the otolith membrane to contain numerous holes into which the ciliary bundles of the sensory hair cells appear to fit (Dale 1976, Popper 1977, Jenkins 1979). The function of the otolith membrane has not yet been explored in any detail, but several investigators have suggested that it keeps the otolith in place near the sensory epithelium by providing a means of attachment between the microvilli on the supporting cells surrounding the sensory cells and the rough surface of the otolith (Dale 1976, Popper 1977, Jenkins 1979).

Several points should be made regarding the relationship between the otoliths and sensory epithelia, since they may affect the relational movements between the two structures and thus the stimulation of the sensory hair cells. First, the extent of the otolithic chambers filled by the otoliths, particularly of the *pars inferior,* varies considerably among different species. The range of variation extends from deep-sea lantern fishes (fam. Myctophidae), where the otolith may fill less than one-quarter of the saccular chamber and cover only the posterior half of the sensory epithelia (Popper 1977), to the acanthurids (Fig. 1-1B) and salmonids (Fig. 1-1C) in which the otoliths fill the chambers and cover more than 90% of the maculae (e.g., Popper 1978a, 1978b). The second point to be made is that the otoliths in both the sacculus and lagena do

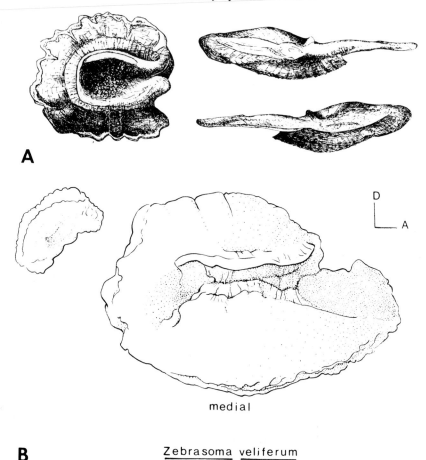

A

D
└ A

medial

B   Zebrasoma veliferum

Figure 1-2. Saccular (right) and lagenar otoliths. (A) An ostariophysan, *Phoxinus laevis;* top is lateral view, bottom is medial view. Note the fluting on the saccular otolith and the deep groove, or sulcus, on the lagenar otolith (from Wohlfahrt 1939). (B) Medial view of the otoliths from *Zebrasoma*. The deeper sulcus is on the saccular otolith in non-ostariophysans.

not necessarily cover the whole sensory epithelium, although in virtually all species the otolith membrane extends out from the otolith to cover the entire sensory epithelium. The extent of sensory epithelium uncovered by the saccular otolith may be from 10% to 50% in different species. A small region on the dorsal tip of the lagenar macula is uncovered in all species studied.

There is also substantial interspecific variation in the shapes of the sensory maculae in the *pars inferior* and particularly in the sacculus (Fig. 1-3). In ostariophysans, the saccular macula is relatively narrow and elongate (Fig. 1-3A), while in non-ostariophysan teleosts it is broader at the rostral than the caudal end (Figs. 1-3B and C). The pattern is also variable in primitive fishes such as several chondrosteans (Fig. 1-3D).

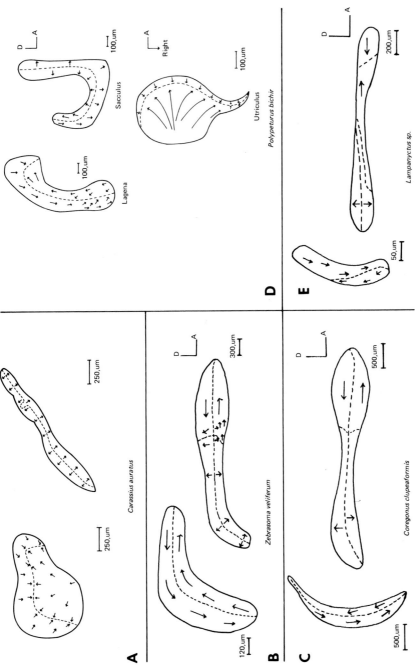

Figure 1-3. Hair cell orientation patterns in the sacculus (right) and lagena from a number of different fishes. The tips of the arrows in each figure indicate the side of the sensory cells in that region on which the kinocilium is found. (A) The ostariophysan, *Carassius auratus* (goldfish) (redrawn from Platt 1977); (B) *Zebrasoma valiferum*, (C) *Coregonus clupeaformis* (or lake whitefish), (D) The bichir or reed fish, *Polypterus bichir* (from Popper 1978b); (E) The myctophid (lantern-fish) *Lampanyctus* (redrawn from Popper 1977).

The lagenar macula is substantially less variable in shape than that of the sacculus. The only major differences, other than those seen in the primitive fishes (Fig. 1-3D), are found in comparing the non-ostariophysans (Figs. 1-3B and C) with ostariophysans (Fig. 1-3A).

The sensory maculae contain sensory hair cells that are surrounded by supporting cells. The number of sensory cells on a typical teleost saccular macula may number ten thousand or more, and it may even exceed the number of sensory cells in the auditory regions of any tetrapod ear. Further, there is some evidence that the number of sensory cells varies between species (see Platt 1977, Popper 1978a) and, more significantly, Platt (1977) has shown that in the goldfish (*Carassius auratus*), the number of sensory cells increases with the size of the animal. There is also some evidence that the density of sensory cells varies in different regions of the same macula (Platt 1977).

The sensory hair cells have an apical ciliary bundle that consists of a single, eccentrically positioned kinocilium and a larger number (40 to 70 in different species) of stereocilia. The stereocilia are usually graded in size, with the longest being found next to the kinocilium. The supporting cells are covered with short microvilli (Fig. 1-4A) that may attach to the otolith membrane (Dale 1976, Popper 1977).

The patterns of the ciliary bundles may vary in different regions of the sensory maculae and between different teleost species. The names for the ciliary bundles are as yet tentative, and different authors have given various names to what appear to be similar bundles. For the sake of convenience, terms used by Popper (1977, 1978b) will be used to refer to bundles. Note that similar, though not necessarily the same, types of bundles have been seen in other vertebrate classes (e.g., Lewis and Li 1975).

The most ubiquitous of the ciliary bundles found in fishes is the F1 bundle (Fig. 1-5A), which contains a series of stereocilia and a single kinocilium one to two times the length of the longest stereocilia. The F1 bundle is found over the bulk of the saccular macular in many ostariophysans and non-ostariophysans as well as in chondrosteans (Platt 1977, Popper 1976, 1977, 1978a, 1979). The F2 bundle contains short, and frequently nongraded, stereocilia and a long kinocilium (Fig. 1-4B). This type of ciliary bundle is most often found on the very margin of the saccular and lagenar maculae and may consist of one or a few cell rows (Fig. 1-4C). The F3 ciliary bundle (Fig. 1-4D) is the most difficult to define morphologically. It looks very much like the F1 bundle except that all of the cilia are very long. The F3 bundle has been found in epithelial regions not covered by the otolith in salmonids, over the whole lagenar maculae in holocentrids (but not in many other non-ostariophysans) (Popper 1977), over the complete saccular macula in myctophids (Popper 1977), and completely covering the saccular, lagenar, and utricular maculae in several herring-like fishes of the family Clupeidae (Platt and Popper 1979).

The functional significance of the different ciliary bundle types is yet unknown, although some physiological evidence has led to the suggestion that longer ciliary bundles are associated with vibrational senses while shorter bundles may be involved with audition (see Lowenstein et al. 1964, Platt 1977). Bundles similar to the F2 bundles of fishes are also found in sharks and may be in regions of macular growth (Corwin 1977).

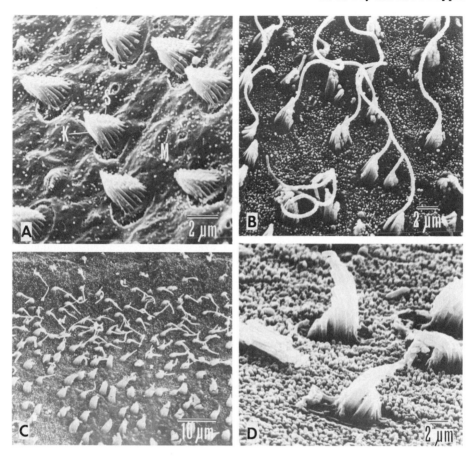

Figure 1-4. Sensory hair cells from teleost fishes. (A) Type I ciliary bundles with the kinocilium just slightly longer than the longest stereocilia. (B) Type II ciliary bundles having very long kinocilia. (C) Border of the saccular macula showing Type II ciliary bundles on the macula margin grading into type I bundles more medially. (D) Type III ciliary bundles with longer cilia than the type I. Letter designations within the figures are as follows: K-kinocilium; M-microvilli; S-stereocilia (from Popper 1977).

2.3 Innervation

While not extensive, data for several teleosts including *Carassius auratus* (Hama 1969), *Lota lota* (the burbot), and the moray eel, *Gymnothorax* (Popper 1979), suggest that the saccular macula is innervated by both afferent and efferent fibers of the eighth nerve. Studies of the three otolithic organs of the holostean, *Amia calva* (Popper and Northcutt unpublished), and of the utriculus of *Lota lota* (Flock 1964), indicate that there are many more sensory cells than innervating nerve fibers. However, the ratio between afferent and efferent fibers is not known. A similar pattern has been found in elasmobranches by Corwin (1977), who suggests that the large number of sensory cells synapsing on fewer nerve fibers may provide a mechanism for enhancement of detection or signal averaging in the ear (also see Bullock, Chapter 16). More central projections of the 8th nerve are largely unknown, though recent studies by McCormick (1978) and by Northcutt (Chapter 3) have shown that there are several medullary nuclei associated with the 8th nerve, as well as projections to mesencephalic, diencephalic, and telencephalic regions in several fish species (also see Grozinger 1967, Page 1970, Piddington 1971, Maler, Karten, and Bennett 1973a,b, Knudsen 1977).

2.4 Hair Cell Orientation

Of particular interest in recent studies of the ultrastructure of the sensory regions of the ears of fishes has been the orientation patterns of the sensory hair cells. In fishes, as in most other vertebrates (e.g., Miller, Chapter 6), the sensory cells on the various otic maculae are organized into groups, each containing hair cells having their kinocilium located on the same side of the cell (Figs. 1-3, 1-4A and D).

While data are still somewhat limited, it is beginning to appear that hair cell orientation patterns, particularly on the sacculus and lagena, can be divided into two major types based on taxonomic relationships among fishes and into several lesser types of as yet unknown taxonomic or functional significance. One of these types is found in the Ostariophysi while the other is in the non-ostariophysan teleosts. Patterns in primitive fishes differ somewhat from teleost patterns.

The hair cell orientation pattern in the ostariophysan sacculus has some similarity to the patterns found in the tetrapod sacculus (Lindeman 1969, Lewis and Li 1975, Miller, Chapter 6). The saccular macula in ostariophysans contains two hair cell orientation groups (e.g., Fig. 1-3A) with dorsally oriented cells on the dorsal side of the macula and ventrally oriented cells on the ventral side. While only 6 or 7 species of ostariophysans have been studied to date (Hama 1969, Platt 1977, Jenkins 1979; Popper unpublished), this basic pattern appears to exist throughout the ostariophysan taxa. The only variation in this pattern has been reported in several species of catfish (order Siluiformes) where Jenkins (1979) has found some alteration of dorsally and ventrally oriented cells at the anterior end of the saccular macula. The orientation pattern on the ostariophysan lagena is only known for the goldfish (Platt 1977) and one catfish, *Arius felis* (Popper unpublished). Both species have two oppositely orientated groups of hair cells (Fig. 1-3A) as in the sacculus. However, in the lagena, the considerable curvature of the macula results in a wider range of presumed "best directions" of stimulation. (See, however, physiological data on directional sensitivity in Section 6).

The second basic saccular orientation pattern is found in the non-ostariophysan teleosts where, in addition to having dorsally and ventrally oriented cells on the caudal region of the macula, there are also horizontally oriented cells on the rostral portion (Figs. 1-3B, C, and E). Horizontally oriented cell groups are found in almost all of the major non-ostariophysan taxa (e.g., Dale 1976, Enger 1976, Jorgensen 1976, Popper 1976, 1977, 1978a, 1978b) with the exception of a mormyrid (Popper unpublished). In general, the basic pattern for the horizontal cell groups is to have the posteriorly oriented cells located dorsal to the anterior cell group (Figs. 1-3A and B). However, there is substantial interspecific variation in this pattern in a wide range of teleosts.

In some cases, one of the cell groups will be located rostrad to the others, as in mycto-phids (Fig. 1-3E) and several holocentrids (Popper 1977), while in the moray eel and several relatives, the two groups alternate as to which is dorsal to the other (Popper 1979 unpublished). There is substantially less variation involving the vertically oriented groups in non-ostariophysans, although a number of species including the cod, *Gadus morhua* (Dale 1976), and several gobies (Popper unpublished) have horizontally oriented cells in the ventro-caudad macula region.

The non-ostariophysan lagenar pattern, with a few exceptions, does not vary sub-stantially among different species. In most species investigated so far, there is a group of dorsally oriented cells rostrad to ventrally oriented cells (Figs. 1-3B and C). A signif-icant exception to this pattern is seen myctophids that have a very small lagenar macu-la (100 to 200 hair cells), and where the pattern seen in other fishes is reversed (see Fig. 1-3E). As in the ostariophysans, the cells on the non-ostariophysan lagena are actually oriented in a wide range of directions as a result of macula curvature.

Totally different patterns than those in teleosts are found in several non-teleosts, including chondrosteans (Popper 1978b) and the holostean, *Amia calva* (Popper and Northcutt unpublished). Each of these species has two major orientation groups on the saccular macula, with cells actually being oriented in four directions due to macula curvature (Fig. 1-3D). This differs from the pattern in non-ostariophysans where there is a distinct separation between the four hair cell orientation groups. The lagenar orien-tation pattern in these species also is somewhat different from non-ostariophysans (Fig. 1-3D) although no single pattern is yet apparent.

2.5 Auxiliary Auditory Structures

In addition to the inner ear, many species of fish have one or two additional structures that are associated with audition. The best known of these is the gas-filled swimblad-der, which is located in the abdominal cavity, ventral to the vertebral column (Fig. 1-5A). The swimbladder also functions as a hydrostatic organ (Steen 1970) and, in a more limited number of species, in sound production (Tavolga 1971, Demski, Gerald, and Popper 1973).

The physical relationship between the swimbladder and the inner ear most likely plays a considerable role in determining the hearing sensitivity, and possibly hearing range, of different teleosts. A wide range of adaptations are found among teleosts for improving coupling between the swimbladder and the ear. In the clupeids (herring-like fishes), the swimbladder has a thin anterior tube leading to an expanded gas-filled chamber that terminates directly in the auditory bulla close to the utriculus (O'Connell 1955, Denton and Blaxter 1976, Blaxter and Tytler 1978). The mormyrids (elephant-nose fishes) have a small swimbladder, presumably broken off from the main swim bladder during development (Stipetic 1939) within a region surrounded by the semi-circular canals, while the anabantids, or bubble-nest builders, maintain a bubble of air in the buccal cavity, which early experiments showed (Schneider 1941) improves hearing.

These adaptations have led to the suggestion that the intimacy of the swimbladder and ear would affect auditory sensitivity, and supporting data are found in a recent

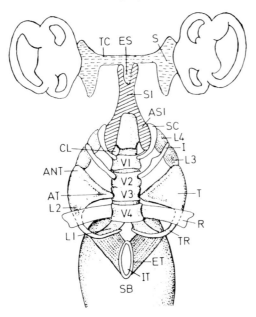

Figure 1-5. Schematic dorsal view of the inner ear, Weberian ossicles, and swim bladder in an ostariophysan fish, demonstrating the connections between the three organs. Letter designations within the figures are as follows: Ant-anterior process of tripus; ASI-atrium sinus impar; AT-articular process of tripus; CL-claustrum; ES-endolymphatic sac; ET-tunica externa of the swim bladder; I-intercalarium; IT-tunica interan of swim bladder, L1, L2, L3, L4-ligaments connecting Weberian ossicles to one another; R-ribs; S-sacculus of inner ear; SB-swimbladder; SC-scaphium; SI-sinus impar; T-Tripus; TC-transverse canal; TR-transformator process of tripus; V1, V2, V3, V4-first four vertebrate. Redrawn from Popper (1971) and Chranilov (1927) (from Henson 1974).

study of the family Holocentridae (squirrelfish). One holocentrid group has a swimbladder that terminates some distance from the ear (Figs. 1-6A, B, and C), another group has the swimbladder terminating close to the ear (Figs. 1-6D, E, and F), while a third group has a swimbladder that terminates on the auditory bulla (Figs. 1-6G, E, and H) (Nelson 1955). Behavioral data (see Fig. 1-7) show that a species of holocentrid with specialized connections between the swimbladder and inner ear, *Myripristis kuntee,* has substantially better sensitivity and range of hearing than *Adioryx xantherythrus,* a species with no such intimacy (Coombs and Popper 1979). It is important to note, however, that species differences in the frequency range of hearing cannot simply be attributed to structures peripheral to the ear. For example, Fay and Popper (1974, 1975) studied microphonic responses from three species (*Tilapia macrocephala, Ictalurus nebulosus,* and *Carassius auratus*) to direct vibratory stimulation, thereby bypassing the swimbladder and ossicular system, and showed that the species variation in hearing bandwidth is fully reflected in the response of the inner ears themselves.

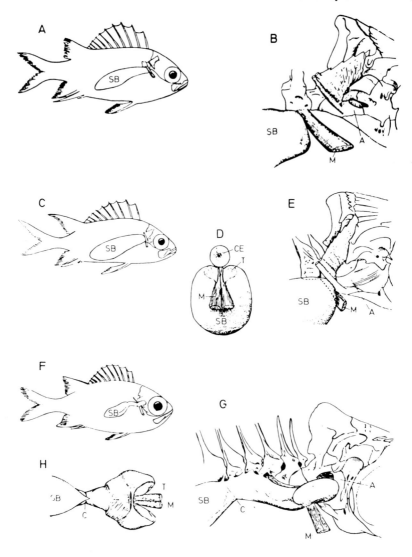

Figure 1-6. Swimbladder-inner ear arrangements characteristic of three groups of squirrelfish: (A and B) subfamily Holocentrinae, genera *Adioryx* and *Flammeo;* (C, D, and E) subfamily Holocentrinae, genus *Holocentrus;* (F, G, and H) subfamily Myripristinae, several genera including *Myripristis.* (A), (C), and (F) show the position and relative size of the swimbladder *in situ.* Lateral view in (B), (E), and (G) delineate the relationship between the anterior portion of the swimbladder and the posterior auditory region of the skull. The anterior end of the swimbladder is depicted in anterior (D) and ventral (H) views. Letter designations within the figures are as follows: A-auditory region of the skull; C-constriction between anterior and posterior chambers of swimbladder; CE-centrum of second vertebra; M-retractor muscle of upper pharyngeal jaws; SB-swimbladder; T-thinned portion of swimbladder membrane where it meets the auditory area of the skull. From Nelson (1955), courtesy Field Museum of Natural History.

While the swimbladder in general enhances sound detection, its precise role is not clear. A number of recent experiments have shown that the swimbladder may amplify signals (Chapman and Sand 1974) and that the swimbladder response is flat within the range of hearing of several species (Sand and Hawkins 1973, Popper 1974). Other evidence leads to the suggestion that the swimbladder may have differential responses in different regions along their lengths (Vaitulevich and Ushakov 1974, Clarke, Popper, and Mann 1975) and may show a preferential response to sound from different directions (Tavolga 1977).

Perhaps the most unique of the teleost adaptations for audition are the Weberian ossicles (Fig. 1-5), a series of bones found only in the Ostariophysi, which connect the swimbladder to the inner ear. Presumably, movements of the swimbladder walls induce motion in the Weberian ossicles, which in turn cause fluid movements of the inner ear fluids (Chranilov 1927). Fluid motion in the ear "catch" the fluted otoliths (Fig. 1-2A), resulting in a shearing action on the sensory hair cells (von Frisch 1938). It has been widely suggested that the presence of the Weberian ossicles as a coupling mechanism enhances hearing sensitivity and bandwidth, and behavioral data (see Section 4) support this argument. However, it is questionable whether the Weberian ossicles provide any better acoustic coupling than occurs when the swimbladder terminates intimately with the otolithic bulla as in *Myripristis* (Coombs and Popper 1979) (Fig. 1-7). Of course, it is also possible that the Weberian ossicles have other functions in addition to coupling, such as filtering or amplifying certain signals (e.g., Alexander 1962, von Bergeijk 1967a). Direct experimental verification of the function of the Weberian ossicles is limited to a single study by Poggendorf (1952) which showed that breaking the ossicular chain in a catfish (*Amiurus*) caused a rather flat 35 dB to 40 dB loss of sound pressure sensitivity.

3 Inner Ear Stimulation

There are three major areas of inquiry regarding inner ear stimulation including (a) the general mode of stimulation of the hair cells, (b) the pathways of sound to the ear, and (c) the types of analyses that occur in the ear due to the interactions between the hair cells and the otoliths. In the case of the first and third areas, few data are available directly related to inner ear function in fishes. Most speculation is based on studies of the anatomy and ultrastructure of the fish ear and knowledge of what occurs in the lateral line organs and in the ears of mammals. Investigations have provided some insight into the second question, the pathways of sound to the ear, and it now appears that such pathways are somewhat more complex than previously thought.

It is generally agreed that the relevant stimulus for a sensory hair cell is a shearing action that causes bending of the ciliary bundle (see von Békésy 1960, Flock 1971, 1977). In the teleost ear, the otolith lies in close proximity to the sensory cells and, presumably, provides a shearing stimulus on the ciliary bundle (Pumphrey, 1950, Dijkgraaf 1960). Two mechanisms by which sound may cause this relative shearing have been recently studied. In several ostariophysans (Fay and Popper 1974, 1975), the cod, *Gadus morhua* (Sand and Enger 1973), and probably in a wide range of other species where the swimbladder is used in audition, motions of the swimbladder wall

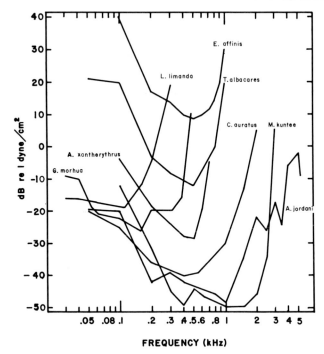

Figure 1-7. Behavioral sound pressure audiograms for eight teleost species, illustrating the wide variation in absolute sensitivity, bandwidth, and the upper frequency limit of hearing. Data for *Gadus morhua* from Chapman and Hawkins (1973), *Limanda limanda* from Chapman and Sand (1974), *Euthymus affinis* from Iversen (1969), *Thunnus albacares* from Iversen (1967), *Adioryx xantherythrus* and *Myripristis kuntee* from Coombs and Popper (1979), *Astyanax jordani* from Popper (1970), and *Carassius auratus* from a composite of several audiograms taken from Fay (1978a).

occurring in response to sound pressure fluctuations are conducted to the inner ear. As described earlier, this motion is coupled to the otolith and the relative motion between the otolith and the apical end of the hair cells causes deformation of the cilia. A second, more general pathway probably operates in all fishes (but not at all frequencies in all species) and involves direct stimulation of the ear by the impinging sound field (Wever 1969, 1971). In this system, the fish's body, which is about the same density as that of water, moves with the impinging sound field, while the far denser otoliths may move at a different amplitude and phase, resulting in relative motion between the sensory maculae and the otoliths. This inertial mode of stimulation would be expected to exert a similar effect on all otolithic organs of the ear. Thus, a distinction between an auditory and vestibular organ becomes difficult in some cases and may depend on the kind of neural information processing that is associated with the input from each organ. Whether an organ would actually respond in this way to sound may possibly be determined by the nature of the ciliary attachment of the otolith (which may vary in different parts of the same sensory macula), as well as the coupling of the otolith to the macula. The swimbladder route of sound to the ear is thought to be relatively more efficient at the higher frequencies (above 200 Hz or so) and to cause stimulation in proportion to the sound pressure level of the impinging stimulus. The tissue conduction route, on the other hand, is more important at the lower frequencies, where particle motion is greater at a given sound pressure level and where the near field extends greater distances. In this case, auditory stimulation occurs in proportion to a vectorial component of the impinging sound (e.g., displacement) which contains directional information.

The precise patterns of movement between the otolith and sensory macula is unknown as yet, although there are very limited data leading to the suggestion that the response pattern is more complex than a simple sliding action between the two tissues (Sand and Michelsen 1978, Fay and Olsho 1979). A number of factors would, more than likely, affect such motion and the way in which hair cells on different macular regions are stimulated, thus leading to the suggestion that different macula regions may be stimulated by different signals according to a "place" principle (see van Bergeijk 1967b, Sand 1974, Popper and Clarke 1976, Sand and Michelsen 1978, Fay and Olsho 1979). This spatial transformation could be affected by the shape of the otoliths and maculae, the drag on the otoliths during stimulation, and the connections between otolith and macula. In addition, the interspecific variation in each structure (Section 2.3) leads to the suggestion that different species may perform different types of sound analysis, or alternatively, that a variety of different mechanisms are found within fishes to ultimately do the same type of signal analysis.

4 Hearing Sensitivity and Bandwidth

Audiograms comprise, by far, the greatest amount of comparative data we have on the function of fish auditory systems. In recent years, most of these data have been reviewed and presented several times (Popper and Fay 1973, Tavolga 1976, Fay 1978a, Popper 1980) and will not be presented in detail again here. Instead, audiograms will be reviewed for several selected species and present understanding of those factors that determine the sensitivity and hearing bandwidth for any given species will be discussed.

Audiograms for eight species are plotted together in Fig. 1-7. In general, the species represented range from some of the least (*Euthynnus affinis*) to most (*Myripristis kuntee*) sensitive tested, and from some with the narrowest (*Limanda limanda*) to the widest (*Astyanax jordani*) bandwidths tested. Sensitivity at best frequency ranges over 60 dB and the frequency above which sensitivity declines ranges over four octaves. There are many possible sources of variation in the determination of these functions other than species differences, however. The most important of these are the level of ambient noise, and whether or not the systems studied responded to the pressure (scalar) or a motional (vector) component of the stimulus.

One nearly universal feature of sound pressure audiograms for fishes is the decrease in sensitivity below several hundred Hz, which parallels the shape of typical ambient noise spectra (e.g., Wenz 1964) and suggests that the low frequency portions of most of these audiograms are masked. Apparent variation in low frequency sensitivity between species may thus be due to different ambient noise levels existing in different experimental set-ups. Note that only the study of Chapman and Hawkins (1973) on *Gadus morhua* confirmed that this was not the case for their thresholds (see Fig. 1-7). It is also possible that this parallel between sensitivity and ambient noise spectra represents a real feature of auditory systems evolved in a noise environment. It would make little biological sense, for example, to extend hearing sensitivity to a point very far below the limits of the usual ambient noise levels. In any case, the region of declining sensitivity toward the higher frequencies is probably not determined by ambient noise levels and is more certainly a species characteristic.

Sound pressure audiograms such as those in Fig. 1-7 are valid descriptions of auditory sensitivity only for systems that are pressure sensitive. Some species, such as the flatfish, *Limanda limanda*, lack a swimbladder and have been shown to respond to particle motion and not to sound pressure (Chapman and Sand 1974). The ratio of pressure to particle velocity (impedance) may vary at different points in a sound field due to near field effects and standing waves. The differences in sound pressure sensitivity between *Limanda limanda* and a tuna, *Euthynnus affinis*, (also lacking a swimbladder) may thus be due to the impedance characteristics of the sound fields in which they were tested. Unfortunately, without extensive impedance measurements these differences remain uninterpretable. This same problem arises for characterizing the sensitivity of species such as the goldfish, *Carassius auratus*, which have been shown to be pressure sensitive at the higher frequencies but displacement or velocity sensitive at lower frequencies (Enger 1966, Fay and Popper 1974, 1975) when tested in small tanks in the laboratory. This dual sensitivity probably exists for all species possessing a swimbladder but is practically impossible to analyze quantitatively in small airbounded tanks in which the acoustic impedance is inevitably low and approaches that of air (Parvulescu 1964). Progress in the comparative study of low frequency sensitivity thus seems most likely to occur in situations where impedance can be measured and manipulated, such as in large open fields (e.g., Chapman and Hawkins 1973) in well controlled standing waves (Cahn, Siler, and Wodinsky 1969, Fay and Popper 1974, Hawkins and MacLennan 1976), or in using direct vibration of the animal's head (e.g., Sand 1974, Fay and Popper 1975, Fay and Olsho 1979) in both physiological and behavioral studies (Fay and Patricoski 1979). The latter technique is most promising in that the direction of motion can be precisely controlled and measured.

At frequencies above 100 Hz to 200 Hz, measures of sound pressure sensitivity probably become more valid for several reasons. Typical ambient noise spectra continue to roll off, particle displacement amplitudes roll off for equivalent sound pressure levels, and the extent of the near field shrinks in proportion to wavelength. As indicated above, the data for *Limanda limanda* and *Euthynnus affinis* (Fig. 1-7) are probably not meaningfully plotted in units of sound pressure since both lack a swimbladder and are mostly likely responding to particle motion throughout their range. The curves for *Thunnes albacares* (a species of tuna with a swimbladder), *Gadus morhua*, and *Adioryx xantherythrus* are typical for fishes with swimbladders but without special connections to the inner ears. *Gadus* was shown in a free field situation to be pressure sensitive down to 50 Hz (Chapman and Hawkins 1973), and it is likely that at least the high frequency portions of the functions for *Thunnus albacares* and *Adioryx xantherythurs* represent valid sound pressure thresholds. The curves for the goldfish, another squirrelfish, *Myripristis kuntee,* and for the blind Mexican cave fish, *Astyanax jordani,* are typical for fishes with swimbladders and with special connections to the ears. The goldfish and *Astyanax jordani* are both Ostariophysi, having Weberian ossicles, while *Myripristis kuntee* has anterior projections of the swimbladder ending close to the auditory bulla and saccule (see Section 2.5). Note here that *Myripristis kuntee* and *Adioryx xantherythrus* were tested in the same sound field using identical methods (Coombs and Popper 1979). These differences in hearing sensitivity and bandwidth are the best behavioral demonstration to date illustrating the difference between specialized and apparently unspecialized routes of sound conduction between the swimbladder and the ear.

While it is clear that species differ considerably in hearing sensitivity (60 dB) and bandwidth (4 octaves), and that these differences appear to be brought about by variation in the mechanical response of the ear and more peripheral structures, clear correlations with other aspects of behavior and inner ear ultrastructure are not apparent. In fact, several species that are well known as sound producers, such as *Adioryx xantherythrus* (Salmon 1967) and *Opsanus tau* (see Fine, Winn, and Olla 1978) do not appear to be acoustically specialized and hear quite poorly (Fish and Offutt 1972, Coombs and Popper 1979). The lesson here may be simply that the adaptive significance of teleost hearing in general is not well understood and that a search for specialized sound detection-production relationships may take us away from rather than toward a general understanding of the use of sound by fishes.

5 Auditory Processing

Interest in functional analysis of auditory processing by fishes has arisen less from questions concerning intrafamily or intraclass comparison than from questions of the rather gross relationships between structure and function within the vertebrates as a whole. The early work of von Frisch (1938), Stetter (1929), Dijkgraaf and Verheijen (1950) and others on capacities for frequency discrimination, for example, was motivated in large part by the simple observation that fishes appear to lack a basilar membrane and other biomechanical mechanisms for a "place" analysis of frequency in the inner ear. A demonstration that frequency discrimination was possible thus provided evidence for the operation of a "volley"-like principle in which temporal rather than spatial neural patterns could form a code for a sensory quality possibly analogous to "pitch." Furthermore, the values for the just-noticeable differences for frequency in fishes could help to define an "existence region" for pitch-like phenomena mediated entirely by a temporal code.

5.1 Frequency Discrimination

Figure 1-8 presents frequency discrimination limens for several teleost species compared with the range of values for a number of mammalian and avian species, excluding man. The data suggest that *Carassius auratus* and *Phoxinus laevis* (both Ostariophysi) are more sensitive to frequency differences than several non-ostariophysans.

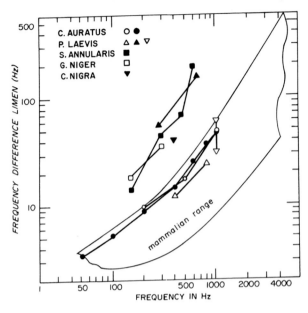

Figure 1-8. Auditory frequency discrimination thresholds for fishes compared with those for certain mammals, excluding man (Fay 1974c). Filled circles, *Carassius auratus* (Fay 1970a); open circles, *Carassius auratus* (Jacobs and Tavolga 1968); inverted open triangles, *Phoxinus laevis* (Wohlfahrt 1939); filled triangles, *Phoxinus laevis* (Stetter 1929); open triangles, *Phoxinus laevis* (Dijkgraaf and Verheijen 1950); filled squares, *Sargus annularis;* open squares, *Gobius niger,* inverted filled triangles, *Corvina nigra* (Dijkgraaf 1950).

While this difference may be correlated with differences in inner ear structure and with overall hearing sensitivity and bandwidth, a likely hypothesis for its physiological basis has not been forthcoming. Before this difference can be taken more seriously, more comparative studies should be done, preferably using identical techniques and signals at comparable levels above threshold. The major point of Fig. 1-8 is that the ostariophysans, at least, are not unusual among vertebrates in their frequency discrimination capacities. A tentative conclusion here is that the mechanisms for frequency discrimination may be the same for all vertebrates, at least for frequencies below about 1 kHz (Fay 1973). However, while many have assumed that this mechanism is based on temporal rather than spatial neural codes, critical evidence has been lacking. Dudok van Heel (1956) provided indirect evidence for this idea by showing that the upper frequency at which frequency discrimination could be made was extended upward by a rise in ambient temperature. Since the upper frequency limit for a neural frequency following response was previously shown to rise with temperature (Adrian, Craik, and Sturdy 1938), Dudok van Heel concluded that discriminations were based on a frequency-following coding principle.

Behavioral data suggesting a temporal analysis of frequency have recently received support from a physiological study by Fay (1978b) showing that just discriminable differences in stimulus period were approximately equal to the standard deviations of interspike-interval distributions of a phase-locked response. These data lead to the suggestion that frequency discrimination in *Carassius auratus* depends on the temporal accuracy with which spikes are phase-locked in saccular neurons. These results are consistent with a simple signal detection theory model of discrimination behavior stating that frequency discrimination decisions are based on estimates of the temporal intervals between neural spikes and are limited by the accuracy with which these intervals are represented in the nervous system.

5.2 Simultaneous Frequency Analysis and Masking

While the evidence cited indicates that frequency discrimination behavior may be based on temporal neural patterns, recent results from a variety of experiments have suggested that at least some aspects of simultaneous frequency analysis are based on a spatial or "place" representation of frequency at the periphery.

5.2.1 Physiological Studies

Sand (1971) and Enger (1973) studied the masking of signals by noise in medullary neurons of goldfish and found that masking was an inverse function of the frequency separation between signal and masker, illustrating the frequency selectivity of central neurons. Sand (1974) and Fay and Olsho (1979) have also demonstrated that the relative response of the goldfish saccule to stimulation in different directions is to some degree frequency dependent. This is an indication that the direction of otolith movement, and thus the makeup of the population of active neurons, may be frequency dependent. Other evidence for this notion comes from attempts to measure saccular otolith movement in *Perca fluviatilis* (Sand and Michelsen 1978). Although fragmentary, the results could be interpreted as an indication that the otolith develops complex rotational movements around a horizontal axis, the position of which shows a slight frequency dependence. Finally, Popper and Clarke (1976) investigated the frequency-dependent fatigue effects of intense tones on subsequent tonal thresholds and found that sensitivity was impaired most in the frequency region of the fatiguing tone.

More direct evidence for a peripheral frequency analysis comes from studies of the tuning characteristics of saccular neurons in goldfish (Furukawa and Ishii 1967, Fay 1978c) and the sculpin, *Cottus scorpius* (Enger 1963). Representative tuning curves for these two species are plotted together in Fig. 1-9. Clearly, these neurons are not homogeneous with respect to sensitivity, best frequency, or bandwidth, and in addition show great diversity in spontaneous discharge patterns and rates of adaptation. Furukawa and Ishii (1967) and Fay (1978c) have observed that saccular neurons tuned to the lower frequencies appeared to originate from the posterior portion of the macula, while the higher frequency types appeared to innervate the anterior end of the macula. While this suggestion requires confirmation by more systematic study, it points to a crude form of tonotopic organization within the saccular macula. The origin of the tuning illustrated here, however, is not at all clear. Fay (1978c) has

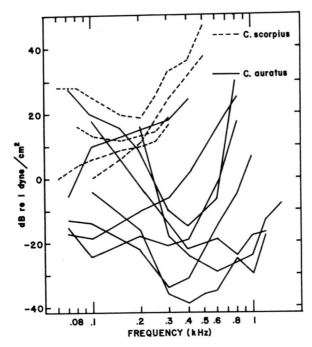

Figure 1-9. Tuning curves for representative saccular neurons from *Cottus scorpius* (Enger 1963) and *Carassius auratus* (Fay 1978c). The data for *Cottus scorpius* are based on impulse rate criteria while those for *Carassius auratus* are based on phase-locking synchronization criteria.

suggested that the low frequency neurons are stimulated by a direct tissue-conducted route to the ear while the higher frequency fibers receive input from the swimbladder, at least in the goldfish. Frequency analysis, in this case, could simply be a reflection of the different frequency response characteristics associated with these two quite different pathways to the ear. In fact, under the conditions of recording, the low frequency saccular neurons were shown to be displacement sensitive and to have the same sensitivity and bandwidths as lagenar neurons (Fay and Olsho 1979). The possibility that the variation in tuning seen here is due to receptor cell tuning of the type observed by Hopkins (1976) in electroreceptors cannot be ruled out and certainly deserves further experimental attention. It is thus clear from the electrophysiological data that there are more complex and varied processes operating in the fish saccule than was originally expected from the gross morphology of the ear and that some sort of frequency analysis probably takes place peripherally. Whether, and to what extent, this limited "place" principle is used by the organism in frequency analysis is taken up in the next section.

5.2.2 Psychophysical Studies

Psychophysical studies of auditory masking in fishes have been motivated in part by the notion that a peripheral analysis of frequency may be revealed through a demonstration of filter-like processes. One such measure for which comparative data exist is the ratio between the level of a tonal signal at threshold and the level of a wide-band noise masker (the critical masking ratio or CR). The CR values measured systematically in several teleost species are plotted together in Fig. 1-10. Note that the data illustrated fall within the total range of values determined for a larger number of teleost species (Tavolga 1974, Chapman 1973, Buerkle 1969) and that these values fall within the range of mammalian variation (except below 200 Hz where the fishes show generally greater sensitivity than mammals) (Fay 1978a). There are several important features of these data that require comment. The CR functions of frequency are similar to those seen in mammals in that they have a positive slope of approximately 3 dB/octave. The data for one ostariophysan (goldfish) do not differ systematically from those for several non-ostariophysan species, except in that the function extends to higher frequencies in goldfish. One of the most interesting features of these data is that the critical ratio depends on the azimuthal angular separation between the signal (tone) and masker sources. When the two sources are both located at the same azimuthal angle (dashed lines), the CRs are about 7 dB greater (showing 7 dB more masking) than when the noise and signal are separated by 85° (dotted lines). This effect has been demonstrated several times in different species (Chapman 1973, Chapman and Johnstone 1974, Hawkins and Sand 1977) and is analogous to the "cocktail party effect" and the binaural masking level difference (see Green and Yost 1975 for a recent review) that has been demonstrated in man and other mammals (see Gourevitch, Chapter 12). In terrestrial animals, this effect depends primarily on differences in the interaural phase values between signal and masker, while in fishes, it presumably arises from the directional characteristics of the ears themselves (see Schuijf and Buwalda, Chapter 2, and Section 6.0 of this chapter). In any case, making the assumptions that a rectangular band of noise centering on the signal frequency produces masking, and that the power of this band is equal to the power of the signal at threshold, the value of the critical ratio may be used to calculate the width of the hypothetical "critical" bands that produce the masking. Clearly, the widths of these critical ratio bandwidths are quite narrow for fishes and indicate a remarkable degree of frequency analysis. On the other hand, since one of the basic assumptions underlying the calculation of critical ratio bandwidths is that signal detection is accomplished through a spectral filtering mechanism, the results themselves cannot be used as evidence for the notion that such a filtering mechanism exists in the teleost auditory system.

More direct measures of the auditory critical band in fishes have been made for goldfish (Tavolga 1974) and for *Gadus morhua* (Hawkins and Chapman 1975). The threshold for a 500 Hz tone was measured for the goldfish as a function of the width of a masking noise band centered on the signal frequency. Sensitivity to the pure tones became poorer as the bandwidth of the noise was increased up to but not beyond 200 Hz, indicating that beyond this critical value, noise power was not adding to the masking effect. This critical band estimate is comparable to that for the monkey, measured at

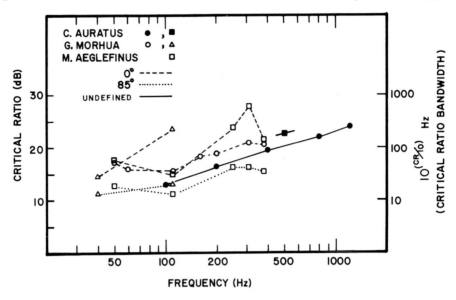

Figure 1-10. Critical masking ratio values for three teleost species. Filled circles, *Carassius auratus* (Fay 1974a); filled square, *Carassius auratus* (Tavolga 1974); open circles, *Gadus morhua* (Chapman and Hawkins 1973); open triangles, *Gadus morhua* (Chapman 1973); open squares, *Melanogrammus aeglefinus* (Chapman 1973). The solid lines indicate that the angular separation between the signal and masker was undefined. The dashed and dotted lines indicate a 0° and 85° azimuthal separation, respectively, between the signal and masker sources.

500 Hz (Gourevitch 1970). In *Gadus morhua,* the critical band was measured to be about 100 Hz at 380 Hz and 60 Hz at 160 Hz.

Other direct measures of auditory filter characteristics come from studies of the masking effects of tones in *Carassius auratus* (Tavolga 1974, Fay, Ahroon, and Orawski 1978) and in *Gadus morhua* (Buerkle 1968, 1969; Hawkins and Chapman 1975). At a qualitative level, the data from each experiment agree in showing that the masking effect is generally an inverse function of the frequency separation between the signal and the masker, suggesting the operation of a set of filter-like mechanisms with a continuous distribution of best frequencies throughout the hearing range. Quantitative comparisons among the functions are difficult to make, however, because of the differences in the experimental paradigms used in each case. Some of the data for *Carassius auratus* (Fay, Ahroon, and Orawski 1978) and *Gadus morhua* (Hawkins and Chapman 1975) are plotted together in Fig. 1-11. The Hawkins and Chapman (1975) curves for *Gadus* are perhaps the most remarkable in showing very narrowly tuned filter characteristics. Indeed, the authors concluded that central (and presumably temporally based) mechanisms rather than peripheral-mechanical mechanisms were operating, since the tuning appeared to far exceed that to be expected from the ear itself. While the exact nature of such hypothetical central analyzing mechanisms has not been explicitly discussed in the literature, it is worth noting here that the process of autocorrelating the

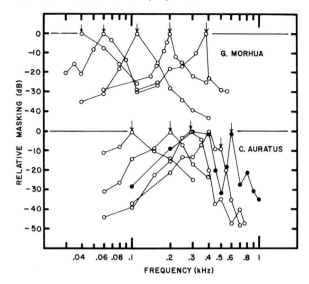

Figure 1-11. Psychophysical tonal masking functions for *Gadus morhua* (Hawkins and Chapman 1975) and for *Carassius auratus* (Fay, Ahroon, and Orawski 1978). In each case, a tonal or narrow band noise signal (frequency indicated by arrows) was masked by tones of various frequencies, and the amount of masking is shown relative to the maximum amount of masking observed at a given signal frequency. For *Gadus morhua*, masking was measured by determining the signal threshold in the presence of maskers of fixed intensity. For *Carassius auratus*, masking was measured by determining the level of the masker necessary to render inaudible a signal fixed at 15 dB above quiet threshold (the psychophysical tuning curve paradigm).

stimulus waveform in time is equivalent to the determination of the shape of its power spectrum, as through narrow band filtering, for example. Again, the results of masking studies by themselves do not allow one to decide whether the system under study analyses in the frequency or time domain.

In any case, the question raised by Hawkins and Chapman (1975) regarding the quantitative relationships between the filtering achieved by the ear, and that which can be demonstrated psychophysically, is an important one. In an attempt to clarify precisely how tonal masking patterns would compare to the neural tuning originating in the ear of the same species, Fay, Ahroon, and Orawski (1978) measured psychophysical tuning curves in the goldfish for comparison with neural tuning curves (Fay 1978c) obtained under comparable acoustic conditions for the same species.

Psychophysical tuning curves were generated by conditioning animals to a tone or narrow noise band probe signal at 10 dB above threshold and then measuring the level of a tonal masker that just masked the probe as a function of the masker frequency. In comparable experiments with birds (see Saunders 1976, Dooling, Chapter 9) and mammals (McGee, Ryan, and Dallos 1976), these masking functions resemble neural tuning curves for 8th neurons, leading to the hypothesis that both procedures measure similar aspects of peripheral filtering in the auditory system. For goldfish, the greatest correspondence between the neural tuning curves (Fig. 1-9) and the psychophysical tuning curves (Fig. 1-11) occurs for signal frequencies below about 400 Hz where the slopes of the upper and lower "legs" of the "V"-shaped psychophysical tuning curve correspond to the slopes of the low frequency and high frequency neural tuning curves respectively. Note, however, that the psychophysical tuning curves for signal frequencies at and below 300 Hz have peaks corresponding to the signal frequencies while there are no neural tuning curve peaks in the same range. A possible explanation for this is that signal frequencies in this range activate both high frequency and low frequency neurons and that detection decisions are based on input from those neurons with the most favorable signal-to-noise ratios (S/N). In this case, a masking peak would always occur at the frequency of the signal, giving the impression of the operation of a set of fixed filters with a continuous distribution of best frequencies. This same effect is apparent in comparing psychophysical tuning curves measured for cutaneous vibration in man (Labs, Gescheider, Fay, and Lyons 1979) with neural curves for monkeys, and it may also play a role in determining the shapes of the tonal masking functions for *Gadus morhua* (Fig. 1-11).

For signal frequencies above about 400 Hz, the psychophysical tuning curve often shows two peaks (solid symbols in Fig. 1-11). The first always occurs in the 300 Hz to 400 Hz range while the second peak tends to occur at the frequency of the signal, with a steep roll off in the masking effect extending about 200 Hz above and below the signal frequency. Neglecting these regions of reduced masking for the moment, it is clear that the overall shape, bandwidth, and particularly the roll-off rate below 400 Hz of higher frequency psychophysical tuning curves correspond to those of the high frequency neural curves. In some respects, then, features of the psychophysical tuning curve are predictable from the forms of the neural tuning curve, and it appears that the goldfish is capable of using a limited degree of peripheral frequency analysis in enhancing the detectability of masked signals.

It is equally clear, however, that processes other than peripheral neural tuning operate to produce the steep masking roll-offs above and below signal frequencies in the range above 400 Hz to 500 Hz (see filled symbols for the goldfish in Fig. 1-11). It appears that tones placed between 100 Hz and 200 Hz of the signal frequency are particularly ineffective as maskers. One explanation for this effect is that beat frequencies (or bands) are particularly highly detectable in the 100 Hz to 200 Hz range. In this case, the neural patterns evoked by the continuous masker may be modified by the amplitude modulations produced by the addition of the signal, even at rather high masker-to-signal ratios. This idea is developed further in the next section.

5.3 Temporal Processing

There have been few studies on fishes directly addressing the question of temporal processing. Using a stimulus generalization paradigm, Fay (1972) found evidence for the existence of periodicity pitch in goldfish by demonstrating a "perceptual" similarity between a tonal signal and a 1 kHz tone modulated in amplitude at a rate corresponding to the frequency of the signal. More recent studies of the detection of amplitude modulation for tones and noise were carried out by Fay (1977, in press).

Here, goldfish were conditioned to respond when a continuous signal such as a tone or wideband noise (carrier signal) began to be modulated in amplitude at a given rate (modulation-frequency, f_m) and at a given degree of amplitude change (modulation depth, m) (Fig. 1-13 illustrates an amplitude modulated waveform and shows the definition of m). The value of m that produces a just-detectable difference between a modulated and unmodulated signal is measured as a function of modulation-frequency.

In psychophysical experiments with human observers, using a white noise carrier, this function has a low-pass characteristic (Viemeister 1977, Rodenburg 1977) with a 3 dB-down point for $f_m = 55$ Hz (Viemeister 1977). One interpretation of this function is that some portion of the auditory system effectively filters out (attenuates) neural representations of sound amplitude modulations above a certain modulation frequency. The value of this critical frequency (the 3 dB-down point) characterizes the hypothetical low pass filter in the frequency domain, and the associated time constant $[2\pi F]^{-1}$, where F = frequency at the −3-dB point on the modulation function) characterizes the system in the time domain. The time constant for human observers is about 3 msec, a value comparable to that obtained using other procedures to measure the "minimum integration time" of the human auditory system (Green 1973).

The modulation functions for goldfish using an 800 Hz tone and wideband noise

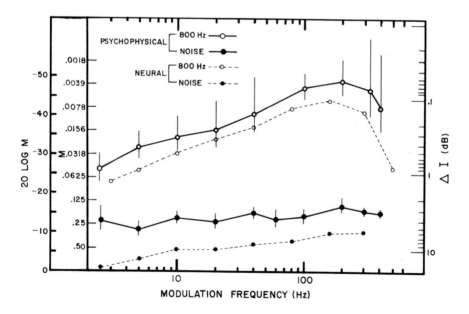

Figure 1-12. Temporal modulation functions for goldfish measured psychophysically (solid lines) and for single saccular neurons (dashed lines) using a wide band noise carrier (filled symbols) and an 800 Hz tonal carrier (open symbols). The psychophysical data were determined by measuring the degree of sinusoidal amplitude modulation (modulation depth, m), which is just detectable when impressed upon a continuous carrier signal at 35 dB above quiet threshold. The neural data show the value of m, which produces a small but statistically reliable degree of phase-locking between the neural response and the modulation envelope (from Fay 1979).

carrier are plotted together in Fig. 1-12. The function for the noise carrier is relatively flat at a value of m corresponding to a peak-to-trough intensity difference of 3 to 4 dB. This intensity difference threshold (delta I) is in the range of those reported for the goldfish (Jacobs and Tavolga 1967) and for *Gadus morhua* (Chapman and Johnstone 1974) using tonal signals and successive discrimination procedures, and is well within the range of values reported for a variety of terrestrial vertebrates (see Dooling, Chapter 9). The noise modulation function for goldfish is remarkable in that it shows no signs of rolling off below 400 Hz. The goldfish thus has a minimum integration time below 0.4 msec and appears to be clearly superior to human observers in its ability to detect high frequency envelope fluctuations.

The function for the 800 Hz carrier shows that sensitivity to modulation increases steadily with frequency up to about 200 Hz and then declines. Significantly, the modulation frequency range of greatest detectability corresponds to the frequency difference between masker and probe signal (beat frequency range) at which maskers are least effective in the psychophysical tuning curve experiment (see filled symbols of Fig. 1-11). This shows that the failure of the masker to interfere with the signal in this case could be due to the highly detectable amplitude modulations (beats) occurring in the 100 Hz to 200 Hz range. The modulation function for the 800 Hz tone is remarkable also in that the amplitude variation, which is just detectable at 200 Hz, is about .07 dB (m = 0.004), a value considerably smaller than the intensity discrimination ability measured for the goldfish by more traditional methods (Jacobs and Tavolga 1967). In fact, these data predict that an 800 Hz tone (masker) presented simultaneously with a 600 Hz or 1000 Hz tone (signal) will produce detectable beats (m = .004) even where the masker-to-signal ratio is as great as 49 dB. (Note that −20 log m in Fig. 1-12 corresponds to the amplitude ratio of two sinusoids which when added produce a modulated (beating) signal with the corresponding modulation depth.) It is not known whether this exquisite sensitivity to amplitude modulation remains for carrier frequencies well below 800 Hz, but to the extent that it does, rather steeply skirted tonal masking functions resembling auditory "filters" such as those of Fig. 1-11 would be expected.

It is possible that these differences between the tonal and noise modulation functions rest on the spectral differences between the two stimuli. While the long term noise spectrum remains flat and unaffected by the modulation, the spectrum of the modulated tone contains three lines: one at the carrier frequency (fc) and one each at frequencies fc + fm and fc − fm. An accurate frequency domain analyzer could thus detect temporal modulation as a given spectral pattern. However, since the deviation between the tonal and noise modulation functions begins to occur at values of fm that would produce spectral side band separations considerably smaller than critical bandwidth estimates (Tavolga 1974) or even frequency discrimination thresholds (Fay 1970a), it is unlikely that a spectral analyzing mechanism is at work here. Preliminary results of neurophysiological studies on the response of saccular neurons to the same modulated stimuli are illustrated in Fig. 1-13, and representative neural modulation functions are plotted in Fig. 1-12. These neural functions parallel the psychophysical curves quite well and show that behavioral modulation detection could be based on the degree to which neural responses are entrained by the modu-

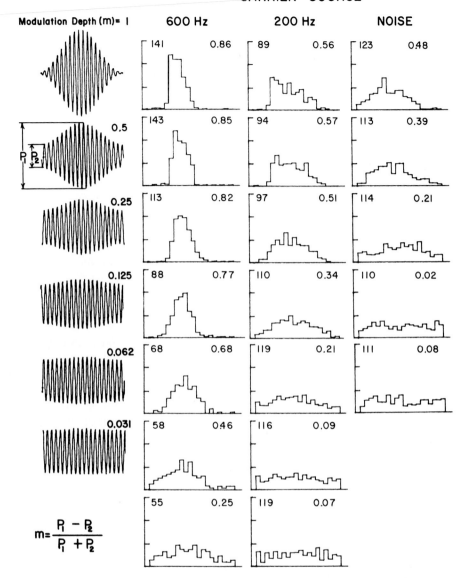

Figure 1-13. Amplitude modulated signals and modulation period histograms for a saccular neuron (M00605) responding to the AM signals presented about 35 dB above behavioral threshold. The waveforms on the left illustrate an 800 Hz tone (carrier) modulated at 50 Hz at a variety of modulation depths. The number associated with each waveform is the value of modulation depth (m) as defined at bottom left. The neural period histograms display the relative frequency distributions of impulses within individual cycles of modulation. Each vertical division on the ordinate is equal to 10% of the total number of impulses counted. The number at the left of each period histogram is the neuron's average firing rate in impulses per second, and the number on the right is the coefficient of synchronization as defined in Fay (1978c). Each column illustrates the response of the same neuron to a different carrier source.

lation envelope. Note in Fig. 1-13 that an amplitude variation in the stimulus of considerably less than 1 dB (m = 0.031) can cause a significant degree if neural synchronization to the stimulus envelope. This high degree of "amplification" can be thought of as an example of temporal contour enhancement which equals or exceeds that measured for neurons of the mammalian cochlear nucleus (Møller 1973). This effect appears not to be due to neural interactions but rather to some property of the hair cell receptors or their synapses onto saccular neurons.

Other studies of temporal processing include two attempts to measure temporal summation at threshold, or the "maximum" integration time (Green 1973), in the goldfish (Offutt 1967, Popper 1972). Unfortunately, these two studies present conflicting data. Although both investigations show that threshold does not depend on the duration of tone pulses in a train to be detected, Offutt (1967) found that sensitivity increases with the percent of time the signal is "on" (duty-cycle), while Popper (1972) found that threshold is independent of duty cycle. Popper's data indicate that the maximum integration time is probably shorter than the shortest signal used (10 msec), while Offutt's data suggest that the integration time is probably longer than the longest signal used (500 msec). Note that neither study measured the integration time.

A recent study of forward and backward masking in goldfish (Popper and Clarke 1979) complicates the view of temporal processing further. Here it was found that the masking effect of a brief noise burst extended both forward and backward in time for as much as 250 msec, with most masking occurring within 50 msec of the masker. Again, signal detectability did not depend on signal duration but was an inverse function of the silent interval between signal and masker. These results indicate that the effects of stimulation persist for rather large intervals. This is consistent with the data of Offutt (1967) and with the notion of a rather long minimum integration time but is qualitatively inconsistent with the modulation detection data (Figs. 1-12, 1-13), which show a high degree of temporal resolution. In any case, more work needs to be done before sense can be made of the rather complex temporal effects of hearing in the goldfish. It would also seem that introducing another variable here (such as species differences) would not make much sense until the relations between stimulus variables in one species are more fully understood.

In spite of these problems, the tentative conclusion has been drawn that the auditory system of the goldfish is relatively well adapted for the coding and analysis of temporal envelope patterns as well as for waveform fine structure. This conclusion is in accord with three other sets of observations: 1) saccular neurons of the goldfish and other species are less highly tuned than those of birds and mammals, thus allowing for a more faithful temporal representation of complex waveforms within individual channels; 2) fishes possess multidirectional hair cell orientation patterns, while mammals and birds do not; thus, complex asymmetrical waveforms may be coded in greater detail in fishes; 3) temporal rather than spectral acoustic patterns appear to carry biologically relevant information for at least some fishes (Fine, Winn, and Olla 1978).

6 Directional Characteristics of the Ear

The striking patterns of hair cell orientation in the ears of fishes (e.g., Popper 1977) suggest that the directional characteristics of relative otolith movement is represented in the neurally coded output of the ear and is likely to be of some biological significance. It is now quite clear that the ears of most species may receive acoustic input both directly, through the tissue conduction of particle motion, and indirectly, through the vibratory response of the swimbladder to sound pressure fluctuations. Furthermore, the relative effect of these two conduction modes is frequency dependent, with the tissue conducted particle motion component increasing in relative importance toward the lower frequencies. While this creates an enormous stimulus specification problem for the worker interested in measuring absolute hearing sensitivity, it is of potentially great value to the fish in acquiring high quality information about its acoustic environment. In combination with specializations for conducting swimbladder motion to the ear in many species, hair cells oriented so as to respond maximally to this input provide the animal with high sensitivity, wide hearing bandwidth, and information on the fine structure and phase of the sound pressure waveform. Otolithic organs not as well coupled to the swimbladder and hair cells that are not oriented to respond maximally to swimbladder input, provide information on the amplitude and phase of the displacement (or velocity) waveform reaching the animal directly. The information represented in this way is thought to possibly subserve the determination of sound source location (Schuijf and Buwalda, Chapter 2), enhanced capacities for signal detection in noise and possibly frequency analysis. In addition, a comparison of the relative amplitudes and phase of the pressure and displacement (or velocity) waveform could allow the fish access to information about the impedance of an impinging signal and thus to information about the size, vibrational modes and range of the signal's source. Whether fishes do in fact process this information is not known. The suggestion that they do is not far fetched, however, since analogous source characteristics of electrical signals appear to be processed in modified acoustico-lareralis systems of the electro-sensitive fishes (Heiligenberg 1977).

Since Schuijf and Buwalda (Chapter 2) treat mechanisms for sound localization in detail, it will only be pointed out in passing here that fishes do appear to be capable of acoustic localization (Moulton and Dixon 1967, Chapman 1973, Popper, Salmon, and Parvulescu 1973, Chapman and Johnstone 1974, Schuijf 1975, 1976a, 1976b, Hawkins and Sand 1977), with minimum audible angles as small as 16° in the vertical plane (Hawkins and Sand 1977) and 22° in the horizontal plane (Chapman and Johnstone 1974) for *Gadus morhua*. Two intact labyrinths appear to be necessary for horizontal localization (Schuijf 1975) but may not be necessary for vertical localization (Hawkins and Sand 1977; Popper in press). It is presumed that much of the information necessary for localization is contained in the axes of water particle movement set up by sound, which are then coded by the ear by virtue of the directional orientation patterns of hair cells within the ear's maculae and the orientation of the paired maculae themselves.

One of the consequences of the variation in directional sensitivity of the maculae and receptor cells within individuals is that signals should interfere with (mask) each other primarily to the extent that they have axes of particle motion in common. In field studies of tonal detection in the presence of broad band noise, Chapman (1973), Chapman and Johnstone (1974) and Hawkins and Sand (1977) have shown that the signal-to-noise ratios at threshold may be improved by up to 7 dB or so by increasing the angular separation between signal and masker sources up to 90° (see Fig. 1-10). This is clearly analogous to the "cocktail party" effect well known in human hearing research but is presumably due to peripheral rather than central neural directional filtering. Since the detectability of most signals in the usual environment is most probably determined by levels of ambient noise, the improvement in detectability afforded by this directional analyzing capability is undoubtedly of great biological significance; it is greater, for example, than simple gains in absolute sensitivity.

Evidence that the ears of fishes do indeed respond in a directional manner comes from experiments on *Perca fluviatilis* (Sand 1974) and *Melanogrammus alglefinus* (Enger, Hawkins, Sand, and Chapman 1973) in which microphonic potentials were recorded as a function of the angle of head vibration in the horizontal plane. The results of Sand (1974) are clearest in showing that the response of the saccule is a cosine function of the angle between the long axis of the macula and the axis of stimulation. A comparison of the relative outputs of the two nonparallel sacculi would thus provide clear information on the axis of particle motion, and in addition, restrict the primary "view" of the auditory system to sources located in front of and behind the animal. Sand (1974) also observed that the relative amplitudes of saccular microphonics to vertical and horizontal vibration was a function of recording location along the macula, with the vertical component predominating in the posterior portion. The fact that these relative amplitudes were also frequency dependent led Sand to suggest that frequency as well as direction may be coded in the neural output of the saccule.

The only study of neural coding of directional and frequency information comes from a study by Fay and Olsho (1979) on the goldfish in which the activity of single saccular and lagenar nerve fibers was recorded in response to vibrational stimuli in three orthogonal directions. The sensitivity and stimulus-response phase angle of each neuron was measured for stimulation in the three directions. The sensitivity measures were used to calculate "best directions" of stimulation in the saggital and horizontal planes (Fig. 1-14). Neurons from the saccule and lagena are practically indistinguishable in every respect. Best directions in the horizontal plane (20° to 30° from midline) correspond to the microphonic data of Sand (1974) and presumably reflect the orientation of the saccule in the head. Best directions in the saggital plane are rather widely dispersed, with a modal point about 50° from vertical. While this modal point corresponds to the sensitivity axes of saccular hair cells in goldfish (Platt 1977) and several other ostariophysans (Jenkins 1979), the dispersion of the distributions does not correspond to the rather narrow range of "best directions" of hair cell orientations expected from the saccule. The reasons for this variation are not clear, but the data support the suggestion that otolith movement patterns are rather complex (see Section 2.2) and that neurally coded output of the ear may not be predictable from hair cell orientation maps alone (see also Popper 1978b, 1980). The distributions of stimulus-response phase angles are similarly diffuse and do not correspond to the rather narrow bimodal distributions to be expected from directionally sensitive receptors. These data, in combination with those of Sand (1974) and Sand and Michelsen (1978), could mean that otolith movements in response to vibration are nontranslatory and generally

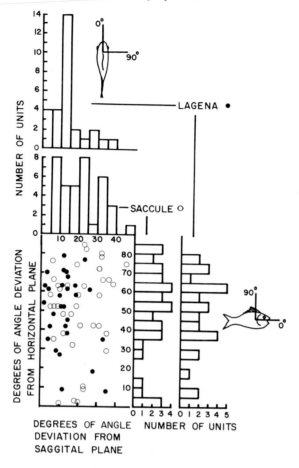

Figure 1-14. Directional characteristics of single saccular (open symbols) and lagenar (closed symbols) neurons of goldfish. For each neuron shown, displacement sensitivity based on phase-locking criteria was measured for direct vibration of the fish's head in three orthogonal directions (vertical, lateral, and anterior-posterior). The axis of greatest sensitivity (the angle of a resultant vector) was then calculated for each neuron in the saggital plane (histograms projected to the right) and in the horizontal plane (histograms projected above). Figure reprinted with permission from Comp. Biochem. Physiol. Vol. 62A, R. Fay and L. Olsho, "Discharge Patterns of Saccular and Lagenar Neurons," 1979, Pergamon Press, Ltd.

complex. The observations of Popper (1977) and Platt (1977) that the modes of attachment between the otolith and hair cell cilia generally are variable suggest, in addition, that even the relationship between otolith movement and hair cell stimulation itself is likely to vary among different points on the macula. While all this variation may be disconcerting to the investigator, this particular electrophysiological approach to the functional analysis of species specific hair cell orientation patterns is probably most valuable. Perhaps the careful selection of species for further study will reduce some of the present uncertainty.

7 Conclusions

Since the time we collaborated on an earlier review on this topic (Popper and Fay 1973), the literature has grown substantially and fruitfully in several areas. Perhaps the greatest increase has occurred in our knowledge of the structure and ultrastructure of the ear. It is now clear that there is enormous interspecific variation in the relative sizes and shapes of the otolithic organs, including their maculae and otoliths, in the relationships between the otoliths, the otolithic membrane, and the underlying maculae, and in the patterns of hair cell orientation and ciliary types within the maculae. When we realize that the number of extant teleost species is greater than the number of all other vertebrate species combined, the task of gaining a realistic general understanding of the dimensions of variation and their functional correlates appears to have no end. The diversity of questions we could ask, and conceivably answer, appears similarly limitless, and we are placed in the position of having to select or formulate the ones that are most likely to bring a wide range of descriptive data under some sort of theoretical "control." Two very general approaches to this problem arise from the respective traditions of our scientific training; that of comparative psychology and that of zoology.

The tradition of comparative psychology emphasizes, at the same time, comparison in a gross sense, that is, across vertebrate classes, and the belief that structure-function relationships worked out in detail for a limited number of species in one class will provide a significant and general framework for understanding the auditory system within that class. In combination with similar detailed analysis in other vertebrate classes, in which the same variables, experimental techniques and theoretical constructs are used, this approach promises to reveal a set of common principles of auditory function operating throughout the vertebrates. Much of the psychophysical work on discriminative capacities in fishes and the accompanying use of hypothetical constructs such as the critical band and other notions of frequency analysis fit into this model. In order for this approach to be as fruitful as possible, we must look with the finest experimental grain at the structural, neurophysiological, and neuroanatomical mechanisms underlying psychophysically defined capabilities of the auditory system until we are satisfied that we understand them in at least one species. An understanding of the mechanisms of frequency and time analysis in the goldfish and codfish is beginning to come together under this approach, although there is clearly much to be done, particularly in determining the central neural mechanisms involved. The analysis of directional hearing in one or two species treated by Schuijf and Buwalda (Chapter 2) is another example of the potential value of this approach. Of course, the limitations inherent in this paradigm cannot be overlooked. It is always possible that the species chosen for study are highly specialized or otherwise inappropriate as general models. In addition, the focus on psychophysical behavior in highly controlled and often unusual acoustic environments may both fail to reveal the "important" auditory functions from a biological point of view and, perhaps, mislead us with epiphenomena. For example, while it is clear that fishes can be trained to make tonal frequency discriminations, it is not clear that we are studying something analogous to pitch perception in man. Another danger in this type of approach is that the application of hypothetical constructs such as the critical band from work on mammals may not be appropriate for other vertebrate classes. While the fishes may exhibit behavior consistent with the critical band concept, the underlying mechanisms may be quite different from those operating in mammals to the point that the behavioral consistency is little more than coincidence. Of course, one of the values of this type of approach is that the generality of such hypothetical constructs can be assessed.

The second general approach, the zoological one, focuses on comparison within class and family and emphasizes the diversity of structures and functions in a wide variety of species. This paradigm promises to reveal the dimensions of structure-function variation. Here, we are less likely to be misled with epiphenomena or to waste too much time with functions of little or no biological significance. In order for this approach to succeed, of course, there must be experimental designs which focus on the kinds of functions which clearly have survival value; for example, unconditioned behaviors such as the jamming avoidance response of electric fishes or the head turning response of owls toward sound sources (Knudsen, Chapter 10). Unfortunately, there are few such responses of fishes to sound, and this has clearly limited our progress in bringing the morphological data under control. This area would probably profit from some careful observation of fishes in their usual environments, perhaps with manipulations of their auditory systems through lesions or with manipulation of the sonic environment such as through the introduction of high levels of masking noise.

One of the troublesome aspects of the study of hearing in fishes is that few species clearly communicate using sound, and many that do have no apparent specialization for hearing and in fact are quite insensitive to sound. Thus, the type of zoological approach so valuable in the study of anuran hearing (see Capranica and Moffat, Chapter 5) will probably not be as valuable in the study of fish hearing. This fact, however, in combination with the wide variation in peripheral auditory morphology in fishes leads to the suggestion that factors other than vocal communication can have large effects on the evolution of the ear. We speculate that one of the keys to understanding auditory processing by fishes may simply lie in the special acoustic characteristics of the underwater environment and the fishes' relationships to it. The relative incompressibility of water produces complex amplitude and phase relationships between the pressure and motional components of acoustic disturbances that extend and vary within considerable distances from sound sources. As Schuijf and Buwalda (Chapter 2) show, an analysis of the temporal relationships between these components may be quite important for, among other things, auditory localization. The auditory systems of fishes may thus be understood as general acoustic signal processing systems that are possibly adapted for temporal as opposed to spectral processing and for listening within regions of three-dimensional space characteristic of the widely divergent adaptive zones of individual teleost species.

Acknowledgments. We thank Ms. Sheryl Coombs for her critical reading of the manuscript. Portions of the work discussed here were supported by grants BNS 76-02031 from NSF (National Science Foundation) and NS-15268-01 from NINCDS (National Institute of Neurological and Communicative Disorders and Stroke) to R.R.F. and by grants BNS 78-22411 from NSF, NS-15090 from NINCDS, and a Research Career Development Award from NINCDS (NS-00312) to A.N.P.

References

Adrian, E. D., Craik, K. J. W., Sturdy, R. S.: The electrical response of the auditory mechanism in cold-blooded vertebrates. Proc. Roy. Soc. Sec. B. 125, 435-455 (1938).

Alexander, R. McN.: The structure of the Weberian apparatus in the Cyprini. Proc. Zool. Soc. Lond. 139, 451-473 (1962).

Blaxter, J. H. S., Tytler, P.: Physiology and function of the swimbladder. In: Advances in Comparative Physiology and Biochemistry, Vol. 7. New York: Academic Press, 1978, pp. 311-367.

Buerkle, U.: Relation of pure tone thresholds to background noise level in the Atlantic cod (Gadus morhua). J. Fish. Res. Bd. Canada. 25, 1155-1160 (1968).

Buerkle, U.: Auditory masking and the critical band in Atlantic cod (Gadus morhua). J. Fish. Res. Bd. Canada, 26, 1113-1119 (1969).

Cahn, P. H. (ed.).: Lateral Line Detectors. Bloomington, Indiana: University Press (1967).

Cahn, P. H., Siler, W., Wodinsky, J.: Acoustico-lateralis system of fishes: tests of pressure and particle-velocity sensitivity in grunts, Haemulon sciurus and Haemulon parrai. J. Acoust. Soc. Amer. 46, 1572-1578 (1969).

Carlstrom, D.: A crystallographic study of vertebrate otoliths. Biol. Lull. 124, 441-463 (1963).

Chapman, C. J.: Field studies of hearing in teleost fish. Helgolander wiss. Meeresunters. 24, 371-390 (1973).

Chapman, C. J., Hawkins, A. D.: A field study of hearing in the cod, Gadus morhua L. J. Comp. Physiol. 85, 147-167 (1973).

Chapman, C. J., Johnstone, A. D. F.: Some auditory discrimination experiments on marine fish. J. Exp. Biol. 61, 521-528 (1974).

Chapman, C. J., Sand, O.: Field studies of hearing in two species of flatfish Pleuronectes platessa (L.) and Limanda limanda (L.) (Family Pleuronectidae). Comp. Biochem. Physiol. 47A, 371-385 (1974).

Chranilov, N. S.: Beiträge zur Kenntnis des Weberschen Apparates der Ostariophysi. 1. Vergleichend-anatomische Übersicht über die Knochenelemente des Weberschen Apparates bei Cypriniformes. Zool. Jb. (Anat. Ont.). 49, 501-597 (1927).

Clarke, N. L., Popper, A. N., Mann, Jr., J. A.: Laser light-scattering investigation of the teleost swimbladder response to acoustic stimuli. Biophysical J. 15, 307-318 (1975).

Coombs, S., Popper, A. N.: Hearing differences among Hawaiian squirrelfish (family Holocentridae) related to differences in the peripheral auditory system. J. Comp. Physiol. 132, 203-207 (1979).

Corwin, J. T.: Morphology of the macula neglecta in sharks of the genus Carcharhinus. J. Morph. 152, 341-361 (1977).

Dale, T.: The labyrinthine mechanoreceptor organs of the cod Gadus morhua L. (Teleostei: Gadidae). Norw. J. Zool. 24, 85-128 (1976).

Demski, L. S., Gerald, J. W., Popper, A. N.: Central and peripheral mechanisms of teleost sound production. Amer. Zool. 13, 1141-1167 (1973).

Denton, E. J., Blaxter, J. H. S.: The mechanical relationships between the clupeid swimbladder, inner ear and lateral line. J. Mar. Biol. Assoc. U. K. 56, 787-807 (1976).

Dijkgraaf, S.: Über die Auslösung des Gasspuckreflexes bei Fischen. Experientia. 6, 188-190 (1950).

Dijkgraaf, S.: Über die Schallwahrnehmung bei Meeresfischen. Z. vergl. Physiol. 34, 104-122 (1952).

Dijkgraaf, S.: Hearing in bony fishes. Proc. Roy. Soc. Lond. Ser. B. 152, 51-54 (1960).

Dijkgraaf, S., Verheijen, F.: Neue Versuche über das Tonunterscheidungsvermögen der Elritze. Z. vergl. Physiol. 32, 248-256 (1950).

Dudock Van Heel, W. H.: Pitch discrimination in the minnow (*Phoxinus laevis*) at different temperature levels. Experientia. 12, 75-76 (1956).

Enger, P. S.: Single unit activity in the peripheral auditory system of a teleost fish. Acta. Physiol. Scand. 59, Suppl. 3, 9-48 (1963).

Enger, P. S.: Acoustic thresholds in goldfish and its relation to the sound source distance. Comp. Biochem. Physiol. 18, 859-868 (1966).

Enger, P. S.: Masking of auditory responses in the medulla oblongata of goldfish. J. Exp. Biol. 59, 415-424 (1973).

Enger, P. S.: On the orientation of haircells in the labyrinth of perch (*Perca fluviatilis*). In: Sound Reception in Fish. Schuijf, A., Hawkins, A. D. (eds.). Amsterdam: Elsevier, 1976, pp. 49-62.

Enger, P. S., Hawkins, A. D., Sand, O., Chapman, C. J.: Directional sensitivity of saccular microphonic potentials in the haddock. J. Exp. Biol. 59, 425-433 (1973).

Fay, R. R.: Behavioral audiogram for the goldfish. J. Aud. Res. 9, 112-121 (1969).

Fay, R. R.: Auditory frequency discrimination in the goldfish (*Carassius auratus*). J. Comp. Physiol. Psychol. 73, 175-180 (1970).

Fay, R. R.: Perception of amplitude-modulated signals in the goldfish. J. Acoust. Soc. Amer. 52, 660-666 (1972).

Fay, R. R.: Auditory frequency discrimination in vertebrates. J. Acoust. Soc. Amer. 56, 206-209 (1973).

Fay, R. R : Masking of tones by noise for the goldfish (*Carassius auratus*). J. Comp. Physiol. Psychol. 87, 708-716 (1974a).

Fay, R. R.: Sound reception and processing in the carp: Saccular potentials. Comp. Biochem. Physiol. 46(A), 29-42 (1974b).

Fay, R. R.: Auditory temporal modulation transfer function for the goldfish. J. Acoust. Soc. Amer. 62(Suppl. 1), 588 (1977).

Fay, R. R.: Sound detection and sensory coding by the auditory systems of fishes. In The Behavior of Fish and Other Aquatic Animals. Mostofsky, D. (ed.). New York: Academic Press, 1978a, pp. 197-236.

Fay, R. R.: Phase-locking in goldfish saccular nerve fibers accounts for frequency discrimination capacities. Nature. 275, 320-322 (1978b).

Fay, R. R.: The coding of information in single auditory nerve fibers of the goldfish. J. Acoust. Soc. Amer. 63, 136-146 (1978c).

Fay, R. R.: Psychophysics and neurophysiology of temporal factors in hearing by the goldfish: amplitude modulation detection. J. Neurophysiol. (in press).

Fay, R. R., Ahroon, W. A., Orawski, A. A.: Auditory masking patterns in the goldfish (*Carassius auratus*): Psychophysical tuning curves. J. Exp. Biol. 74, 83-100 (1978).

Fay, R. R., Patricoski, M. M.: Sensory mechanisms for low frequency vibration detection in fishes. U. S. Geological Survey open file report on Abnormal Animal Behavior Prior to Earthquakes. Buskirk, R. (ed.). (1979).

Fay, R. R., Olsho, L.: Discharge patterns of lagenar and saccular neurons of the goldfish eighth nerve: Displacement sensitivity and directional characteristics. Comp. Biochem. Physiol. 62A, 377-386 (1979).

Fay, R. R., Popper, A. N.: Acoustic stimulation of the ear of the goldfish (*Carassius auratus*). J. Exp. Biol. 61, 243-260 (1974).

Fay, R. R., Popper, A. N.: Modes of stimulation of the teleost ear. J. Exp. Biol. 62, 379-388 (1975).

Fine, M., Winn, H., Olla, B.: Communication in Fishes. In How Animals Communicate. Sebeok, T. A. (ed.). Bloomington, Indiana: Indiana University Press, 1978, pp. 472-518.

Fish, J. F., Offutt, G. C.: Hearing thresholds from toadfish, *Opsanus tau,* measured in the laboratory and field. J. Acoust. Soc. Amer. 51, 1318-1321 (1972).

Flock, A.: Structure of the macula utriculi with special reference to directional interplay of sensory responses as revealed by morphological polarization. J. Cell Biol. 22, 413-431 (1964).

Flock, A.: Sensory transduction in hair cells. In: Handbook of Sensory Physiology, Vol. II. Lowenstein, W. R. (ed.). Berlin: Springer-Verlag, 1971, pp. 396-441.

Flock, A.: Physiological properties of sensory hairs in the ear. In: Psychophysics and Physiology of Hearing. Evans, E. F., Wilson, J. P. (eds.). London: Academic Press, 1977, pp. 15-25.

Furukawa, T., Ishii, Y.: Neurological studies on hearing in goldfish. J. Neurophysiol. 30, 1377-1403 (1967).

Gourevitch, G.: Detectability of tones in quiet and in noise by rats and monkeys. In: Animal Psychophysics: The Design and Conduct of Sensory Experiments. Stebbins, W. C. (ed.). New York: Appleton-Century-Crofts, 1970, pp. 67-97.

Green, D. M.: Minimum integration time. In: Basic Mechanisms in Hearing. Møller, A. (ed.). New York: Academic Press, 1973, pp. 829-846.

Green, D. M., Yost, W. A.: Binaural analysis. In: Handbook of Sensory Physiology, Vol. V, Part 2. Keidel, W., Neff, W. (eds.). New York: Springer-Verlag, 1975, pp. 461-480.

Grozinger, B.: Elektro-physiologische Untersuchungen an der Hörbahn der Schleie (*Tinca tinca L.*). Z. vergl. Physiol. 57, 44-76 (1967).

Hama, K.: A study on the fine structure of the saccular macula of the goldfish. Z. Zellforsch. 94, 155-171 (1969).

Hawkins, A. D., Chapman, C. J.: Masked auditory thresholds in the cod *Gadus morhua L.* J. Comp. Physiol. 103A, 209-226 (1975).

Hawkins, A. D., MacLennan: An acoustic tank for hearing studies on fish. In: Sound Reception in Fish. Schuijf, A., Hawkins, A. D. (eds.). Amsterdam: Elsevier, 1976, pp. 149-169.

Hawkins, A. D., Sand, O.: Directional hearing in the median vertical plane by the cod. J. Comp. Physiol. 122, 1-8 (1977).

Heiligenberg, W.: Principles of Electrolocation and Jamming Avoidance in Electric Fish. In: Studies of Brain Function, Vol. 1. Braitenberg, V. (ed.). Berlin: Springer-Verlag (1977).

Henson, O. W.: Comparative anatomy of the middle ear. In: Handbook of Sensory Physiology, Vol. V, Part 1. Keidel, W., Neff, W. (eds.). New York: Springer-Verlag, 1974, pp. 38-110.

Hopkins, C. C.: Stimulus filtering and electroreception: Tuberous electroreceptors in three species of gymnotid fish. J. Comp. Physiol. 111, 171-206 (1976).

Iversen, R. T. B.: Response of the yellowfin tuna (*Thunnus albacares*) to underwater sound. In: Marine Bio-Acoustics II. Tavolga, W. N. (ed.). Oxford: Pergamon Press, 1967, pp. 105-121.

Iversen, R. T. B.: Auditory thresholds of the scombrid fish *Euthynnus affinis*, with comments on the use of sound in tuna fishing. FAO Conference on Fish Behaviour in Relation to Fishing Techniques and Tactics. FAO Fisheries Rep. No. 62 (3), 849-859 (1969).

Jacobs, D. W., Tavolga, W. N.: Acoustic intensity limens in the goldfish. Anim. Behav. 15, 324-335 (1967).

Jacobs, D. W., Tavolga, W. N.: Acoustic frequency discrimination in the goldfish. Anim. Behav. 16, 67-71 (1968).

Jenkins, D. B.: A light microscopic study of the saccule and lagena in certain catfishes. Amer. J. Nat. 150, 605-630 (1977).

Jenkins, D. B.: A transmission and scanning electron microscopic study of the saccule in five species of catfishes. The Amer. J. of Anat. 154, 81-101 (1979).

Jorgensen, J. M.: Hair cell polarization in the flatfish inner ear. Acta Zool. 57, 37-39 (1976).

Knudsen, E. I.: Distinct auditory and lateral-line nuclei in the midbrain of catfishes. J. Comp. Neurol. 173, 417-432 (1977).

Labs, S. M., Gescheider, G. A., Fay, R. R., Lyons, C. H.: Psychophysical tuning curves in vibrotaction. Sensory Processes 2, 231-247 (1979).

Lewis, E. R., Li, C. W.: Hair cell types and distributions in the otolithic and auditory organs of the bullfrog. Brain Res. 83, 35-50 (1975).

Lindeman, H. H.: Regional differences in structure of the vestibular sensory regions. J. Laryngol. Otol. 81, 1-17 (1969).

Lowenstein, O.: The labyrinth. In: Fish Physiology, Vol. 5. Hoar, W. W., Randall, D. J. (eds.). New York: Academic Press, 1971, pp. 207-240.

Lowenstein, O., Osborne, M. P., Wersäll, J.: Structure and innervation of the sensory epithelia of the labyrinth in the thornback ray (*Raja clavata*). Proc. Roy. Soc. Lond. B. 160, 1-12 (1964).

Maler, L., Karten, H. J., Bennett, M. V. L.: The central connections of the posterior lateral line nerve of *Gnathonemus petersii*. J. Comp. Neur. 151, 57-66 (1973a).

Maler, L., Karten, H. J., Bennett, M. V. L.: The central connections of the anterior lateral line nerve of *Gnathonemus petersii*. J. Comp. Neur. 151, 67-84 (1973b).

McCormick, C.: Central projections of the lateralis and eighth nerves in the bowfin, *Amia calva*. Ph.D. Thesis, Univ. of Michigan (1978).

McGee, T., Ryan, A., Dallos, P.: Psychophysical tuning curves of chinchillas. J. Acoust. Soc. Amer. 60, 1146-1150 (1976).

Møller, A.: Coding of amplitude modulated sounds in the cochlear nucleus of the rat. In: Basic Mechanisms in Hearing. Møller, A. (ed.). New York: Academic Press, 1973, pp. 593-619.

Moulton, J. M., Dixon, R. H.: Directional hearing in fishes. In: Marine Bio-Acoustics, Vol. II. Tavolga, W. N. (ed.). Oxford: Pergamon Press, 1967, pp. 182-232.

Nelson, E. M.: The morphology of the swimbladders and auditory bulla in Holocentridae. Fieldiana Zool. 37, 121-137 (1955).

O'Connell, C. P.: The gas bladder and its relation to the inner ear in *Sardinops caerulea* and *Engraulis mordax*. Fishery Bull. 56, 505-533 (1955).

Offutt, G. C.: Integration of the energy in repeated tone pulses by man and the goldfish. J. Acoust. Soc. Amer. 41, 13-19 (1967).

Offutt, G. C.: Response of the tautog (*Tautoga onitis*, teleost) to acoustic stimuli measured by classically conditioning the heart rate. Conditional Reflex 6, 205-214 (1971).

Page, C. H.: Electrophysiological study of auditory responses in the goldfish brain. J. Neurophysiol. 33, 116-127 (1970).

Parvulescu, A.: Problems of propagation and processing. In: Marine Bio-Acoustics, Tavolga, W. N. (ed.). Oxford: Pergamon Press, 1964, pp. 87-100.

Platt, C.: Hair cell distribution and orientation in goldfish otolith organs. J. Comp. Neurol. 172, 283-298 (1977).

Platt, C., Popper, A. N.: Otolith organ receptor morphology in herring-like fish. In: Vestibular Function and Morphology. Gualterratti, G. (ed.). New York: Springer-Verlag (1979).

Piddington, R. W.: Central control of auditory input in the goldfish - II. Evidence of action in the free-swimming animal. J. Exp. Biol. 55, 585-610 (1971).

Poggendorf, D.: Die absoluten Hörschwellen des Zwergwelses (*Amiurus nebulosus*) und Beiträge zur Physik des Weberschen Apparatus der Ostariophysen. Z. vergl. Physiol. 34, 222-257 (1952).

Popper, A. N.: Auditory capacities of the Mexican blind cave fish (*Astyanax jordani*) and its eyed ancestor (*Astyanax mexicanus*). Anim. Behav. 18, 552-562 (1970).

Popper, A. N.: The effects of size on auditory capacities of the goldfish. J. Aud. Res. 11, 239-247 (1971).

Popper, A. N.: Auditory threshold in the goldfish (*Carassius auratus*) as a function of signal duration. J. Acoust. Soc. Amer. 52, 596-602 (1972).

Popper, A. N.: The response of the swimbladder of the goldfish (*Carassius auratus*) to acoustic stimuli. J. Exp. Biol. 60, 295-304 (1974).

Popper, A. N.: Ultrastructure of the auditory regions in the inner ear of the lake whitefish. Science. 192, 1020-1023 (1976).

Popper, A. N.: A scanning electron microscopic study of the sacculus and lagena in the ears of fifteen species of teleost fishes. J. Morph. 153, 397-418 (1977).

Popper, A. N.: Scanning electron microscopic study of the otolithic organs in the bichir (*Polypterus bichir*) and shovel-nose sturgeon (*Scaphirhynchus platorynchus*). J. Comp. Neurol. 181, 117-128 (1978a).

Popper, A. N.: A comparative study of the otolithic organs in fishes. Scanning Electron Microscopy. II, 405-416 (1978b).

Popper, A. N.: The ultrastructure of the sacculus and lagena in a moray eel (*Gymnothorax* sp.). J. Morphol. 161, 241-256 (1979).

Popper, A. N.: Organization of the inner ear and auditory processing. In: Fish Neurobiology and Behavior. Northcutt, R. G., Davis, R. E. (eds.). Ann Arbor: Univ. of Michigan Press (in press).

Popper, A. N., Clarke, N. L.: The auditory system of the goldfish (*Carassius auratus*); effects of intense acoustic stimulation. Comp. Biochem. Physiol. 53A, 11-18 (1976).

Popper, A. N., Clarke, N. L.: Simultaneous and non-simultaneous auditory masking in the goldfish, *Carassius auratus*. J. Exp. Biol. 83, 145-158 (1979).

Popper, A. N., Fay, R. R.: Sound detection and processing by teleost fishes: a critical review. J. Acoust. Soc. Amer. 53, 1515-1529 (1973).

Popper, A. N., Salmon, M., Parvulescu, A.: Sound localization by the Hawaiian squirrelfishes, *Myripristis berndti* and *M. argyromus*. Anim. Behav. 21, 86-97 (1973).

Pumphrey, R. J.: Hearing. In Physiological Mechanisms in Animal Behavior. Symp. Soc. Exp. Biol. 4, 1-18 (1950).

Retzius, G.: Das Gehörorgan der Wirbelthiere. Vol. I. Stockholm: Samson and Wallin (1881).

Rodenburg, M.: Investigations of temporal effects with amplitude modulated signals. In: Psychophysics and Physiology of Hearing. Evans, E., Wilson, J. (eds.). New York: Academic Press, 1977, pp. 429-437.

Salmon, M.: Acoustical behavior of the menpachi, *Myripristis berndti*, in Hawaii. Pacific Sci. 21, 364-381 (1967).

Sand, O.: An electrophysiological study of auditory masking of clicks in goldfish. Comp. Biochem. Physiol. 40A, 1043-1053 (1971).

Sand, O.: Directional sensitivity of microphonic potentials from the perch ear. J. Exp. Biol. 60, 881-899 (1974).

Sand, O., Enger, P. S.: Evidence for an auditory function of the swimbladder in the cod. J. Exp. Biol. 59, 405-414 (1973).

Sand, O., Hawkins, A. D.: Acoustic properties of the cod swimbladder. J. Exp. Biol. 58, 797-820 (1973).

Sand, O., Michelsen, A.: Vibration measurements of the perch saccular otolith. J. Comp. Physiol. 123, 85-89 (1978).

Saunders, J.: Psychophysical analysis of pure tone masking in the parakeet. In: Hearing and Davis. Hirsh, S., Eldridge, D., Hirsh, I., Silverman, S. (eds.). St. Louis: Washington University Press, 1976, pp. 199-212.

Schneider, H.: Die Bedeutung der Atemhöhle der Labyrinthfische für ihr Hörvermögen, Z. vergl. Physiol. 29, 172-194 (1941).

Schuijf, A.: Directional hearing of cod (*Gadus morhua*) under approximate free field conditions. J. Comp. Physiol. 98, 307-332 (1975).

Schuijf, A.: The phase model of directional hearing in fish. In: Sound Reception in Fish. Schuijf, A., Hawkins, A. D. (eds.). Amsterdam: Elsevier, 1976a, pp. 63-86.

Schuijf, A.: Timing analysis and directional hearing in fish. In: Sound Reception in Fish. Schuijf, A., Hawkins, A. D. (eds.). Amsterdam: Elsevier, 1976b, pp. 81-112.

Schuijf, A., Hawkins, A. D.: Sound Reception in Fish. Vol. 8. Amsterdam: Elsevier (1976).

Steen, J. B.: The swimbladder as a hydrostatic organ. In: Fish Physiology, Vol. IV. Hoar, W. S., Randall, D. J. (eds.). New York: Academic Press, 1970, pp. 413-443.

Stetter, H.: Untersuchungen über den Gehörsinn der Fische besonders von *Phoxinus laevis* L. und *Amiurus nebulosus* Raf. Z. vergl. Physiol. 9, 339-477 (1929).

Stipetic, E.: Über das Gehörorgan der Mormyriden. Z. vergl. Physiol. 26, 740-752 (1939).

Tavolga, W. N.: Sound production and detection. In: Fish Physiology, Vol. V. Hoar, W. S., Randall, D. J. (eds.). New York: Academic Press, 1971, pp. 135-205.

Tavolga, W. N.: Signal/noise ratio and the critical band in fishes. J. Acoust. Soc. Amer. 55, 1323-1333 (1974).

Tavolga, W. N.: Recent advances in the study of fish audition. In: Sound Reception in Fishes. Tavolga, W. N. (ed.). Benchmark Papers in Animal Behavior, Vol. 7. Stroudsburg, Penna.: Dowden, Hutchinson and Ross, Inc., 1976, pp. 37-52.

Tavolga, W. N.: Mechanisms for directional hearing in the sea catfish (*Arius felis*). J. Exp. Biol. 67, 97-115 (1977).

Tavolga, W. N., Wodinsky, J.: Auditory capacities in fishes. Pure tone thresholds in nine species of marine teleosts. Bull. Am. Mus. Nat. Hist. 126, 177-240 (1963).

Vaitulevich, S. F., Ushakov, M. N.: Holographic study of swim bladder vibrations in *Cyprinus carpio*. Biofizika. 19, 528-533 (1974).

van Bergeijk, W. A.: Directional and non-directional hearing in fish. In: Marine Bio-Acoustics. Tavolga, W. N. (ed.). Oxford: Pergamon Press, 1964, pp. 281-299.

van Bergeijk, W. A.: The evolution of vertebrate hearing. In: Contributions to Sensory Physiology. Neff, W. D. (ed.). New York: Academic Press, 1967a, pp. 1-49.

van Bergeijk, W. A.: Discussion. In: Marine Bio-Acoustics II. Tavolga, W. N. (ed.). Oxford: Pergamon Press, 1967b, pp. 244-245.

Viemeister, : Temporal factors in audition: a systems analysis approach. In: Psychophysics and Physiology of Hearing. Evans, E., Wilson, J. (eds.). New York: Academic Press, 1977, pp. 419-427.

von Békésy, G.: Experiments in Hearing. New York: McGraw-Hill (1960).

von Frisch, K.: The sense of hearing in fish. Nature. 141, 8-11 (1938).

Wenz, G. M.: Curious noises and the sonic environment in the ocean. In: Marine Bio-Acoustics. Tavolga, W. N. (ed.). Oxford: Pergamon Press, 1964, pp. 101-119.

Werner, C. F.: Das Gehörorgan der Wirbeltiere und des Menschen. Leipzig: G. Thieme (1960).

Wersäll, J., Flock, A., Lundquist, P-G.: Structural basis for directional sensitivity in cochlear and vestibular sensory receptors. Cold Spring Harbor Symp. Quant. Biol. 30, 115-145 (1965).

Wever, E. G.: Cochlear stimulation and Lempert's mobilization theory. Principles and methods. Arch. Otolaryng. 909, 68-73 (1969).

Wever, E. G.: The mechanics of hair-cell stimulation. Trans. Amer. Otol. Soc. 59, 89-107 (1971).

Wohlfahrt, T. A.: Untersuchungen über das Tonunterscheidungsvermögen der Elritze (*Phoxinus laevis* Agass). Z. vergl. Physiol. 26, 570-604 (1939).

Chapter 2

Underwater Localization–A Major Problem in Fish Acoustics

ARIE SCHUIJF[1] AND ROBBERT J. A. BUWALDA[2]

1 General Introduction

Only recently have data on acoustic localization by aquatic vertebrates become available. It has quickly become apparent that it is not possible to extrapolate localization mechanisms from terrestrial vertebrates to fishes, whereas this appears possible for certain pinniped mammals (Moore and Au 1975).

At present the amount of data on localization by fishes is still very limited as compared with those on localization by terrestrial vertebrates (e.g., see Knudsen, Chapter 10; Capranica and Moffat, Chapter 5; Gourevitch, Chapter 12).

Sound detection by fishes has been the subject of many reviews of and discussions in research papers (e.g., Dijkgraaf 1952, Griffin 1955; van Bergeijk 1967, Popper and Fay 1973; Hawkins 1973, Fay 1978a; Fay and Popper, Chapter 1). While considerations on acoustic localization in fish are usually present in these papers this subject is more extensively discussed in specialized reviews (e.g., von Frisch and Dijkgraaf 1935, Reinhardt 1935, van Bergeijk 1964, Moulton and Dixon 1967, Popper, Salmon, and Parvulescu 1973, Schwartz 1973, Schuijf 1974, 1976a, Schuijf and Buwalda 1975, Myrberg, Gordon, and Klimley 1976, Sand 1976, Tavolga 1976, 1977).

The aim of this chapter is, first, to provide an up to date critical review on acoustic localization in fishes and, second, to place the various models of the detection system in a framework. By excluding trivial mechanisms, like the acoustic kinesis (gradient seeking), from this review, acoustic localization is understood to imply some form of directional hearing. While acoustic localization then also implies distance perception, this aspect is only touched on speculatively, in view of the complete lack of pertinent data.

Two more topics will be omitted. The ability of some surface dwelling fishes, like the topminnow (*Aplocheilus delineatus*), to locate sources of surface waves has been adequately analyzed and documented by Schwartz (1965, 1973); few new data have been obtained since. "Echolocation" in the sea catfish (*Arius felis*) (Tavolga 1971,

[1] Sections 1, 2 and 4.
[2] Sections 2 and 3.
[1,2] Laboratory of Comparative Physiology, State University of Utrecht, Jan van Galenstraat, 40, Utrecht, The Netherlands.

1976), on the other hand, is still virtually *terra incognita* as regards the mechanisms involved. Both echolocation and surface wave detection are highly specialized forms of acoustic localization found in only a few species. Discussing the mechanisms involved is unlikely to be of much value in placing the various models of the detection system in a framework.

1.1 Differences between Auditory Localization in Fishes and in Man

In theory, fish might utilize the same binaural differences for directional cues as does man. However, the close proximity of the fish's ears and the high speed of underwater sound propagation produce differences in stimulus timing that arise from unequal sound path lengths several orders of magnitude smaller than in man. Moreover, most fishes are small in comparison to a wavelength, and their bodies are about as dense as water. Both of these factors reduce possible binaural differences in stimulus strength to levels beyond discrimination. The physical separation of the left and right ears, of prime importance in human directional hearing, is therefore unlikely to provide fish with a localization cue.

Fish, as opposed to man, possess a lateral line system for detecting mechanical stimuli in addition to their labyrinths. Until now it has not been provided conclusively that the lateral line participates in localizing sound sources. Consequently, models of acoustic localization in fish can be divided into: those that involve the lateral line system; those that involve the labyrinths; or a combination of both sensory organs.

In 1964 van Bergeijk proposed his influential model of directional hearing asserting that only the lateral line system can convey the directional information contained in the incident sound wave. Binaural directional hearing in the "horizontal" plane with the bilateral labyrinths would be impossible, according to van Bergeijk, because the pulsating swimbladder would mask any difference between the labyrinths in response to the particle motions in the incident wave. The underlying notion is that the acoustic pressure, a nondirectional quantity, forces the swimbladder to pulsate independently of the direction of incidence in a process that is generally much more effective in transferring vibrations to the otolith organs than directly through inertia. This mechanism, in addition to the factors already mentioned, should effectively reduce the two ears to a single, nondirectional receptor. Van Bergeijk concluded that acoustic localization is only possible in the near-field of the sound source where, at normal intensities, the water displacements can exceed the threshold for the lateral-line receptors.

After van Bergeijk argued that directional hearing would be impossible in the far-field of a sound source, discussions arose regarding where to fix the boundary of the near-field. Acoustic theory shows that the "radius" of the near-field should be measured in wavelength units. In other words, the extent of the near-field should be expressed in terms of the acoustic distance to the sound source: $kr = 2\pi r/\lambda$ where r = radial distance to the sound source, λ = wavelength of the pure tone, and k = wave number of the sound wave. The wave number equals the phase difference in radians between two points at unit distance along the propagation direction of a plane sound wave. The wave number thus equals $2\pi/\lambda$. The near-field/far-field transition is agreed on by most workers in fish acoustics to occur at $kr = 1$ for harmonically pulsating (breathing) spheres, the so-called monopole sources. The gradients of the acoustic pressure in the near-field are large compared to those in the far-field. These gradients are directly responsible for the magnitude of the particle motions in the sound wave. Since only the motions in the near-field are adequate to stimulate the lateral-line system of the fish, the acoustic distance will be one of the factors (in van Bergeijk's theory) controlling the fish's ability to localize the sound source.

An analysis of the merits of this model (Schuijf 1976a) pointed to two weak or wrong assumptions: one is that the coupling of the swimbladder is such that it precludes directional hearing with the labyrinths and the second is that the lateral line is involved in acoustic localization. Dijkgraaf (1964) criticized the model on the latter point. The fact that Harris and van Bergeijk (1962) demonstrated the directional response of the lateral-line microphonics to the near-field of a vibrating object is insufficient evidence for the use of the lateral-line system by the fish for locating sound sources.

In a model for acoustic localization that only involves the labyrinths, proposed by Dijkgraaf (1960), there is, in principle, no distance limit to directional hearing. Dijkgraaf suggested that the propagation direction would be detected in the "horizontal" plane through a bilateral pair of otolith organs that do not lie in parallel planes. In this view the otolith organs form two inherently directional acceleration detectors having different axes of optimal sensitivity. Determination of the angular position of a sound source is performed in this model by using the ratio between the effective stimulus strengths for the two detectors. The directional cue here is the orientation of the imaginary line along which the particle accelerations in the incident wave act. An auditory system based on three or more of such detectors with noncoplanar directivity axes can, in theory, serve for detection of elevation as well as azimuth of a sound source.

The theoretical considerations above indicate that studies in acoustic localization by fish should start by regarding the directional hearing system as a black box whose functional organization and input variables are for the greater part unknown and, moreover, likely to exhibit considerable interspecific variation.

An analysis of this black box typically starts by acquiring data on the performance of the entire system. Lack of knowledge concerning the adequate input conditions dictates that the acoustic environment imposes no a priori restrictions on the potential localization cues. In practice this demand is only met by experimenting in the field. The value of reports on the most important single feat of performance, i.e., whether acoustic localization is possible at all, is determined largely by such experiments. The experimental evidence for directional hearing in fishes is discussed in Section 2.

Manipulation of the stimulus conditions is necessary to test which cues are relevant for localization. The results of such analyses allow the stimulus situation to be reduced to the essentials and enable one to produce appropriate directional stimuli in the laboratory. It is then feasible to probe the black box in suitably contrived experiments, thus to acquire a better understanding of its organization and—the last step in the black-box approach—the relationship between function and morphology. Section 3 deals with these aspects. The conclusions from the data in Sections 2 and 3 can, under certain assumptions, be framed into an organizational model of the operation of the entire detection system (see Section 4). By then, it will be evident that acoustic localization, involving virtually all aspects of fish auditory physiology and morphology, is indeed a major topic in fish acoustics.

2 Behavioral Evidence for Directional Hearing

2.1 Introduction

Although the need for experimentation in the field in studies of acoustic localization by fishes was recognized as early as in 1935 by von Frisch and Dijkgraaf, the first successful application was not possible until the development and availability of the U.S. Navy Underwater Sound Reference Laboratory type J9 sound projector and similar electrodynamic designs that enable the production of free field conditions of a deep mid-water environment.

The evidence for directional hearing that has accumulated during the past 15 years is reviewed in this section from comparative and methodological points of view. The first aspect is centered on the presence or absence of a swimbladder and further auditory specializations, since these factors are of considerable importance both in fish audition in general (see Fay and Popper, Chapter 1) and in acoustic localization (see Section 1). The methodological aspects relevant to evaluating the strength of the evidence include considerations of the acoustic environment used in the experiment and the related problems of the "quality" of the directional stimuli, the experimental design and nature of the behavioral response, and possible controls guaranteeing that only the directional aspects of the stimulus were responded to.

2.2 Fish with Swimbladders

2.2.1 Fish without Specialized Swimbladder-labyrinth Connections

A straightforward way of testing directional detection is to study the ability of fish to discriminate between differently positioned, but equidistant and equally loud, sound sources.

Such an approach has been attempted successfully with cod (*Gadus morhua*) by Olsen (1969a) (also cited in Sand and Enger 1974) and Schuijf (1975). Both were working from rafts moored in the middle of a Norwegian fjord in order to optimize the acoustic conditions. Both employed a two-alternative food-rewarded conditioning paradigm, in which the experimental subject was required to indicate the active one of two (left and right) sound projectors by swimming to either of two opposing corners of its confining cage. Although Olsen provides no data on the possible occurrence of different discrimination cues other than direction of sound incidence, Schuijf reports a gradual disappearance of discrimination concomitant with diminishing the angular separation between the two sound sources. The identity of the sources was apparently not relevant for positive discrimination in Schuijf's experiment. However, even virtually identical sound sources are subject to the effects of asymmetries in the acoustic environment. A thorough analysis of the prevailing acoustic conditions, such as those described by Schuijf (1975), is therefore needed to exclude such position effects.

Symmetrical acoustics are also required to corroborate the validity of interchanging the sound sources as a control for determining position. Such a control was applied by Schuijf and Buwalda (1975) in a variation of Schuijf's (1975) discrimination experiment. This variation was contrived to test the ability of cod to discriminate between sound waves coming from the direction of the fish's head or tail. The positive results obtained in this binaurally symmetrical situation indicate that binaural differences are not always required for determination of the direction of the sound source.

An experimental task that is even less demanding than absolute discrimination is the detection of a change in the direction of sound incidence. Again using a free-field situation, Schuijf, Baretta, and Wildschut (1972) have shown through the food conditioning of a wrasse (*Labrus bergylta*), that the fish will readily respond when the sound is repeatedly switched from one sound projector to another in an otherwise uninterrupted train of pulses when the speakers have an angular separation of 71°. This is not so, however, when this separation is minimized. A similar result was found for cod and haddock (*Melanogrammus aeglefinus*) using a classical heart rate conditioning paradigm, a technique that seems especially suited for such discrimination experiments. These fishes were confined to a small opaque, cylindrical cage to prevent "scanning" head or body movements and to keep them from seeing the sound sources. The cage was suspended 6 m from the sea bottom in 25 m deep water. Both these species proved capable of directional discrimination in the horizontal plane (Chapman and Johnstone 1974) as well as in the median vertical plane (Hawkins and Sand 1977). In a similar set-up Buwalda, Schuijf, and Hawkins (in ms.) ascertained a cod's ability to discriminate between diametrically opposed sound sources in the median vertical and transverse vertical planes, and found that at least some fishes can discriminate in situations that are ambiguous or otherwise confusing for man and for animals with similar directional hearing mechanisms.

Chapman (1973), Chapman and Johnstone (1974), and Hawkins and Sand (1977) have likewise shown that a directional masking phenomenon exists in fishes. They found the threshold for detecting a low frequency pure tone in broad band noise to be a function of the angular separation between noise source and signal source. The signal-to-noise ratio at threshold was highest when both signal and noise came from the same direction and dropped about 7 dB at an angular separation of 20°, both in the horizontal and in the median vertical plane. Such an effect illustrates the working of one of the two basic techniques for improving the signal-to-noise ratio employed in sensory systems: the tuning of a spatial or a spectral window to the desired signal. The fish's auditory system apparently utilizes both techniques (see Fay and Popper, Chapter 1 on critical bands in fish).

Having thus established the simpler manifestations of directional hearing, the results of Schuijf and Siemelink (1974) and Schuijf (1975), who demonstrated that cods are capable of acoustic orientation and not merely of a left-right or fore-aft discrimination can logically be discussed. In both studies elaborate controls were employed in a four-alternative, food-rewarded choice conditioning paradigm (i.e., making sure of symmetrical free-field conditions, interchanging the sources, and denying visual orientation to the sources) and it was ascertained that the fish was only responding to directional acoustic cues. It is necessary to point out, however, that while a conditioned response toward food dispersers that are in line with an equal number of sound sources is proof for recognition of a directional cue, it is not proof for acoustic orientation in the sense of true direction detection. The proper experimental design for showing direction detection would be to include control stimulus directions that are new for the fish and to observe whether the directional responses correspond to the actual physical stimulus direction. Until such an experiment has been performed, the results of Schuijf (1975) remain the best and most complete proof for directional detection in fish.

2.2.2 Hearing Specialists

The swimbladder-labyrinth connection is decisive in separating fish into hearing specialists and nonspecialists (cf. Fay and Popper, Chapter 1). Although a fish should properly be called a hearing specialist only if its superior hearing capacities are shown to be dependent on the presence of specializations such as the swimbladder-labyrinth connection, all fish possessing swimbladder adaptations apparently useful for better hearing will be accepted in this section. All such specializations operate on the principle of optimizing the coupling between the swimbladder and labyrinths. Thus van Bergeijk's argument regarding masking of potentially directional information to the labyrinths by the essentially nondirectional information from the swimbladder should apply here.

While the data on directional hearing by specialists are still limited in quantity and quality, the originally negative evidence (von Frisch and Dijkgraaf 1935, Reinhardt 1935) is now being replaced by more positive results.

Moulton and Dixon (1967) have shown that the polarity of the tail flip reflex of the goldfish (*Carassius auratus*) depends on the side of incidence of the sound stimulus. Although the strength of this evidence is severely marred by the complex acoustics prevailing in the test conditions, the experiments are relevant in showing that the sacculi are involved in the discrimination.

Substantially better acoustic conditions were used in the studies by Olsen (1969b) with herring (*Clupea harengus*) and Popper et al. (1973) with squirrelfishes of the genus *Myripristis*. In both cases a directive influence on the locomotion patterns of unconditioned groups of fishes confined in a cage was found: the squirrelfishes tended to turn toward the active one of a pair of sound projectors; the herring turned away.

Following these authors in using a field situation but, unlike them, providing a control by irregularly interchanging the two sources, Schuijf, Visser, Willers, and Buwalda (1977) succeeded in demonstrating the ability of a conditioned ide (*Leuciscus idus*) to discriminate between sound waves coming from the direction of its head and those impinging on its tail.

It then appears that the close association of swimbladder and labyrinths does not preclude (coarse) angular discrimination in hearing specialists, contrary to the suggestion of van Bergeijk (1964).

2.3 Fish without Swimbladders

Convincing evidence for acoustic localization in fish without swimbladders is, so far, only available for sharks. For example, Nelson's (1967) demonstration of conditioned angular discrimination in the lemon shark (*Negaprion brevirostris*) is similar in design to many of the above experiments. However, Nelson's experiments may be less reliable owing to the acoustical shortcomings of the shallow pool in which the experiments were conducted.

The bulk of data on localization in sharks stems from acoustic attraction experiments, conducted in relatively deep water with, mostly, carcharhinid and sphyrnid species (e.g., Nelson and Gruber 1963, Nelson and Johnson 1972, Myrberg et al. 1976). While such experiments are of considerable interest, there are a number of problems in infering acoustic localization from acoustic attraction experiments. These include knowing neither the range of attraction *a priori* nor the number of sharks present in this area before the onset of attraction. These data are necessary for a statistically valid account of an induced aggregation of sharks (Myrberg et al. 1976) and can perhaps best be obtained in aerial observations (Nelson and Gruber 1963). From such observations the range of attraction sometimes appears to exceed 200 yards. Another factor complicating the statistics is the effect of animals following one another, the contribution of which to aggregation is difficult to estimate. Finally it must be realized that the visual presence of the often very conspicuous sound projector may provide a steering stimulus. It is not sufficient to demonstrate that a silent projector is not attractive, because the absence of attractive sound as a releasing stimulus also eliminates the need for a steering stimulus.

3 Properties of the Detection System

3.1 Introduction

However beneficial van Bergeijk's influence may have been on the acoustic insights of the biologists studying fish hearing, his very authority has long prevented an open view on the problem of acoustic localization. Very few of the many studies instigated by his theoretical work have dealt with this subject directly. They have concentrated primarily on the major task of proving van Bergeijk's conclusions wrong by unequivocally demonstrating the directional hearing ability in fishes (see Section 2). Data of a more descriptive character on the directional detection system involved are consequently still rather scanty. This section will review some properties of the detection system: the operating system's parameters, its functional organization, and the morphological and functional properties of its constituent parts.

3.2 System Parameters: Bounds for Angular Detection and Discrimination

3.2.1 Angular Resolving Power

As argued in Chapter 1 the sense of hearing is very important in providing fishes with information on their surroundings. The accuracy and, concomitantly, the usefulness of the "acoustic image" of the outer world depends on the spatial resolving power of the detection system. In this respect, the fish's hearing system is no match for some of the more advanced terrestrial and aquatic mammals, i.e., man and bottlenose porpoise

(*Tursiops truncatus*). Whereas these can boast of an angular resolution better than one or two degrees of arc (Howard and Templeton 1966, Renaud and Popper 1975), the few experiments conducted with fish (all of them using pure tones in the low frequency range) indicate a just noticeable difference (j.n.d.) greater by one order of magnitude. As compared with many other vertebrates (cf. Gourevitch, Chapter 12), however, fish do not perform that badly.

The actual j.n.d. values for angular resolution by fish seem very dependent on experimental design. While Schuijf (1975), employing a rather permissive reward conditioning paradigm, estimated angular resolution in cod to be no better than 45° at 75 Hz, Nelson (1967) found the appreciably smaller value of 19° (mean from pooled data at 40 Hz, 80 Hz, and 320 Hz) in a two-alternative forced choice conditioning experiment with lemon sharks. As the experimental task contained elements of acoustic orientation, Nelson's direct estimate of the mean orientation error, which amounts to the even lower value of 7.1°, may also be relevant. A negative reinforcement regimen stimulates cod and haddock to a comparable performance. Chapman and Johnstone (1974) and Hawkins and Sand (1977) demonstrated that at about 100 Hz and with high intensity levels, these species can readily be conditioned to switch between two identical sound projectors 20° apart in the horizontal plane and 16° apart in the median vertical plane. Discrimination deteriorated with lower sound levels. In a similar heart rate conditioning paradigm, Buwalda (unpublished results) diminished the angular separation between the sources in a stepwise fashion and found the angular j.n.d. for cod to range from 8° to 21° (mean 13.6°) at 105 Hz under optimal conditions. Buwalda subsequently established that the magnitude of the j.n.d. depends on the signal-to-noise ratio rather than on the signal level itself (see Fig. 2-1).

Apart from giving an idea of the directional acuity of the piscine hearing system, the magnitude of the j.n.d. is of rather limited value in providing an insight in the mechanisms involved. It is a fallacy to suppose that animals with comparable directional acuity must have similar directional hearing systems. For example, comparing the limited data on fish with those of pinniped mammals (Gentry 1967, Terhune 1974, Moore 1975, Moore and Au 1975) shows these animals to have comparable directional acuity. It is physically and physiologically improbable, however, that fish use the binaural difference cues employed by pinnipeds, whose directional hearing system is akin to that of other mammals (Moore and Au 1975).

It appears more rewarding to study the angular resolving power for fishes as a function of stimulus conditions, such as acoustic impedance, signal-to-noise ratio, etc. Information pertaining to the nature of the mechanism involved might be obtained through a detailed analysis of the spatial distribution of angular resolution. It is a textbook matter now that for man the just noticeable angular difference is a function of the orientation of the sound source relative to the listener's own frame of reference, with best resolution occurring in front or behind and relatively poor separation of sources from the side. Such a result is to be expected, in view of man's directional hearing system operating on binaural sound path length differences. As a general rule it can be stated that the function describing the relation between the sound source position and the magnitude of the directional cue will be characteristic for a given system. Consequently the angular derivative of this function, i.e., the distribution of angular resolution, will also be characteristic and may thus become a valuable tool in

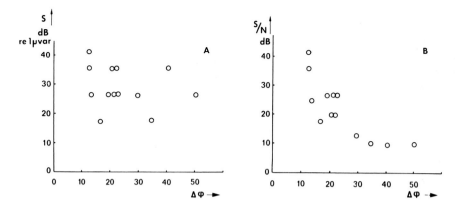

Figure 2-1. Angular discrimination threshold data (horizontal axis) for cod plotted as a function of (A) signal intensity level expressed in dB re 1 μvar (1 μvar = 1 μbar/ρc = 6.7 $\times$ 10^{-8} m/s) and of (B) signal-to-noise ratio. Only in the latter case is a relation apparent between the just noticeable angular difference $\Delta\phi$ and the ordinate variable. The data indicate a minimum j.n.d. of about 13° (at 105 Hz in the horizontal plane) and a minimally required signal to noise ratio for directional discrimination of about + 10 dB.

identifying the underlying mechanism. Relevant data are, however, still sadly lacking for fish. Future research might thus well be directed towards this potentially rewarding subject.

3.2.2 Intensity Threshold Phenomena

In absolute threshold determinations under free-field conditions, the detection of discrimination of sound direction fails at an appreciably higher stimulus level than detection of sound presence (Chapman and Johnstone 1974, Schuijf 1975, Hawkins and Sand 1977). This phenomenon is consistent with the notion that two different subsystems are operative within the fish's hearing system: one responsible for directional detection and therefore restricted to input variables carrying directional information, the other free to use any available input, directional or not.

In point of fact, some fishes, when faced with a signal detection task, apparently respond to either the acoustic pressure or to a kinetic variable, whichever is prevailing in the given acoustic conditions (Cahn, Siler, and Wodinsky 1969, Chapman and Hawkins 1973, Buwalda, Portier, and Schuijf, in ms.). Directional detection, on the other hand, seems dependent on a kinetic variable as shown by the findings that the angular j.n.d. is influenced (masked) by the particle motion component of noise but not by the pressure component (Buwalda, unpublished results) and by observations that the intensity threshold for directional discrimination is apparently best expressed in terms of a kinetic variable (Chapman and Johnstone 1974). Expressed as acoustic displacement, this threshold varies from about 2×10^{-10} m to 10^{-11} m and thus falls in the range of the calculated displacement sensitivity of the otolith organs of fishes (Chapman and Hawkins 1973, Chapman and Sand 1974). A corollary of these facts is that under suitable stimulus conditions the threshold for directional discrimination should approach the threshold for signal detection.

3.2.3 Distance Range

Since van Bergeijk's assumption (that acoustic localization is impossible in the acoustic far-field) is the crux of his model of directional hearing with the lateral line, the acoustic distance to the sound source at which localization is still possible becomes an important parameter in deciding on a mechanism of directional hearing in fish. In the majority of the experiments mentioned in Section 2, the distance limit to directional detection was set by the mechanical limitations of the set-up rather than by the inability to localize the distant objects on the part of the experimental subjects. As the acoustical distances ($kr = 2\pi r/\lambda$) to the source exceeded unity in a number of cases (Olsen 1969a, 1969b, Chapman 1973; Chapman and Johnstone 1974, Schuijf and Siemelink 1974, Schuijf 1975, Schuijf and Buwalda 1975, Hawkins and Sand 1977), directional hearing is apparently not restricted to the near-field. Only the experiments of Popper et al. (1973) provide evidence against this notion, although Popper and Fay (1973) later ascribed the reported disappearance of the directive influence on locomotion patterns at sound source distances over about 2 m to 3 m to reluctance to respond rather than to a sensory inability to react to sources outside of the near-field.

The thesis that directional hearing is not subject to a distance limen provided that the signal-to-noise ratio is sufficiently high is corroborated by the results of Olsen (1976), who was able to condition a school of saithe (*Pollachius virens*) to feed near an active source (emitting 150 Hz pure tones) and to choose correctly and immediately between sound sources some 80 m apart that were placed in such a fashion as to exclude position learning effects. Pertinent evidence is also provided by the acoustic attraction experiments on sharks, which can apparently be lured from hundreds of yards in a straight course to the sound source (cf. Myrberg et al. 1976).

3.2.4 Frequency Range

At a fixed distance from the source, the ratio of acoustic pressure to the motional variables is affected by sound frequency because the near-field extent is frequency dependent (cf. Siler 1969). Moreover, the pressure-to-displacement transforming effect of the swimbladder provides a gain over the displacement sensitivity of the otolith organs that is, theoretically, proportional to frequency (Sand and Hawkins 1973). Both effects add to the swimbladder's efficiency with increasing frequency.

While directional hearing at very low frequencies might be possible regardless of the validity of van Bergeijk's assumption on the swimbladder's masking of directional information, there should, as a consequence to this theory, be an upper-frequency limit to directional hearing that would coincide with the gain of the swimbladder effect exceeding unity and the fish becoming effectively pressure sensitive. Whereas this point is reached at about 50 Hz to 100 Hz in cod (Chapman and Hawkins 1973, Sand and Hawkins 1973, Sand and Enger 1973, Buwalda and van der Steen 1979), Olsen (1969a) and Chapman and Johnstone (1974) found directional discrimination to persist up to at least 300 Hz to 400 Hz in cod, which is then completely comparable to the lemon shark without any swimbladder (Nelson 1967). It is possible that the upper limit of directional hearing falling short of the upper limit of sound detection in cod was more due to subliminal displacement stimuli at the highest frequencies than to a fundamental failing of the directional detection system.

3.3 Functional Organization: A Multiple Input Detector

3.3.1 The Effect of Phase Shift between Acoustic Pressure and Particle Motion

It will be clear by now that several acoustic variables, whether inherently directional or not, are available to the auditory system of fishes with a swimbladder. Directional hearing is apparently based on a motional variable, which is considered to be the particle acceleration $a(t)$, for reasons discussed in Section 4. As follows from the previous section, directional detection persists under a variety of conditions with widely ranging pressure-to-motion ratios. It is tempting to explain such a result by assuming that at least part of the motion-sensitive detectors constituting the directional detection system are not affected by the pressure-to-motion transforming action of the swimbladder; there should be only a weak coupling of pressure into the directional hearing system. This, indeed, was van Bergeijk's rationale for attributing directional hearing to the lateral line. (He apparently overlooked, though, that the lateral line, too, might respond to the near-field set up by the swimbladder pulsations.) Such weak coupling would of course optimize the transfer of directional information to the Central Nervous System (CNS) of the fish. It appears, however, that the acoustic pressure, the $p(t)$ input, does exert control over the directional choice of fish, in addition to the motional variables in the sound field.

Schuijf and Buwalda (1975) discovered that the directional choice of cod could be affected by a phase inversion of $p(t)$ with respect to the unaffected particle acceleration $a(t)$. The subject was trained to discriminate sounds of opposing propagation direction by swimming toward the direction of the sound (a 75 Hz pure tone) source. However, when the active source was located behind the animal, for instance, and the accompanying acoustic pressure $p(t)$ was artificially shifted $180°$, the cod swam forward. As the oscillatory motions were not affected by the phase inversion in $p(t)$, the implication is that the phase relationship between $p(t)$ and $a(t)$ is relevant for part of the processes in acoustic localization and that both $p(t)$ and $a(t)$ are necessary for directional detection. Buwalda et al. (in ms) have shown similar results for the cod in the median and transverse vertical planes as well.

A phase analysis between $p(t)$ and $a(t)$ also occurs in acoustic localization by Ostariophysi. Buwalda et al. (in ms. a) showed this in a laboratory experiment with *Leuciscus idus* using methods analogous to the field study of Schuijf and Buwalda (1975).

They could make the conditioned fish turn away from the sound source when p(t) in the original wave was shifted 180° with respect to a(t).

Subsequently Buwalda et al. (in ms. a) studied the choice behavior of the conditioned animal to nonreinforced probe stimuli with variable phase differences between a(t) and p(t) but with a constant amplitude ratio between these variables (see Fig. 2-2). The clear structure of the data, with rather sharp transition regions occurring at phase differences ψ(a,p) between a(t) and p(t) of 0 and $\pm \pi$ radians, suggested to Buwalda et al. that a lead/lag detector, comparing its two inputs a(t) and p(t) for precedence, might be the underlying principle of the phase analysis between a(t) and p(t) governing the discrimination of sources 180° apart. The data of Fig. 2-2 should be compared with the phase cues existing near monopole sources in an unbounded medium (see Fig. 2-7).

In contrast, then, to van Bergeijk's opinion, the acoustic pressure p(t) is not a disadvantage imposed on directional detection, but its presence is even necessary. The conclusion can be that the multiple-input character of the total hearing system of fishes with a swimbladder is maintained down to the level of the subsystems involved in acoustic localization. Apparently the directional sensitivity of this system is related to a vector input, while the p-input is required to enable unambiguous *orientation* with such a directionally sensitive sensory subsystem (see Section 4 for a further discussion).

3.3.2 The Irrelevance of Spatial Differences in Phase

Experiments by the authors have disclosed that both normally propagating and standing sound waves may be employed in testing directional hearing in fish provided that, at the position of the fish, the acoustic variables satisfy certain conditions as to amplitudes, direction (of oscillation), and phase relations.

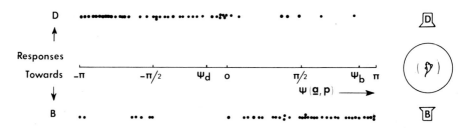

Figure 2-2. The dependence of the directional choice of an ide (*Leuciscus idus* L.) on the phase relationship ψ(a,p) between particle acceleration a(t), and acoustic pressure p(t). The fish, kept in a circular netting cage with a central arena, was trained to orient itself toward sound projectors B and D. The phase difference ψ(a,p) between a(t) and p(t) characteristic for the training situations are indicated by ψ_b and ψ_d. Dots indicate prompt, unconditioned responses toward either sound projector on administration of a probe stimulus in which ψ(a,p) could be varied artificially from $-\pi$ to π rads. Note the clear structure in the data, with response reversals occurring at 0 and $\pm \pi$ rads. The acceleration was measured to be positive when directed toward D, that is rostrad for the fish when orientated as in the inset figure.

Employing standing waves as opposed to propagating waves for locally producing adequate stimulus conditions has some interesting implications for the mechanism of directional hearing in fish. First, there is no such thing as wave front propagation in a standing wave and the inherent time difference cues between separated receptors are consequently absent. Second, a perfect simulation of the normal situation as to amplitude, direction, and phase of the involved acoustic variables is often only possible within a very restricted volume of space (sometimes only containing part of the body of big fishes like cod), and rather large differences, with respect to the normal propagating wave situation, may exist outside this volume. As directional detection remains unaffected in a standing wave (Schuijf and Buwalda 1975, for cod; Buwalda et al. in ms. a, for *Leuciscus idus*), and as standing and propagating waves are not even discriminated (Buwalda et al. in ms., for cod), it must be concluded that propagation of a characteristic spatial distribution of the acoustic variables—in short, spatial phase differences—is irrelevant as opposed to the phase differences between the acoustic variables (cf. Section 3.3.1).

3.3.3 Separate Processing of Acoustical Inputs in Fish?

When sensory processing in a given unknown system requires comparison of information from two or more inputs, finding the level in the system at which interaction between these inputs occurs is of primary importance in gaining insight into the functional organization of the system.

In the case of acoustic localization in fish, involving a phase comparison of acoustic pressure and particle motion, this interaction could well occur at the extreme periphery of the hearing system. Mechanical superposition processes between the "normal" (direct) particle motions and the "indirect" motions due to swimbladder pulsations might result in nontranslatory (rotational) oscillations characteristic of any given stimulus situation. A detector capable of resolving such oscillations might then provide the CNS with the information conveyed by a particular oscillation pattern. Such a mechanism is not readily reconciled with the notion of a weak coupling of pressure and motion apparent from the data in Section 3.2.

On the other hand, if separate reception and processing of the different inputs occurs, integration presumably takes place at a higher level, as in the CNS.

To provide a first answer to such questions Buwalda et al. (in ms. b) undertook to study the degree of perceptual segregation between the nondirectional p-input and the particle motion by means of cross-modality masking. The degree of segregation can be inferred from the efficiency of masking of one input by another. Masked auditory thresholds as a function of stimulus composition were determined for cod using classical heart rate conditioning. The stimulus generating system employed horizontal and vertical standing waves, permitting independent control of acoustic pressure (p), horizontal (u), and vertical (w) particle velocity for both signal (105 Hz pure tone) and masker (50 Hz wide-noise band, centered at 105 Hz), over a wide range. The fish's masked threshold appeared to depend on the ratios of p, u, and w in the signal and masker. The threshold for a signal with a high p/u or p/w ratio was determined by the p-level in the masker. The same was true, *mutatis mutandis,* for u and w. It proved impossible, within apparatus' limits, to mask a given sound variable (e.g., p) with another (e.g., u), at signal-to-noise ratios well below -10 dB to -30 dB. Such results clearly point to separate processing of pressure and motion in the cod's hearing system.

These experiments are strongly reminiscent of the studies by Cahn, Siler and Auwarter (1970) on *Haemulon parrai* (a grunt), both in the use of standing waves to control the p/v ratio and in employing the cross-modality masking paradigm. Their findings, implicating a masking of velocity detection by pressure, but not the converse, are very difficult to assess, however, partly because of a lack of description of their data and partly because the range of control of the p/v ratio was limited to about 25 dB for noise and 35 dB for pure tones (as opposed to 50 dB and 70 dB respectively in these authors' experiments). Nevertheless, their results indicate that the distinct perceptual segregation of acoustic inputs found in cod may not be characteristic for all or even most fish.

3.4 Considerations on the Involved Sensory Organs

3.4.1 Which Receptors are Involved?

3.4.1.1 *Lateral Line versus Labyrinth.* Now that the multiple input character of the directional hearing system of fish has been established, it becomes opportune to have a closer look at which receptors mediate directional hearing.

The obvious approach for assessing the involvement of a receptor in a given sensory task is to look for the effect of its elimination. Application of Dijkgraaf's (1973) method for severing the appropriate cranial nerve roots, which obviates many of the drawbacks of straightforward surgical extirpation, demonstrated that unilaterally severing the saccular and lagenar nerves deprives the cod of its acoustic localization ability, but not its acoustic detection ability (Schuijf 1975). Thus, it appears that the combination of one intact ear with intact lateral line is not sufficient for localization. This observation does not, however, rule out the possibility of the lateral line being involved in directional hearing. For a decision in that matter the reverse experiment is needed: eliminating the lateral line and observing the fish's directional detection remaining unimpaired.

A more subtle method for eliminating the contribution of a receptor is the creation of stimulus conditions known to be inadequate for the particular receptor. For instance, the lateral line is not stimulated in the pressure node of a standing wave, notwithstanding even extremely high particle motion levels (Cahn, Siler, and Fujiya 1973), evidently because the whole fish is carried along with the particle oscillations, and the non mass-loaded neuromasts fail to register such a motion. Such results, surprising as they were to some fish acousticians, have a hydrodynamic parallel in the failure of the lateral line to detect massive water currents displacing the entire fish (cf. Dijkgraaf 1963). The lateral line system, of course, can detect such stimuli as small squirts of water or (the acoustic parallel) the oscillations in the near-field of a sound source (Harris and van Bergeijk 1962) with its inherent steep gradients owing to divergence.

Whereas the lateral line is not stimulated in a standing wave, otolith organs are (Fay and Popper 1974, 1975, Hawkins and McLennan 1976, Buwalda and van der Steen 1979). The experiments conducted by Schuijf and Buwalda (1975) and Buwalda et al. (in ms., in ms. a) demonstrate the feasibility of standing waves in directional detection studies. Therefore, the conclusion seems justified that the lateral line, though directionally sensitive (Harris and van Bergeijk 1962, Horch and Salmon 1973, Tavolga 1977) and in this respect meeting the demands on an input channel processing the directional information conveyed by particle motion, is not an essential part of the detection system responsible for directional hearing. The directional detector is apparently confined to the otolith organs and associated structures.

3.4.1.2 *The Role of the Macula Neglecta in Sharks.* Although sharks are among the first fishes for which directional hearing has been demonstrated convincingly (see Section 2.3), rather less is known of the receptors involved than is the case for teleosts. Sharks lack a swimbladder, or comparable structures, and are consequently only sensitive to particle motion (Banner 1967, Kelly and Nelson 1975) (see, however, Section 4.4).

In addition to the sacculus and lagena, which may function as mass-loaded detectors of the accelerations in a sound wave, the macula neglecta has recently been credited with a function in acoustic detection in sharks. While the original evidence was largely of a morphological nature (Tester, Kendall, and Milisen 1972), both Fay, Kendall, Popper, and Tester (1974) and Bullock and Corwin (1979) succeeded more or less convincingly in demonstrating the vibration sensitivity of the macula neglecta and the crucial role of the fossa parietalis in conducting vibrations to that part of the labyrinth via an opening in the otic capsule (the fenestra ovalis). The macula neglecta, with its unloaded cupula, appeared to respond to a velocity component in the vibrations (Fay et al. 1974).

Circumstantial evidence for an acoustic functioning is found in a comparative morphological study of the macula neglecta in six elasmobranch species (Corwin 1978). The macula neglecta appeared best developed both as to relative magnitude and to internal organization of the hair cell fields in carcharhinid and other free-swimming species. Since these species show best responding in acoustic attraction experiments, there is a conspicuous correlation between the importance of acoustic localization for a shark and the development of its macula neglecta.

Findings such as these, led Corwin (1977) to suggest that the macula neglecta of carcharhinid sharks might be involved in directional hearing. As the perfectly parallel alignment of the hair cells should guarantee that the macula neglecta is optimally sensitive to vibrations incident parallel to the posterial canal duct containing the two macular hair cell patches, and as the left and right posterial canal ducts are approximately perpendicular to each other, Corwin (1977, personal communication) feels that both maculae neglectae constitute an orthogonal vector detector, somewhat comparable to Dijkgraaf's (1960) model of directional hearing with the otolith organs.

The problem in assessing Corwin's proposed mechanism for acoustic localization is that the functional organization of the non mass-loaded macula neglecta and associated structures is more suggestive of a typical near-field acoustic receptor. The experimental evidence for vibration sensitivity does not suggest otherwise, since only vibrators and short-range sound sources were used (Fay et al. 1974, Bullock and Corwin

1979). A decision on the role of the highly interesting macula neglecta will therefore have to await more physiological data, preferably obtained under realistic acoustic conditions.

3.4.2 Functional Properties Related to Directional Detection

3.4.2.1 *A Functional Organization of the Labyrinth.* The tendency for specialization within the labyrinth is apparent from the outset in the fact that in most fishes the receptors for position sense and for hearing reside in the pars superior and pars inferior of the labyrinth, respectively. As this gross morphological partitioning is tied to a segregation between perceptual modalities, so might the more subtle functional organization of the system involved with sound detection be correlated with some form of organization within the pars inferior. Indeed, the very complexity of this structure strongly suggests a regional functional differentiation. Not only do the three pairs of maculae containing the total complement of receptor cells in the fish ear, differ in dimensions, in spatial orientation, in being loaded with more or less massive otoliths or otoconial mass (or even with a cupula only) in a more or less close association with accessory structures, etc., but studies at the microstructural level, particularly those employing the scanning electron microscope, have revealed that regional differentiation, even within a given macula, is present as evidenced by hair cell orientation patterns and differences in kinocilium length, etc. (Lowenstein, Osborne, and Thornhill 1968, Dale 1976, Enger 1976, Jørgensen 1976, Popper 1976a, 1976b, 1977, 1978a, 1978b, Platt 1977, Fay and Popper, Chapter 1).

It is obviously relevant for the functional organization of the acoustic localization system whether such structural differentiation could result in separate detectors for acoustic pressure and for particle motion. Let us therefore try to assess, in general terms, the effect of some aspects of structural variation on input selectivity.

(a) Because a fish is about as dense as water and small compared to a wavelength, the particle oscillations in the acoustic far-field can only be detected by employing an inertial system, i.e., hair cells coupled to a calcareous otolith or otoconial mass. On the other hand, a group of haircells associated with an unloaded cupula-like structure, or coupled very loosely to an otolith, will respond only in rather strong spatial *gradients* of particle motion, such as can be found in the secondary near-field of the swimbladder and might thus indirectly be quite selective to pressure input in all but the most extreme near-field conditions. The degree of association with or coupling to an otolith may thus determine the p-selectivity.

(b) While fish tissues are generally considered to be essentially transparent to propagated sound waves, they might well behave differently to the near field of the swimbladder. Both the physical proximity of a particular receptor field to the swimbladder and the properties of the intervening tissues may influence the ratio of the field's "direct" particle motion sensitivity with respect to its sensitivity for the swimbladder vibrations. Whereas mass loaded hair cells within the skull are not shielded from the near-field motions of an external sound source because the rigid skull as a whole will be carried along with the particle motions, this need not be true for the induced motions from the swimbladder that forms an internal, secondary sound source. The crucial difference in excitation is that the skull as a rigid body is elastically constrained in its motion relative to the swimbladder's center, whereas this is not the case for the external, primary sound source (for arguments see Schuijf 1976b). Screening by rigid skull structures will then promote a specificity for the "direct" incident particle motion by reducing the effect of the secondary near-field, whereas association with near-field transparent, "acoustic window"-like structures could improve a coupling to the swimbladder and promote p-selectivity by enhancing the input via the "indirect" swimbladder route over the direct input.

(c) Where hair cell polarization is assumed to underly directional sensitivity in the acoustico-lateralis system, the clear receptor cell orientation patterns found in the sacculus and lagena suggest yet another mechanism for separating between inputs. Because of the fixed spatial relationship between swimbladder and labyrinths, the indirect vibrations reaching the otolith organs will have a fixed direction in the fish's reference frame. A macular or intra-macular hair cell field having an orientation pattern or sensitivity axis perpendicular to this direction would thus be insensitive to pressure and would make a perfectly selective detector for particle motion albeit in one direction only. Sound waves can, however, come from all directions and a realistic particle motion detector can therefore not rely on this sole principle. This same argument excludes directional sensitivity as a sufficient means for establishing p-specificity.

It thus appears from this certainly not exhaustive discussion, that there is quite a variety of mechanisms or principles, by which a certain part of the labyrinth may acquire a specificity for one of the available input variables. Some of these can be recognized in the arrangements found in Ostariophysi. In this group of fish the sacculus, containing a delicate fluted otolith (the sagitta), is functionally connected to the swimbladder via a hydraulo-mechanic system consisting of the endolymphatic transverse canal, the perilymphatic sinus impar and the Weberian ossicles (see Fay and Popper, Chapter 1). No such connection is present for the relatively large lagena. Whereas the sacculus will be stimulated very efficiently via its direct connection with the swimbladder, the lagena might even be screened from the latter's near field. The resulting differential sensitivity to the swimbladder vibrations might thus make the sacculus essentially a p-receptor, while the lagena could constitute the much-sought-after detector of that carrier of directional information, the particle motion. Fragmentary evidence for such a view was first provided by Furukawa and Ishii (1967), noting that while first order saccular neurons could be stimulated very efficiently in their stimulus generating system, the lagenar neuron's responses were correlated with the fish holder vibrating at its resonance frequency (200 Hz). Very recently the issue has been studied systematically by Fay and Olsho (1979). Sensitivity thresholds (at 100 Hz) in goldfish lagenar neurons, defined as the stimulus level producing a given degree of phase locking, varied when expressed in terms of acoustic pressure but clustered when expressed in terms of a motional variable, i.e., displacement (3 to 10×10^{-10} m in the most sensitive preparations). Previous measurements involving recording of microphonic potentials had shown the sacculus of Ostariophysi to respond to pressure at a sound intensity level some 40 dB lower than the level at which it would respond to particle motion, with deflated swimbladder (Fay and Popper

1974, 1975). The sacculus should thus be quite selective to pressure. However, Fay and Olsho (1979) found a group of saccular neurons that were similar in most respects to lagenar afferents. These so-called LF (low best frequency) saccular neurons are apparently mainly afferents from the posterior macula sacculi.

While Ostariophysi certainly possess pressure specific receptors in the anterior part of the sacculi, such a statement cannot be made with certainty for particle motion specific receptors. The lagena and posterior sacculus were motion specific in the acoustic conditions of Fay and Olsho's experiments, i.e., at pressure-to-motion ratios some 30 dB below the far-field ratio. Although it is certainly possible that lagena and/or posterior saccular macula will remain motion specific in conditions more closely resembling a far-field situation, calling these structures motion specific receptors will have to wait for experimental proof.

Whereas Ostariophysi may well possess functionally and physically separate pressure and particle motion detectors, the situation in nonspecialized fish with a swimbladder is less clear. Experiments in which the ratio of direct to indirect stimulation was somehow manipulated have been reported for only a few species. These experiments indicate that the sacculus may be motion specific (Fay and Popper 1975, for the mouth breeder *Tilapia*) or respond to motion below a certain frequency and to pressure above this frequency (Sand and Enger 1973, for the cod).

Such results, apart from showing profound interspecific differences are in fact worthless for the issue of functional differentiation, because they represent the characteristics of one labyrinthine part (i.e., the sacculus) as a whole. A functional differentiation within this part, or a different behavior of another structure, may thus go unnoticed. Pertinent data can only be obtained through a detailed analysis.

The (preliminary) results of the only study along such lines thus far for nonspecialists are reported by Buwalda and van der Steen (1979) for cod. Using standing waves for stimulus control and a microphonic null response as an indicator, they found complete cancellation of direct by indirect stimulation in the anterior and middle part of the saccular macula at a p/v ratio 10 dB to 15 dB below the far-field value, at 120 Hz. The apparent p-specificity of these parts should extend further down in the cod's hearing range (cf. Chapman and Hawkins 1973). The horizontally sensitive posterior part of the macula, however, proved 8 dB to 10 dB less sensitive to p than the rest of the sacculus, although it seems optimally situated to receive the swimbladder's nearfield. As this result implies a screening by the surrounding structures, nearby labyrinthine parts such as the lagena might even be less susceptible to swimbladder vibrations.

Attractive as this notion may be in its correspondence to the Ostariophysian condition, the only evidence yet is for a regional variation in differential sensitivity to pressure and motion. This, on the other hand, already provides a basis for further (neural) segregation of the directional and nondirectional inputs in cod.

3.4.2.2 *Directional Sensitivity*. Evidence that the fish ear satisfies the basic requirements of a directional detector was first provided by Enger, Hawkins, Sand, and Chapman (1973). Reasoning that vibrating a fish in air gives an adequate simulation of the kinetic part of underwater sound stimulation, they recorded saccular microphonic potentials in haddock (*Melanogrammus aeglefinus*) clamped to a horizontally vibrating table. The amplitude of the microphonics appeared to depend on the angle between the fish's longitudinal axis and the driving direction, thus confirming the directional sensitivity of the fish labyrinth to a motional variable implicit in the model for directional hearing in fish proposed by Dijkgraaf (1960) (see Section 1.3). Sand (1974) followed up these studies in the perch (*Perca fluviatilis*) with improved techniques and found a regional variation of directional sensitivity in the pars inferior. Whereas both the lagena and the sacculus responded to horizontal vibrations, the posterior part of the saccular macula was also quite sensitive to vertical stimulation and the lagena even more so. These results tally well with hair cell orientation patterns in the perch labyrinth: they are mainly horizontal in the anterior macula sacculi, changing gradually into a more vertical orientation in the posterior parts, including the lagena (Enger 1976, Popper 1977). Horizontal vibrations were most effective both in sacculus and lagena when incident at about $\pm$ 20° (and $\pm$ 200°) to the fish's mediosagittal plane, i.e., when incident along the intersections of the left and right macular planes with the horizontal plane.

A similar broad agreement between electrophysiologically determined directional sensitivity and morphological data was demonstrated by Fay and Olsho (1979) in goldfish. However, where Sand (1974) and Enger et al. (1973) employed the technique of recording microphonic potentials with its inherent spatial integrating effect, Fay and Olsho studied the directional characteristics of first order saccular and lagenar afferents, thus obtaining a much better spatial resolution (see Fig. 1-14 in Fay and Popper, Chapter 1). Some scattering in their results, then, is not completely surprising, since minor local deviations (less than 20°, Platt 1977) from the average orientation direction are to be expected. Fay and Olsho feel, on the other hand, that the quite large deviations encountered in some conditions (particularly in the sacculus) could be explained as being the result of complex behavior of the otolith in response to translational oscillations. Such complex behavior has also been suggested by Sand (1974) on the basis of a frequency dependent shift in directional sensitivity along the perch saccular macula and has subsequently been demonstrated by means of interferometric techniques (Sand and Michelsen 1977).

However this may be, it does not alter the fact that in both goldfish and in perch, a basis for directional hearing in three spatial dimensions is present in the bilaterally symmetrical arrangement of two detectors having axes of optimal sensitivity in the horizontal plane subtending an angle of about 30° to 40°, and having both an intramacular and intermacular variation in sensitivity to vertical vibrations. Exactly how the directional information is extracted from the output of the various directionally sensitive parts has not yet been demonstrated conclusively. The stimulus component in the horizontal plane is apparently found by comparing the outputs of the horizontally sensitive parts of the left and right ears, as proposed in the original model of Dijkgraaf (1960). Vertical components in the respective macular planes might be assessed monaurally on the basis of the outputs of more and less vertically orientated hair cell fields (Sand 1974). A combination of monaural and binaural interaction could thus determine the actual sound direction (with 180° ambiguity). On the other hand, the vertically sensitive parts of left and right ear might equally well partake in binaural "vector weighing" as the horizontally sensitive parts, defining position vector components in the fish's transversal and horizontal plane, respectively. In this case a

central processing effecting a vector addition of the results of more peripheral binaural interactions would determine the actual direction of the instantaneous particle
motion. Such mechanisms are compatible with the recent findings indicating a neural
topological representation of auditory space in vertebrates (Knudsen and Konishi
1978, Pettigrew et al. 1978, Knudsen, Chapter 10). However, such considerations
will remain largely speculative until more is known of the CNS of fishes and about
binaural processing in particular.

The very usefulness of directional sensitivities demonstrated through vibrating a
fish in air remains a matter of speculation as long as the influence of sound pressure
is not assessed. In goldfish the portion of the auditory system stimulated in such experiments may constitute particle motion specific receptors (see Section 3.4.2.1) but
such cannot be held for perch *a priori*. In contrast to the above studies, then, Buwalda
and van der Steen (1979) stimulated cod in an underwater sound field, in which the
ratio of particle motion to pressure could be varied from +50 dB to −10 dB relative
to the far field value by employing standing waves. The results of stimulation in both
extreme situations (at 120 Hz) can be seen in Fig. 2-3. A perfect cosine-dependence of

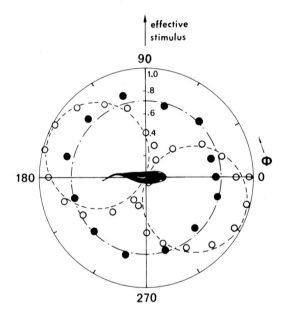

Figure 2-3. The sensitivity of the cod sacculus to directional and nondirectional stimuli.
The microphonic potentials measured in the anterior part of the right sacculus were
converted into effective input stimulus strengths via a stimulus response calibration
curve. Open circles represent data obtained on rotating the fish in an underwater sound
field with a ratio of pressure to horizontal particle velocity more than 50 dB below the
far-field ratio. The two tangent circles allow a postulated perfect cosine dependence of
the directional sensitivity to be compared with the data. The filled circles give the results obtained with a p/v· ratio of +10 dB (f = 122 Hz, as above). The deviations from
the circle with a radius corresponding to the average input strength are very slight, indicating that directionality is lost and, thus, testifying to the nondirectional character
of the pressure input (see text). From Buwalda and van der Steen (1979).

the directional sensitivity was observed only when relatively little acoustic pressure was present; the addition of pressure to more realistic levels progressively destroyed this pattern until at high p-levels the directionality was lost completely. Similar results were obtained for middle and posterior parts of the saccular macula, although in the latter case the effect of pressure was less pronounced (see Section 3.4.2.1). Directional sensitivity is then severely reduced in cod (and probably in most fishes with swimbladders) in many acoustic conditions, unless either a receptor more specific for motion or a neural mechanism for restoring the directional sensitivity can be found. Since the contribution of p should be invariant with direction, such a mechanism may well rely on rather simple operations.

A final point should be made concerning the influence of p on directionality. It is generally assumed, and indeed demonstrated in cod (see Fig. 2-3), that p constitutes nondirectional input for the fish's hearing system. However, Tavolga (1977) has shown in the sea catfish (*Arius felis*) that the amplitude of the swimbladder wall vibrations as measured with various transducers varies with the animal's orientation relative to the sound source, with the highest amplitudes occurring for sounds impinging on the head. It is not certain, though probable, that such behavior results in differential stimulation of the labyrinths. Anyway, more data are needed to know whether the occurrence of this phenomenon is restricted to *Arius felis* with its highly specialized swimbladder and accessory structures or whether it has a more general significance.

3.4.2.3 *Timing.* An analysis of the phase of the p(t) input relative to the a(t) input, necessary for eliminating the 180° ambiguity in determining sound source position from the direction of **a**, requires that the time structure of both inputs be adequately encoded in the afferent information flux up to the level of the auditory system at which the actual analysis is performed. This demand is, in principle, easily met. Phase locking of neural activity, ensuring that the time structure is not only encoded, but actually preserved (within certain physiological limits), seems to be an inherent property of most acousticolateralis afferents and has been observed in the VIIIth nerve of fishes by several authors (Enger 1963, Grözinger 1967, Furukawa and Ishii 1967, Fay 1978b, 1978c, Fay and Olsho 1979).

Recently Fay (1978c) has provided evidence that phase locking in goldfish first order saccular neurons is (in the best-locking neurons) accurate enough to account for the observed frequency discrimination performance in goldfish, which indicate a just noticeable difference in stimulus period of about 3% to 5%. Although a period length discrimination is not a sensory task entirely analogous to a phase comparison between two acoustic inputs, it should be noted in Fig. 2-2 that, for the ide studied by Buwalda et al. (in ms. a), in the transition regions around the values 0° and 180° for the phase angle between p(t) and a(t) a phase shift of some 10° (i.e., about 3% of the period length) can result in a response reversal in acoustic orientation. That the limits to both sensory abilities are so comparable suggests that they may be set by the same process: variability in phase locking.

Representing the stimulus period more or less faithfully in the interspike interval is only one aspect of phase locking. The phase relationship between the neural activity and the stimulus is at least as important, because a proper operation of the phase analyzer relies heavily on preservation of the phase angle between the p and the a input throughout the afferent path. Fay and Olsho (1979) provide evidence that the across-fiber distribution of phase angles between input and neural activity both in goldfish saccular and lagenar neurons, does not always conform to the theoretical ideal which is sharply bimodal with the peaks spaced 180° (to account for opposingly oriented hair cell fields). It will be clear that such a divergence from the ideal condition presents grave problems for the phase analyzer, if it is to compare the phase of p to **a** on the basis of a multifiber input.

A somewhat comparable situation exists for periodicity detection. Multifiber input with large interfiber phase variation would result in a gross input for the period detector in which the time structure is largely lost. However, goldfish seem to base decisions in period length discrimination on a small population of the best phase locking neurons rather than on total input (Fay 1978c). Similarly, the phase comparator eliminating a 180° ambiguity in the directional hearing of fish may well select from among the many available inputs the small population of neurons having an equal phase (shift) with respect to their a(t) or their p(t) input and, thus, a constant phase relationship to each other.

4 A Model Description of Acoustic Localization

4.1 Outline of the Problem

In the preceding sections the localization system of the fish was regarded as a black box. The present problem is to construct a model of the auditory space perception in fish on the basis of the psychophysical data and the other facts known about the sytem. A successful model should frame all observed relations between the physically defined stimulus—from which direction the sound impinges onto the fish— and the subjective direction experienced by the fish. This output variable is a nonobservable quantity, but in an appropriate experiment it can be inferred from the responses elicited in the fish. The functional organization of the model must be compatible with the morphological facts, but a handicap is not knowing which structures are part of the system and where they fit in the organization diagram. Lack of empirical data prevents such a model synthesis at any level other than the auditory periphery.

A hypothesis must be constructed about the way the (angular) position of a sound source is detected before the procedure of constructing the model can be started. The hypothesis here is that the line of action of the particle motion—the directive cue for the detection—is assessed by a number of receptors having different axes of directional sensitivity. Dijkgraaf's model of directional hearing with the labyrinths (see Section 1) forms a special case. The hypothesis above is compatible with the observed fact that time-of-arrival differences such as those that occur in travelling waves, are irrelevant for directional detection (see Section 3.3.2). However, in theory, an alternative mechanism with a single receptor that is directionally sensitive can also furnish directional information, provided that a reference indicative of the absolute stimulus magnitude is available to overcome the inherent intensity-direction ambiguity. But in this case a "cone of confusion," coaxial with the receptor axis, would result; any direction of incidence at a given inclination with the receptor axis would be indistinguishable. Again, to eliminate this kind of directional ambiguity, more directional receptors with differently directed axes would be required. Both hypothetical mechanisms deserve consideration. It will turn out below that for localization in the "horizontal" plane of the fish both hypothetical mechanisms require a minimum of three receptors—two directional receptors and one nondirectional receptor.

Synthesis of the model also requires that the input quantities of the system be known. The difference in density and compressibility between the body of aquatic animals and their medium is small with the exception of the swimbladder. A fishes body will therefore be carried along with the oscillatory particle motions of the incident wave. As the wavelength of the sound will be large compared with the dimensions of the body of the fish this will imply that the tissues of the fish oscillate throughout with the local particle acceleration equalling those of the external wave. This also implies that an enclosed inertial mass like an otolith finds itself in an accelerated reference frame. De Vries (1950) accordingly recognized that the particle acceleration should be regarded as the input quantity for the otolith-haircell system. The otolith mechanics transfers the acceleration into a displacement of the otolith relative to the resting position of a haircell unit. Well above the natural frequency of the otolith suspension, about 22 Hz in the ruff *Acerina cernua* (de Vries 1950), this displacement output lags $180°$ after displacement in the sound wave and is proportional to it. Nevertheless the acceleration should be considered the real input.

A second input variable affects directional detection; the acoustic pressure $p(t)$ will induce volume oscillations in the swimbladder. It follows that Fig. 2-4A represents the directional detection system of fishes with a swimbladder in block diagram.

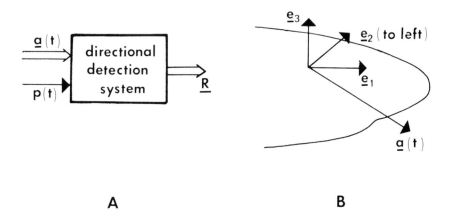

A   B

Figure 2-4. Definition of the external quantities for the black box formed by the directional detection system. (A) External quantities in vector form are either represented by double arrows or by as many single arrows as there are spatial components. Vector **R** denotes the subjective direction experienced by the fish. (B) The unit vectors e_1, e_2, and e_3 form the basis of a rectangular, body-fixed reference system necessary to decompose the input acceleration **a**(t) into components. Vector e_2 points to the left side of the fish.

4.2 The Model of Vectorial Weighing

The present models of directional localization with the labyrinths (Sand 1974, Schuijf 1976a, 1976b) assume the participation of a very limited number of directionally sensitive detectors with differently directed axes. The directions of the detector configuration's symmetry axes are not yet known with respect to the body of the fish, but these directions need not be specified for clarifying the principle of vectorial weighing.

An adequate description of how the particle accelerations of the external sound act on the detection system of the fish requires that the instantaneous acceleration vector $a(t)$ be specified in a reference frame moving with the fish (Fig. 2-4B). The particle acceleration $a(t)$ in the unperturbed sound field (that is, if the fish were absent) can be decomposed into components along the body-fixed coordinate axes, defined by the mutually perpendicular vectors e_1, e_2 and e_3 of unit length. This forms a complete description of the $a(t)$ input of the black box of Fig. 2-4A.

Vectorial weighing is a special hypothetical model of the operation of the directional detection system. The essence of vectorial weighing is readily explained for the case of directional detection in the plane of two acceleration detectors (i.e., otolith organs). Consider a bilateral pair of such detectors that enclose some angle β (Fig. 2-5). Then the skew coordinate axes e_1 and e_2 are advantageously chosen along the detector axes

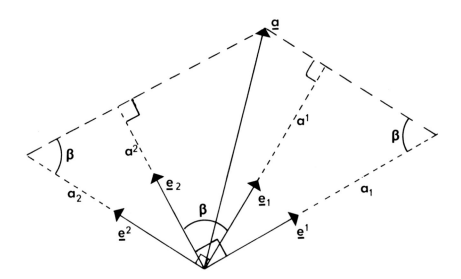

Figure 2-5. Vectorial weighing for two types of projections. The unit vectors e_1 and e_2 indicate the positive directions of the detector axes. The normal (= perpendicular) projections of a onto e_1 and e_2 are denoted by a^1 and a^2, respectively, to conform to the mathematical convention of using superscripts (not powers!) for this kind of vector decomposition, differing from the usual one parallel to the coordinate axes in which a forms the diagonal of a parallelogram with sides a_1 and a_2. The two imaginary axes e^1 and e^2 (superscripts, again) are normal to e^2 and e^1, respectively. From geometrical similarities it follows that $a_2/a_1 = a^2/a^1$.

instead of perpendicular to each other. For the mechanics of the otoliths, the perpendicular projections of a onto e_1 and e_2 are decisive; these are a^1 and a^2, respectively (see Fig. 2-5). The quotient a^1/a^2, the outcome of vectorial weighing (Schuijf and Buwalda 1975), determines uniquely the orientation of the line—in the plane spanned by e_1 and e_2—along which the particles in the sound wave oscillate. Encoding of the propagation direction of a sound wave is unambiguous in vectorial weighing with respect to amplitude as, for instance, a doubling of $a(t)$ does not affect the quotient.

It should be pointed out that:

1. The components a^1 and a^2 oscillate like $a(t)$, hence the time-dependence may be written as $a^1(t)$, etc.
2. The relationship between the instantaneous signs of $a^1(t)$ and $a^2(t)$—same sense or opposite—is essential for detecting whether the particles move within the interior angle formed by e_1 and e_2 or within one of its supplements. This is reflected in a sign change in the quotient as well. This aspect of vectorial weighing lends significance to the existence of hair cell fields polarized in opposing directions.
3. A detection process like vectorial weighing, based on detection of the instantaneous particle acceleration, shows a 180° directional ambiguity, since propagation direction and the direction of $a(t)$ in a traveling wave are alternatingly common and opposite (Schuijf and Buwalda 1975). This problem in the modeling is overcome by the phase model (see Section 4.3).

4.3 The Phase Model of Directional Hearing in Fish

According to the phase model of directional hearing, fish might use the phase shift between the particle acceleration and the acoustic pressure to cope with the inherent 180° ambiguity of a directional detection based solely on the instantaneous direction of the particle acceleration. This phase shift for different locations of the sound source with respect to the fish will be treated.

Schuijf (1976a) treated this problem by means of analytical methods for an arbitrary bearing of the sound source in the "horizontal" plane of the fish. The detector configuration in this phase model consisted of a bilateral pair of acceleration detectors plus, effectively, a single pressure detector. The present ideas on the organization of such a detection system are illustrated in Fig. 2-6.

In Schuijf's theory, the phase comparison is distributed over two unilateral processes simultaneously operative. Subsequently the information from both ears would be integrated (see also Section 3.4.2.2), and all phase comparisons would occur in the auditory space detector of Fig. 2-6. The empirical fact that the principle of phase analysis holds in cod for particle motions along the rostro-caudal and dorso-ventral body axes (Schuijf and Buwalda 1975, Buwalda et al., in ms.) is not at variance with unilateral phase comparisons, since binaural differences do not occur in these situations. Binaural differences do occur between the situation of a sound traveling one way along the fish's transverse axis and the situation of a sound traveling in the opposite direction, but, again, this does not preclude the presence of unilateral subsystems for a phase comparison between $a(t)$ and $p(t)$.

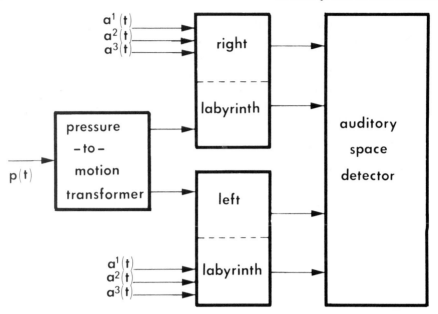

Figure 2-6. Model of the functional organization of the acoustic localization system of a fish for detection in the "horizontal plane." In the inertial mode the acceleration components along the detector axes a^1, a^2, and a^3 cannot be shielded from any of the labyrinth parts, whereas pressure-induced swimbladder oscillations may effectively reach only part of the labyrinths (see Section 3.4.2). After a possible peripheral segregation in the labyrinth of both kinds of stimuli (indicated by the dotted line), further processing for eliminating direction/intensity ambiguity and $180°$ directional ambiguity would occur in the auditory space detector.

Theory and experiment both suggest that essential elements of phase analysis are retained in a special case that will be more closely analyzed: the discrimination of frontally vs. caudally incident sound waves.

Examine the cue for phase analysis in this simple case of wave propagation in one dimension. It is a convenient shorthand to denote the propagation direction of a wave by the vector n (normal to the wave front). In the present case this implies that n can either be directed in the $+e_1$ direction or in the $-e_1$ direction of the fish (compare Fig. 2-4B). For a single frequency, far-field traveling wave the acoustic theory shows that the acceleration leads $p(t)$ by $90°$ when the wave impinges caudally onto the fish and lags $90°$ after $p(t)$ in the opposite case. This assumes that $+e_1$ is the positive direction for measuring the accelerations. The phase shift between $a(t)$ and $p(t)$ will now be denoted as $\psi(a,p)$. The acoustic rule above is written in mathematical terms as:

$$\psi(a,p) = \begin{cases} -\pi/2 \text{ radians if } x \to +\infty \\ +\pi/2 \text{ radians if } x \to -\infty \end{cases}$$

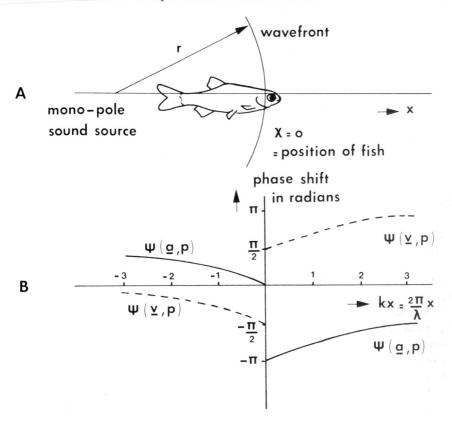

Figure 2-7. Model of the phase information available to a fish in the sound field of a pulsating sphere (monopole). (A) Geometry used for the derivation of the graphs in the lower figure. (B) Phase difference $\psi(a,p)$ between the acceleration a(t) with respect to the $+e_1$, direction of the fish (compare Fig. 2-4B) and p(t), the acoustic pressure, as function of coordinate position x of the sound source in the fish's reference frame. In fact this position is specified in terms of acoustic distances kx for general validity of the figure.

where x denotes the position coordinate of the sound source along the e_1 axis (compare Figs. 2-4B and 2-7). This forms the principle of phase analysis for discriminating opposing directions.

This result may be generalized when the source is at some finite distance from the fish, where spherical spreading of waves occurs under free field conditions. The results of such a calculation are shown in Fig. 2-7B. Note that the absolute value of kx corresponds to the acoustic distance to the source kr (see Section 1). Comparison with the choice behavior of *Leuciscus idus* in a sound field in which $\psi(a,p)$ could be varied from $-\pi$ to $+\pi$ independently of distance (Buwalda et al., in ms. a, see Section 3.3.1) via standing waves shows that response reversals occurred at $\psi(a,p) = 0$ and $\psi(a,p) = \pm\pi$ (see Fig. 2-2), corresponding to the phase jump at zero distance to the source in Fig. 2-7.

The present efforts in this field are directed toward gaining insight into the problem of the physiological realization of phase analysis and in particular into the mechanisms operative in segregating the necessary directional and nondirectional information.

4.4 Speculations on Phase Analysis in Fish without Swimbladder

This review has emphasized that bony fish, like cod and ide, use the acoustic pressure to overcome the 180° directional ambiguity in their detection with the labyrinths. Sharks show acoustic attraction from afar but the limited psychoacoustic studies seem to confirm that the sensitivity threshold of these fish is determined by particle motion (Banner 1967, Kelly and Nelson 1975), as was anticipated due to their lack of a swimbladder. On the other hand, it has not been demonstrated that sharks are completely insensitive to acoustic pressure, and the use of p(t) as a coherent reference cannot be excluded; an inefficient transformer of p(t) into motion for the hair cells would be sufficient. Potential candidates in this respect are the large, oil-containing liver, which may be more compressible than water, and the parietal fossa, which can hardly dilate but is sheared easily and is covered by a tough skin (cf. Tester et al. 1972).

Even if sharks are completely insensitive to pressure, another type of timing analysis is conceivable. The cue proposed by Schuijf (1975) to explain directional hearing in sharks is illustrated in Fig. 2-8. The direct wave results at some instant t in a particle motion as indicated by the vector. Similarly the indirect wave results through reflection at the soft surface in a displacement that is usually opposite the propagation direction because of the phase inversion at the surface (only p and not ξ, the vertical displacement component changes sign). The displacement is, moreover, delayed owing to

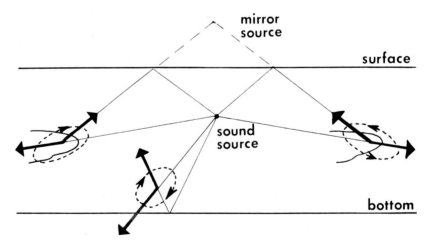

Figure 2-8. In principle, sharks might use the rotational sense of particle motion to remove directional ambiguity when inferring the propagation direction from motion to and from the sound source. Reflections at surface and bottom are essential in generating the cue. The detection comes down to relative timing between differently oriented hair cell fields.

its longer path. The combined motions, at the position of the shark's head, result in rotation along a flattened elliptical orbit. The cause is that both contributing waves do not propagate in exactly the same direction and form a Lissajous figure.

One can prove that the rotational sense of the motion depends on the position of the sound source with respect to the fish. If the sound source is caudad for the fish, as for the shark on the right in Fig. 2-8, then the rotational sense is counterclockwise and, for symmetry reasons, clockwise for the shark on the left. The same principle holds for reflections at a hard bottom and also for moderately and very shallow water layers. There are limitations to the reliability of this cue with sine waves if the reflected wave is delayed by more than one-fourth of a period. This can happen when the sound source is distant by more than $\lambda/4$ from both bottom and surface.

The available directional cue would be the orientation of the major axis of the ellipse, while the remaining $180°$ ambiguity can be resolved through the rotational sense. This cue is, for a given situation, invariant with respect to space. The shark, however, will resolve the sense of the rotation within its own frame of reference (e.g., by decomposition into oscillations along the main detector axes and subsequent analysis of their relative timing) and may even experience a reversal of the rotational sense, for instance under a roll of $180°$. Hence, an extra invariant orientational cue like gravity is needed to inform the shark of its orientation in space.

It should be noted that in fishes with a swimbladder, the interaction of the particle motions in the incident wave and the swimbladder pulsations may also lead to rotational motions in the labyrinths (see Section 3.3.3).

The similarity of these phenomena in fish with and without swimbladder suggests a basis for an evolution of directional hearing in fish that does not necessarily parallel the fishes' evolution proper. Primitive hearing systems, sensitive only to motion and capable of unambiguous directional detection in the manner depicted above for sharks, may have evolved into more sensitive systems through pressure detection via the swimbladder while retaining the original directional detectors and the principle of analysis of rotational sense. Further development may have been either in the direction of a more efficient, peripheral mechanical separation of the directional and the ever more important nondirectional inputs, or in the direction of a more sophisticated neural processing of the various inputs. In this way, the evolution of sensitive hearing need not have impaired the ability of acoustic localization.

5 Concluding Remarks

Evidence for acoustic localization has been proffered for representatives of all of the three major groups (in a bioacoustic sense) of fishes, i.e., specialized and nonspecialized fish with swimbladders and fish without swimbladders. Data on the mechanisms involved have nearly all been obtained, however, from only a few gadoid and cyprinid species. More comparative studies are therefore required to substantiate the generalizations made in this paper.

The data obtained thus far are consistent with the notion of highly evolved directional hearing systems, either on a peripheral level (as may be true for Ostariophysi) or in the sense of sophisticated processing of directional data (as may be postulated for cod). Hence, research into a more primitive directional detection system seems especially promising. The outcomes of such studies might bear on a hypothesis put forward here on the evolution of directional hearing in fishes. In this view even the most primitive ear was capable of directional detection free from ambiguities; subsequent development was directed toward sensitive hearing without losing the benefit of acoustic localization.

Whichever the way of evolution, the outcome for modern teleost fish appears to be a detection system in which acoustic pressure and particle acceleration constitute the basic input variables and are both required. Any future analysis of this system should therefore consider both pressure and motion to do justice to its multiple input character. A major aim for such an analysis is to find out how fish segregate these physical input quantities to isolate the directional cues. It is not clear as to whether the process of converting translatory accelerations into complex otolith motions (Sand 1974, Sand and Michelsen 1977, Fay and Olsho 1979) either plays an essential role in this segregation, is irrelevant, or does not even exist. One thing is certain—cues provided by the normal propagation of sound are not needed for directional detection. The system apparently does not discriminate between the various ways in which a particular set of input conditions can be produced, even if they are as artificial as the use of standing waves. Provided that adequate manipulation of acoustic conditions are used by fish acousticians, this fact should greatly increase the experimental possibilities.

Even more fragmentary than the data on the mechanisms involved is the knowledge concerning the biological significance of directional hearing. Apart from the studies on acoustic attraction of sharks, hardly any attention has been paid to the role of sound coming from distant sources in evoking directional responses and even less to their directive effects on schools of fish (Olsen 1976).

Directional hearing seems at least useful for fishes such as the cod, which produces sound with its swimbladder through drumming muscles (Hawkins and Rasmussen 1978). By being capable of directional discrimination in horizontal and vertical planes, pelagic fishes like cod seem eminently adapted to their 3-dimensional habitat. Distance perception, which would complete one of the most all-round acoustic localization systems found among vertebrates, has not been demonstrated yet, but fishes seem well equipped to detect the distance cues present both in the pressure to motion ratios and in the relative timing relations between these input variables.

Many fishes, however, are not pelagic and live in acoustic conditions that might be adverse to directional hearing. A thorough survey of acoustic localization abilities in special environments such as shallow water layers, or near reflecting boundaries such as the water surface, should therefore be the finishing touch to a really comprehensive study of acoustic localization.

Acknowledgments. We acknowledge Dr. A. N. Popper and Dr. R. R. Fay for valuable and expert criticisms and their helpful suggestions, extending beyond the call of editorial duty, in revising this chapter.

We further wish to thank Prof. Dr. F. J. Verheijen for critically reading parts of the manuscript, Miss Selma van Cornewal and the Stichting Film en Wetenschap for preparing the figures, and Miss Renée Enklaar for skillfully working out and typing the various versions of the manuscript.

References

Banner, A.: Evidence of sensitivity to acoustic displacements in the lemon shark *Negaprio brevirostris* (Poey). In Lateral Line Detectors. Cahn, P. (ed.). Bloomington: Indiana University Press, 1967, pp. 265-273.

van Bergeijk, W. A.: Directional and nondirectional hearing in fish. In: Marine Bioacoustics. I. Tavolga, W. N. (ed.). Oxford, etc.: Pergamon Press, 1964, pp. 281-299.

van Bergeijk, W. A.: The evolution of vertebrate hearing. In: Contributions to Sensory Physiology. 2. Neff, W. (ed.). New York: Academic Press, 1967, pp. 1-49.

Bullock, T. H., Corwin, J. T.: Acoustic evoked activity in the brain in sharks. J. Comp. Physiol. 129, 223-234 (1979).

Buwalda, R. J. A., Portier, J. A., Schuijf, A.: Independent detection of acoustic pressure and particle motion by cod. (in prep. b).

Buwalda, R. J. A., Schuijf, A., Hawkins, A. D.: On the discrimination, in three dimensions, of sound waves from opposing directions by the cod. (in ms.).

Buwalda, R. J. A., Schuijf, A., Visser, C., Willers, A. F. M.: Experiments on acoustic orientation in an Ostariophysian fish. (in prep. a).

Buwalda, R. J. A., Steen, J. van der: Sensitivity of the cod sacculus to directional and nondirectional stimuli. Comp. Biochem. Physiol. 64A, 467-471 (1979).

Cahn, P. H., Siler, W., Auwarter, A.: Acoustico-lateralis System of Fishes: Cross-modal coupling of signal and noise in the Grunt *Haemulon parrai*. J. Acoust. Soc. Am. 49, 591 (1970).

Cahn, P. H., Siler, W. Fujiya, M.: Sensory detection of environmental changes by fish. In: Responses of Fish to Environmental Changes. Chavin, W. (ed.). Springfield, Ill.: Charles C. Thomas, 1973, ch. XIII.

Cahn, P. H., Siler, W., Wodinsky, J.: Acoustico-lateralis System of Fishes: Tests of pressure and particle-velocity sensitivity in Grunts, *Haemulon sciurus* and *Haemulon parrai*. J. Acoust. Soc. Am. 46, 1572-1578 (1969).

Chapman, C. J.: Field studies of hearing in teleost fish. Helgoländer wiss. Meeresunters. 24, 371-390 (1973).

Chapman, C. J., Hawkins, A. D.: A Field Study of Hearing in the Cod, *Gadus morhua* L. J. Comp. Physiol. 85, 147-167 (1973).

Chapman, C. J., Johnstone, A. D. F.: Some auditory discrimination experiments on marine fish. J. Exp. Biol. 61, 521-528 (1974).

Chapman, C. J., Sand, O.: Field studies of hearing in two species of flatfish *Pleuronectes platessa* (L.) and *Limanda limanda* (L.) (fam. Pleuronectidae). Comp. Biochem. Physiol. 47, 371-386 (1974).

Corwin, J. T.: Morphology of the macula neglecta in sharks of the genus *Carcharhinus*. J. Morphol. 152, 341-362 (1977).

Corwin, J. T.: The relation of inner ear structure to the feeding behavior in sharks and rays. Scan. Electr. Microsc. II, 1105-1112 (1978).

Dale, T.: The labyrinthine mechanoreceptor organs of the cod *Gadus morhua* L. (Teleostei: Gadidae). Norw. J. Zool. 24, 85-128 (1976).

Dijkgraaf, S.: Über die Schallwahrnehmung bei Meeresfischen. Z. vergl. Physiol. 34, 104-122 (1952).

Dijkgraaf, S.: Hearing in bony fishes. Proc. Roy. Soc. B 152, 51-54 (1960).

Dijkgraaf, S.: The functioning and significance of the lateral-line organs. Biol. Rev. 38, 51-105 (1963).

Dijkgraaf, S.: The supposed use of the lateral line as an organ of hearing in fish. Experientia 20, 586 (1964).

Dijkgraaf, S.: A method for complete and selective surgical elimination of the lateral-line system in the Codfish, *Gadus morhua*. Experientia 29, 737-738 (1973).

Enger, P. S.: Unit activity in the fish auditory system. Acta Physiol. Scand. 59, suppl., 1-48 (1963).

Enger, P. S.: On the orientation of haircells in the laybrinth of perch (*Perca fluviatilis*). In: Sound Reception in Fish. Schuijf, A., Hawkins, A. D. (eds.). Amsterdam: Elsevier, 1976, pp. 49-62.

Enger, P. S., Hawkins, A. D., Sand, O., Chapman, C. J.: Directional sensitivity of saccular microphonic potentials in the haddock. J. Exp. Biol. 59, 425-434 (1973).

Fay, R. R.: Sound detection and sensory coding by the auditory systems of fishes. Ch. 5 in: The Behavior of Fish and Other Aquatic Animals. New York, etc.: Academic Press, 1978a, pp. 197-236.

Fay, R. R.: Coding of information in single auditory-nerve fibers of the goldfish. J. Acoust. Soc. Am. 63, 136-146 (1978b).

Fay, R. R.: Phase-locking in goldfish saccular nerve fibres accounts for frequency discrimination capacities. Nature 275, 320-322 (1978c).

Fay, R. R., Kendall, J. I., Popper, A. N., Tester, A. L.: Vibration detection by the macula neglecta of sharks. Comp. Biochem. Physiol. 47, 1235-1241 (1974).

Fay, R. R., Popper, A. N.: Acoustic stimulation of the ear of the goldfish (*Carassius auratus*). J. Exp. Biol. 61, 243 (1974).

Fay, R. R., Popper, A. N.: Modes of stimulation of the teleost ear. J. Exp. Biol. 62, 379-387 (1975).

Fay, R. R., Olsho, L. W.: Discharge patterns of lagenar and saccular neurones of the goldfish eighth nerve: displacement sensitivity and directional characteristics. Comp. Biochem. Physiol. A 62, 377-387 (1979).

von Frisch, K., Dijkgraaf, S.: Können Fische die Schallrichtung wahrnehmen? Z. vergl. Physiol. 22, 641-655 (1935).

Furukawa, T., Ishii, Y.: Neurophysiological studies on hearing in Goldfish. J. Neurophysiol. 30, 1377-1403 (1967).

Gentry, R. L.: Underwater auditory localization in the California Sea Lion (*Zalophus californianus*). J. Aud. Res. 7, 187-193 (1967).

Griffin, D. R.: Hearing and acoustic orientation in marine animals. Deep Sea Res., suppl. 3, 406-417 (1955).

Grözinger, B.: Elektro-physiologische Untersuchungen an der Hörbahn der Schleie (*Tinca tinca* (L.)). Z. vergl. Physiol. 57, 44-76 (1967).

Harris, G. G., van Bergeijk, W. A.: Evidence that the lateral-line organ responds to near-field displacements of sound sources in water. J. Acoust. Soc. Am. 34, 1831-1841 (1962).

Hawkins, A. D.: The sensitivity of fish to sounds. Oceanogr. Mar. Biol. Ann. Rev. 11, 291-340 (1973).

Hawkins, A. D., MacLennan, D. N.: An acoustic tank for hearing studies on fish. In: Sound Reception in Fish. Schuijf, A., Hawkins, A. D. (eds.). Amsterdam: Elsevier, 1976, pp. 149-169.

Hawkins, A. D., Rasmussen, K. J.: The calls of gadoid fish. J. Mar. Biol. Ass. U.K. 58, 891-911 (1978).

Hawkins, A. D., Sand, O.: Directional hearing in the Median Vertical Plane by the Cod. J. Comp. Physiol. 122, 1-8 (1977).

Howard, I. P., Templeton, W. B.: Human spatial orientation. London, etc.: J. Wiley & Sons, 1966, 533 pp.

Horch, K., Salmon, N.: Adaptations to the acoustic environment by the squirrelfishes *Myripristis violaceous* and *M. pralinus*. Mar. Behav. Physiol. 2, 121-137 (1973).

Jørgensen, J. M.: Hair cell polarization in the Flatfish Inner Ear. Acta Zool. (Stockh.) 57, 37-39 (1976).

Kelly, J. C., Nelson, D. R.: Hearing thresholds of the horn shark *Heterodontus francisci*. J. Acoust. Soc. Am. 58, 905-909 (1975).

Knudsen, E. I., Konishi, M.: A neural map of auditory space in the owl. Science 200, 795-797 (1978).

Lowenstein, O., Osborne, M. P., Thornhill, R. A.: The anatomy and ultrastructure of the labyrinth of the lamprey (*Lampetra fluviatilis* L.). Proc. Roy. Soc. B 170, 113-134 (1968).

Moore, P. W. B.: Underwater localization of click and pulsed pure-tone signals by the California sea lion (*Zalophus californianus*). J. Acoust. Soc. Am. 57, 406-410 (1975).

Moore, P. W. B., Au, W. W. L.: Underwater localization of pulsed pure tones by the California sea lion (*Zalophus californianus*). J. Acoust. Soc. Am. 58, 721-727 (1975).

Moulton, J. M., Dixon, R. H.: Directional hearing in fishes. In: Marine Bioacoustics. 2. Tavolga, W. N. (ed.). Oxford, etc.: Pergamon Press, 1967, pp. 187-232.

Myrberg, A. A., Gordon, C. R., Klimley, A. P.: Attraction of free ranging of sharks by low frequency sound, with comments on its biological significance. In: Sound Reception in Fish. Schuijf, A., Hawkins, A. D. (eds.). Amsterdam: Elsevier, 1976, pp. 205-228.

Nelson, D. R.: Hearing thresholds, frequency discrimination and acoustic orientation in the Lemon shark, *Negaprio brevirostris* (Poey). Bull. Mar. Sci. 17, 741-768 (1967).

Nelson, D. R., Gruber, S. H.: Sharks: attraction by low-frequency sounds. Science 142, 975-977 (1963).

Nelson, D. R., Johnson, R. H.: Acoustic attraction of Pacific reef sharks: effect of pulse intermittency and variability. Comp. Biochem. Physiol. 42, 85-95 (1972).

Olsen, K.: Directional hearing in cod (*Gadus morhua* L.). 8th working group for fishing technology. I. F. meeting. Lowestoft, England, 1969a, (mimeo), 18 pp.

Olsen, K.: Directional responses in herring to sound and noise stimuli. ICES paper CM 1969/B20. Gear and behaviour Committee, 1969b.

Olsen, K.: Evidence for localization of sound by fish in schools. In: Sound Reception in Fish. Schuijf, A., Hawkins, A. D. (eds.). Amsterdam: Elsevier, 1976, pp. 257-270.

Pettigrew, A., Chung, S. H., Anson, M.: Neurophysiological basis of directional hearing in amphibia. Nature 272, 138-142 (1978).

Platt, C.: Hair cell distribution and orientation in goldfish otolith organs. J. Comp. Neurol. 172, 283-298 (1977).

Popper, A. N.: The ultrastructure of the auditory regions in the inner ear of the Lake Whitefish. Science 192, 1020-1022 (1976a).

Popper, A. N.: Patterns of hair-cell orientation in the sacculus and lagena of two species of teleost fish. J. Acoust. Soc. Am. 59, Suppl. 1, 91 (1976b).

Popper, A. N.: A scanning electron microscopic study of the sacculus and lagena in the ears of fifteen species of Teleost fishes. J. Morphol. 153, 397-418 (1977).

Popper, A. N.: Scanning electron microscopic study of the otolithic organs in the Bichir (*Polypterus bichir*) and Shovel-nose Sturgeon (*Scaphirhynchus platorynchus*). J. Comp. Neurol. 181, 117-128 (1978a).

Popper, A. N.: A comparative study of the otolithic organs in fishes. Scanning El. Microsc. II, 1405-1416 (1978b).

Popper, A. N., Fay, R. R.: Sound detection and processing by teleost fishes: a critical review. J. Acoust. Soc. Am. 53, 1515-1529 (1973).

Popper, A. N., Salmon, M., Parvulescu, A.: Sound Localization by the Hawaiian squirrelfishes *Myripristis berndti* and *M. argyromus*. Anim. Behav. 21, 86-97 (1973).

Reinhardt, F.: Über Richtungswahrnehmung bei Fischen, besonders bei der Elritze (*Phoxinus laevis* L.) und beim Zwergwels (*Amiurus nebulosus* Raf.). Z. vergl. Physiol. 22, 570-603 (1935).

Renaud, D. L., Popper, A. N.: Sound localization by the bottlenoise porpoise *Tursiops truncatus*. J. Exp. Biol. 63, 569-586 (1975).

Sand, O.: Directional sensitivity of microphonic potentials from the perch ear. J. Exp. Biol. 60, 881-899 (1974).

Sand, O.: Microphonic potentials as a tool for auditory research in fish. In: Sound Reception in Fish. Schuijf, A., Hawkins, A. D. (eds.). Amsterdam: Elsevier, 1976, pp. 27-48.

Sand, O., Enger, P. S.: Evidence for an auditory function of the swimbladder in the cod. J. Exp. Biol. 59, 405-414 (1973).

Sand, O., Enger, P. S.: Possible mechanisms for directional hearing and pitch discrimination in fish. Abh. Rheinisch-Westfälische Akad. Wiss. 53, Symposium Mechanoreception, 223-242 (1974).

Sand, O., Hawkins, A. D.: Acoustic properties of the cod swimbladder. J. Exp. Biol. 58, 797-820 (1973).

Sand, O., Michelsen, A.: Vibration measurements of the perch saccular otolith. J. Comp. Physiol. 123, 85-89 (1978).

Schuijf, A.: Field studies of directional hearing in marine teleosts. Thesis, University of Utrecht, Utrecht: Elinkwijk, 1974.

Schuijf, A.: Directional hearing of cod (*Gadus morhua*) under approximate free field conditions. J. Comp. Physiol. 98, 307-332 (1975).

Schuijf, A.: The phase model of directional hearing in fish. In: Sound Reception in Fish. Schuijf, A., Hawkins, A. D. (eds.). Amsterdam: Elsevier, 1976a, pp. 63-86.

Schuijf, A.: Timing analysis and directional hearing in fish. In: Sound Reception in Fish. Schuijf, A., Hawkins, A. D. (eds.). Amsterdam: Elsevier, 1976b, pp. 87-112.

Schuijf, A., Baretta, J. W., Wildschut, J. T.: A field investigation on the discrimination of sound direction in Labrus berggylta (Pisces: Perciformes). Neth. J. Zool. 22, 81-104 (1972).

Schuijf, A., Buwalda, R. J. A.: On the mechanism of directional hearing in cod (*Gadus morhua* L.). J. Comp. Physiol. 98, 333-344 (1975).

Schuijf, A., Siemelink, M.: The ability of Cod (*Gadus morhua*) to orient towards a Sound Source. Experientia 30, 773-774 (1974).

Schuijf, A., Visser, C., Willers, A. F. M., Buwalda, R. J. A.: Acoustic localization in an Ostariophysian Fish. Experientia 33, 1062-1063 (1977).

Schwartz, E.: Bau und Funktion der Seitenlinie des Streifenhechtlings (*Aplocheilus lineatus* Cuv. u. Val.). Z. vergl. Physiol. 50, 55-87 (1965).

Schwartz, E.: Zur Lokalisation akustischer Reize von Fischen und Amphibien. Fortschr. Zool. 21, 121-135 (1973).

Siler, W.: Near- and far-fields in a marine environment. J. Acoust. Soc. Am. 46, 483-484 (1969).

Tavolga, W. N.: Acoustic orientation in the sea catfish *Galeichthys felis.* Ann. N. Y. Acad. Sci. 188 (1971).

Tavolga, W. N.: Acoustic obstacle detection in the sea catfish (*Arius felis*). In: Sound Reception in Fish. Schuijf, A., Hawkins, A. D. (eds.). Amsterdam: Elsevier, 1976, pp. 185-204.

Tavolga, W. N.: Mechanisms for directional hearing in the sea catfish (*Arius felis*). J. Exp. Biol. 67, 97-115 (1977).

Terhune, J. M.: Directional hearing of a harbor seal in air and water. J. Acoust. Soc. Am. 56, 1862-1865 (1974).

Tester, A. L., Kendall, J. I., Milisen, W. B.: Morphology of the ear of the shark genus *Carcharhinus* with particular reference to the macula neglecta. Pac. Sci. 26, 264-274 (1972).

de Vries, H. L.: The mechanics of a labyrinth otolith. Acta oto-lar. 38, 262-273 (1950).

Chapter 3

Central Auditory Pathways in Anamniotic Vertebrates

R. GLENN NORTHCUTT*

1 Introduction

The late nineteenth and early twentieth centuries witnessed rapid growth in descriptive neuroanatomy. This period of intensive study of nervous systems in a wide variety of vertebrates resulted in several hypotheses concerning the origin and subsequent evolution of the otic and lateralis systems. These hypotheses possess two common features: they are based on descriptive anatomical material and were not tested experimentally as the appropriate techniques did not yet exist; and they reflect certain supposed anatomical relationships among anamniotic vertebrates that were believed to form a linear series of increasingly complex groups.

Anamniotes comprise approximately 55% of the living vertebrate species and are so termed because of their reproductive strategy: they do not possess an amniotic egg that allowed amniotic vertebrates to successfully invade a wide range of terrestrial niches. There are at least four anamniotic radiations, and each has a separate phylogenetic history for the last 400 million years (Fig. 3-1).

One radiation, the agnathans, are represented today by lampreys and hagfishes, collectively termed the cyclostomes. These species possess neither jaws nor paired fins and are relatively restricted regarding size and variety of their prey and in their locomotor efficiency. The living agnathans are believed to be evolved from early ostracoderms; they have lost an external dermal armor, elongated the trunk, and radically reorganized many of the head structures for their specialized feeding habits. Thus, many morphological characters in living cyclostomes are highly derived, and these taxa represent a separate radiation, paralleling the other anamniotic radiations, not simply an ancestral stock from which other vertebrates evolved.

Gnathostomes (jawed) vertebrates (Fig. 3-1) represent the second anamniotic radiation. They occur slightly later in the fossil record and are usually assumed to have arisen from some group of early agnathans (Romer 1966, Hotton 1976). The earliest gnathostomes apparently radiated very rapidly into three distinct groups: placoderms,

*Division of Biological Sciences, University of Michigan, Ann Arbor, Michigan 48109.

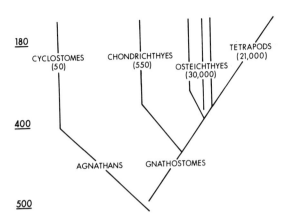

Figure 3-1. Phylogeny of vertebrates illustrating time of origin of major radiations in millions of years (ordinate) and approximate number of living species (in parentheses).

chondrichthians, and osteichthians. All three groups possess jaws and paired fins. Jaws allow increased efficiency in feeding and offer a wider range of prey; paired fins increase locomotor stability and maneuverability. These developments underlay the evolution of new and more active predators.

There are no extant placoderms, and this group of moderate to large-sized vertebrates is thought to have been replaced by chondrichthians. The chondrichthians are are primarily a marine radiation, whereas osteichthians have evolved in both marine and freshwater systems. The chondrichthians, or cartilaginous fishes, and the osteichthians, or bony fishes, represent long-separate anamniotic radiations, each with unique organisms representing distinct grades of organization. For example, the cartilaginous skates and rays arose at approximately the same time as the osteichthian teleosts. Thus, phylogenetically, skates and rays are no older than the most recently evolved group of bony fishes or, for that matter, birds and mammals. Cartilaginous fishes are not an earlier and simpler group of fishes, ancestral to other gnathostomes; instead they are a separate radiation with their own long history characterized by extensive changes and the evolution of new groups paralleling bony fishes and land vertebrates.

The osteichthians, or bony fishes, are frequently divided into three major groups: actinopterygians (ray-finned fishes), dipnoans (lungfishes), and crossopterygians (lobe-finned fishes). The affinities of these groups are presently in dispute, and each has a long evolutionary history. The ray-finned fishes constitute most of the extant bony fishes, dipnoans are represented by only three genera, and crossopterygians have a single extant genus, *Latimeria*.

Early in their history, the crossopterygians are thought to have given rise to the fourth anamniotic radiation, the amphibians. Amphibians arose from bony fishes in the Devonian (Fig. 3-1) and thus have an evolutionary history almost as long as that of the bony fishes themselves. Again, living bony fishes should not be considered simpler than living amphibians. Both groups may retain some primitive characteristics inherited from their common ancestors, but both radiations have long, separate histories, and both have evolved organisms reflecting different grades of organization. For example, both have given rise to new major groups—teleosts in the case of bony fishes and true land vertebrates (amniotes) in the case of amphibians.

New interpretations of the fossil record have invalidated the earlier assumption that anamniotes represent a linear series of ever increasing complexity. A more accurate picture is that of four distinct anamniotic radiations, separated from a common ancestor for at least 400 million years, evolving at different rates and each producing new groups. This view of anamniotic affinities necessitates a very different approach by modern anatomists addressing questions about the origin and evolution of structures. It is impossible to elucidate the origin of a structure by asking if it first occurs in, for example, agnathans, chondrichthians, or bony fishes. Rather, one must ask if a given structure is found in all three groups with a similarity greater than chance. If so, the structure was most likely found in their common ancestor. In other words, all anamniotic radiations must be analyzed for the presence or absence of particular structures; their patterns of distribution can then be used to determine their probable evolutionary origin and subsequent phylogenetic history.

Experimental anatomical studies on the central projections of the octavus and lateralis nerves and their higher order pathways in anamniotes are limited to the last seven years (Gregory 1972, Mehler 1972, Maler, Karten, and Bennett 1973a, 1973b, Campbell and Boord 1974, Maler 1974, Maler, Finger, and Karten 1974, Rubinson 1974, Boord and Campbell 1977, McCormick 1978, Northcutt 1979a, 1979b). The results of these studies, and recent unpublished observations on cartilaginous and bony fishes, are summarized in this review along with the data used to examine two of the early hypotheses regarding the otic and lateralis systems.

In 1892 Ayers first formulated the acousticolateralis hypothesis, which proposed that the inner ear of vertebrates arose as an elaboration of a portion of the head lateral line system that had sunk below the ectodermal surface. This hypothesis is based primarily on the observation that both the inner ear and lateral line organs arise ontogenetically from a series of dorsolateral placodes (Fig. 3-2; Mayser 1882, Beard 1884, Wilson 1889, Wilson and Mattocks 1897). The acousticolateralis hypothesis was reiterated by van Bergeijk (1966, 1967) who argued in its support based on three lines of evidence: (1) the lateral line and inner ear arise from the same ectodermal tissue; (2) both otic sense organs and neuromasts are composed of identical sensory cells (hair cells); (3) both systems project to the same medullar nuclei. Wever (1976) rejected the acousticolateralis hypothesis on the basis that embryological data do not support the contention that both otic and lateralis systems arise from a single placode; rather, they appear to arise from closely associated but separate structures. Experimental anatomical data generated over the past four years indicate that the otic and lateralis systems are clearly separate systems with parallel central nervous system pathways.

Van Bergeijk (1966, 1967) and Wever (1976) argue that the inner ear arose as a vestibular organ and only secondarily evolved auditory functions. This idea (labyrinth hypothesis) was originally based on the assumption that audition in anamniotes is present, or well developed, only in some teleosts and amphibians. The labyrinth hypothesis drew support from early descriptive anatomical studies which stated that most anamniotes possess a single medullar nucleus, which receives the entering octavus

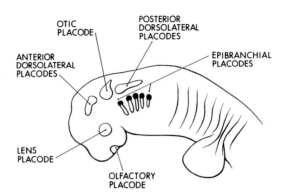

Figure 3-2. Diagrammatic lateral view of an anamniotic embryo illustrating ectodermal placodes involved in formation of special sense organs.

nerve fibers, and that this single nucleus is homologous to the vestibular complex of amniotic vertebrates (Larsell 1967). More recently, behavioral and physiological studies indicate that all anamniotes (except agnathans, which have not been examined) possess auditory abilities that cannot be dismissed as lateral line responses to low frequency sounds (see Fay and Popper, Chapter 1 and Capranica and Moffat, Chapter 5). Specific auditory nuclei have not yet been identified, but recent experimental anatomical data reveal that at least four distinct octavus nuclei, rather than a single nucleus, occur in most anamniotes. These data may not provide sufficient evidence for rejecting the labyrinth hypothesis, but they certainly cast serious doubt on its validity.

New experimental anatomical data allow us to reexamine older hypotheses in a new light, but they also pose new problems. All living bony fishes, including *Latimeria,* possess more octavus nuclei than the living amphibians, as well as different nuclear topology. Earlier descriptive anatomical studies concluded that auditory nuclei in amphibians arose by "capture" of lateralis nuclei and that the emergence of amphibians was thus characterized by a larger number of octavus nuclei. The new anatomical data force a reexamination of amphibian audition and its relationship to that of bony fishes and reptiles.

2 Results

The octavolateralis area in anamniotic vertebrates constitutes a large portion of the dorsal alar plate of the hindbrain and is the primary target of three pairs of cranial nerves: anterior lateral line, octavus, and posterior lateral line (Herrick 1899, Johnston 1902, Maler 1974, McCready and Boord 1976, McCormick 1978). The organization and central pathways of both the octavus and lateral line systems will be discussed as the two systems are traditionally considered to be closely related. However, the lateral line nuclei and pathways will be covered only in sufficient detail to place the octavus system in perspective and to reject the acousticolateralis hypothesis that the inner ear arose from lateral line organs.

2.1 Cyclostomes

There are two families of living cyclostomes, myxinoids (hagfishes) and petromyzontids (lampreys). There are no experimental data on the central nervous system of hagfishes; details of their octavolateralis organization are based solely on older descriptive anatomy (Jansen 1930, Larsell 1947, 1967). Hagfishes appear to possess poorly developed anterior and posterior lateral line nerves and an otic organ comprised of a single saccus communis that gives rise to a single canal. At either end of the canal, small ampullae are assumed to represent anterior and posterior semicircular organs. As in other anamniotic vertebrates, the hagfish octavus nerve consists of anterior and posterior rami, which are believed to innervate utricular and saccular divisions of the inner ear, respectively (Larsell 1967).

The octavolateralis area in hagfishes is less differentiated than in lampreys, and only a single nucleus has been recognized (Ayers and Worthington 1908, Jansen 1930, Larsell 1967). Both lateral line and octavus nerves are said to terminate within a single octavolateral nucleus. There is no general agreement on whether hagfishes possess a cerebellum, thus no agreement on whether vestibulocerebellar projections exist (Larsell 1967).

In lampreys (Figs. 3-3, 3-4), the octavolateralis area consists of dorsal, medial, and ventral nuclei (Pearson 1936a, Rubinson 1974, Northcutt 1979a). Rostrally, the dorsal nucleus begins immediately caudal to the cerebellar plate and continues caudally in the medulla, ending slightly behind the entry of the anterior lateral line nerve (Fig. 3-4D, E). Rostrally the medial nucleus, unlike the dorsal nucleus, is continuous with the cerebellar plate. Caudally it can be traced into the medulla where it ends just rostral to the obex (Fig. 3-4E). The dorsal and medial nuclei are similarly organized: both are capped laterally by the cerebellar crest, a layer of unmyelinated fibers; both consist of centrally situated neuropils supplied by the lateral line nerves (Fig. 3-3B); and both consist of prominent periventricular cell plates (Fig. 3-3B,C). In the silver lamprey, *Ichthyomyzon unicuspis*, the anterior lateral line nerve projects to both the dorsal and medial nuclei, whereas the posterior lateral line nerve projects only to the medial octavolateralis nucleus (unpublished observations). Neither lateral line nerve projects to the ventral octavolateralis nucleus.

Rostrally, the ventral octavolateralis nucleus forms the lateral edge of the cerebellar plate, and it continues caudally to obex levels (Figs. 3-3A, 3-4). The ventral nucleus is bordered laterally by the octavus nerve, and the ventral nuclear cells are more scattered than those of the dorsal and medial octavolateralis nuclei. The ventral nucleus contains three distinct populations of large neurons—the anterior, intermediate, and posterior octavomotor nuclei (Figs. 3-3A,C, 3-4). The ventral nucleus is thus a long column of scattered small and medium-sized cells, within which three aggregations of much larger neurons can be identified.

It has been claimed that octavus fibers project only to the ventral nucleus (Heier 1948), to both the medial and ventral nuclei (Johnston 1902), and primarily to the medial nucleus (Rubinson 1974). In addition, most older descriptions claim that primary octavus fibers project to the ipsilateral cerebellum, as well as to the contralateral cerebellum, after decussating in the cerebellar commissure (Heier 1948, Larsell 1967).

Experimental determination of the central projections of the octavus nerve in adult

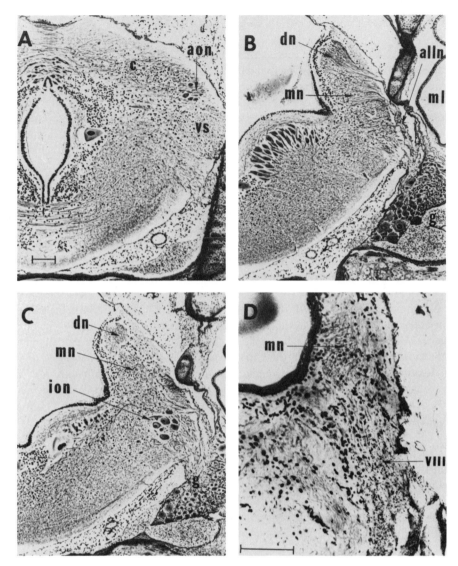

Figure 3-3. Photomicrographs of lamprey hindbrain. (A) Transverse section through the cerebellum and isthmus at the level of the anterior octavomotor nucleus. (B) Transverse section through the medulla at the level of the entry of the anterior lateral line nerve. (C) Transverse section of the octavolateralis area at the level of entry of the octavus nerve. (D) Transverse section at the level of entry of the octavus nerve showing the degenerating octavus fibers. Bar scales equal 100 μm. Magnifications of (A) through (C) are identical. Abbreviations: alln-anterior lateral line nerve; aon-anterior octavomotor nucleus; c-cerebellum; dn-dorsal octavolateralis nucleus; g-octavo-facial ganglion; ion-intermediate octavomotor nucleus; ml-membranous labyrinth; mn-medial octavolateralis nucleus; vs-trigeminal sensory root; VIII-octavus root.

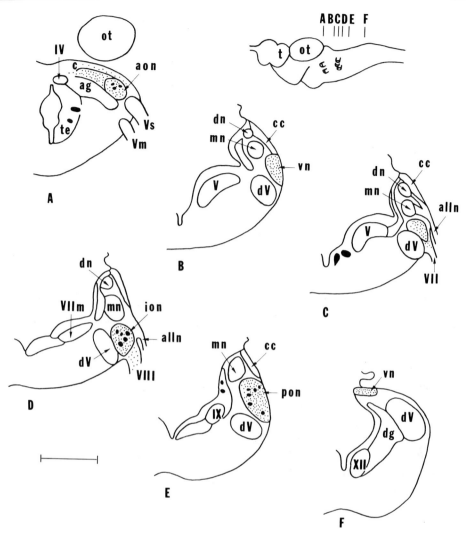

Figure 3-4. Charting of the degenerating octavus afferents following removal of the membranous labyrinth. (A-F) Transverse sections through the hindbrain at levels indicated in the lateral view of the brain. Fine stippling indicates degenerating fibers and terminals, large solid black circles indicate individual large neurons. Bar scale equals 500 μm. Abbreviations: ag-alar gray; alln-anterior lateral line nerve; aon-anterior octavomotor nucleus; c-cerebellum; cc-cerebellar crest; dg-dorsal gray; dn-dorsal octavolateralis nucleus; dV-descending trigeminal tract and nucleus; ion-intermediate octavomotor nucleus; mn-medial octavolateralis nucleus; ot-optic tectum; pon-posterior octavomotor nucleus; t-telencephalon; te-tegmentum; vn-ventral octavolateralis nucleus; IV-trochlear motor nucleus; V-trigeminal motor nucleus; Vm-trigeminal motor root; Vs-trigeminal sensory root; VIIm-facial motor nucleus; VII-facial root; VIII-octavus root; IX-glossopharyngeal motor nucleus; XII-hypoglossal motor nucleus.

silver lampreys (Fig. 3-4; Northcutt 1979a) reveals that the degenerating octavus fibers enter the ipsilateral ventral octavolateralis nucleus where they form ascending and descending limbs (Figs. 3-3D, 3-4). The octavus fibers of the ascending limb course rostrally through the ventral nucleus (Fig. 3-4A-C), where most of the fibers terminate in the lateral two-thirds of the nucleus. Octavus fibers continue rostrally, terminating on cell bodies of the anterior octavomotor nucleus (Fig. 3-4A), and at this level turn medially to terminate in both the cellular and molecular layers of the cerebellum (Fig. 3-4A). No degenerating octavus fibers were seen in the cerebellar commissure or in the contralateral cerebellum.

The octavus fibers of the descending limb course caudally through the entire extent of the ventral nucleus (Fig. 3-4D-F), including the cell bodies of the intermediate and posterior octavomotor nuclei (Fig. 3-4D,E). Degenerating octavus fibers were traced caudally to obex levels (Fig. 3-4F) where the ventral nucleus ends in a dorsomedial position in the medullar wall.

Thus in the silver lamprey the ventral and octavomotor nuclei of the octavolateralis area are the primary, if not sole, medullar targets of the primary octavus fibers. Rubinson (1974) reported octavus fibers projecting to the medial nucleus in larval lampreys; however, interpretation of these results are complicated by two factors: (1) the medulla of larval lampreys is not well differentiated, and Rubinson's cytoarchitectural boundaries appear to be very different than those of other workers (Pearson 1936a, Northcutt 1979a); and (2) Rubinson's experimental results may include projections of part of the anterior lateral line nerve as well as the octavus nerve. In lampreys, the ganglion of the anterior lateral line nerve is located immediately ventral to the otic capsule; its entering roots pass through a foramen in the floor of the otic capsule, then course medially to the octavus ganglion to enter the lateral wall of the medulla (Fig. 3-3B; Johnston 1905). Thus, damage to the medial wall of the otic capsule not only destroys the octavus ganglion, but also interrupts the roots of the anterior lateral line nerve.

Higher order octavus projections in cyclostomes have not been demonstrated experimentally, even though projections to a midbrain area, the torus semicircularis, and to the thalamus were claimed in an earlier descriptive study of lampreys (Heier 1948). Further experimental studies of lampreys are clearly needed to determine higher order projections and to resolve the question of auditory functions. Lowenstein, Osborne, and Thornhill (1968) and Thornhill (1972) established that lampreys possess a macula neglecta, an otic organ suspected of mediating hearing in other anamniotes, but there are no functional studies to demonstrate the presence or absence of its auditory function in lampreys.

2.2 Chondrichthyes

The cartilaginous fishes comprise at least two groups, the holocephalons (chimeras) and the elasmobranchs (sharks, skates, and rays). For several years the inner ear in cartilaginous fishes was assumed to possess only vestibular functions. However, recent studies (Lowenstein and Roberts 1951, Nelson and Gruber 1963, Tester, Kendall, and Milisen 1972, Fay, Kendall, Popper, and Tester 1974, Corwin 1977, 1978, Popper and

Fay 1977) indicate that sharks, at least, detect underwater sound at frequencies up to 1000 Hz. Tester et al. (1972), Fay et al. (1974), and Corwin (1977, 1978) argue that the macula neglecta is likely involved in sound detection, but this does not preclude the involvement of other maculae or the additional involvement of the lateral line system, particularly for low frequency sounds.

There has been no previous experimental anatomy of the primary—or higher order—projections of the octavus nerve in cartilaginous fishes. Thus the octavus nerve in the thornback skate, *Platyrhinoidis triseriata*, was transected unilaterally in order to trace its primary projections by the Fink-Heimer method for degenerating axons and terminals (Fig. 3-5).

Like other cartilaginous fishes, *Platyrhinoidis* possesses an octavolateralis area (Fig. 3-5C,D) composed of dorsal, medial, and ventral columns (Larsell 1967, McCready and Boord 1976, Smeets and Nieuwenhuys 1976, Northcutt 1978). The dorsal column, or anterior lateral line lobe, consists of a dorsal nucleus (Fig. 3-5C,D), composed of large Purkinje-like cells and smaller triangular cells, capped dorsally by a thick layer of granule cells terms the lateral granular layer (see LG, Fig. 3-5C). Laterally, the dorsal nucleus is covered by a layer of fine fibers, the cerebellar crest, that is pierced by entering dorsal root fibers of the anterior lateral line nerve. The dorsal nucleus begins caudally at mid-medullar levels (Fig. 3-5D) and continues rostrally under the lateral edge of the corpus of the cerebellum where it ends without fusing with the cellular layers of the cerebellum. Its dorsal cap of granule cells continues rostrally and fuses with a similar granule layer covering the dorsolateral edge of the medial nucleus. These two granule populations thus form the lateral portion of the lower leaf of the cerebellar auricle (see LG, Fig. 3-5A,B). The axons of the granule cell caps form the cellebellar crest and thus terminate on the distal portions of the Purkinje-like cells of the dorsal and medial nuclei (Boord 1977).

In *Platyrhinoidis,* the medial column, or posterior lateral line lobe, is organized much like the dorsal column. It consists of a medial nucleus of large neurons capped dorsolaterally by a granule layer (Fib. 3-5B) and covered laterally by the cerebellar crest. The cerebellar crest is pierced by the ventral root of the anterior lateral line nerve as well as the entering fibers of the posterior lateral line nerve (Fig. 3-5C,D).

Earlier studies (Kappers 1947, Larsell 1967) recognize a ventral column of the octavolateralis area that receives primary octavus projections, but there is no agreement concerning the boundaries or the number of nuclei that compose the ventral column. Analysis of the ventral column in *Platyrhinoidis* reveals four octavus nuclei rostrocaudally: anterior, magnocellular, descending, and posterior nuclei (Fig. 3-5).

The anterior nucleus consists of medium-sized fusiform neurons located ventromedial to the medial nucleus. The rostral border of the anterior nucleus is rostral to that of the medial nucleus and occurs as the anterior nucleus is replaced by the granule cells of the lower leaf of the auricle. Caudally the anterior nucleus is replaced by the magnocellular nucleus (Fig. 3-5C), composed of large polygonal cells whose dendrites are oriented laterally and medially. The laterally directed dendrites extend into the entering octavus root, as well as more dorsally into the neuropil of the medial nucleus of the posterior lateral line lobe. The medially directed dendrites of the magnocellular nucleus extend into the neuropil of the reticular formation. Laterally and caudally the

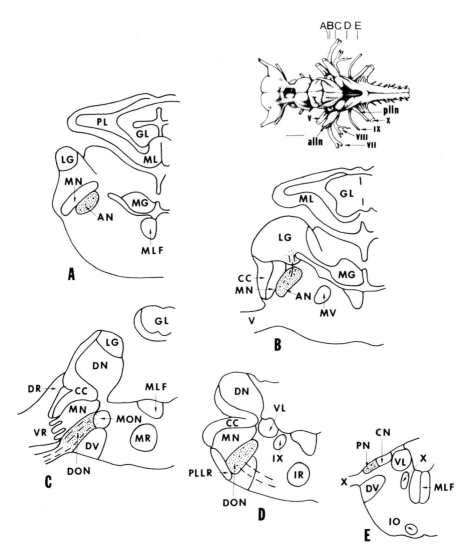

Figure 3-5. Charting of the degenerating octavus afferents in the thornback skate following transection of the octavus nerve proximal to the ganglion. (A-E) Transverse sections through the hindbrain at levels indicated in the dorsal view of the brain. Dashed lines indicate degenerating fibers and fine stippling indicates degenerating terminals. Bar scale equals 5 mm. Abbreviations: alln-anterior lateral line nerve; AN-anterior octavus nucleus; CC-cerebellar crest; CN-caudal octavolateralis nucleus; DN-dorsal octavolateralis nucleus; DON-descending octavus nucleus; DR-dorsal root of anterior lateral line nerve; DV-descending trigeminal nucleus and tract; GL-granule cell layer; IO-inferior olive; IR-inferior reticular formation; LG-lateral granular layer; MG-medial granular layer of auricle; ML-molecular layer; MLF-medial longitudinal fasciculus; MN-medial octavolateralis nucleus; MON-magnocellular octavus nucleus; MR-medial reticular formation; MV-trigeminal motor nucleus; PL-Purkinje cell layer; plln-posterior lateral line nerve; PLLR-root of the posterior lateral line nerve; PN-posterior octavus nucleus; VL-vagal lobe; VR-ventral root of anterior lateral line nerve; V-trigeminal nerve; VII-facial nerve; VIII-octavus nerve; IX-glossopharyngeal nerve and nucleus; X-vagal nerve and nucleus.

magnocellular nucleus is replaced by smaller bipolar and polygonal cells of the descending octavus nucleus (Fig. 3-5C,D). The descending nucleus is the largest single octavus nucleus and extends the length of the medulla; it is replaced at obex levels by a posterior octavus nucleus formed by small spherical and bipolar neurons (Fig. 3-5E).

Earlier descriptive studies of the distribution of the primary octavus fibers (Kappers 1947, Larsell 1967) claimed that primary fibers enter the medulla and turn caudally. Many of the fibers were said to cross the midline at the level of the abducens nucleus, terminating in the contralateral reticular formation. Other fibers were said to continue caudally and ipsilaterally to spinal cord levels. Ascending octavus fibers were believed to project to the medial nucleus of the posterior lateral line lobe, as well as more rostrally to the granule cells of the lower leaf of the auricle and the cerebellar nuclei. Collaterals of the ascending fibers of the anterior and posterial lateral line nerves were also believed to terminate in the ventral octavus column (Larsell 1967).

Recently Boord and Campbell (1977) determined the primary projections of the anterior and posterior lateral line nerves in *Mustelus*. The dorsal root of the anterior lateral line nerve consists of afferent fibers that innervate head electroreceptive ampullary organs; the dorsal root fibers terminate in the dorsal nucleus and granular cap of the anterior lateral line lobe. The ventral root of the anterior lateral line nerve, and the posterior lateral line nerve, consist of afferents that innervate the neuromasts of the head and trunk, respectively, and terminate in the medial nucleus and granular cap of the posterior lateral line lobe. Similar projections also characterize the lateral line nerves of *Platyrhinoidis* (unpublished observations). Thus primary lateral line afferents do not terminate on any of the octavus nuclei, with the possible exception of the magnocellular nucleus whose dendrites extend into the medial nucleus and may thus receive lateral line input.

Unilateral transection of the octavus root in *Platyrhinoidis* reveals ascending and descending primary fibers within the ipsilateral ventral column of the octavolateralis area, but no primary fibers crossing the midline (Fig. 3-5). Primary octavus fibers terminate on the distal dendrites of neurons of the magnocellular nucleus, but few terminals occur on the more proximal portions of these dendrites or on cell bodies (Fig. 3-5C). Descending fibers continue caudally, terminating heavily among the cells of the descending octavus nucleus (Fig. 3-5C,D). Primary octavus fibers turn medially throughout the rostrocaudal extent of the descending nucleus and pass through the descending trigeminal tract to terminate on the distal portions of dendrites from the reticular formation. These medially coursing fibers likely represent the octavus fibers claimed in earlier descriptive studies to be decussating to the contralateral medulla. Ipsilaterally, degenerating octavus fibers continue caudally to terminate in the posterior octavus nucleus at obex levels (Fig. 3-5E). Primary fibers were not traced more caudally, and the primary octavospinal projections of earlier studies were not confirmed.

Ascending octavus fibers course rostrally, terminating throughout the extent of the anterior nucleus (Fig. 3-5A,B). A portion of these fibers continue beyond the anterior nucleus, turning dorsally and caudally to terminate in the medial third of the granule layer of the lower leaf of the cerebellar auricle (Fig. 3-5B). Primary afferents were not observed to terminate in the upper leaf of the auricle, nor in the cerebellar nucleus as claimed in earlier descriptive studies. At present it is impossible to determine which, if any, of the octavus nuclei may possess auditory functions. Although the eighth nerve in *Platyrhinoidis* possesses anterior and posterior rami, these rami fuse prior to the ganglion, and the saccular and neglecta branches are not amenable to degeneration studies. The central projections of individual otic receptors must be resolved by other anatomical tracing methods.

Anatomical data on higher order octavus projections are presently lacking, but Bullock and Corwin (1979) report that auditory stimuli produce evoked potentials in the cerebellum, midbrain, and telencephalon of carcharhinid and triakid sharks. The responsive loci were not anatomically marked, but Bullock and Corwin report that the locus of best response was distinctly different for auditory, photic, and electric stimuli. These results strongly suggest separate and distinct anatomical pathways for auditory and lateralis inputs to telencephalic levels.

2.3 Osteichthyes

The Osteichthyes, or bony fishes, comprise three major radiations—the actinopterygians (ray-finned fishes), the dipnoans (lungfishes), and the crossopterygians (lobe-finned fishes)—with problematical phyletic relationships (Fig. 3-1). All three radiations appear during the Devonian and no one radiation can be considered phylogenetically older than another. Traditionally, the dipnoans and crossopterygians are considered to be more closely related to one another than to the actinopterygians (Romer 1966); however, many systematists believe that all three radiations are equally distinct (Schaeffer and Rosen 1961, Schaeffer 1969, Moy-Thomas and Miles 1971). Experimental anatomical data on the octavolateralis area are presently available only for some ray-finned fishes, and even descriptive studies on the lungfishes and crossopterygians are meager.

2.3.1 Actinopterygians

Three grades of organization related to more effective feeding and locomotor mechanisms characterize the three groups of ray-finned fishes: Chondrostei, Holostei, and Teleostei. The chondrosteans consist of the sturgeons and paddle fishes, as well as two distinctly different genera, *Polypterus* and *Calamoichthys,* so abberant that many workers separate them from chrondrosteans and assign them to a fourth group of actinopterygians, the polypteriforms, or even to a fourth osteichthian radiation. There are numerous derived brain characters that separate *Polypterus* and *Calamoichthys* from the chondrosteans, but these genera also possess sufficient derived actinopterygian characters to warrant retaining them as a grade within ray-finned fishes (Northcutt and Braford 1978).

The gars and the bowfin, *Amia,* are the only surviving holostean genera. These forms are restricted to North and Central American freshwater systems, and represent relict populations closely related to a broad radiation of holosteans that gave rise to teleosts late in the Mesozoic.

The teleosts form the single largest group of living vertebrates and may have arisen from holosteans a number of times (Greenwood, Rosen, Weitzman, and Myers 1966).

Four major groups of teleosts are presently recognized: osteoglossomorphs, clupeomorphs, elopomorphs, and euteleosts (Greenwood 1973). The osteoglossomorphs and clupeomorphs appear to be more closely related to one another than to the elopomorphs and euteleosts, which also appear to be sister groups. The euteleosts include the vast majority of the living teleosts and are clearly the most varied and successful group.

2.3.1.1 *Chondrosteans and Polypteriforms.* There have been no experimental studies on the octavolateralis area or primary octavus projections in these fishes. Descriptive studies indicate that the octavolateralis area is organized like that of Cyclostomes and Chondrichthyes (Johnston 1901, Larsell 1967, McCormick 1978). All chondrosteans and polypteriforms possess dorsal, medial, and ventral columns within the octavolateralis area and all possess dorsal and ventral roots of the anterior lateral line nerve. In each case, the dorsal nucleus is associated with the dorsal root of the anterior lateral line nerve, and all of these taxa are known to possess electroreceptive ampullary organs (Pfeiffer 1968, Jørgensen, Flock and Wersäll 1972, Roth 1973, Teeter and Bennett 1976). Thus the earliest actinopterygians, like chondrichthians, appear to have been electroreceptive, and electroreception appears to be the sole function of the anterior lateral line lobe (dorsal nucleus) in anamniotes. Likewise, all chondrosteans and polypteriforms possess a posterior lateral line lobe (medial nucleus) that receives the ventral root of the anterior lateral line nerve and the posterior lateral line nerve, which are likely concerned only with mechanoreceptive information from the head and trunk neuromasts.

Earlier studies recognized only a single octavus nucleus, ventral to the medial nucleus and usually termed nucleus ventralis or area vestibularis (Hocke Hoogenboom 1929, Larsell 1967). However, McCormick (1978) reported that the sturgeon, *Scaphirhynchus,* and *Polypterus* possess a ventral octavolateralis column that is divided into anterior, magnocellular, descending, and posterior octavus nuclei as in elasmobranchs. If future experimental studies confirm McCormick's observations, it will be reasonably established that primitive actinopterygians possess a pattern of octavolateralis organization identical to that of elasmobranchs, and that this pattern is the ancestral gnathostome pattern.

2.3.1.2 *Holosteans. Amia* and *Lepisosteus* possess an octavolateralis area in which the dorsal nucleus and dorsal root of the anterior lateral line nerve have been lost. Thus, the octavolateralis area in these taxa consists of a medial nucleus, capped by the cerebellar crest, and a more ventral octavus column (Figs. 3-6, 3-7). McCormick (1978) reports that unilateral transection of the anterior and posterior lateral line nerves in *Amia* reveals projections to the ventral and dorsal halves, respectively, of the medial nucleus, as well as projections to a caudal nucleus at obex levels, and a more rostral projection to the eminentia granularis—a granular cell mass forming lateral lobes closely associated with the cerebellum. The anterior lateral line nerve projects to the lateral third of the eminentia, while the posterior lateral line nerve projects to the middle third of the eminentia. These projections are identical to lateral line projections to the lateral granular ridge of the lower leaf of the elasmobranch auricle, and strongly suggest that the eminentia granularis of actinopterygians is homologous to the lower leaf of the elasmobranch auricle.

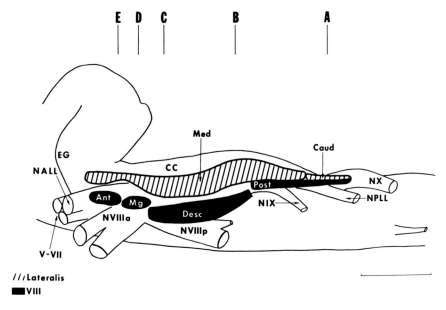

Figure 3-6. Lateral view of the brainstem of the holostean *Amia calva* illustrating the position of the octavus nuclei and the medial and caudal octavolateralis nuclei. (A-E) indicate the levels of the transverse sections of Fig. 3-7. Abbreviations: Ant-anterior octavus nucleus; CC-cerebellar crest; Caud-caudal octavolateralis nucleus; Desc-descending octavus nucleus; EG-eminentia granularis; Med-medial octavolateralis nucleus; Mg-magnocellular octavus nucleus; NALL-anterior lateral line nerve; N IX-glossopharyngeal nerve; NPLL-posterior lateral line nerve; N VIIIa-anterior ramus of the octavus nerve; N VIIIp-posterior ramus of the octavus nerve; N X-vagus nerve; Post-posterior octavus nucleus; V-VII-trigeminal and facial nerves. Bar scale equals 1 mm (after McCormick 1978).

McCormick (1978) observed no lateral line recipient areas in the ventral octavus column, with the possible exception of the magnocellular octavus nucleus (Fig. 3-7D) whose dendrites extend into the neuropil of the medial nucleus and may receive lateral line input.

Four octavus nuclei are recognized in *Amia* (McCormick 1978) and all receive primary octavus fibers, as experimentally determined. A magnocellular nucleus (Figs. 3-6, 3-7D) lies medial to the entering octavus fibers and consists of large (90 μ) multipolar neurons. Caudally, the magnocellular nucleus is replaced by medium-sized bipolar and polygonal cells forming a descending octavus nucleus (Fig. 3-7B,C). At obex levels, the descending nucleus is replaced by a population of small cells (5 μ) forming the posterior octavus nucleus (Fig. 3-7A). An ascending octavus limb projects to a fourth, anterior octavus nucleus, which consists of a small and medium-sized bipolar neurons lying ventral to the medial octavolateralis nucleus (Figs. 3-6, 3-7E). Ascending primary octavus fibers continue beyond this level and terminate in the medial third of the eminentia granularis, but no other cerebellar related projections were seen. McCormick

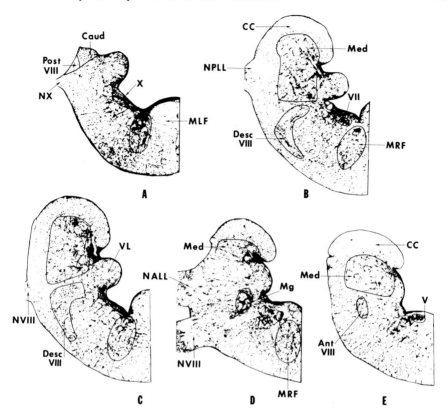

Figure 3-7. Photomicrographs of transverse sections through the octavolateralis area of the medulla of *Amia*. Abbreviations: Ant VIII-anterior octavus nucleus; Caud-caudal octavolateralis nucleus; CC-cerebellar crest; Desc VIII-descending octavus nucleus; Med-medial octavolateralis nucleus; Mg-magnocellular octavus nucleus; MLF-medial longitudinal fasciculus; MRF-medial reticular formation; NALL-anterior lateral line nerve; NPLL-posterior lateral line nerve; N VIII-octavus nerve; N X-vagus nerve; Post VIII-posterior octavus nucleus; VL-vagal lobe; V-trigeminal motor nucleus; VII-facial motor nucleus; X-vagal motor nucleus (after McCormick 1978).

(1978) did not report a descending octavus projection to any part of the reticular formation, but fibers do extend far ventrally and medially in the descending octavus nucleus (Fig. 3-7B,C), and dendrites of the reticular neurons could conceivably extend this far laterally as in elasmobranchs.

Holosteans thus appear to be characterized by the same number of octavus nuclei as chondrosteans and elasmobranchs. The primary difference in the organization of their octavolateralis area and its primary afferent projections is the loss of ampullary organs with the concomitant loss of the dorsal root of the lateral line nerve and anterior lateral line lobe. With the possible exception of the magnocellular octavus nucleus, there is no overlap in the primary terminal sites of the lateral line and octavus nerves. The higher order octavus pathways in holosteans have not been experimentally examined in detail.

2.3.1.3 *Teleosts.* Analysis of the octavus system and the octavolateralis area in teleosts is compounded by several factors: (1) this is the single largest group of bony fishes, and very few species have been examined; (2) experimental data exist for both lateral line and octavus nerves in only a single teleost species; and (3) several teleost families appear to have independently evolved electroreceptors, with a concomitant hypertrophy of part of the octavolateralis area (McCormick 1978).

Most teleost species do not possess electroreceptors, and the lateralis portion of the octavolateralis area is identical to that in holosteans (Figs. 3-7, 3-8). The anterior lateral line nerve consists of a single root, homologous to the ventral root of chondrosteans, which enters the medial nucleus as does the posterior lateral line nerve (Figs. 3-8, 3-9). Thus the medial nucleus, capped by the cerebellar crest, is the most dorsal element of the octavolateralis area. The medial nucleus is replaced rostrally by the eminentia granularis (Figs. 3-8A, 3-9A-C) and caudally by a caudal nucleus of small granule cells (Figs. 3-8D, 3-9G) as in other actinopterygians and in elasmobranchs. Descriptive studies have suggested that the medial nucleus is the primary terminal site of the lateralis nerves; among nonelectric teleosts, experimental evidence exists only for the anterior lateral line nerve of *Cyprinus* (Luiten 1975), but would tend to confirm the descriptive work. In *Cyprinus,* the anterior lateral line nerve does project ipsilaterally to the medial nucleus, and not to more ventral portions of the octavolateralis area.

In electroreceptive teleosts, the lateralis portion of the octavolateralis area is extremely complex, and there is no agreement regarding homologies of the electroreceptive portions with areas in other teleosts or other anamniotes. Many, if not all, siluriform teleosts have developed electroreceptors, as have all mormyrids and gymnotoids. Thus electroreception has likely evolved independently at least two if not three times from nonelectroreceptive holosteans. These electroreceptive teleosts have separate electroreceptive and mechanoreceptive regions within the octavolateralis area (Maler et al. 1973a, 1973b, Maler et al. 1974, Maler 1974, Knudsen 1977). In mormyrids and gymnotids, hypertrophy results in distinct lobes termed the anterior and posterior lateral line lobes (Franz 1911, Berkelback van der Sprenkel 1915, Maler et al. 1974). At present, there exist two distinct interpretations regarding the homologies of these lobes with portions of the octavolateralis area in other teleosts and anamniotes: (1) the lateral line lobes of electroreceptive teleosts are directly homologous to the lateral line lobes of primitive actinopterygians and elasmobranchs (Larsell 1967, Maler 1974); and (2) the electroreceptive lobes of these specialized teleosts are independently evolved from a portion of the generalized teleost medial nucleus and are not homologous to the dorsal nucleus of chondrosteans and elasmobranchs (McCormick 1978). This author's own work and a survey of existing data and arguments strongly support the second interpretation. That is, electroreceptors and the dorsal octavolateralis nucleus of primitive gnathostomes were probably lost in holosteans, and both teleost electroreceptors and electroreceptive areas in the octavolateralis area have most likely developed independently from portions of the medial octavolateralis nucleus in generalized teleosts. However, far more teleosts must be surveyed for electroreceptors, and more intensive studies must be carried out on the living holosteans to confirm this theory.

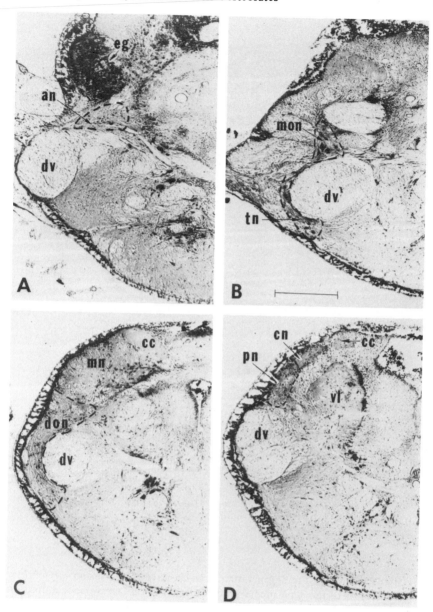

Figure 3-8. Photomicrographs of transverse sections from rostral (A) to caudal (D) of the octavolateralis area of the medulla of the teleost *Gillichthys*. Bar scale equals 400 μm. Abbreviations: an-anterior octavus nucleus; cc-cerebellar crest; cn-caudal octavolateralis nucleus; don-descending octavus nucleus; dv-descending trigeminal tract; eg-eminentia granularis; mn-medial octavolateralis nucleus; mon-magnocellular octavus nucleus; pn-posterior octavus nucleus; tn-tangential octavus nucleus; vl-vagal lobe.

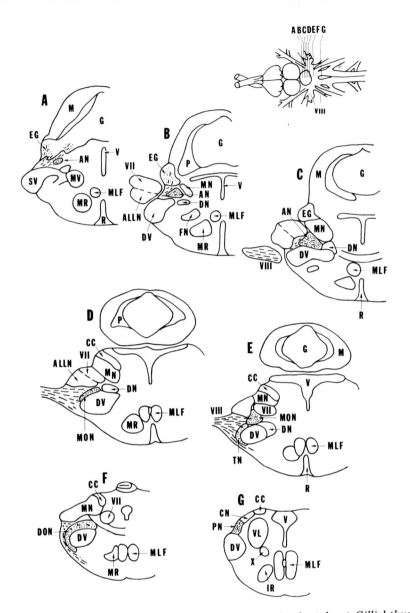

Figure 3-9. Charting of the degenerating octavus nerve in the teleost *Gillichthys* following transection of the nerve medial to the ganglia. (A-G) Transverse sections through the hindbrain at levels indicated in the dorsal view of the brain. Dashed lines indicate degenerating fibers, and stippling indicates degenerating terminals. Abbreviations: ALLN-anterior lateral line nerve; AN-anterior octavus nucleus; CC-cerebellar crest; CN-caudal octavolateralis nucleus; DN-nucleus of the descending trigeminal tract; DON-descending octavus nucleus; DV-descending trigeminal tract; EG-eminentia granularis; FN-facial motor nucleus; G-granular layer of cerebellum; IR-inferior reticular formation; M-molecular layer of cerebellum; MLF-medial longitudinal fasciculus; MN-medial octavolateralis nucleus; MON-magnocellular octavus nucleus; MR-medial reticular formation; MV-trigeminal motor nucleus; P-Purkinje layer of cerebellum; PN-posterior octavus nucleus; R-raphe nucleus; SV-sensory root of trigeminal nerve; TN-tangential octavus nucleus; V-ventricle; VL-vagal lobe; VII-facial sensory root; VIII-octavus nerve; X-vagal lobe.

The literature on the octavus portion of the octavolateralis area is as confusing as that on the lateralis portion. Ramon y Cajal (1908) divided the ventral column into four nuclei: Deiters, tangential, dorsal, and descending. The nucleus of Deiters is described as lying medial to the entering octavus fibers and consists of large oval or fusiform neurons. Pearson (1936b) includes these cells as part of his ventral nucleus. Maler et al. (1973b) recognized the same group of large cells in *Gnathonemus* but traced no octavus fibers to this group with the Fink-Heimer method. However, Korn, Sotelo, and Bennett (1977) demonstrated both electronic and chemical junctions between octavus fibers and the magnocellular nucleus in the toadfish, *Opsanus*.

Both Ramon y Cajal (1908) and Pearson (1936b) recognized a tangential nucleus of large cells lying lateral to the descending trigeminal tract. These neurons begin at the caudal level of the magnocellular nucleus and extend caudally just beneath the lateral surface of the medulla. Pearson (1936b) and McCormick (1978) have argued that this octavus nucleus is seen only in teleosts and cannot be recognized in other actinopterygians; Maler et al. (1973b) have reported octavus projections to the tangential nucleus in *Gnathonemus*.

The dorsal octavus nucleus of Ramon y Cajal (1908) is described as a group of small cells lying dorsal to Deiters' nucleus and beneath the cerebellar crest. It is said to be continuous with the descending nucleus, which continues posterior to Deiters' nucleus caudally and merges with the ventral border of the medial octavolateralis nucleus. The dorsal and descending octavus nuclei of Ramon y Cajal probably correspond to cell groups that Pearson (1936b) and McCormick (1978) describe as parts of the medial octavolateralis nucleus and their descending octavus nuclei. Maler et al. (1973b) have not described either dorsal or descending nuclei as such, but at least a portion of these cells are probably included in the cell groups they term nucleus octavus.

McCormick (1978) examined a number of teleost species and argued that the ventral octavolateralis column consists of the same anterior, magnocellular, descending, and posterior nuclei that she had seen in *Amia*. She also recognized a tangential nucleus in teleosts, the only "new" octavus nucleus she could discern in teleosts.

In an attempt to resolve some of the earlier discrepancies concerning the number of octavus nuclei and the projections of the primary octavus fibers in teleosts, this author experimentally examined the primary octavus projections in a nonelectroreceptive teleost, *Gillichthys mirabilis* (Northcutt 1979b).

The ventral octavolateralis column of *Gillichthys* is divided into anterior, magnocellular, tangential, descending, and posterior nuclei (Figs. 3-8, 3-9). These nuclei and their positions correspond closely to McCormick's description of the octavus nuclei in other teleosts. The anterior octavus nucleus consists of small bipolar or fusiform cells located beneath the eminentia granularis (Fig. 3-8A). These cells are replaced more caudally by large polygonal cells (Fig. 3-8B), usually described as the magnocellular nucleus. A third group of octavus neurons, the tangential nucleus, occurs at the same level (see tn, Fig. 3-8B) as a ventrolateral extension of the magnocellular nucleus. These more laterally situated cells are generally slightly smaller than the magnocellular cells and stream around the lateral edge of the descending trigeminal tract, embedded among the entering octavus fibers. The tangential nucleus is rapidly replaced by much smaller fusiform neurons, forming the descending octavus nucleus that is embedded in the descending octavus limb (Fig. 3-8C). This nucleus continues to obex levels where it is replaced by even smaller granule cells of the posterior octavus nucleus (Fig. 3-8D).

Fink-Heimer analyses of the degenerating primary octavus fibers and terminals in *Gillichthys* reveal that all five of the octavus nuclei receive primary octavus fibers. The dorsal half of the octavolateralis area (nucleus medius and the cerebellar crest) do not receive octavus projections (Fig. 3-9). However, primary octavus projections do reach three additional neural regions in the hindbrain: medial reticular formation (Fig. 3-9F), a rostral portion of the eminentia granularis (Fig. 3-9A,B), and a lateral portion of the granule layer of the cerebellar corpus (Fig. 3-9A). Some octavus fibers in the descending limb continue to course ventrally and medially, beyond the boundaries of the descending octavus nucleus (Fig. 3-9E,F), and ramify among the distal portions of the medial reticular dendrites, as in elasmobranchs.

Octavus fibers in the ascending limb continue beyond the anterior nucleus and divide into lateral and medial bundles, terminating in the eminentia granularis and lateral granule layer of the cerebellum respectively (Fig. 3-9A). The projection to the eminentia granularis in *Gillichthys* is similar to that seen in *Amia* and *Platyrhinoidis,* but in these taxa only a single octavus projection is seen this far rostrally. The more medial granule population receiving octavus input is most likely homologous to the flocculus of tetrapods—although the eminentia has been claimed to be homologous to the elasmobranch auricle or tetrapod flocculus, particularly if the eminentia granularis, like the lateral granular ridge of elasmobranchs, projects back onto the dorsal and medial octavolateralis nuclei.

There are few experimental data on the higher order projections of the octavolateralis area in teleosts (Knudsen 1977). Unilateral lesions of the dorsolateral portion (lateral line lobe) of the octavolateralis area in *Ictalurus* result in bilateral ascending projections to the lateral half of the midbrain, the torus semicircularis. The torus in *Ictalurus* is divided into a medially positioned auditory nucleus (nucleus centralis) and a more lateral area (nucleus lateralis). Multiple unit recordings (Knudsen 1977) indicate that the lateral area is involved with the lateral line. Knudsen was able to further divide the lateral nucleus into a lateral portion, consisting predominantly of electroreceptive units and a medial portion consisting of mechanoreceptive units. These data demonstrate that the three modalities normally processed by the octavolateralis area are maintained as separate channels at midbrain levels; considered in the light of recent results in elasmobranchs (Bullock and Corwin 1979), the data strongly suggest that this separation will be maintained to telencephalic levels.

2.4 Dipnoi

The lungfishes are represented by three living genera relegated to two families: Lepidosirenidae and Ceratodontidae. The lepidosirenids consist of the African genus *Protopterus* and the South American genus *Lepidosiren.* The Ceratodontidae comprise a single Australian species, *Neoceratodus fosteri,* considered the most primitive living lungfish (Moy-Thomas and Miles 1971).

There are no experimental anatomical studies of dipnoans, and the lateralis and octavus nerves and octavolateralis area have been described only for *Neoceratodus* (Holmgren and van der Horst 1925). These workers noted that the anterior lateral line nerve of *Neoceratodus* possesses both dorsal and ventral roots, that the dorsal root enters a dorsomedially situated nucleus they termed the dorsal lateral line lobe, and that the ventral root of the anterior lateral line nerve enters a more ventrolateral cell group they termed the ventral lobe nucleus. Holgren and van der Horst did not recognize a separate, more ventrally located octavus column but claimed that the eighth nerve enters the ventral lobe nucleus. Larsell (1967) reinterpreted Holmgren's and van der Horst's description and equated their dorsal lateral line lobe to the dorsal octavolateralis nucleus of other bony fishes. Larsell believed that the medial nucleus received both octavus and lateralis inputs, thus he accepted the earlier claim that *Neoceratodus* does not possess a distinct ventral octavus column. Given the data now available for other bony fishes, it is very likely that lungfishes do possess a separate and distinct octavus column, but further observations are needed to confirm this probability.

This description of the octavolateralis area in lungfishes is far from adequate, but the available information does suggest that these taxa likely retain the ancestral gnathostome pattern of dorsal and medial nuclei capped by a cerebellar crest. The presence of a dorsal octavolateralis nucleus further suggests that these taxa should be electroreceptive. Roth (1973) reported that lungfishes possess ampullary organs, and evoked potentials have been recorded from midbrain in response to weak electric field stimuli in *Lepidosiren* (Northcutt and Bodznick, unpublished observations).

2.5 Crossopterygii

The crossopterygians comprise two major groups: the rhipidistians and the coelacanths. The rhipidistians became extinct during the Permian period but prior to that time gave rise to the amphibians. The coelacanths are thought to have evolved from the rhipidistians during the Silurian and are represented today by a single species, *Latimeria chalumnae*. *Latimeria* is thus the sole living crossopterygian and represents a radiation temporally remote from the crossopterygian lineage that gave rise to land vertebrates.

Several studies describe the general course of the cranial nerves and external brain morphology in *Latimeria* (Millot and Anthony 1965, Lemire 1971, Nieuwenhuys, Kremers, and van Huijzen 1977, Northcutt, Neary, and Senn 1978), but histological details of the octavolateralis area have not been published.

The organization of the octavolateralis area and cerebellum of *Latimeria* is illustrated in Fig. 3-10, which is based on a series of transverse sections prepared from an immature female caught in Iconi, Grande Comoro, on March 22, 1972 by a French-British-American expedition. Brain sections of this specimen were prepared in two series: one stained to reveal the cell bodies only (cresyl violet) and alternate sections stained by the Klüver-Barrera method to demonstrate myelinated fibers as well as cell bodies.

Maximal development of the octavolateralis area occurs immediately caudal to the cerebellum (Fig. 3-10B). At this level the octavolateralis area consists of a dorsal nucleus, a medial nucleus, and a more ventral octavus column as in chondrosteans and elasmobranchs. The anterior lateral line nerve has dorsal and ventral roots that enter

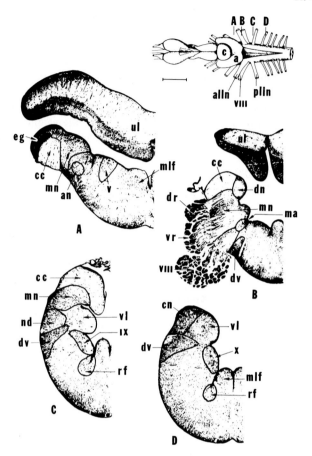

Figure 3-10. Photomicrographs of transverse sections through the octavolateralis area of the medulla of the coelacanth, *Latimeria chalumnae*. The levels of the various sections are indicated in the dorsal view of the brain. Abbreviations: a-auricle; alln-anteral lateral line nerve; an-anterior octavus nucleus; c-corpus of cerebellum; cc-cerebellar crest; cn-caudal octavolateralis nucleus; dn-dorsal octavolateralis nucleus; dr-dorsal root of anterior lateral line nerve; dv-descending trigeminal tract; eg-eminentia granularis; ma-magnocellular octavus nucleus; mlf-medial longitudinal fasciculus; mn-medial octavolateralis nucleus; nd-descending octavus nucleus; plln-posterior lateral line nerve; rf-inferior reticular formation; ul-upper leaf of auricle; vl-vagal lobe; vr-ventral root of anterior lateral line nerve; V-trigeminal motor nucleus; VIII-octavus nerve; IX-glossopharyngeal motor nucleus; X-vagal motor nucleus. Bar scale = 10 mm.

the dorsal and medial nuclei, respectively (Fig. 3-10B). The posterior lateral line nerve enters the medial nucleus further caudally. Both the dorsal and medial octavolateralis nuclei are closely associated with a cerebellar crest (Fig. 3-10A-C) that becomes continuous with the molecular layer of the cerebellum at rostral levels. The dorsal octavolateralis nucleus, like that of elasmobranchs and chondrosteans, is not continuous rostrally with the cerebellum; it ends before the medial nucleus, which is continuous with the lower leaf of the auricle. The distinct cap of granule cells seen in elasmobranchs is not associated with the dorsal and medial nuclei of *Latimeria,* but granule cells do replace the medial nucleus at cerebellar levels (Fig. 3-10A). In fact, the posterior two-thirds of the lower leaf of the auricle consists primarily of granule cells and is likely homologous to the eminentia granularis of other fishes.

The Purkinje-like cells of the medial nucleus are replaced at obex levels by a smaller population of granule cells termed the caudal nucleus (Fig. 3-10D). This caudal population, located above the descending trigeminal tract, is extremely uniform but may actually consist of a posterior octavus nucleus and a caudal lateralis nucleus as seen in other fishes.

Latimeria possesses a well developed eighth nerve (Fig. 3-10B) and, as in other fishes, a distinct magnocellular octavus nucleus lies medial to the entering octavus fibers (Fig. 3-10B). A population of medium-sized fusiform and smaller polygonal cells replaces the magnocellular neurons caudally (Fig. 3-10C). This population is embedded in the descending octavus fibers and is comparable to the descending octavus nucleus of other fishes. It can be traced caudally to obex levels where it is replaced by the caudal nucleus.

An exact boundary between the entering octavus and anterior lateral line nerve fibers is impossible to establish (Fig. 3-10B) without experimental data, thus the ascending octavus fibers cannot be traced with any accuracy. However, a population of smaller fusiform neurons, rostral to the magnocellular nucleus, is in a position comparable to the anterior octavus nucleus of other fishes (Fig. 3-10A).

This analysis of the octavus column in *Latimeria* suggests a pattern of organization comparable to that of chondrosteans, dipnoans, and elasmobranchs. Thus, the extant coelacanth likely retains the primitive gnathostome pattern. The presence of two anterior lateral line roots and a dorsal octavolateralis nucleus also strongly suggests that *Latimeria* is electroreceptive. Together the data argue that the early rhipidistians possessed similar octavial organization and were likely electroreceptive, which suggests that the earliest amphibians, many of which were aquatic throughout life, might have retained electroreceptive ability as they retained the ordinary lateral line system (Olson 1971).

2.6 Amphibia

Living amphibians comprise three different orders: Anura (frogs and toads, approximately 2600 species), Urodela (salamanders, approximately 300 species), and Apoda (caecilians, approximately 150 species). Each order is distinguished by extensive structural differences and each is characterized by a distinct lifestyle. The three orders are easily discernible from their earliest appearance in the fossil record, but they also share a number of derived features that suggest common ancestry. Members of all three amphibian orders possess teeth with a weak, uncalcified segment between the base and crown (pedicellate teeth), and similarities in the middle ear bones and vertebral-skull articulation (Parsons and Williams 1963). Based on these characteristics, Parsons and Williams concluded that modern amphibians represent a monophyletic group, the Lissamphibia. Most workers employ this model of amphibian origin; however, additional data are needed to accept this hypothesis (Thomson 1968, Estes and Reig 1973) particularly for the caecilians.

2.6.1 Caecilians and Salamanders

To date there are no experimental data on the central auditory pathways of these animals. In fact, it is probable that no descriptions of the primary octavus nuclei in caecilians even exist.

Herrick (1930, 1948) described the primary octavus area of the salamanders *Necturus* and *Ambystoma* as an obscure group of cells forming a single vestibular nucleus; he claimed that entering octavus fibers form ascending and descending branches that terminate on this nucleus and on cells of the superior trigeminal nucleus, the lateral line lobe, the cerebellum, and the spinal trigeminal and dorsal funicular nuclei. Herrick (1948) assumed that the inner ear of salamanders has no auditory function, and that the octavus region of the medulla is characterized by extremely "primitive" lateral line, vestibular, and trigeminal regions with extensive overlap of all modalities.

Larsell (1967) largely accepted Herrick's observations on salamanders and further claimed that the octavolateralis area in salamanders and larval anurans consists of dorsal, medial, and ventral nuclei. Larsell thus interpreted the octavolateralis area of salamanders as being identical to that of primitive bony fishes. He claimed that both dorsal and medial octavolateralis nuclei receive input from anterior and posterior lateral line nerves and that a more ventral octavolateralis nucleus receives lateral line and primary vestibular input.

Herrick's and Larsell's theories of octavolateralis organization were generated when it was generally assumed that only land vertebrates possess auditory functions, and evolution proceeds in a linear fashion from simple to complex. They envisioned an octavolateralis area receiving only lateral line and vestibular inputs before the origin of land vertebrates with whom audition was thought to represent a new modality with a concomitant loss of the lateral line system. The primary nuclei that received lateral line inputs were thought to be retained and invaded by the "new" auditory fibers; thus an older system was believed to be utilized in a new way. Larsell (1934, 1967) claimed that metamorphosing anurans recapitulated, in a real sense, this supposed phylogenetic sequence. He further claimed that the dorsal octavolateralis nucleus of tadpoles loses its innervation by the dorsal root of the anterior lateral line nerve at metamorphosis, and that cells of the dorsal nucleus are invaded by newly developed auditory fibers, thus transforming the old dorsal lateral line nucleus into the dorsal acoustic nucleus of adult frogs. Larsell also claimed that the medial octavolateralis nucleus loses its innervation by the ventral root of the anterior lateral line nerve and the posterior lateral line nerve, and that its cells become part of the ventral or magnocellular octavus nucleus.

Experimental data for salamanders and larval anurans do not exist to evaluate Larsell's hypothesis of octavolateralis organization, but it is highly unlikely that this organization is as Larsell suggested. It is now known that all jawed fishes possess distinctly separate lateralis and octavus columns, and that most of these taxa also possess auditory functions. Thus most, if not all, jawed vertebrates possess separate octavus and lateralis systems, and there is no reason to believe amphibians would present an exception. Furthermore, the lungs in anuran tadpoles serve as pressure receptors and are coupled to the perilymphatic fluid of the inner ear by bronchial columella, thus enabling sensitive detection of waterborne sounds (Witschi 1949, Capranica 1976). These data suggest that experimental anatomical studies will likely reveal separate central auditory and lateralis systems in larval amphibians, even as they exist in adult anurans (Campbell and Boord 1974).

2.6.2 Anurans

Experimental data on the octavolateralis area and higher order pathways are available for *Xenopus,* a permanently aquatic genus, and *Rana,* one of the most advanced anurans with a terrestrial adult stage. Most of the existing data are for *Rana,* and it is the only anamniotic genus in which the central auditory pathways are experimentally identified to telencephalic levels (Fig. 3-11).

There is far less information on *Xenopus,* but these data are equally important as *Xenopus* retains a lateral line system in the adult, and its octavolateralis area can be directly compared to larval anurans and other anamniotes. *Xenopus* has anterior and posterior lateral line nerves and receptors (Russell 1976) and an inner ear that functions as an underwater auditory receptor. The auditory receptors of the inner ear of *Xenopus* are similar in number and position to those of *Rana* (Witschi, Bruner, and van Bergeijk 1953), and there is no reason to assume that their functions or central projections differ from those of ranids.

The octavolateralis area of *Xenopus* (Fig. 3-12) is a dorsomedial nucleus extending the length of the medulla. This nucleus consists of a medial cell plate and a more lateral neuropil formed by the laterally directed dendrites of the medial cell plate and the terminals of the posterior lateral line nerve (Campbell and Boord 1974). Campbell and Boord reported that there is no overlap between the octavus and posterior lateral line nerves and that the dorsomedial nucleus and neuropil appear to be homologous to the medial octavolateralis nucleus of fishes. These workers traced octavus fibers to more ventrolaterally situated nuclei and divided the octavus column into dorsal and ventral nuclei (Fig. 3-12), as in *Rana* (Fig. 3-11D). Campbell and Boord did not experimentally trace the central projections of the anterior lateral line nerve in *Xenopus,* but their data suggest that the octavolateralis area consists of a dorsal column of cells receiving lateral line input, and a ventral column of cells receiving octavus input. This pattern is remarkably similar to that seen in holosteans and nonelectroreceptive teleosts; a medial octavolateralis nucleus receives lateral line input, but no dorsal nucleus is apparent as in *Latimeria* or primitive ray-finned fishes. The ventral octavus column of *Xenopus* also appears to be organized differently than in other anamniotes; only two nuclei have been identified, rather than four and five as in other gnathostomes. Furthermore, the two octavus nuclei of *Xenopus* are oriented dorsoventrally, rather than forming a rostrocaudal longitudinal series as in fishes. These relationships pose a number of problems in homologizing the octavus nuclei

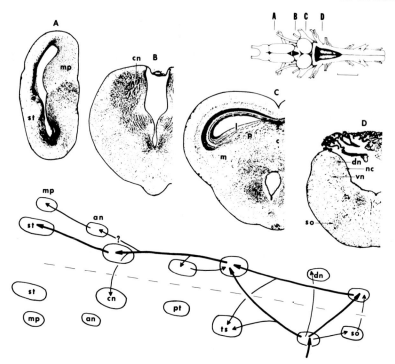

Figure 3-11. Photomicrographs of transverse sections through the brain of the bullfrog, *Rana catesbeiana*, illustrating major auditory centers and their interconnections. The levels of the various sections are indicated in the dorsal view of the brain. Bar scale equals 10 mm. Abbreviations: an-anterior thalamic nucleus; c-commissural toral nucleus; cn-central thalamic nucleus; dn-dorsal octavus nucleus; l-laminar toral nucleus; m-magnocellular toral nucleus; mp-medial pallium; nc-caudal octavus nucleus; p-principal toral nucleus; pt-pretectum; so-superior olive; st-striatum; ts-torus semicircularis; vn-ventral octavus nucleus.

of amphibians to those of other anamniotes. These problems will be discussed further in the last section.

Acoustic signals play a dominant role in anuran social behavior and are involved in mate selection, formation and maintenance of breeding aggregations, and insuring reproductive isolation (Blair 1958). Not surprisingly, the anuran auditory system is specialized to detect species specific calls, particularly mating calls (Frishkopf, Capranica, and Goldstein 1968, Capranica 1976).

Anurans possess auditory receptors in two separate inner ear sensory papillae: the basilar papilla, located in a diverticulum of the posterior saccular wall and a larger amphibian papilla, located in a chamber opening into the saccular medial wall. Both the basilar and amphibian papillae contribute axons to the octavus nerve via its posterior ramus. The anuran octavus nerve consists of an anterior ramus, which innervates horizontal and anterior semicircular canals, the utricle, and a part of the saccule and a posterior ramus, which innervates the remaining part of the saccule, the lagena, the

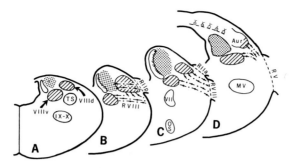

Figure 3-12. Chartings of the central projections of the octavus and posterior lateral line nerves in the South African clawed toad, *Xenopus laevis*. (A) through (D) represent transverse sections from the caudal medulla (A) to the cerebellum (D). Degenerated axons are indicated by beaded dashes and dots, terminal sites of the posterior lateral line fibers by stippling, and terminal sites of the octavus fibers by diagonal lines. Abbreviations: Aur-auricle; MV-trigeminal motor nucleus; VII-facial motor nucleus; VIIId-dorsal octavus nucleus; VIIIv-ventral octavus nucleus; IX-X-glossopharyngeal and vagal motor nuclei; OS-superior olive; Rv-trigeminal root; RVIII-octavus root; RIIa-anterior lateral line nerve root; RIIp-posterior lateral line nerve root; TS-solitary tract (after Campbell and Boord 1974).

posterior semicircular canal, and the two auditory papillae (Gregory 1972). Thus, the anuran octavus nerve is divided into an anterior vestibular ramus and a mixed auditory and vestibular posterior ramus.

In the bullfrog, *Rana catesbeiana,* octavus nerve fibers are divided into three classes according to their best excitation frequencies: 200 Hz to 300 Hz, 600 Hz to 700 Hz, and a broad band from 1000 Hz to 1700 Hz (Feng, Narins, and Capranica 1975). The low frequency units differ from the other two classes in that they are inhibited by sounds in the midfrequency range. This inhibitory phenomenon led Frishkopf and Goldstein (1963) to classify peripheral auditory units in *Rana* as simple or complex (inhibitable). Simple and complex peripheral units are now known to exist in many other anurans, as are three frequency populations (see Capranica 1976 for review). Feng et al. (1975) demonstrated that the high frequency simple units arise from the basilar papilla, whereas the simple midfrequency and complex low frequency units arise from the more complex amphibian papilla. Thus, the two papillae are specialized to receive nonoverlapping frequencies.

The anuran periphal auditory system acts as a feature detector with each class of auditory units responding selectively to a component of a species specific call. In bullfrogs, the mating call is characterized by energy concentrated in a narrow peak at 1500 Hz and a broader band at 200 Hz to 300 Hz (Capranica 1968). These energy peaks correspond well to the best excitatory frequencies of the high and low frequency auditory fibers. Other calls, such as the male territorial call, warning call, and release call, possess midfrequency energy ranges (500 Hz to 700 Hz) that inhibit complex unit firing (Capranica 1968).

The anuran anterior and posterior octavus rami enter the medulla separately. The anterior ramus enters ventrally and is termed the ventral octavus root, whereas the posterior ramus enters dorsally and is termed the dorsal octavus root. The octavus roots distribute to the primary octavus nuclei, traditionally divided into a dorsal acoustic nucleus and a ventral vestibular nucleus (Fig. 3-11D; Larsell 1967, Gregory 1972, Mehler 1972). Thus the amphibian and basilar papillae project only to the dorsal octavus nucleus.

Opdam, Kemali, and Nieuwenhuys (1976) recently described a third nucleus (nucleus caudalis, Fig. 3-11D) closely associated with the dorsal and ventral nuclei in the ranid medulla. Rostrally, the caudal nucleus occupies a periventricular position, medial to both dorsal and ventral nuclei, but it continues beyond the caudal limit of the dorsal acoustic nucleus. Gregory's (1972) and Mehler's (1972) chartings of the central course of the octavus nerve indicate that the caudal nucleus receives primary octavus input; it is not known whether this nucleus receives auditory and/or vestibular terminals. Wilczynski (1978) reported that cells of the caudal nucleus project bilaterally to the torus, but this is also true for some cells of the ventral vestibular nucleus. Thus, without further information, it is impossible to determine whether the caudal nuclear ascending pathway to the torus is auditory, vestibular, or both.

The dorsal acoustic nuclei are interconnected by an extensive commissural system (Fig. 3-11; Grofova and Corvaja 1972). Approximately half the dorsal acoustic cells are binaurally driven; of these, most are excited by contralateral and inhibited by ipsilateral stimulation, although some dorsal acoustic cells are excited by both. Most of the monaural cells are excited by ipsilateral tones, but 20% are excited exclusively by contralateral tones (Feng 1975, Feng and Capranica 1976).

The dorsal acoustic nucleus projects bilaterally to the superior olive and torus semicircularis (Fuller and Ebbesson 1973), with most of the fibers terminating contralaterally (Fig. 3-11). The anuran superior olive (Fig. 3-11D), in turn, projects heavily to the ipsilateral torus and sparsely to the contralateral torus (Rubinson and Skiles 1975). Physiological data on the olivary cells in *Rana* are not available, but Feng and Capranica (1978) have investigated these cells in *Hyla*. In this genus, the olivary cells are similar to those of the dorsal acoustic nucleus, with approximately half the units exhibiting binaural sensitivity. However, a few olivary cells, termed complex cells, have very different properties. The complex cells are excited by contralateral ear stimulation, but simultaneous ipsilateral stimulation, with tones close to the cells' best excitatory frequency, inhibits these cells. More remote ipsilateral tones, however, facilitate these units.

Dorsal acoustic and superior olivary inputs converge on the midbrain torus semicircularis (Fig. 3-11C). The torus is the expanded caudal roof of the midbrain, homologous to the mammalian colliculus. The anuran torus is divided into five nuclei: commissural, laminar, magnocellular, principal, and subependymal midline nuclei (Potter 1965a). The ascending dorsal nuclear input terminates mainly in the principal nucleus; the superior olivary input terminates in the principal nucleus and, more sparsely, in the commissural and magnocellular nuclei (Rubinson and Skiles 1975). Auditory terminals are not seen among the cell bodies of the laminar nucleus, but its cells send dendrites ventrally into the principal nucleus, and Potter (1965b) has confirmed auditory activity in the laminar nucleus as well as in the principal and magnocellular nuclei. The laminar nucleus also receives hypothalamic (Neary and Wilczynski 1977), spinal

(Ebbesson 1976), and dorasl column (Wilczynski, Neary, and Andry 1977) inputs. Additional toral inputs from the medial and lateral medullary reticular formations and the midbrain (tegmental) reticular nuclei are known (Wilczynski 1978). The medullary reticular cells that project to the torus are closely associated with the medullary auditory centers, whereas the tegmental cells appear to arise from two distinct populations: a caudal portion of the superficial isthmal nucleus, which may be homologous to the mammalian nuclei of the lateral lemniscus (Wilczynski 1978), and a rostral portion of the superficial isthmal nucleus, which receives afferents from the telencephalic striatum and projects to the laminar and principal toral nuclei. Thus, the torus appears to receive additional ascending pathways from medullar and caudal tegmental cell groups, and these inputs may represent additional auditory pathways as well as input from the telencephalon. Intertoral connections also exist with the commissural, magnocellular, and principal nuclei (Wilczynski 1978).

Most toral auditory units are phasically active, and both sharply tuned and broad band units are known (Potter 1965b). Frishkopf et al. (1968) discovered biomodally tuned units, which represent the first central convergence of the two auditory papillae; these cells respond equally to both high and low frequency stimulation. Similar convergence is now known for a variety of anurans (Ewert and Borchers 1971, Loftus-Hills 1971, Feng 1975).

Single unit studies of the anuran torus have not indicated tonotopic or spatiotopic maps; however, an evoked potential study (Pettigrew, Chung, and Anson 1978) revealed both types of maps in *Rana esculenta*. Pettigrew et al. reported that the rostral torus is most sensitive to sounds located in front of the animal, whereas the caudal torus is most sensitive to caudally located sounds. A tonotopic toral map also indicated high frequencies located rostrally and low frequencies located caudally.

The torus in anurans has both ascending and descending pathways; Ascending toral fibers project to the ventral half of the ipsilateral lateral thalamic nucleus and to the more rostrally located central thalamic nucleus (Neary 1974, Wilczynski 1978). The toral projection to the central thalamic nucleus (Fig. 3-11B) is bilateral, with the contralateral fibers decussating in the suprachiasmatic commissure. The lateral thalamic nucleus forms part of the anuran pretectum and is divided into dorsal and ventral subdivisions. The dorsal subdivision receives input from the optic tectum and projects back onto the tectum, whereas the ventral subdivision receives toral input and projects back onto the torus (Wilczynski 1978). Ewert (1970) and Ingle (1973) demonstrated that the visual pretectum normally exerts a strong inhibitory effect on tectal neurons involved in certain visual behaviors. Trachtenberg and Ingle (1973) and Wilczynski and Northcutt (1977) have suggested that the dorsal subdivision of the lateral nucleus mediates this pretectal inhibition of the tectum. Given the similarity of reciprocal connections between the torus and the ventral subdivision of the lateral nucleus, it is possible there exists an analogous auditory system between pretectum and torus, as between pretectum and tectum.

Descending toral efferents originate from the laminar and principal nuclei, and a completely ipsilateral pathway terminates in the lateral isthmal reticular formation, the superior olive, and the medial medullary reticular formation (Wilczynski 1978). Direct toral feedback to the dorsal and caudal octavus nuclei does not appear to exist, and descending toral efferents have not been traced caudal to the obex in *Rana*.

Two ascending auditory thalamo-telencephalic pathways appear to exist in anurans (Fig. 3-11). The central thalamic nucleus projects, via the lateral forebrain bundle, to the ipsilateral striatum (Fig. 3-11A,B; Kicliter and Northcutt 1975, Wilczynski 1978, Kicliter 1979). Mudry, Constantine-Paton, and Capranica (1977) have physiologically confirmed the presence of auditory units in the central thalamic nucleus of *Rana pipiens* and *Rana catesbeiana*. Units of the central thalamic nucleus respond preferentially to the simultaneous presentation of high and low frequency components of mating calls. In addition, Mudry and Capranica (1978) reported auditory activity in the striatum, but the response properties of these units must still be characterized.

Mudry and Capranica also reported auditory activity in the medial pallium of the telencephalon. This activity does not result from striatal projections onto the medial pallium, as no such projections appear to exist (Wilcynski and Northcutt 1979, Ronan and Northcutt 1979). However, Ronan and Northcutt reported that the anterior thalamic nucleus projects bilaterally on the medial pallium. Although toral efferents do not terminate within the confines of the anterior thalamic nucleus, it is possible that some dendrites of the anterior thalamic nuclear cells project caudally into the central thalamic nucleus. It is also possible that collaterals or some cells of the central thalamic nucleus project to the anterior thalamic nucleus.

Auditory information clearly reaches telencephalic levels in anurans, but far more data are needed to characterize these projections. Physiological study of the anterior thalamic nucleus should determine whether this nucleus does receive auditory input. If so, two separate thalamic auditory pathways project to the telencephalon. Further physiological characterization of medial pallial and striatal auditory units should prove particularly interesting. The anterior thalamic nucleus receives visual, hypothalamic, somatosensory, and possibly auditory input, which projects bilaterally on the medial pallium. The medial pallium also receives input from most second order telencephalic olfactory centers, and projects to the anterior thalamic nucleus, and preoptic and posterior hypothalamus (Ronan and Northcutt 1979). Thus the medial pallium may affect the torus, optic tectum, and tegmentum via its preoptic and hypothalamic connections (Wilczynski 1978). Given the variety of sensory convergence in the medial pallium, bilateral input from the medial pallium into the preoptic nucleus, and preoptic bilateral input into the laminar nuclei of the torus, it is likely that the medial pallium functions as an arousal system.

The striatum is a more likely candidate for the highest neural center involved in conspecific recognition in anurans. This structure receives a well-developed auditory thalamic pathway, as well as a retino-tecto-thalamic pathway (Gruberg and Ambros 1974). The striatum, in turn, projects to the superficial isthmal nucleus, anterior entopeduncular nucleus, and caudal tegmental nuclei (Wilczynski 1978, Wilczynski and Northcutt 1979). These connections give the striatum access to the ipsilateral optic tectum, torus, and medullar reticular formation. Thus the striatum possesses anatomical connections that allow complex feedback circuits involving lower ascending sensory centers, as well as connections that tie it with the major integrating centers mediating motor responses.

3 Discussion and Conclusions

3.1 Anamniotic Octavus Patterns

Among anamniotes there are three distinct patterns of octavus organization: agnathan, piscine, and amphibian. The agnathan octavus column that receives primary octavus terminals consists of a ventral nucleus (Figs. 3-3, 3-4) of small and medium-sized cells, within which three aggregations (octavomotor cells or nuclei) are embedded. The piscine pattern occurs in cartilaginous and bony fishes and consits of a primary octavus column divided into anterior, magnocellular, descending, and posterior nuclei (Figs. 3-5—3-10). Teleosts have an additional octavus nucleus, the tangential nucleus, which may be an elaboration of part of the magnocellular nucleus of other fishes. The amphibian pattern consists of two primary octavus nuclei, a dorsal auditory nucleus and a ventral vestibular nucleus, as well as the more medially situated caudal nucleus whose function is uncertain (Figs. 3-11, 3-12).

In all three anamniotic groups, primary octavus fibers also project ipsilaterlly to one or more cerebellar related cell populations, in addition to the medullar octavus nuclei. Octavus fibers terminate in the cerebellum of agnathans (Fig. 3-4A), in the lower leaf of the auricle of cartilaginous fishes (Fig. 3-5B), in the eminentia granularis and auricle of teleosts (Fig. 3-9A), and in the auricle of amphibians (Fig. 3-12D). Comparisons of these cerebellar related populations receiving octavus input is complicated by differences in the number and position of such populations in anamniotes.

The eminentia granularis does not appear to be homologous to the auricle; rather, it appears to be a separate cell population closely related to the lateral line lobes, as its axons form part of the cerebellar crest capping the lateral line lobes. The extent of octavus projections to the cerebellum indicate that this structure in agnathans may be homologous only to the eminentia granularis and auricle in other anamniotes. In other vertebrates, the corpus of the cerebellum does not receive primary octavus input. The "auricle" of cartilaginous fishes consists of lower and upper leaves. The lower leaf is composed primarily of granule cells and is likely homologous to the eminentia granularis of bony fishes; the upper leaf is homologous to the auricle of bony fishes and the flocculonodular lobe of tetrapods. It is not presently known whether amphibians that retain the lateral line system also possess an eminentia granularis and cerebellar crest.

Phyletic comparisons of the three anamniotic octavus patterns are also complicated because we do not know whether agnathans possess auditory abilities, or whether fishes possess separate auditory and vestibular nuclei. At present it is possible to state only that there is no one-to-one nuclear correspondence among these three octavus patterns. We must have experimental data indicating whether agnathans possess auditory end organs and, if so, where such organs terminate within the octavus column in order to decide among the following hypotheses concerning the agnathan octavus column:

1. The agnathan octavus pattern is the most primitive among living vertebrates and is solely vestibular.
2. The agnathan octavus column is solely vestibular but has lost auditory functions and is, therefore, secondarily simple rather than primitive.

3. The agnathan octavus column is divided into auditory and vestibular portions and is the most primitive among vertebrates.
4. The agnathan octavus column is divided into auditory and vestibular portions but has evolved independently of jawed vertebrates.

Similar problems also plague octavus comparisons among fishes and tetrapods. Auditory functions are now established for many cartilaginous and bony fishes (Fay and Popper, Chapter 1), but the exact otic end organs and their termination sites among the octavus nuclei are unknown. Until experimental data provide answers to these questions, it is impossible to evaluate the following hypotheses:

1. Auditory end organs and medullar nuclei have evolved independently in fishes and tetrapods.
2. Auditory end organs and their medullar nuclei are homologous in fishes and tetrapods.
3. There is phyletic continuity between auditory end organs and medullar nuclei in fishes and some tetrapods; i.e., between fishes, extinct primitive amphibians, and living reptiles, but not living amphibians.

Until these questions are answered, it will be impossible to evaluate the evolution of higher order auditory pathways, even though experimental studies of several anamniotes have established such pathways to midbrain (Knudsen 1977) and telencephalic levels (Mudry and Capranica 1978, Wilczynski 1978, Bullock and Corwin 1979).

3.2 Evaluation of Earlier Hypotheses

Ayers (1892) and van Bergeijk (1966, 1967) argued that the vertebrate inner ear arose phylogenetically as an elaborated portion of the head lateral line system. Three lines of evidence are used to support this acousticolateralis hypothesis: (1) the lateral line and inner ear both arise from ectodermal placodes; (2) the sensory cells of the inner ear and lateral line are identical; and (3) the octavus and lateral line nerves project to the same medullar nuclei. While it is true that both the inner ear and lateral line organs arise from ectodermal placodes (Fig. 3-2), so do taste buds (epibranchial placodes, Fig. 3-2), the olfactory organ, and parts of the eye. The embryological data allow us to conclude only that all special sense organs are related ontogenetically by arising wholly, or in part, from ectodermal placodes that may be induced by neural crest (Balinsky 1975). If we use ontogenetic evidence to argue for certain phylogenetic sequences, we could just as validly argue that the taste buds or olfactory organs arose phylogenetically from the lateral line or any other placodally derived organ. The ontogenetic data are consistent with a phylogenetic hypothesis that all vertebrate special sense organs are derived from some type of primitive ectodermal sense organ in early chordates; but the ontogenetic data are also consistent with a phylogenetic hypothesis that all vertebrate special sense organs arose at the same time by ectodermo-neural crest interactions and that no special sense organ is phylogenetically older than any other. In this context, the earliest fossil vertebrates (ostracoderm agnathans of the Ordovician and Silurian) possessed both inner ears and lateral line organs (Moy-Thomas and Miles 1971).

The similarity between hair cells of the inner ear and lateral line neuromasts is striking; however, many of these features, including the apical cilia, are common to sensory epithelial cells in general. Again, these similarities indicate only that vertebrate special sense organs are ontogenetically and, probably, phylogenetically related to ectodermal tissues. They do not allow us to conclude that any given sensory system preceded any other.

The third line of evidence used to argue for the acousticolateralis hypothesis is that the lateral line and octavus nerves project to the same medullar nuclei; this is now known to be false. In all anamniotes thus far examined experimentally, there is little or no overlap in the terminations of the lateralis and octavus nerves (Maler et al. 1973b, Campbell and Boord 1974, McCormick 1978).

In summary, the embryological evidence indicates that all vertebrate special sense organs arise ontogenetically through ectodermo-neural crest interactions, and that all living anamniotes possess separate medullar columns receiving lateral line and octavus inputs. There is no phylogenetic evidence that any special sensory system is older than any other. The data are consistent with two different phylogenetic hypotheses: (1) all vertebrate special sense organs arose from some unspecified ectodermal sense organ(s) of the ancestral chordates; and (2) all vertebrate special sense organs arose from a type of ectodermo-neural crest interaction that first occurred with the origin of vertebrates, and all special sense organs are equally old.

Van Bergeijk (1966, 1967) and Wever (1976) argue that the inner ear arose phylogenetically as a vestibular organ and only secondarily evolved auditory functions (labyrinth hypothesis). This hypothesis is based on the supposed distribution of audition in anamniotes and the belief that most anamniotes possess a single medullar octavus nucleus, homologous to the vestibular complex of amniotic vertebrates (Larsell 1967). Recent behavioral and physiological studies indicate that at least some members of all jawed anamniotic groups possess auditory abilities (for review, see Popper and Fay 1977 and Chapter 1).

At present it is not known whether agnathans possess audition. However, Lowenstein et al. (1968) and Thornhill (1972) indicate that lampreys possess otic end organs homologous to those of gnathostomes. If audition is demonstrated in agnathans, and the otic organs mediating audition project to comparable areas of the octavus column in agnathans and fishes, the labyrinth hypothesis must be rejected. However, if agnathans do not possess audition, the labyrinth hypothesis cannot be accepted on this basis alone. It could be argued that living agnathans possess a secondarily simple inner ear in which auditory functions have been lost; however, this hypothesis cannot be tested. Future studies may reject the labyrinth hypothesis, or it may not be possible to establish the phylogenetic origin of audition.

3.3 Audition and the Emergence of Land Vertebrates

Evaluation of peripheral and central auditory systems has proceeded largely independently in most vertebrates. Until recently it was assumed that only amphibians and teleosts, among anamniotes, possessed audition. Behavioral and experimental evidence now indicates that audition occurs more widely in anamniotes, but there is still no agreement regarding which inner ear organs respond to sound and the phylogeny of these organs. Wever (1976) argues that audition and the otic organs mediating this modality have evolved independently in teleosts, living amphibians, and amniotes. His argument is based primarily on differences in the stimulating mechanisms of inner ear receptors in each of these groups, and the belief that any given stimulating mechanism could not have evolved into any other. Lombard (Chapter 4) raises similar questions regarding the evolution of middle ear structures in amphibians and amniotes. Far more data regarding which otic organs respond to sound, their embryology, their homologies, and their central projections are clearly needed to illuminate the phylogeny of peripheral auditory mechanisms in vertebrates.

Not surprisingly, the phylogeny of the octavus nuclei is equally clouded. Until recently, it was assumed that fishes possess only a single octavus nucleus receiving both lateral line and vestibular inputs (Larsell 1967). The dorsal octavus nucleus of amphibians, which receives auditory input, was believed to arise by capture of the dorsal nucleus of the lateral line system (Larsell 1967). We now know that some amphibians that possess a lateral line system also possess both dorsal acoustic and more ventral vestibular nuclei (Campbell and Boord 1974). Thus, it is highly unlikely that the dorsal acoustic nucleus of amphibians arose from a portion of the lateral line system. Furthermore, both cartilaginous and bony fishes possess a larger number of octavus nuclei (4 to 5) than do amphibians (3). Not only is the number larger in fishes than in amphibians, but the octavus nuclear column in fishes is organized as a single rostrocaudal series while that of amphibians is oriented dorsoventrally (Figs. 3-6, 3-11D, 3-12). The reptilian octavus pattern is more similar to that of fishes than to that of living amphibians. Most reptiles possess two anterodorsally located auditory nuclei and four posteroventrally located vestibular nuclei (Campbell and Boord 1974, Leake 1974). Further experimental studies on the central projections of piscine otic organs are needed to determine the degree of similarity between fishes and reptiles. However, the differences in number and topography of the octavus nuclei in fishes, amphibians, and reptiles suggest that audition may have arisen a number of times in vertebrates or that modern amphibians may have reorganized, or even lost, parts of the ancestral tetrapod auditory system.

Acknowledgments. Original research for this chapter was supported in part by NIH grants NS11006 and EYNS02485 and by a Rackham Faculty Research Grant from the University of Michigan. Mary Sue Caudle Northcutt assisted in many phases of the research and in the preparation of the manuscript. The author is also indebted to the Guggenheim Foundation for their support during the period in which this manuscript was written.

References

Ayers, H.: Vertebrate cephalogenesis. II. A contribution to the morphology of the vertebrate ear, with a reconsideration of its function. J. Morphol. 6, 1-360 (1892).
Ayers, H., Worthington, J.: The finer anatomy of the brain of *Bdellostoma dombeyi.* I. The acoustico-lateralis system. Amer. J. Anat. 8, 1-16 (1908).
Balinsky, B. I.: An Introduction to Embryology, 4th ed. Philadelphia: Saunders, 1975.

Beard, J.: On the segmental sense organs of the lateral line, and on the morphology of the vertebrate auditory organ. Zool. Anz. 7, 123-126 (1884).

Berkelbach van der Sprenkel, H.: The central relations of the cranial nerves in *Siluris glanis* and *Mormyrus cashive*. J. Comp. Neurol. 25, 5-65 (1915).

Blair, W. F.: Mating call in the speciation of anuran amphibians. Amer. Naturalist 92, 27-51 (1958).

Boord, R. L.: Auricular projections in the clearnose skate, *Raja eglanteria*. Amer. Zool. 17, 887 (1977).

Boord, R. L., Campbell, C. B. G.: Structural and functional organization of the lateral line system of sharks. Amer. Zool. 17, 431-441 (1977).

Bullock, T. H., Corwin, J. T.: Acoustic evoked activity in the brain in sharks. J. Comp. Physiol. 129, 223-234 (1979).

Campbell, C. B. G., Boord, R. L.: Central auditory pathways in nonmammalian vertebrates. In: Handbook of Sensory Physiology (Vol. V/1: Auditory System. Keidel, W. D., Neff, W. D., eds.). Berlin-Heidelberg-New York: Springer-Verlag, 1974, pp. 337-362.

Capranica, R. R.: The vocal repertoire of the bullfrog (*Rana catesbeiana*). Behavior 31, 302-325 (1968).

Capranica, R. R.: Morphology and physiology of the auditory system. In: Frog Neurobiology. Llinas, R., Precht, W. (eds.). Berlin-Heidelberg-New York: Springer-Verlag, 1976, pp. 551-575.

Corwin, J. T.: Morphology of the macula neglecta in sharks of the genus *Carcharhinus*. J. Morphol. 152, 341-362 (1977).

Corwin, J. T.: The relation of inner ear structure to the feeding behavior in sharks and rays. Scanning Electron Microscopy II, 1105-1112 (1978).

Ebbesson, S. O. E.: Morphology of the spinal cord. In: Frog Neurobiology. Llinas, R., Precht, W. (eds.). Berlin-Heidelberg-New York: Springer-Verlag, 1976, pp. 679-706.

Estes, R., Reig, O. A.: The early fossil record of frogs: A review of the evidence. In: Evolutionary Biology of the Anurans. Vial, J. L. (ed.). Columbia: University of Missouri Press, 1973, pp. 11-63.

Ewert, J.-P.: Neural mechanisms of prey-catching and avoidance behavior in the toad (*Bufo bufo*, L.). Brain, Behav. Evol. 3, 36-56 (1970).

Ewert, J.-P., Borchers, H. W.: Reaktionscharakteristic von Neuronen aus dem Tectum opticum und Subtectum der Erdkröte *Bufo bufo* (L.). Z. vergl. Physiol. 71, 165-189 (1971).

Fay, R. R., Kendall, J. I., Popper, A. N., and Tester, A. L.: Vibration detection by the macula neglecta of sharks. Comp. Biochem. Physiol. 47A, 1235-1240 (1974).

Feng, A. S.: Sound localization in anurans: An electrophysiological and behavioral study. Doctoral Thesis, Cornell University, (1975).

Feng, A. S., Capranica, R. R.: Sound localization in anurans. I. Evidence of binaural interaction in dorsal medullary nucleus of bullfrogs (*Rana catesbeiana*). J. Neurophysiol. 39, 871-881 (1976).

Feng, A. S., Capranica, R. R.: Sound localization in anurans. II. Binaural interaction in superior olivary nucleus of the green tree frog (*Hyla cinerea*). J. Neurophysiol. 41, 43-54 (1978).

Feng, A. S., Narins, P. M., Capranica, R. R.: Three populations of primary auditory fibers in the bullfrog: their peripheral origins and frequency sensitivities. J. Comp. Physiol. 100, 221-229 (1975).

Franz, V.: Das Mormyridenhirn. Zool. Jahrb., Abt. f. Anat. 32, 401-464 (1911).

Frishkopf, L. S., Capranica, R. R., Goldstein, M. H.: Neural coding in the bullfrog's auditory system—A teleological approach. Proc. IEEE 56, 969-980 (1968).

Frishkopf, L. S., Goldstein, M. H.: Responses to acoustic stimuli from single units in the eighth nerve of the bullfrog. J. Acoust. Soc. Amer. 35, 1219-1228 (1963).

Fuller, P. M., Ebbesson, S. O. E.: Projections of the primary and secondary auditory fibers in the bullfrog (*Rana catesbeiana*). Program, Soc. Neurosci. Ann. Meet.: 333 (1973).

Greenwood, P. H.: Interrelationships of osteoglossomorphs. In: Interrelationships of Fishes. Greenwood, P. H., Miles, R. S., Patterson, C. (eds.). London: Academic Press, 1973, pp. 307-332.

Greenwood, P. H., Rosen, D. E., Weitzman, S. H., Myers, G. S.: Phyletic studies of teleostean fishes, with a provisional classification of living forms. Bull. Am. Mus. Nat. Hist. 131, 339-346 (1966).

Gregory, K. M.: Central projections of the eighth nerve in frogs. Brain, Behav., Evol. 5, 70-88 (1972).

Grofova, I., Corvaja, N.: Commissural projections from the nuclei of termination of the VIIIth cranial nerve in the toad. Brain Res. 42, 189-195 (1972).

Gruberg, E. R., Ambros, V. R.: A forebrain visual projection in the frog (*Rana pipiens*). Exp. Neurol. 44, 187-197 (1974).

Heier, P.: Fundamental principles in the structure of the brain; a study of the brain of *Petromyzon fluviatilis*. Acta Anat. 5, (Suppl. 8) 1-213 (1948).

Herrick, C. J.: The cranial and first spinal nerves of *Menidia*: a contribution upon the nerve components of the bony fishes. J. Comp. Neurol. 9, 153-455 (1899).

Herrick, C. J.: The medulla oblongata of *Necturus*. J. Comp. Neurol. 50, 1-96 (1930).

Herrick, C. J.: The Brain of the Tiger Salamander. Chicago: University of Chicago Press, 1948.

Hocke Hoogenboom, K. J.: Das Gehirn von *Polydon folium*. Z. Mikr. Anat. Forsch. 18, 311-392 (1929).

Holmgren, N., van der Horst, C. J.: Contribution to the morphology of the brain of *Ceratodus*. Acta Zool. 6, 59-165 (1925).

Hotton, N.: Origin and radiation of the classes of poikilothermous vertebrates. In: Evolution of Brain and Behavior in Vertebrates. Masterton, R. B., Bitterman, M. E., Campbell, C. B. G., Hotton, N. (eds.). Hillsdale, NJ: Lawrence Erlbaum Assoc.'s, 1976, pp. 1-24.

Ingle, D.: Disinhibition of tectal neurons by pretectal lesions in the frog. Science 180, 422-424 (1973).

Jansen, J.: The brain of *Myxine glutinosa*. J. Comp. Neurol. 49, 359-507 (1930).

Johnston, J. B.: The brain of *Acipenser*. Zool. Jahr. Anat. und Ontogenie 15, 59-260 (1901).

Johnston, J. B.: The brain of *Petromyzon*. J. Comp. Neurol. 12, 1-106 (1902).

Johnston, J. B.: The cranial nerve components of *Petromyzon*. Morph. Jahrb. 34, 149-203 (1905).

Jørgensen, J. M., Flock, Å., Wersäll, J.: The Lorenzian ampullae of *Polydon spatula*. Z. Zellforsch. 130, 362-377 (1972).

Kappers, C. U. A.: Antomie comparée du Système Nerveux. Haarlem: De Erven F. Bohn (Paris: Masson et Cie), 1947.

Kicliter, E.: Some telencephalic connections in the frog, *Rana pipiens*. J. Comp. Neurol. 185, 75-86 (1979).

Kicliter, E., Northcutt, R. G.: Ascending afferents to the telencephalon of ranid frogs: an anterograde degeneration study. J. Comp. Neurol. 161, 239-254 (1975).

Knudsen, E. I.: Distinct auditory and lateral line nuclei in the midbrain of catfishes. J. Comp. Neurol. 173, 417-432 (1977).

Korn, H., Sotelo, C., Bennett, M. V. L.: The lateral vestibular nucleus of the toadfish *Opsanus tau*: Ultrastructural and electrophysiological observations with special reference to electrotonic transmission. Neuroscience 2, 851-884 (1977).

Larsell, O.: The differentiation of the peripheral and central acoustic apparatus in the frog. J. Comp. Neurol. 60, 473-527 (1934).

Larsell, O.: The cerebellum of myxinoids and petromyzonts, including developmental stages in the lampreys. J. Comp. Neurol. 86, 395-445 (1947).

Larsell, O.: The Comparative Anatomy and Histology of the Cerebellum from Myxinoids through Birds. Minneapolis: University of Minnesota Press, 1967.

Leake, P. A.: Central projections of the statoacoustic nerve in *Caiman crocodilus*. Brain, Behav. Evol. 10, 170-196 (1974).

Lemire, M.: Étude architectonique du rhombencéphale de *Latimeria chalumnae* Smith. Bull. Musée Nat. d'Histoire Naturelle, sér. 3, no. 2, 41-95 (1971).

Loftus-Hills, J. J.: Neural correlates of acoustic behavior in the Australian bullfrog *Limnodynastes dorsalis* (Anura: Leptodactylidae). Z. vergl. Physiologie 74, 140-152 (1971).

Lowenstein, O., Osborne, M. P., Thornhill, R. A.: The anatomy and ultrastructure of the labyrinth of the lamprey (*Lampetra fluviatilis* L.). Proc. R. Soc. Lond. B. Biol. Sci. 170, 113-134 (1968).

Lowenstein, O., Roberts, T. D. M.: The localization and analysis of the responses to vibration from the isolated elasmobranch labyrinth. A contribution to the problem of the evolution of hearing in vertebrates. J. Physiol. (Lond.) 114, 471-489 (1951).

Luiten, P. G. M.: The central projections of the trigeminal, facial, and anterior lateral line nerve in the carp (*Cyprinus carpio* L.). J. Comp. Neurol. 160, 399-418 (1975).

Maler, L.: The acousticolateral area in bony fish and its cerebellar relations. Brain, Behav. and Evol. 10, 130-145 (1974).

Maler, L., Karten, H. J., Bennett, M. V. L.: The central connections of the posterior lateral line nerve of *Gnathonemus petersi*. J. Comp. Neurol. 151, 57-66 (1973a).

Maler, L., Karten, H. J., Bennett, M. V. L.: The central connections of the anterior lateral line nerve of *Gnathonemus petersi*. J. Comp. Neurol. 151, 67-84 (1973b).

Maler, L., Finger, T., Karten, H. J.: Differential projections of ordinary lateral line receptors and electroreceptors in the gymnotid fish, *Apteronotus* (*Sternarchus*) *albifrons*. J. Comp. Neurol. 158, 363-382 (1974).

Mayser, P.: Vergleichend anatomische Studien über das Gehirn der Knochenfische mit besonderer Berücksichtigung der Cyprinoiden. Z. Wiss. Zool. 36, 259-364 (1882).

McCormick, C. A.: Central projections of the lateralis and eighth nerves in the bowfin, *Amia calva*. Doctoral Thesis, University of Michigan, (1978).

McCready, P. J., Boord, R. L.: The topography of the superficial roots and ganglia of the anterior lateral line nerve of the smooth dogfish, *Mustelus canis*. J. Morphol. 150, 527-538 (1976).

Mehler, W. R.: Comparative anatomy of the vestibular nuclear complex in submammalian vertebrates. In: Basic Aspects of Central Vestibular Mechanisms. Brodal, A., Pompeiana, O. (eds.). Progress in Brain Research 37. Amsterdam: Elsevier, 1972, pp. 55-67.

Millot, J., Anthony, J.: Anatomie de *Latimeria chalumnae*, t. II, Système Nerveux et Organes des Sens. Paris: C. N. R. S., 1965.

Moy-Thomas, J. A., Miles, R. S.: Paleozoic Fishes, 2nd ed. Philadelphia: Saunders, 1971.

Mudry, K. M., Capranica, R. R.: Electrophysiological evidence for auditory responsive areas in the diencephalon and telencephalon of the bullfrog, Rana catesbeiana. Soc. for Neurosci. Abstracts 4, 101 (1978).

Mudry, K. M., Constantine-Paton, M., Capranica, R. R.: Auditory sensitivity of the diencephalon of the leopard frog Rana p. pipiens. J. Comp. Physiol. 114, 1-13 (1977).

Neary, T. J.: Diencephalic efferents of the torus semicircularis in the bullfrog, Rana catesbeiana. Anat. Rec. 178, 425 (1974).

Neary, T. J., Wilczynski, W.: Autoradiographic demonstration of hypothalamic efferents in the bullfrog, Rana catesbeiana. Anat. Rec. 187, 665 (1977).

Nelson, D. R., Gruber, S. H.: Sharks: Attraction by low-frequency sounds. Science 142, 975-977 (1963).

Nieuwenhuys, R., Kremers, J. P. M., van Huijzen, Chr.: The brain of the crossopterygian fish Latimeria chalumnae: A survey of its gross structure. Anat. Embryol. 151, 157-169 (1977).

Northcutt, R. G.: Brain organization in the cartilaginous fishes. In: Sensory Biology of Sharks, Skates, and Rays. Hodgson, E. S., Mathewson, R. F. (eds.). Arlington: Office of Naval Research, Department of the Navy, 1978, pp. 117-193.

Northcutt, R. G.: Central projections of the eighth cranial nerve in lampreys. Brain Res. 167, 163-167 (1979a).

Northcutt, R. G.: Primary projections of VIII nerve afferents in a teleost, Gillichthys mirabilis. Anat. Rec. 193, 638 (1979b).

Northcutt, R. G., Braford, M. R., Jr.: New observations on the organization and evolution of the telencephalon of actinopterygian fishes. In: Comparative Neurology of the Telencephalon. Ebbesson, S. O. E. (ed.). New York: Plenum, in press.

Northcutt, R. G., Neary, T. J., Senn, D. G.: Observations on the brain of the coelacanth Latimeria chalumnae: External anatomy and quantitative analysis. J. Morphol. 155, 181-192 (1978).

Olson, E. C.: Vertebrate Paleozoology. New York: Wiley, 1971.

Opdam, P., Kemali, M., Nieuwenhuys, R.: Topological analysis of the brain stem of the frogs Rana esculenta and Rana catesbeiana. J. Comp. Neurol. 165, 307-332 (1976).

Parsons, T. S., Williams, E. E.: The relationships of the modern amphibia. Quart. Rev. Biol. 38, 26-53 (1963).

Pearson, A. A.: The acoustico-lateral centers and the cerebellum, with fiber connections, of fishes. J. Comp. Neurol. 65, 201-294 (1936a).

Pearson, A. A.: The acoustico-lateral nervous system in fishes. J. Comp. Neurol. 64, 235-273 (1936b).

Pettigrew, A., Chung, S.-H., Anson, M.: Neurophysiological basis of directional hearing in amphibia. Nature 272, 138-142 (1978).

Pfeiffer, W.: Die Fahrenholzchen Organe der Dipnoi und Brachiopterygii. Z. Zellforsch. 90, 127-147 (1968).

Popper, A. N., Fay, R. R.: Structure and function of the elasmobranch auditory system. Amer. Zool. 17, 443-452 (1977).

Potter, H. D.: Mesencephalic auditory region of the bullfrog. J. Neurophysiol. 28, 1132-1154 (1965a).

Potter, H. D.: Patterns of acoustically evoked discharges of neurons in the mesencephalon of the bullfrog. J. Neurophysiol. 28, 1155-1184 (1965b).

Ramon y Cajal, S.: Sur un noyau special du nerf vestibulaire des poissons et des oiseaux. Trabl. Lab. Invest. Biol. Univ. Madr. 6, 1-20 (1908).

Romer, A.: Vertebrate Paleontology, 3rd ed. Chicago: University of Chicago Press, 1966.

Ronan, M. C., Northcutt, R. G.: Afferent and efferent connections of the bullfrog medial pallium. Soc. for Neurosci. Abstracts 5, 146 (1979).

Roth, A.: Electroreceptors in Brachiopterygii and Dipnoi. Die Naturwissenschaften 60, 106 (1973).

Rubinson, K.: The central distribution of VIII nerve afferents in larval *Petromyzon marinus*. Brain, Behav. Evol. 10, 121-129 (1974).

Rubinson, K., Skiles, M. P.: Efferent projections of the superior olivary nucleus in the frog, *Rana catesbeiana*. Brain, Behav. Evol. 12, 151-160 (1975).

Russell, I. J.: Amphibian lateral line receptors. In: Frog Neurobiolog. Llinas, R., Precht, W. (eds.). Berlin-Heidelberg-New York: Springer-Verlag, 1976, pp. 513-550.

Schaeffer, B.: Adaptive radiation of the fishes and fish-amphibian transition. Ann. N. Y. Acad. Sci. 167, 5-17 (1969).

Schaeffer, B., Rosen, D. E.: Major adaptive levels in the evolution of the actinopterygian feeding mechanism. Amer. Zool. 1, 187-204 (1961).

Smeets, W. J. A. J., Nieuwenhuys, R.: Topological analysis of the brain stem of the sharks *Squalus acanthias* and *Scyliorhinus canicula*. J. Comp. Neurol. 165, 333-368 (1976).

Teeter, J. H., Bennett, M. V. L.: Ampullary electroreceptors in sturgeon. Soc. for Neurosci. Abstracts 2, 185 (1976).

Tester, A. L., Kendall, J. I., Milisen, W. B.: Morphology of the ear of the shark genus *Carcharhinus* with particular reference to the macula neglecta. Pacific Sci. 26, 264-274 (1972).

Thomson, K. S.: A critical review of the diphyletic theory of rhipidistian-amphibian relationships. In: Current Problems of Lower Vertebrate Phylogeny. Ørvig, T. (ed.). New York: Wiley, 1968, pp. 285-306.

Thornhill, R. A.: The development of the laybrinth of the lamprey (*Lampetra fluviatilis* Linn. 1758). Proc. R. Soc. Lond., B. 181, 175-198 (1972).

Trachtenberg, M. C., Ingle, D.: Thalamo-tectal projections in the frog. Brain Res. 79, 419-430 (1974).

Wever, E. G.: Origin and evolution of the ear of vertebrates. In: Evolution of Brain and Behavior in Vertebrates. Masterton, R. B., Bitterman, M. E., Campbell, C. B. G., Hotton, N. (eds.). Hillsdale, NJ: Lawrence Erlbaum Assoc.'s, 1976, pp. 89-106.

Wilczynski, W.: Connections of the midbrain auditory center in the bullfrog, *Rana catesbeiana*. Doctoral Thesis, University of Michigan, (1978).

Wilczynski, W., Neary, T. J., Andry, M. L.: Somatosensory projections to the thalamus in ranid frogs. Soc. for Neurosci. Abstracts 3, 95 (1977).

Wilczynski, W., Northcutt, R. G.: Afferents to the optic tectum of the leopard frog: An HRP study. J. Comp. Neurol. 173, 219-299 (1977).

Wilczynski, W., Northcutt, R. G.: An HRP study of the anuran striatum. Anat. Rec. 193, 721 (1979).

Wilson, H. V.: The embryology of the sea bass *Serranus atrarius*. U. S. Nat'l Marine Fishes Service Bull. 9, 209-277 (1889).

Wilson, H. V., Mattocks, J. E.: The lateral sensory anlage in the salmon. Anat. Anz. 13, 658-660 (1897).

Witschi, E.: The larval ear of the frog and its transformation during metamorphosis. Z. Naturforsch. 4b, 230-242 (1949).

Witschi, E., Bruner, J. A., van Bergeijk, W. A.: The ear of the adult *Xenopus*. Anat. Rec. 117, 602-603 (1953).

van Bergeijk, W. A.: Evolution of the sense of hearing in vertebrates. Amer. Zool. 6, 371-377 (1966).

van Bergeijk, W. A.: The evolution of vertebrate hearing. In: Contributions to Sensory Physiology, Vol. 2. Neff, W. D. (ed.). Berlin-Heidelberg-New York: Springer-Verlag, 1967, pp. 1-49.

PART TWO

Amphibians

The auditory systems of amphibia have been of great interest and value to those interested in comparative aspects of hearing primarily because of the striking relations between hearing and vocalization in anurans. In these groups, the biological significance of hearing is quite clear to us, and this fact has prompted a significant body of research on all aspects of their auditory systems. Since the work on the relations between hearing and vocalization has been presented several times recently, we concentrate here on the comparative anatomy and physiology of the peripheral auditory systems. In the preceding section, Northcutt (Chapter 3) discussed the structure of the amphibian CNS. In this section, Lombard (Chapter 4) explores the evolution of the amphibian and tetrapod ear, and Capranica and Moffat (Chapter 5) show some of the intriguing relations between the function of amphibian and mammalian systems. Note that this and the following section on reptiles lack treatments of auditory psychophysics. This is not due to a lack of interest in this type of experiment among comparative researchers but to the persistent and frustrating failure to bring these animals' behavior under the control of an auditory stimulus through conditioning.

Chapter 4

The Structure of the Amphibian Auditory Periphery: A Unique Experiment in Terrestrial Hearing

R. ERIC LOMBARD*

1 Introduction

In a period of objective reflection as preparation for writing this chapter, the question arose: What amphibian otic morphology is there to review that is not adequately covered already? Figure 4-1 illustrates the observation behind this thought. In the one-hundred year period, 1880 to 1980, the publication rate of original works on amphibian otic morphology has never been overwhelming. From 1880 the rate increases, peaks prior to WW II, and then declines precipitously to a steady rate of one paper per year over the past thirty years. This latter pace is, by current tenure committee standards, the output expected of about one-half an assistant professor! The arrows indicate the occurrence of major reviews. Using the publication of Retzius' monograph as a start and the publication date for this chapter as a finish, a trend is evident. Apparently, the field is reviewing a declining volume of new observations at an increasing rate!

   In light of this, this chapter will review how researchers in the field view the amphibian auditory periphery. That is to say, how is the amphibian ear perceived in relation to the ears of other vertebrates? The case shall be made that present ideas on the subject are unsatisfactory and that an alternative viewpoint—that the amphibian ear represents a separate experiment in terrestrial hearing unrelated to that of amniotes—is justifiable. Lombard and Bolt (1979) present an overview of the evolution of the auditory periphery in tetrapods in general. This chapter expands the theme developed in that work for the class Amphibia.

2 The Standard View

Presently, the amphibian ear is regarded in one of two ways: (1) it is primitive and/or generalized, or (2) it is derived and specialized. Romer (1970), in his popular textbook, considers the auditory periphery of fishes, reptiles, birds, and mammals and then writes:

*Department of Anatomy, The University of Chicago, 1025 East 57th Street, Chicago, Illinois 60637.

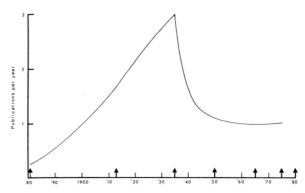

Figure 4-1. Publication rate (curve) and dates of major reviews (arrows) for amphibian otic morphology. The bibliography of original research reports on which this figure is based includes works that are either solely morphological in content or that relate function directly to structure. Physiological papers are not included. The choice of reviews indicated is somewhat subjective. Only half of the reviews considered are reviews by modern criteria, the remainder are monographs or texts included because they summarize preceding works or include large amounts of original data. Though fifteen "reviews" are considered, only seven arrows are indicated on the graph. Reviews published within a year or two are recorded by an "average" date. The seven arrows are: *1881*-Retzius (1881); *1913*-Gaupp (1913); *1934*-Goodrich (1930), deBurlet (1934), deBeer (1937); *1950*-Reinbach (1950); *1965*-Schamlhausen (1968), Olson (1966); *1975*-Baird (1974a, b), Henson (1974), Wever (1974), Wever (1976), Capranica (1976); *1980*-this chapter. A copy of the bibliography upon which the graph is based is available from the author.

> We have omitted any reference to ear construction in living amphibians, for the reason that conditions in these forms are not, in general, primitive. They are specialized and seemingly degenerate in most cases and are, further, extremely varied. (p. 475)[1]

Romer quite clearly characterizes the amphibian ear as specialized and degenerate (degeneracy may also reflect specialization). Wever (1974), in contrast, considering the transmission problems associated with the aquatic-terrestrial transition in the evolution of the middle ear writes:

> This transmission problem was solved by the development of a middle ear apparatus that served as a mechanical transformer, effectively matching the impedance of the ear to that of the air. This development occurred first in amphibians, and then was continued in the reptiles, birds, and mammals. (p. 450)

Wever's use of the word "continued" implies a primitive, and probably generalized,

---

[1]The fifth edition of Romer's text, revised by Parsons (1977), expresses the same view with minor changes in wording.

condition for the amphibian tympanic cavity, columella and tympanum. He continues, addressing the middle ears of all tetrapods in general:

> Although homologies can be found among several of the middle ear structures, . . . there are many differences that indicate separate developments. (p. 450)

Here, a reasonable synonym for Wever's "separate" is "derived" (and probably "specialized"). This last quotation hints at a middle view—that the amphibian ear might contain both general and special features, though Wever does not identify those features he might consider special. These quotations are not intended to imply that these views are either unique to, or original with, Romer or Wever. Rather, they are used simply to indicate instances of the two perspectives in the light of which the amphibian auditory periphery has been interpreted.

These disparate views arise and indeed are dictated by a common perception of general otic evolution in tetrapods. Figure 4-2 illustrates a generally accepted view of tetrapod evolution. The schema indicates that amphibians are derived from fishes (Rhipidistia), amniotes from amphibians, and that modern reptiles and birds, on the one hand, and mammals, on the other, are independently derived from fossil reptiles. The common perspective from which the views outlined above arise presumes that a tympanic middle ear was an early feature in tetrapod evolution. Indeed, one author has suggested (and attempted to adduce evidence) that the "tetrapod tympanic ear" evolved in rhipidistian fishes—the fossil taxon from which the Paleozoic amphibians arose (van Bergeijk 1966). The slash mark in Fig. 4-2 indicates where a tympanic ear might be likely to have arisen under the common perception of otic evolution.

This assumption that the "tetrapod tympanic ear" arose once and very early is necessitated by the belief that the tympana of living tetrapods are homologous (solid circles, Fig. 4-2). That is to say, once evolved, the tympanic ear was maintained throughout the evolution of tetrapods. (see Lombard and Bolt 1979 for a review of the long history of this view.) This notion is based on standard criteria for homology for the tympanum and stapes (columella), as stated by Goodrich (1930):

> It is unlikely, however, that two structures so similar as the tympanum of a frog and a lizard, each situated behind the quadrate and having embedded in it the distal end of the columella, should have been independently developed. (p. 484)

That is, it is the similarity of tympana and their relations that give rise to the notion that the auditory periphery is generally homologous in all tetrapod groups. This view leads in turn to the view that the amphibian auditory periphery is related to that of amniotes, i.e., shares a common heritage. With this perspective, the amphibian auditory periphery is the prototype (historically) and may be thus viewed as primitive. This view is reinforced by the observation that the amphibian ear is generally simpler than that of amniotes. Given this relationship, any "bizarre" amphibian otic features must be viewed as derived specializations added on to the primitive plan.

The commonly held view is that the amphibian auditory periphery is phylogenetical-

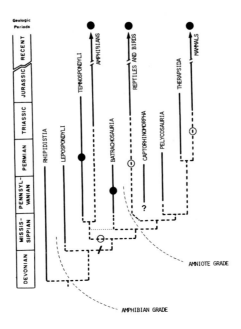

Figure 4-2. Generally accepted relationships of fossil and recent groups discussed in this chapter. The solid lines represent the extent of each group in the fossil record, for example, the Temnospondyli were extant from the late Mississippian to the early Jurassic. The geologic periods to the left are not scaled to their actual time span. The solid circles indicate the presence of a tympanum (most frogs in "Amphibians"; *Sphenodon,* many lizards, turtles, crocodiles, and birds in "Reptiles and Birds"; all mammals, except the whale, *Kogia,* in "Mammals") or evidence for a tympanum (otic notch and dorsolaterally directed stapes in many Temnospondyli; an otic notch in Batrachosauria). The slash mark indicates where the tympanum would have to arise if it were homologous in all tetrapods. The open circles indicate where tympanic ears would arise under the view proposed in this chapter. The independent origins are accurate phylogenetically but only relatively correct with respect to time. The dashed line linking the Batrachosauria to the other amphibians indicates the necessary reordering of the systematics of paleozoic amphibians under the view proposed in this chapter (modified from Lombard and Bolt 1979).

ly related to that of other tetrapods; this view, however, is not supported by the evidence. This contention will be supported by a review of the major structures of the periphery, and the problems that arise when these structures are interpreted from the orthodox point of view will be pointed out. The evidence indicates that the amphibian auditory periphery is neither primitive nor derived with respect to that of anmniotes, but unique in its own right.

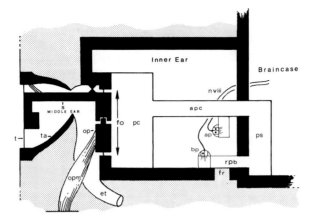

Figure 4-3. Schematic illustration of the major features of the amphibian auditory periphery in a roughly frontal plane. All the features illustrated are not present in many amphibians. Skeletal elements, whether cartilage or bone, are black; non-otic soft tissues are stippled. The illustration is only roughly anatomically correct so that all parts may be shown. The only parts of the otic labyrinth shown are the two recesses containing the sensory papillae. Abbreviations: ap-amphibian papilla; apc-amphibian periotic canal; bp-basilar papilla; et-pharyngotympanic tube; fo-fenestra ovalis; fr-fenestra rotundum; nviii-cranial nerve VIII; op-otic opercular bone; opm-opercular muscle; pc-periotic cistern; ps-periotic sac; rpb-recessus partis basilaris; s-stapes; t-tympanum; ta-tympanic annulus.

3 The Amphibian Auditory Periphery

The major elements of the amphibian ear believed to be involved in the perception of sound are illustrated in Fig. 4-3. The figure represents in a simple manner the most complex state (in terms of number of elements present) achieved in living adult amphibians, namely many anurans (also see Capranica and Moffat, Chapter 5). In the middle ear, the tympanic cavity is broadly confluent with the pharynx via the tympanic tube (et). The cavity is spanned by the stapes (s) that articulates distally with the tympanic membrane (t) and proximally with the fenestra ovalis (fo) of the otic capsule. The fenestra ovalis also contains the otic opercular element (op) that provides insertion for a muscle (opm) originating on the shoulder girdle. The joint between the otic opercular element and the footplate of the stapes is often very complex as schematically illustrated. In the inner ear, the fenestra ovalis and its contained bony elements are directly related to a large periotic space, the periotic cistern (pc). The periotic cistern gives rise to the amphibian periotic canal (apc) that passes to the braincase where it expands into the periotic sac (ps). The periotic system is indirectly related to the auditory sensory epithelia: the basilar and amphibian papillae (ap, bp). These papillae are the only components of the otic labyrinth illustrated. The amphibian papilla is present in all amphibians whereas the basilar papilla is often absent (Lombard 1977). Further, the amphibian papilla is generally an order of magnitude larger in hair

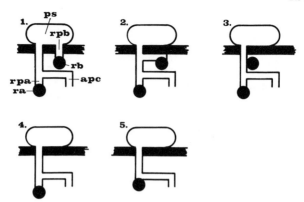

Figure 4-4. Known variation in the relations of the periotic system to the sensory papillae in amphibians in a roughly frontal plane. All five patterns are found in salamanders. Patterns (1), (2), and (4) are found in frogs, and pattern (4) occurs in *Caecilia occidentalis* (which lacks a basilar papilla), the only gymnophionan for which the periotic labyrinth has been examined in detail. Abbreviations: apc-amphibian periotic canal; ps-periotic cistern; ra-recessus amphibiorum; rb-recessus basilaris; rpa-recessus partis amphibiorum; rpb-recessus partis basilaris (from Lombard 1977).

cell count than the basilar papilla. The known variation in the termination of the amphibian periotic canal and its relations to the sensory epithelia is shown in Fig. 4-4. The recesses of the otic labyrinth containing the amphibian and basilar papillae are represented by black dots (ra, rb). That configuration, common to most frogs, is approximated by part 1 of Fig. 4-4. If a "foramen rotundum" occurs, it is located near the basilar papilla and is not related to the middle ear cavity or eustachian tube but rather abuts against the musculature of the head as indicated by (fr) in Fig. 4-3.

4 Problems with the Standard View

Table 1 lists the features of the modern amphibian ear presumed or known to be involved in auditory perception and divides them on the basis of whether they are commonly regarded in the literature as homologs of amniote structures or are derived amphibian features. The tympanum, the tympanic cavity and tube, the articulation of the tympanum and columella, the periotic system, basilar papilla, and perhaps the foramen rotundum are all commonly regarded as homologous in amphibians and amniotes. The otic opercular element, the opercularis muscle, amphibian papilla, and the pattern of the chorda tympani in the region of the ear are considered as derived amphibian features. The impression gained from this table is that indeed the amphibian auditory periphery is closely allied to that of amniotes as most of the features are held in common. Partitioning the features in this manner legitimizes the adjectives "primitive, generalized, derived, and specialized," applied in the standard view. The standard view would have it, for example, that the diminutive basilar papilla of amphibians is

Table 4-1. Presumed homologies of amphibian and amniote ear structures (under the standard view of otic evolution)[1]

Common tetrapod	Derived amphibian
1. Tympanum	
2. Tympanic cavity and tube	
3. Stapes-tympanum articulation	
	4. Otic opercular element
	5. Opercularis muscle
6. Periotic system	
7. Foramen rotundum	
8. Basilar papilla	9. Amphibian papilla
	10. Chorda tympani relations

[1] Items 1, 2, 3, 6, 7, 8 are considered homologous in amphibians and amniotes.

primitive compared to its presumed derivative, the extended cochlea of mammals. On the other hand, the standard view sees the otic opercular apparatus, for example, as a specialization superimposed on a generalized tetrapod ear.

When the otic regions of reptiles and amphibians, both fossil and living, are examined in some detail, five problems may be identified if one takes the view indicated in Table 4-1:

1. location of the tympanum in stem fossil reptiles
2. the orientation and articulation of the stapes in fossil labyrinthodont and living amphibians on one hand, and stem fossil amniotes on the other
3. the orientation and pattern of the periotic system in living amphibians and amniotes
4. the structure-function of the basilar papilla in living amphibians and amniotes
5. the relation of the chorda tympani to the stapes in living tetrapods

Figure 4-5 illustrates the general features of the otic region in a few key fossil groups. In many labyrinthodonts (temnospondyls and batrachosaurs, Fig. 4-2) the posterodorsal margin of the skull is emarginated to form an otic notch. The stapes is directed dorsolaterally from the otic capsule to the otic notch. These two morphological features are regarded as evidence that these organisms possessed a tympanic membrane in the otic notch to which the distal end of the stapes attached. No such otic notch occurs in the early amniote fossils illustrated: the captorhinomorph, representing a morphological stage in the ancestry of modern reptiles plus birds, or the pelycosaur, representing the early ancestors of mammals. Further, the stapes, which is especially massive in pelycosaurs, is directed ventrolaterally from the otic capsule and articulates distally with the quadrate bone, a major component of the jaw articulation. If evolutionary continuity for the tympanum from labyrinthodonts to living amniotes is presumed, where is it placed in the stem amniotes? No evidence exists that these creatures possessed a tympanum, although attempts have been made to hypothesize where one might have been and how it might have been attached to the stapes (e.g., see Romer and Price 1940).

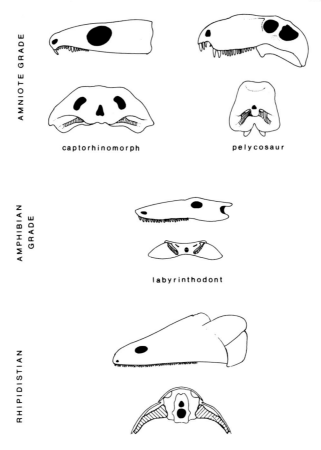

Figure 4-5. Lateral and posterior views of the otic regions of some key fossil vertebrates. The stapes is crosshatched in each. The labyrinthodont illustrated is representative of many of the Temnospondyli and Batrachosauria of Fig. 4-2. In the rhipidistian, captorhinomorph and pelycosaur, the stapes is directed ventrolaterally and articulates distally with the palatoquadrate (rhipidistian) or quadrate (amniotes); the quadrate is derived from the palatoquadrate. In the labyrinthodont, the stapes is directed
dorsolaterally and articulates distally with a tympanum in the otic notch.

The problem of the orientation and articulation of the stapes in labyrinthodont amphibians and stem fossil amniotes is closely related to that of the location of the tympanum. The orientation and distal articulation (in part) of the stapes in the stem amniotes is very much like that in rhipidistian fishes (Fig. 4-5). In both of these groups, the
stapes is directed ventrolaterally and the distal end articulates with the quadrate of the
jaw articulation. However, the orientation of the stapes in the labyrinthodonts is rotated 90° and further, the distal end articulates with a tympanum. If it is presumed
that the labyrinthodont morphology is primitive for tetrapods, then it must be imagined
(1) that the stapes acquired a derived orientation and distal articulation in the evolution
of labyrinthodont amphibians from rhipidistians and then (2) reverted to the primitive

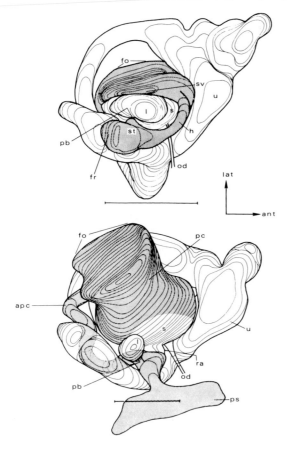

Figure 4-6. Graphic reconstruction of the ventral surfaces of the periotic labyrinths (stippled) and otic labyrinths of an amniote (*Coluber constrictor*, a snake, upper) and an amphibian (*Salamandra salamandra*, lower). The periotic tube (scala vestibuli, helicotrema, and scala tympani in amniotes; amphibian periotic canal in amphibians) passes anterior to the lagena in amniotes and posterior to the lagena in amphibians. The relations of the basilar papilla neuroepithelium to the periotic labyrinth is illustrated for each group in Fig. 4-7. The scalar bars are 1 mm. Abbreviations: apc-amphibian periotic canal; fo-portion of the periotic system related to the fenestra ovalis; fr-portion of the periotic system related to the fenestra ovalis; h-helicotrema; l-lagena; od-otic duct; pb-site of papilla basilaris; pc-periotic cistern; ps-periotic sac; ra-recessus amphibiorum (contains papilla); s-sacculus; st-scala tympani; sv-scala vestibuli; u-utriculus (from Lombard and Bolt 1979).

orientation and distal articulation in the evolution of amniotes from amphibians. Moreover, while these transformations took place in the derivation of amniotes, it must be imagined that the stapes maintained its articulation with a tympanum.

Figure 4-6 illustrates the morphology of the periotic labyrinth in relation to the otic labyrinth of a living amniote and amphibian. In both groups the periotic labyrinth consists of: (1) an enlarged cistern applied to the inner face of the stapedial footplate in the fenestral ovalis (periotic cistern in amphibians; scala vestibuli in amniotes), (2) an enlarged sac within the otic capsule (amniote) or braincase (amphibian), and (3) a tube connecting the two. In the amphibian, the connecting tube (amphibian periotic canal) proceeds from the cistern to the sac by passing *posteriorly* around the lagena. In the amniote the connecting tube (scala vestibuli, helicotrema, and scala tympani) proceeds from the cistern to the sac by passing *anteriorly* around the lagena. This basic difference is found in all living amphibians and amniotes that have been examined (Baird 1960, Lombard 1977). Intermediate conditions do not exist. If it is presumed that the amphibian morphology is primitive and therefore ancestral to that of amniotes, then a transformation from one pattern to the other somewhere in the evolution of amniotes from amphibians must be postulated.

The problem of structure-function in the basilar papilla of living amphibians and amniotes is schematically illustrated in Fig. 4-7. In both groups the relationships of the neuroepithelium to (1) otic fluid space and (2) periotic fluid space and the tectorium are illustrated. In amphibians the neuroepithelium resides in the relatively stable periotic tissue. The apical surface of the epithelium forms part of the wall of the basilar recess of the otic labyrinth fluid space. The tectorium lies over the surface of the hair cells. The periotic labyrinth approaches the region of the neuroepithelium by a recessus partis basilaris from either the amphibian periotic canal or the periotic sac. A thin membrane separates the periotic fluid space from the otic fluid space and nowhere is the periotic fluid space directly related to the neuroepithelium. In amninotes the neuroepithelium rests on a thin membrane overlying a portion of the periotic fluid space (scala tympani). The apical surface forms part of the wall of a recessus basilaris (scala media) as in amphibians and also, as in amphibians, is overlain by a tectorium. Finally, the otic fluid space is again related to a portion of the periotic fluid space

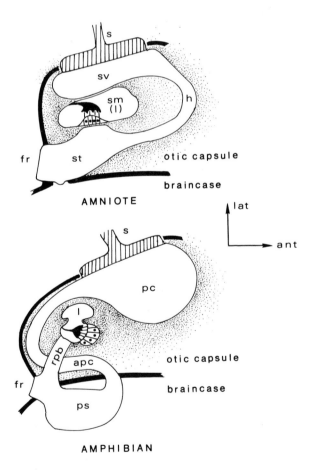

Figure 4-7. Schematic illustration of the relations of the basilar papilla of modern reptiles (above) and amphibians (below) in a roughly frontal plane. The neuroepithelium of the amniote rests on a flexible membrane. The neuroepithelium of the amphibian is embedded in periotic tissue. The periotic system passes anterior to the lagena in amniotes and posterior to the lagena in amphibians. The skeletal elements in the region of the fenestra rotundum are illustrated in black. Abbreviations: apc-amphibian periotic canal; fr-fenestra rotundum; h-helicotrema; l-lagena; pc-periotic cistern; ps-periotic sac; rpb-recessus partis basilaris; s-stapes; sm-scala media; st-scala tympani; sv-scala vestibuli.

(scala vestibuli). Thus, in amniotes the neuroepithelium is intimately related to the periotic fluid space whereas in amphibians it is not. The gross morphology of these relations may be appreciated in Fig. 4-6. Recently, Wever (1974, 1976) has hypothesized that the tectorium moves relative to a stable neuroepithelium in amphibians and that in amniotes the epithelium moves relative to a more stable tectorium. The very different morphology present in each group would indicate that this suggestion has merit. Thus, the mechanism of creating shear at the surface of the neuroepithelium would be very different in each group. If it is presumed that the amphibian morphology and function are primitive and ancestral to that of amniotes, then a radical change in both structure and function in the evolution of amniotes from amphibians must be accepted. Wever, in fact, considers it "almost inconceivable" that the basilar papilla of amphibians could give rise to that of amniotes (Wever 1976, p. 104).

The four problems presented by the standard view described above have been recognized comparatively recently. The first major summary and discussion of the tympanum problem in stem amniotes is that of Romer and Price (1940); of the orientation of the stapes, Watson (1953); of the periotic system, Lombard (1971); and of the basilar papilla, Wever (1974, 1976). The problem presented by the path of the chorda tympani nerve in relation to the stapes, on the other hand, has been recognized for a long time [Gaupp (1913) has a summary of the relevant literature of the nineteenth and early twentieth centuries].

The chorda tympani is a branch of cranial nerve VII and carries afferent taste fibers to the central nervous system. Figure 4-8 illustrates the otic region of rhipidistian fish as seen after the dermal roofing bones have been removed to reveal the neurocranium and some elements of the splanchnocranium. The hyomandibula, a dorsal element of the hyoid arch and the homolog of the stapes is crosshatched. Note that it has five articulations:

1. with the otic capsule (on the lateral ridge formed by the bony covering of the horizontal semicircular canal
2. with the otic capsule (will become the fenestra ovalis)
3. with the ceratohyal (a ventral element of the hyoid skeletal arch)
4. with the quadrate (upper jaw and dorsal element of the mandibular arch)
5. with the opercular dermal bone on the surface of the cheek

The seventh cranial nerve exits the skull anterior to the otic capsule (Fig. 4-8C), passes backwards (except the palatine ramus) in the lateral commissure located between the two articulations of the hyomandibula with the otic capsule (Fig. 4-8B, C), passes behind the hyomandibula and enters its body (Fig. 4-8B), and then passes through a canal to exit on the anterolateral face of the hyomandibula (Fig. 4-8A, C). The chorda tympani (imb) then passes forward medial to the junction between hyomandibula and palatoquadrate. The articulations of the hyomandibula and the path of the seventh nerve are drawn from the works of Eaton (1939) and Jarvik (1954). Under the standard view, the problem created by the path of the chorda tympani becomes apparent when recent tetrapods are examined. The chorda tympani and its relations to the stapes are illustrated in Fig. 4-9 for an amniote and an amphibian. The five rhipidistian articulations are indicated for the "tetrapod ancestor" (fossil amphibian). In the "tetra-

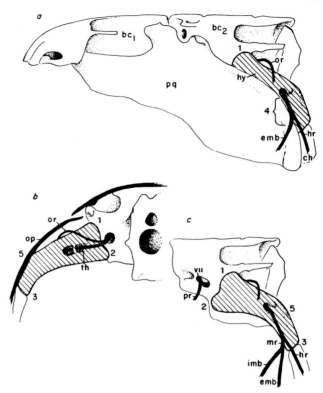

Figure 4-8. Otic region of a rhipidistian fish. Lateral views, (*a*) and (*c*); posterior view, (*b*). In (*a*) the dermal roofing bones, lower jaw and branchial skeleton have been removed. In (*c*) only the caudal portion of the jointed braincase is shown and the palatoquadrate has been removed. In all three views the course of the seventh cranial nerve is shown in relation to the hyomandibula (=stapes of tetrapods). The hyomandibula is cross hatched and its five articulations are indicated (numbers 1-5). Abbreviations: bc_1-rostral braincase; bc_2-caudal braincase; ch-ceratohyal; emb-external mandibular branch of n. VII; hr-hyoid ramus of n. VII; imb-internal mandibular branch of n. VII (Chorda Tympani); op-dermal opercular bone; or-opercular ramus of n. VII; pq-palatoquadrate, pr, palatine ramus of n. VII; th-truncus hyomandibularis of n. VII; VII, exit of n. VII from braincase (from Lombard and Bolt 1979).

pod ancestor," the nerve no longer lies within the shaft of the stapes. The tympanum is indicated to be present on articulation 5 (the old dermal opercular bone articulation), the commonly assumed stapedial-tympanal bone articulation. Note that the amphibian stapes makes use of articulations 1, 2, and 5, the latter with the tympanum, and that the chorda tympani passes ventral to the shaft of the stapes on its way to the tongue. The problem here is, if the amphibian condition is presumed to be primitive to that of amniotes, the chorda tympani must have at some time altered its path to assume its reptilian location running forward (dorsal) to stapes. There is no evidence that this alteration has indeed taken place. This problem of the standard view has generated a fairly large literature over the past century (see Lombard and Bolt 1979). Note

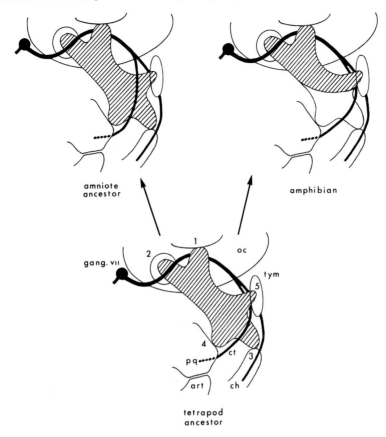

Figure 4-9. Relations of the chorda tympani to the stapes under the standard view. Note that in passing from the tetrapod ancestor to the amniote ancestor, the chorda tympani must alter its relations to the stapes and tympanum to achieve the position illustrated for the amniote. The numbers refer to the five articulations of the rhipidistian hyomandibula. The open circle is the tympanic membrane. Abbreviations: art-articular of the lower jaw (=malleus of mammals); ch-ceratohyal; gang. vii-ganglion of n. VII; oc-otic capsule; pq-quadrate ossification of palatoquadrate (=incus of mammals); tym-tympanic membrane.

that the course of the chorda tympani is a problem only if the tympana of recent tetrapods are assumed to be homologous and the process of the stapes attached to them is assumed to be homologous as well. Further, the first two problems outlined—the tympanum of fossil stem amniotes and the orientation of the stapes—are likewise a result of these initial assumptions.

If it is assumed that the amphibian auditory periphery shares a common ancestral heritage with that of amniotes, we are presented with difficulty in every major morphological feature. Any one of these problems is difficult to solve within the tenets of the standard view and together they present a formidable barrier to its acceptance. These problems may be obviated if we view the amphibian auditory periphery as an independent experiment in terrestrial audition, sharing only its general location with that of amniotes.

5 An Alternative View

Figure 4-10 illustrates an alternative view of the evolution of the tympanum and tympanic articulations with the stapes in tetrapods. The tetrapod ancestor illustrated is identical to that seen in Fig. 4-9 with the important exception that a tympanum is not present. The amphibian otic region is identical in all respects to that seen previously. The illustration of the ancestor of modern amniotes, however, is different. Here no

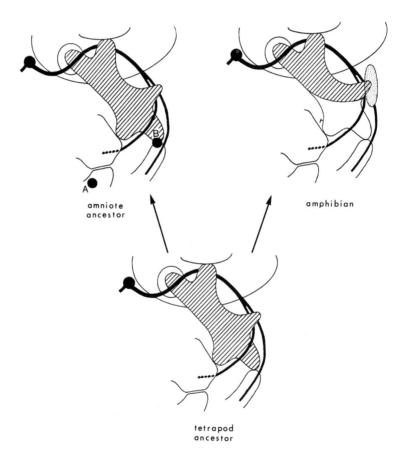

Figure 4-10. Relations of the chorda tympani to the stapes under the view advocated in this paper. Note that in comparison to Fig. 4-9, the tetrapod ancestor and amniote ancestor have no tympanic membrane. The solid circles indicate the locations of the tympana in recent amniotes: (A)-mammals, (B)-reptiles and birds under the view advocated in this chapter. The structures illustrated are identified in Fig. 4-9.

tympanum is indicated. The location of the tympanum by this scheme in recent reptiles plus birds is indicated by the dot lettered "A." If indeed the tympanum does articulate with this process, then the path of the chorda tympani as seen in living reptiles and birds is explained without having to postulate a change in relationship. The basic assumption from which this notion stems is that the seventh cranial nerve maintains constant relationships with the processes of the stapes in tetrapods. (Some birds and all salamanders are exceptional; see Lombard and Bolt 1979.) And this assumption would indicate then that the tympanum of amphibians and of modern reptiles plus birds, for example, are independently derived. The ear of mammals would represent yet another derivation of a tympanic ear in this view as explained in Lombard and Bolt (1979) and illustrated by the dot lettered "B." The three independent origins of tympanic ears advocated by this alternative view are indicated by the open circles in Fig. 4-2.

Table 4-2 indicates the uniqueness of the amphibian ear based on the arguments just presented. Note that in contrast with the standard view illustrated previously, amphibians share with amniotes only the tympanic cavity and tube, and pattern of the chorda tympani—the latter having nothing to do with auditory perception and the former being a general vertebrate characteristic (as a hyoid pouch). The unique amphibian characteristics now include tympanum, tympanum-stapes articulation, otic opercular element, opercular muscle, periotic system, basilar papilla, amphibian papilla, and provisionally, the foramen rotundum. The unique features are all involved in auditory perception. None are homologs with features found in amniotes. The picture presented is of an auditory periphery unique in all the gross features associated with audition. Further, fossil remains interpreted in this light would indicate that the amphibian tympanic ear (1) was the original experiment in terrestrial hearing, (2) preceded that in amniotes by a considerable period of time, and (3) is presently represented in its most complete form only in living frogs.

The ramifications of this alternative view for skull evolution, the origin of

Table 4-2. Suggested homologies of amphibian and amniote ear structures (under the view proposed in this chapter)[1]

Common tetrapod	Unique amphibian
	1. Tympanum
2. Tympanic cavity and tube	
	3. Stapes-tympanum articulation
	4. Otic opercular element
	5. Opercularis muscle
	6. Periotic system
	7. Foramen rotundum
	8. Basilar papilla
	9. Amphibian papilla
10. Chorda tympani relations	

[1] Only items 2 and 10 are considered homologs in amphibians and amniotes. All the structures involved with sound reception are considered uniquely derived within the amphibia.

reptiles, the evolution of hearing within the amniotes, and the systematics of fossil and recent amphibians are briefly explored in a larger work (Lombard and Bolt 1979). The possible ramifications for our views on hearing in the amphibians are appropriate here.

The standard view of the amphibian auditory system implies a general commonalty of function[2] with amniotes, derived from the notion of a common morphological plan. Further, as the amphibian ear is, in a major way, perceived as a primitive prototype for the more "refined" amniote ears under the standard view, a clinal view of auditory function could be considered reasonable. Basically, then, the evolution of auditory function would be perceived as

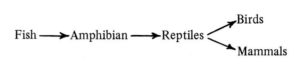

with the implication of common function and clinal abilities (i.e., "improvement" or "refinement" in the direction of the arrows). This global view of tetrapod auditory evolution can be used to support the notion that the amphibian ear may be studied as a simplified version of a mammal's ear, for example, and that such studies will tell us about auditory function in mammals.

The view advocated in this chapter, in contrast, results in a very different picture of auditory evolution. Here the evolution of tetrapod auditory systems would be

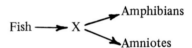

in which each named tetrapod group develops an auditory periphery de novo (no common morphology or function) and in which a cline of "improvement" on the scala natura of tetrapods in clearly not appropriate. This view would indicate that it would be very unlikely that auditory function in amphibians is similar to that in amniotes. Hence, the use of frogs, for example, as "simplified" mammalian substitutes in auditory research could be very misleading (but see Capranica and Moffat, Chapter 5). Further, experimental protocols appropriate to research on mammalian auditory systems (stimulus forms and interpretation of results) may be inappropriate for amphibian systems. How one designs experiments suited to amphibian auditory systems and then interprets the results may indeed be a more difficult problem than we currently understand it to be.

Acknowledgments. This study was supported by grant DEB76-18987 from the National Science Foundation. I would like to acknowledge the large role played by John Bolt in the development of the ideas expressed in this work and the editorial, typing, and illustrative skills of Joan Hives. Special thanks to Herbert Barghusen for his critical comments, both in the preparation of the talk presented in Honolulu and in the preparation of this paper.

[2]Function is used here in the mechanical and perhaps physiological sense, not in the sense of biological role or adaptive significance. (*Sensu* Bock and von Wahlert 1965.)

References

Baird, I. L.: A survey of the periotic labyrinth in some representative recent reptiles. Univ. Kansas Sci. Bull. 41, 891-981 (1960).

Baird, I. L.: Anatomical features of the inner ear in submammalian vertebrates. In: Handbook of Sensory Physiology, Vol. V/1 (Auditory System). W. D. Keidel, Neff, W. D. (eds.). Berlin: Springer-Verlag, 1974a, pp. 159-212.

Baird, I. L.: Some aspects of the comparative anatomy and evolution of the inner ear in submammalian vertebrates. Brain Behav. Evol. 10, 11-36 (1974b).

Bock, W. J., von Wahlert, G.: Adaptation and the form-function complex. Evolution 19, 269-299 (1965).

Capranica, R. R.: Morphology and physiology of the auditory system. In: Frog Neurobiology. Llinas, R., Precht, W. (eds.). Berlin: Springer-Verlag, 1976, pp. 551-575.

deBeer, G. R.: The Development of the Vertebrate Skull. Oxford: Oxford University Press, 1937.

deBurlet, H. M.: Vergleichende Anatomie des stato-akustischen Organs. a) Die innere Ohrsphäre; b) Die mittlere Ohrsphäre. In: Handbuch der vergleichenden Anatomie der Wirbeltiere, Vol. 2. Bolk, L., Göppert, E., Kallius, E., and Lubosch, W. (eds.). Berlin: Urban und Schwarzenberg, 1934, pp. 1293-1432.

Eaton, T. H.: The crossopterygian hyomandibular and the tetrapod stapes. J. Wash. Acad. Sci. 29, 109-117 (1939).

Gaupp, E.: Die Reichertsche Theorie (Hammer-, Ambos- und Kieferfrage). Arch. Anat. Physiol. Abt. Anat.-Entwickl. Suppl.-Band, 1-146 (1913).

Goodrich, E. S.: Studies on the Structure and Development of Vertebrates. London: Macmillan and Company, 1930.

Henson, O. W.: Comparative anatomy of the middle ear. In: Handbook of Sensory Physiology, Vol. V/1 (Auditory System). Keidel, W. D., Neff, W. D. (eds.). Berlin: Springer-Verlag, 1974, pp. 39-110.

Jarvick, E.: On the visceral skeleton in *Eusthenopteron* with a discussion of the parasphenoid and palatoquadrate in fishes. K. Sven. Vetenskapsakad. Handl. 5, 1-104 (1954).

Lombard, R. E.: A comparative morphological analysis of the salamander inner ear. Ph.D. diss., Univ. Chicago (1971).

Lombard, R. E.: Comparative morphology of the inner ear in salamanders (Caudata: Amphibia). Contrib. Verteb. Evol. 2, 1-143 (1977).

Lombard, R. E., Bolt, J.: Evolution of the tetrapod ear: an analysis and reinterpretation. Biol. J. Linn. Soc. 11, 19-76 (1979).

Olson, E. C.: The middle ear—morphological types in amphibians and reptiles. Am. Zool. 6, 399-419 (1966).

Reinbach, W.: Über den schalleitenden Apparat der Amphibien und Reptilien. (Zur Schmalhausenschen Theorie der Gehörknöchelchen.) Zeit. Anat.-Entwickl. 113, 611-639 (1950).

Retzius, G.: Das Gehörorgan der Wirbeltiere, Vols. I and II. Stockholm: Samson and Wallin (1881/1884).

Romer, A. S.: The Vertebrate Body. Philadelphia: W. B. Saunders Company (1970).

Romer, A. S., Price, L. I.: Review of the Pelycosauria. Geol. Soc. Amer., Spec. Pap. 28, 1-538 (1940).

Schmalhausen, I. L.: The Origin of Terrestrial Vertebrates. [English trans. L. Kelso]. New York: Academic Press (1968).

van Bergeijk, W. A.: Evolution of the sense of hearing in vertebrates. Am. Zool. 6, 371-377 (1966).
Watson, D. M. S.: The evolution of the mammalian ear. Evolution 7, 159-177 (1953).
Wever, E. G.: The evolution of vertebrate hearing. In: Handbook of Sensory Physiology, Vol. V/1 (Auditory System). Keidel, W. D., Neff, W. D. (eds.). Berlin: Springer-Verlag, 1974, pp. 423-454.
Wever, E. G.: Origin and evolution of the ear of vertebrates. In: Evolution of Brain and Behavior in Vertebrates. Masterton, R. B., Bitterman, M. E., Campbell, C. B. G., Hotton, N. (eds.). New Jersey: Laurence Erlbaum Association, 1976, pp. 89-105.

Chapter 5

Nonlinear Properties of the Peripheral Auditory System of Anurans

ROBERT R. CAPRANICA and ANNE J. M. MOFFAT*

1 Introduction

The vertebrate ear is highly nonlinear. This is rather surprising since its vibrational amplitudes are so minute in response to normal sound pressures. Generally, one might expect a stable mechanical system to respond linearly when disturbed slightly from its resting state. Thus the nonlinear properties of the peripheral auditory system are of considerable interest inasmuch as they can provide valuable insight into the underlying transduction process in the ear. The two most prominent nonlinear properties are intermodulation distortion and two-tone suppression. Their characteristics have been studied extensively in the mammalian auditory system by a number of investigators. To provide a comparative view, a series of electrophysiological experiments were conducted in order to determine the nonlinear behavior of the anuran's peripheral auditory system. The results have interesting implications regarding the origin of nonlinearities, as well as the mechanical basis for frequency analysis, in the vertebrate inner ear in general. Before presenting these findings, several relevant studies of nonlinearities in the mammalian auditory system are summarized, followed by a brief review of the anatomy of the anuran's ear.

1.1 Intermodulation Distortion Products

When two pure tones of frequency f_1 and f_2 are passed through a nonlinear system, the output generally may contain intermodulation distortion products having frequencies $mf_1 \pm nf_2$ (m and n are positive integers). Psychoacoustic studies (Moe 1942, Plomp 1965) have shown that two of the most prominent distortion products are the cubic difference tone at $2f_1 - f_2$ and the quadratic difference tone at $f_2 - f_1$, where $f_2 > f_1$.

Zwicker (1955) added a third tone at the distortion frequency and, by adjusting its amplitude and phase, was able to cancel the pitch of the $2f_1 - f_2$ difference tone. He

*Section of Neurobiology and Behavior, Langmuir Laboratory, Cornell University, Ithaca, New York 14853.

found that the growth rate of this distortion product as a function of the intensity of the two primary tones was considerably less than predicted by a simple third-order term in a polynomial expansion. Goldstein (1967) used this pitch cancellation technique to quantify the characteristics of the cubic difference tone in greater detail. His study showed that the amplitude of the perceived $2f_1 - f_2$ component increased at the same rate as the amplitude of the two primary tones. Furthermore the level of this distortion product decreased markedly with increasing frequency separation between the primary tones. Goldstein thus concluded that the $2f_1 - f_2$ component does not originate from a saturating polynomial nonlinearity as would be expected from classical auditory theory; instead he claimed it must be generated by an essential mechanical nonlinearity in the cochlea.

Hall (1972a, 1972b) discovered that the quadratic difference tone can be just as prominent as the cubic difference tone. His psychoacoustic measurements show that the effective amplitude of the $f_2 - f_1$ component may be as great as 10 dB below that of the two primary tones, while the effective amplitude of the $2f_1 - f_2$ component can be 15 dB to 20 dB below the level of the primaries. Such high effective amplitudes indicate that the underlying nonlinearity must be very pronounced. Hall further found that the amplitude of the $f_2 - f_1$ component, as in the case of the $2f_1 - f_2$ component, decreased with increasing frequency separation between the primary components. This led him to conclude that the $f_2 - f_1$ distortion product also behaves as an essential nonlinearity.

Electrophysiological studies provide clear evidence that the intermodulation distortion products are generated in the peripheral auditory system. Goldstein and Kiang (1968) recorded the activity of single fibers in the cat's auditory nerve in response to pairs of tones. The tonal frequencies were adjusted so that the combination frequency $2f_1 - f_2$ was approximately equal to the fiber's best excitatory frequency. They found that each fiber fired synchronously to individual cycles of the $2f_1 - f_2$ component. This time-locked activity could be canceled by adding a third tone of appropriate amplitude and phase at the frequency $2f_1 - f_2$. Buunen and Rhode (1978) verified these results and further noted that the firing rate of each fiber decreased as the frequency separation between f_1 and f_2 was increased (while maintaining $2f_1 - f_2$ equal to the fiber's best excitatory frequency). From their electrophysiological results they estimated that the effective intensity of the cubic distortion component was on the average about 20 dB below the level of the two primary tones when their frequencies were below 10 kHz. Greenwood, Merzenich, and Roth (1976), by recording from single cells in the anteroventral cochlear nucleus of the cat, concluded that the distortion product $2f_1 - f_2$ produced maximum firing in auditory nerve fibers when the amplitudes of the two primary tones f_1 and f_2 were approximately equal. If there was an imbalance in the amplitudes of the two tones, then the cubic distortion component was reduced.

Electrophysiological studies have likewise verified the sensitivity of auditory nerve fibers to the quadratic distortion product $f_2 - f_1$. Pfeiffer and Molnar (1974) recorded from single fibers in the auditory nerve of cats in response to pairs of harmonically related tones. The discrete Fourier transform of the resultant period histograms yielded both amplitude and phase of the primary and distortion components in the response patterns. They found that the magnitude of the quadratic distortion product $f_2 - f_1$ was very prominent and comparable to the amplitude of the cubic component

$2f_1 - f_2$ (Pfeiffer, Molnar, and Cox 1974). More recently Smoorenburg, Gibson, Kitzes, Rose, and Hind (1976), by recording from single cells in the cat's anteroventral cochlear nucleus, verified that auditory nerve fibers respond to the distortion product $f_2 - f_1$. They found that the thresholds of the most sensitive units to the $f_2 - f_1$ component were only 0 dB to 40 dB above the thresholds to pure tones (at the corresponding best excitatory frequency). Hence the quadratic differenc tone is quite strong. The amplitude of this distortion component, as in the case of the cubic difference tone, is maximum when the two primary tones are of approximately equal levels.

There is impressive agreement between electrophysiological and psychophysical studies. The evidence is quite convincing that the two intermodulation distortion products $2f_1 - f_2$ and $f_2 - f_1$ originate from nonlinearities within the peripheral auditory system. The obvious question arises: "Where exactly are these nonlinearities?" Von Békésy (1960) measured the radiation of sound from the eardrum and found that the distortion products in the reflected energy were not sufficient to account for the perception of difference tones. He suggested that this distortion must be in the cochlea. Guinan and Peake (1967), in their direct measurements of the vibrational sensitivity of the cat's middle ear, concluded that the ossicles behave as a low-pass, linear system. Rhode (1971) used the Mössbauer technique to measure the mechanical sensitivity of the squirrel monkey's middle ear and found that its ossicles vibrate linearly for sound pressures up to at least 100 dB SPL. And Wilson and Johnstone (1975) measured the movements of the incus in guinea pigs using a capacitive probe technique; the results reveal linearity for pressures up to at least 100 dB SPL.

The logical conclusion is that the primary locus for intermodulation distortion must reside in the inner ear. The distortion product responses behave as if these frequencies were actually delivered externally to the ear. It thus has been proposed (e.g., Hall 1974, Smoorenburg et al. 1976, Kim and Molnar 1976, Siegel, Kim, and Molnar 1977, Buunen and Rhode 1978) that the two primary tones f_1 and f_2 somehow interact nonlinearly in exciting their normal places along the basilar membrane, thus giving rise to distortion products $2f_1 - f_2$ and $f_2 - f_1$. These distortion products then somehow propagate through the cochlea to excite the appropriate sensory regions tuned to the corresponding distortion frequencies. That the distortion products actually propagate from their site of generation to their site of action has been shown recently by the interesting experiments of Siegel et al. (1977). They exposed chinchillas to an intense narrow band of noise which resulted in local damage to the organ of Corti along the basilar membrane. They subsequently found that the distortion components $2f_1 - f_2$ and $f_2 - f_1$ no longer could be detected in the responses of auditory nerve fibers if either of the primary tones fell within the damaged region. Tone pairs having lower frequencies corresponding to the undamaged region continued to produce large, propagated distortion products.

Siegel et al. (1977) proposed that the distortion products likely propagate mechanically along the basilar membrane. However, this is a matter of controversy. Wilson and Johnstone (1972) made direct measurements of basilar membrane motion in the guinea pig using their capacitive probe technique. They were unable to find any significant component of membrane motion at the frequency $2f_1 - f_2$. Rhode (1977) likewise could not detect a mechanical component of basilar membrane motion at this distortion frequency. These direct mechanical measurements suggest that this intermodulation distortion product probably does not propagate as a traveling wave along the basilar membrane. This is puzzling in view of the fact that the distortion component can be totally canceled by adding a third tone of appropriate amplitude and phase.

1.2 Two-Tone Suppression

The first clear finding of inhibitory interaction in response to tonal stimuli, now generally referred to as two-tone suppression, in auditory nerve fibers was reported by Frishkopf (1964) in his study of the little brown bat and by Nomoto, Suga, and Katsuki (1964) in their study of the squirrel monkey. These results clearly demonstrated that the response of a primary fiber to an excitatory tone could be suppressed by the simultaneous presence of a second tone of appropriate amplitude and frequency. This phenomenon has been verified by a number of subsequent investigators in all classes of terrestrial vertebrates: mammals (e.g., Sachs and Kiang 1968, Arthur, Pfeiffer, and Suga 1971, Abbas and Sachs 1976, Javel, Geisler, and Ravidran 1978); birds (Sachs, Young, and Lewis 1974, Gross and Anderson 1976); reptiles (Holton and Weiss 1978); and anurans (e.g., Frishkopf and Goldstein 1963, Liff and Goldstein 1970, Feng, Narins, and Capranica 1975, Capranica and Moffat 1975). Thus, this two-tone nonlinear property seems to occur universally in the vertebrate peripheral auditory system.

In mammals, birds, and reptiles, the excitatory tuning curve of a fiber is intimately flanked on each side by an inhibitory tuning area. As the frequency separation between the excitatory and inhibitory tones is increased, the degree of suppression decreases monotonically (Javel et al. 1978). Abbas and Sachs (1976) have shown that there is a fundamental difference in the attenuating power of suppressive tones in the inhibitory areas on the two sides. For suppressive frequencies greater than the best excitatory frequency, the degree of suppression is related to the intensity ratio of the suppressor and excitor tones. But for tones that fall within the lower inhibitory tuning curve, the degree of suppression depends almost entirely only on the amplitude of the suppressor tone. This may reflect a fundamental difference in the mechanism underlying suppression by frequencies above and below a fiber's best excitatory frequency.

As has already been pointed out, the mammalian eardrum and ossicles respond in a linear fashion for sound pressures up to at least 100 dB SPL. Since two-tone suppression occurs at much lower intensities, the locus of this nonlinearity is not in the middle ear. Furthermore, when the eighth nerve is severed, fibers in the peripheral stump still show tone-on-tone suppression (Frishkopf and Goldstein 1963, Kiang, Watanabe, Thomas, and Clark 1965). Hence efferent inhibition from the central nervous system is not responsible for this nonlinearity. Clearly the mechanism for two-tone suppression must reside in the inner ear.

Several studies have reported interference effects and two-tone suppression in the cochlear microphonic potential (e.g., Wever, Bray, and Lawrence 1940, Legouix, Remond, and Greenbaum 1973, Dallos, Cheatham, and Ferraro 1974). That is, the amplitude of the microphonic potential in response to a tonal stimulus can be reduced by the addition of a second tone. These results have been interpreted as evidence that two-tone suppression is due to a nonlinearity in basilar membrane mechanics (Legouix et al. 1973). A variety of models have attempted to explain how such a nonlinearity could arise in the membrane motion (e.g., Kim, Molnar, and Pfeiffer 1973, Hall 1977b).

The only direct evidence that the basilar membrane may support two-tone suppression comes from the recent studies of Rhode (1977, 1978) using the Mössbauer technique. In measuring vibrations of the squirrel monkey's basilar membrane in response to an excitatory tone, he found that a second, higher frequency tone of sufficient intensity could lead to a reduction in vibrational amplitude of about 10 dB. The degree of suppression diminished with increasing frequency separation between the two tones. Suppression by frequencies below that of the excitatory tone was less pronounced. Overall, these mechanical measurements seem to be in close agreement with the neural correlates of two-tone suppression. But there is a serious problem with such a simple conclusion. The characteristics of two-tone suppression are very similar in other mammals, such as the guinea pig. Several different techniques have been used in measuring the movements of its basilar membrane: Mössbauer method (Johnstone and Taylor 1970, Johnstone and Yates 1974); capacitive probe (Wilson and Johnstone 1972); and speckle illumination (Köhlloffel 1972). The vibration of the guinea pig's basilar membrane was remarkably linear in each of these studies. For example, Wilson and Johnstone found that the amplitude of vibration of its basilar membrane was invariably linear to within 1 dB for sound pressures up to at least 110 dB SPL. It therefore is difficult to reconcile the differences between these mechanical measurements for squirrel monkeys and guinea pigs with regard to a common mechanism for two-tone suppression involving the basilar membrane.

1.3 Anuran Peripheral Auditory System

The anatomy and physiology of the anuran's peripheral auditory system has been described recently in considerable detail (Capranica 1976). The main features will be summarized in order to provide the background for this chapter.

The eardrum of the adult consists of a large, circular membrane exposed directly to the external environment. The middle-ear bones (plectrum, columella and operculum) couple vibrational energy from the eardrum to the fluid-filled inner ear via the oval window; a round window membrane serves as a pressure release. Within the inner ear are two separate, distinct organs—amphibian papilla and basilar papilla—that are specialized for sensitive reception of acoustic stimuli. Each of the papillae possesses its own complement of hair cells with a separate, overlying tectorial membrane. The hair cells rest on stationary, supporting structures (Geisler, van Bergeijk, and Frishkopf 1964, Wever 1973) so that a basilar membrane is totally absent in each organ. Hence only the tectorial membrane plays a role in coupling mechanical energy to the cilia of the receptor cells. The hair cells are innvervated by auditory fibers in the eighth cranial nerve that conduct impulses to the central auditory system.

Recordings from single auditory nerve fibers indicate typical "V" shaped tuning curves with response properties remarkably similar to those in other vertebrate classes. The amphibian papilla gives rise to a distribution of fibers tuned to low and mid frequencies, whereas the basilar papilla is more narrowly tuned to higher frequencies. Based on their studies of the response properties of fibers to tones and clicks, Capranica and Moffat (1977) concluded that a place mechanism must underlie frequency analysis in the amphibian papilla. This conclusion has been verified very recently by the study of Lewis and Leverenz (1979) in which they succeeded in tracing fibers possessing different frequency sensitivities to their origin in the organ. They have found that the low-frequency-sensitive fibers innervate hair cells in the anterior portion of the sensory epithelium of the amphibian papilla, whereas the mid-frequency fibers innervate receptor cells in the posterior region.

In our studies of the response properties of the anuran's amphibian papilla, an interesting result was encountered: many of its fibers are very sensitive to two-tone suppression and intermodulation distortion products (Capranica and Moffat 1974a, 1974b, 1974c, 1976, Moffat and Capranica 1976, 1978, 1979). Since the structure of this organ is so different from the mammalian cochlea, we decided that this sensitivity to nonlinearities should be explored in detail. The remainder of this chapter documents some of our major observations. These results have broad implications regarding the basis for nonlinear processing in the vertebrate peripheral auditory system in general.

2 Methods

American toads (*Bufo americanus*) were collected during springtime in the vicinity of Ithaca, New York, and maintained in laboratory terrariums for subsequent electrophysiological study. Data were obtained from animals weighing 10 g to 50 g, anesthetized with Nembutal (50 μg/g) or immobilized with D-tubocurarine chloride (10 μg/g). Both drugs were given intramuscularly. The only difference in results obtained from the two types of preparations for this species was that the level of spontaneous activity recorded from the anesthetized animals was appreciably lower—many of the units were not spontaneously active. However, thresholds, best excitatory frequency distributions, sensitivity to two-tone suppression, etc., were comparable.

During surgery and the subsequent recording sessions, the animals were placed in a double-walled audiometric room and were covered with damp gauze to aid cutaneous respiration; temperature in the room was maintained at 21°C to 24°C. The VIIIth nerve was exposed through a hole in the roof of the mouth and activity of single fibers was recorded with KCl-filled micropipettes (40 megohms to 80 megohms). The electrode was advanced through the nerve by means of a Kopf hydraulic microdrive located outside the room. Firing rates were displayed on a gated electronic timer-counter and tape recorded for later analysis with a DEC Lab 8E computer.

Acoustic stimuli were generated by a system of audio oscillators (Hewlett-Packard 200 CD), electronic switches (Grason-Stadler 1287), timers (Grason-Stadler 1216) and pulse generators (Ortec 4650 series). Combinations of tones of different frequencies and intensities were controlled by calibrated attenuators and resistive adders. The resultant stimuli were presented to the animal through a PDR-10 earphone enclosed in a metal housing. This housing, which also contained a condenser microphone (B & K 4134) for measurement of sound pressure at the animal's eardrum, was sealed with silicone grease around the outer edge of the tympanic ring to form a closed acoustic system. All sound pressure levels are expressed in dB SPL re 20 μPa. The frequency response of the stimulus delivery system was flat within ± 4 dB over the frequency range of interest (50 Hz to 5000 Hz). The intensities of individual components in complex tones were corrected for the frequency response of the system and therefore reflect absolute sound pressure level. The purity of the tonal stimuli was verified by measuring the output of the condenser microphone amplifier (B & K 2604) with a wave analyzer and graphic level recorder (General Radio 1900-A and 1521B). All distortion products in the tonal combinations were at least 50 dB below the level of the primary tones. Additional details of our experimental setup have been described previously (Capranica and Moffat 1975).

3 Results

Our study is based on 345 auditory nerve fibers recorded from 16 animals. No significant differences in the response properties of animals of different size (and hence presumably different age) or sex were found; nor were there any clear seasonal changes in sensitivity. The results for these animals have, therefore, been combined.

3.1 Frequency Sensitivity

The first information collected for each unit was its excitatory tuning curve in response to pure tones, and three representative curves are shown in Fig. 5-1. The tuning curves have the typical simple "V" shape common to auditory nerve fibers in almost all vertebrate species so far studied. The frequency to which a fiber is most sensitive is termed its best excitatory frequency (BEF) and, based on their BEF's, fibers in the auditory nerve of *Bufo americanus* fall into three groups (Capranica and Moffat 1974a). One group has its BEF's in the low-frequency range below 550 Hz, a second group has its BEF's in the mid-frequency range of 500 Hz to 1000 Hz, and the third group has its BEF's between about 1200 Hz and 1600 Hz. By analogy with the study of Feng et al. (1975) in the bullfrog, *Rana catesbeiana*, we believe that the high frequency fibers originate in the basilar papilla while the low and mid frequency fibers come from the amphibian papilla. In this chapter only properties of amphibian papilla fibers will be discussed and, in fact, the discussion will be largely restricted to the low-frequency-sensitive fibers since only they show the nonlinear properties of sensitivity to intermodulation distortion and two-tone suppression.

3.2 Combination Tones

On measuring a fiber's tuning curve and threshold at best frequency, a variety of tonal pairs was next presented to determine whether it was responsive to intermodulation distortion products. Since all distortion components in the acoustic stimulus were at least 50 dB less intense than that of the two primary tones, evidence of sensitivity to combination tones was accepted only if the level of the primaries was less than 40 dB above the fiber's absolute threshold at its BEF. The only combination tone to which responses could be recorded was the quadratic difference tone $f_2 - f_1$. There were

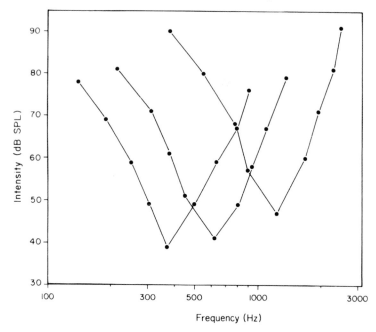

Figure 5-1. Single-tone tuning curves for three representative fibers in the VIIIth nerve of *Bufo americanus*. The rightmost tuning curve (BEF 1390 Hz) is for a fiber from the basilar papilla while the other two (BEF's 620 Hz, 370 Hz) are for fibers from the amphibian papilla.

no responses to higher order difference tones, such as $2f_1 - f_2$, nor to summation tones such as $f_1 + f_2$.

In Fig. 5-2 is shown the sensitivity of a fiber to the difference tone $f_2 - f_1$. This unit had a BEF of 180 Hz and a threshold of 53 dB SPL. The response to a single tone at its best frequency and 10 dB above threshold is seen in Fig. 5-2A. Stimulation by two simultaneous tones, neither of which alone evoked any activity, produced a vigorous response (Fig. 5-2B, C) on condition that their frequency difference was approximately equal to the BEF.

Any tone pair with a frequency difference approximately equal to a unit's best frequency was capable of producing a response, so long as the actual frequencies and intensities of the two tones fell within the sensitivity range of the amphibian papilla. This is illustrated in Fig. 5-3, which shows the tuning curve for a low-frequency fiber as well as a plot of the intensity of a tonal pair, as a function of frequency f_1 of one of the tones (frequency of the second tone $f_2 = f_1 + $ BEF), that evoked a response equivalent to a BEF tone at 10 dB above threshold. Note that the second plot intersects the distortion level in the stimulus (50 dB above the fiber's threshold) at about 1 850 Hz. This frequency corresponds to about the upper limit of the frequency sensitivity of the amphibian papilla (while a.p. units have *best* excitatory frequencies extending only to about 900 Hz, the high-frequency side of their tuning curves extends to around

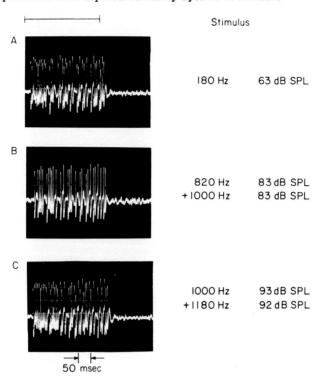

Figure 5-2. Nerve spikes of a low-frequency-sensitive unit to (A) a tone at its BEF, and to tone pairs whose frequency difference is equal to the fiber's BEF: (B) $f_1 = 820$ Hz, $f_2 = 1000$ Hz, and (C) $f_1 = 1000$ Hz, $f_2 = 1\,180$ Hz. The duration of each stimulus was 300 msec, represented by the solid line at the top.

1 700 Hz to 1 800 Hz for intense sounds). The upper limit of sensitivity around 1 800 Hz most likely accounts for the fact that only those units with BEF's below about 550 Hz respond to $f_2 - f_1$. In order to excite higher BEF fibers with this quadratic difference tone, one of the primaries would have to have a frequency greater than the upper cutoff of the amphibian papilla.

The response to $f_2 - f_1$ is greatest when the intensities of the two tones are approximately equal. This is demonstrated in Fig. 5-4 which shows the firing rate of a low-frequency-sensitive fiber (BEF 190 Hz) in response to a tone pair having a frequency difference of 150 Hz ($f_1 = 740$ Hz, $f_2 = 890$ Hz). The intensity of each of the tones relative to the other was varied systematically and the number of spikes evoked during 10 stimulus presentations was counted and averaged. The result shows quite clearly that the response is optimal when the intensities of the two tones are approximately the same.

As shown in Fig. 5-5, the response was also optimal when the frequency difference between the two primaries in the tone pair was close to the fiber's BEF (270 Hz for the fiber in Fig. 5-5). To generate the curve in Fig. 5-5 we fixed one tone f_2 at 970 Hz and 70 dB SPL, and varied the frequency f_1 of the second tone while maintaining its

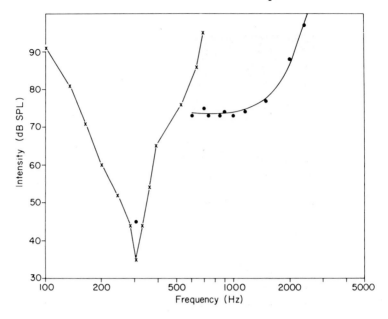

Figure 5-3. Curve on the left (X—X) shows the single-tone tuning curve for a low-frequency-sensitive fiber with a BEF of 305 Hz. On the right (●—●) is a plot of the mean intensity for a pair of tones as a function of f_1 ($f_2 = f_1$ + BEF) that produced a response equal to that evoked by a BEF tone 10 dB above threshold (● in the tuning curve). The level at which distortion products in the acoustic stimulus might begin to excite this unit is 85 dB SPL.

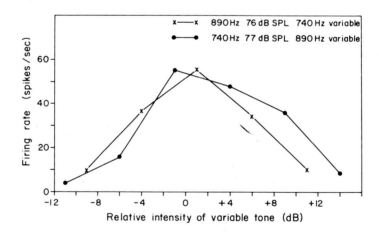

Figure 5-4. Evoked firing rate as a function of relative intensities of the two primaries in a tonal pair. X—X 890 Hz fixed at 76 dB SPL, 740 Hz intensity variable; ●—● 740 Hz fixed at 77 dB SPL, 890 Hz intensity variable. This unit had a BEF of 190 Hz with a threshold of 58 dB SPL.

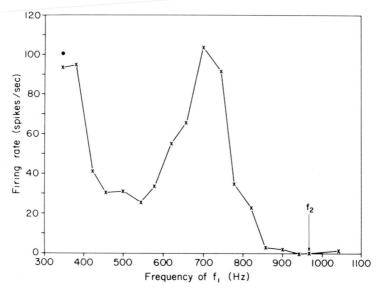

Figure 5-5. Evoked firing rate as the frequency f_1 was varied, with f_2 fixed at 970 Hz. The intensities of the two tones were constant at 70 dB SPL. This unit had a BEF of 270 Hz with a threshold of 38 dB SPL. The increase in firing rate for low frequency values of f_1 is due to the fact that this component by itself began to excite the fiber. The symbol ● indicates the firing rate to a single BEF tone at 10 dB above threshold. The reason that low frequency values of f_1 do not evoke a greater response than ● is because of suppression by tone f_2.

intensity at the same 70 dB SPL level; spikes evoked during 10 presentations of the stimulus were counted and plotted in terms of average firing rate during the stimulus. Note that there is a clear peak in the firing rate when $f_2 - f_1$ = BEF. This rate is approximately equal to that evoked by a single BEF tone at the same intensity. Thus, the underlying nonlinearity responsible for generation of the difference tone must be very pronounced.

The conclusion that the low-frequency fibers from the amphibian papilla are actually responding to energy produced at the difference frequency, and not to some temporal feature in the waveform, is supported by at least two observations. The first is that, as shown in Fig. 5-6, it is possible to generate "difference-tone" tuning curves. Furthest left in this figure is the pure-tone tuning curve for a unit with a BEF of 115 Hz and a threshold of 57 dB SPL. Difference-tone tuning curves, to the right, were obtained by fixing the frequency of one tone at either 600 Hz or 800 Hz and varying the frequency of the second, while changing the intensity of both tones together until a threshold response was obtained. The bandwidths of the difference-tone tuning curves at 10 dB above threshold are all remarkably similar to the corresponding bandwidth of the single tone tuning curve. This result suggests that energy is produced at the difference frequency which acts as if it were a pure tone delivered to the ear at that frequency.

The second piece of evidence that indicates energy must be generated at the difference frequency is illustrated in Figs. 5-7 and 5-8. In Figure 5-7A is shown the response

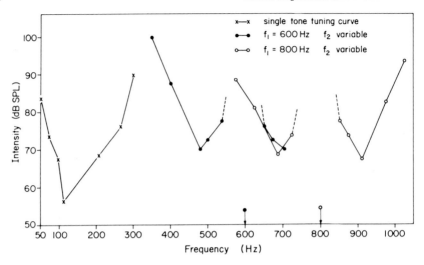

Figure 5-6. Tuning curves for a fiber in response to single tones X—X and to tone pairs: ●—● f_1 = 600 Hz, f_2 variable; ○—○ f_1 = 800 Hz, f_2 variable. The best frequency of this unit was 115 Hz with a threshold of 57 dB SPL. The fiber was lost before the high-frequency side of the tuning curve for f_1 = 600 Hz could be completed.

of a low-frequency-sensitive fiber to a pure tone at its best excitatory frequency 185 Hz at 10 dB above threshold. Figure 5-7B shows the response of the same fiber to a tone pair with a frequency difference of 172 Hz (f_1 = 828 Hz, f_2 = 1000 Hz). Simultaneous presentation of these three tones produced the response shown in Fig. 5-7C. Note that the pattern of spikes occurs in bursts, indicating beating between the 185 Hz tone and the internally generated difference tone—such beating was not present in the acoustic stimulus waveform.

This internal beating is demonstrated more clearly in Fig. 5-8, which shows a sequence of response histograms for a low-frequency-sensitive fiber. The upper histogram indicates the level of activity to a tone of 265 Hz at 10 dB above excitatory threshold; the fiber fired tonically throughout the 35 sec duration of the stimulus. The middle histogram shows the response to a tone pair, f_1 = 700 Hz and f_2 = 965 Hz, presented at 10 dB above the fiber's threshold to this tonal combination. The difference frequency in this stimulus and the frequency of the single tone in the upper histogram differed by only 0.3 Hz. The response to all three of these tones presented simultaneously is shown in the bottom histogram. The cyclical firing probability corresponds to a rate of 0.3 Hz, namely at the beat frequency between the difference tone and the single tone. By careful adjustment of the frequency, phase and amplitude of the tone at 265 Hz, it was possible to completely cancel the response in the bottom histogram. This cancellation did not occur in the stimulus; clearly energy at the difference frequency must be generated within the auditory periphery.

The amount of energy at the difference frequency diminished rapidly with increasing frequency separation between the two primary tones. To explore this dependency,

Stimulus

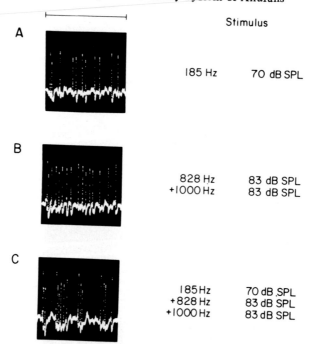

A

185 Hz   70 dB SPL

B

828 Hz   83 dB SPL
+1000 Hz   83 dB SPL

C

185 Hz   70 dB SPL
+828 Hz   83 dB SPL
+1000 Hz   83 dB SPL

Figure 5-7. Responses of a low-frequency-sensitive fiber to: (A) a tone at its BEF at 10 dB above threshold; (B) a tone pair whose frequency difference is close to the fiber's BEF and falls within its tuning curve; (C) the BEF tone and the tone pair presented simultaneously. Bursting in the response indicates beating between energy due to the BEF tone and that generated at the difference frequency of the tone pair; this beating did not occur in the acoustic stimulus. The duration of the stimuli was 300 msec, represented by the solid line at the top.

a series of measurements were conducted in which the thresholds to a tone pair (as a function of frequency separation) were compared with the thresholds to pure tones corresponding to these difference frequencies. By way of example, the procedure leading to the representative result shown in Fig. 5-9 will be described. This fiber had a best excitatory frequency of 290 Hz, threshold 27 dB SPL. First a series of pure tones was presented and the fiber's threshold at each 50-Hz interval beginning at 100 Hz was measured. Following these measurements, thresholds to a pair of equal-amplitude tones for f_2 fixed at 1500 Hz were determined while f_1 was reduced in 50-Hz steps starting at 1400 Hz. This procedure thus identified the minimum sound pressure level of each tone pair necessary to generate threshold excitation energy at its corresponding difference frequency. The difference between these levels and the respective pure tone thresholds provides a measure of the efficiency of difference-tone generation. Fig. 5-9 shows that the amplitude of the difference tone falls off rapidly with its frequency. This conclusion was supported in measurements, using various fixed frequency values of f_2, of a number of other low-frequency-sensitive fibers. In each case the amount of

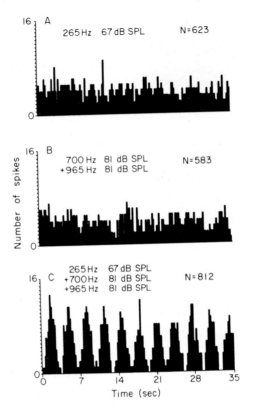

Figure 5-8. PST histograms of the response of a low-frequency-sensitive fiber to: (A) single tone at the fiber's BEF, 10 dB above threshold; (B) tone pair whose frequency difference is equal to the fiber's BEF, 10 dB above threshold for that pair; (C) BEF tone and tone pair presented simultaneously. The cyclical pattern in the bottom histogram reflects the slight difference of 0.3 Hz between the frequency of the single tone and the difference frequency in the tone pair. The duration of the stimulus in each histogram is 35 seconds (bin width = 250 msec).

energy produced at the quadratic difference tone decreased sharply with increasing frequency separation of the primaries.

In order to verify the locus of the underlying nonlinearity, the tympanic membranes were bilaterally removed and the columellae in an anesthetized animal was cut. The earphone was then sealed to the upper jaw and the inner ear was thus stimulated by bone conduction. The low frequency fibers continued to show the same response properties to the difference tone $f_2 - f_1$. These results support the conclusion that the nonlinearity responsible for difference tone excitation must reside within the inner ear.

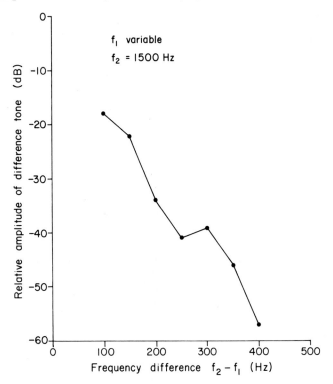

Figure 5-9. Relative amplitudes of quadratic difference tone generated by equal-amplitude primary tones as a function of their frequency separation $f_2 - f_1$.

3.3 Two-Tone Suppression

The phenomenon of two-tone suppression is shown by the sequence of PST histograms in Fig. 5-10. This fiber had a best excitatory frequency of 113 Hz and a threshold of 57 dB SPL. The top histogram shows the response pattern to a 300-msec excitatory tone at 10 dB above threshold. The fiber responded in typical fashion by firing tonically throughout the stimulus. To this excitatory tone, a 200-msec higher frequency (600 Hz) tone was added and gradually its relative amplitude was increased; this second tone was embedded in the middle of the excitatory tone as indicated by the stimulus symbol at the top of the figure. In Fig. 5-10B, the level of the 600 Hz tone was 10 dB greater than the 113 Hz tone. There is an obvious suppression in the excitatory response during the higher frequency tone. When the level of this suppressing tone was increased to 15 dB above that of the excitatory tone, shown in Fig. 5-10C, there were only a few spikes that occurred during its presence. In Fig. 5-10D, the suppression was total when the suppressing tone reached a level of 20 dB above the level of the excitatory tone. Generally, then, the suppression is graded and exhibits little adaptation.

The range of frequencies that will suppress the excitatory response is limited and is defined by the "inhibitory" tuning curve. Figure 5-11 illustrates such a curve for a fiber

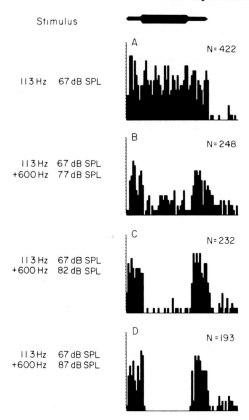

Figure 5-10. Sequence of PST histograms showing tone-on-tone suppression for a representative low-frequency-sensitive fiber. This unit had a best excitatory frequency of 113 Hz and a threshold of 57 dB SPL. (A) Response to a BEF tone at 10 dB above threshold. (B) Effect of adding a second tone of 600 Hz with a relative amplitude 10 dB greater than the excitatory tone. (C) Suppression when the level of the second tone was increased to 15 dB above the excitatory tone level. (D) Total suppression for a relative amplitude difference of 20 dB. The symbol at the top of the figure indicates the time pattern of the two tones: the excitatory tone in each histogram had a duration of 300 msec; the suppressor tone lasted for 200 msec and occurred during the middle of the excitatory tone. N denotes the total number of spikes in each histogram (bin width = 5 msec).

with a BEF of 195 Hz. To generate the inhibitory curve, the unit was stimulated at its BEF 10 dB above threshold (indicated by the point X in the excitatory tuning curve in this figure), and the frequency and intensity of a second tone was varied until the lowest relative intensity of this second tone that would produce complete suppression was found. For the fiber in Fig. 5-11, this tone, defined as the best suppression frequency (BSF), had a frequency of 600 Hz and had to be only 3 dB more intense than the excitatory tone. On varying the frequency and intensity of the suppressor tone, the threshold tuning curve for total suppression was then determined, as indicated by

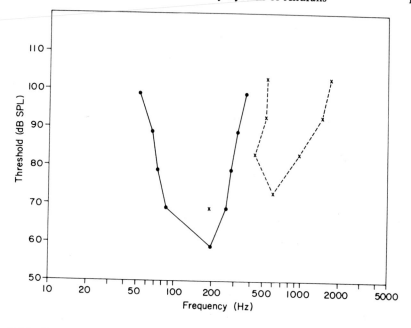

Figure 5-11. Excitatory tuning curve (●——●) for a low frequency fiber and the inhibitory tuning curve (X----X) that produced total suppression of the response to a BEF tone 10 dB above threshold (X in the excitatory tuning curve).

the dashed lines in Fig. 5-11. In general the inhibitory tuning curve always occurred on the high-frequency side of the excitatory tuning curve. No evidence was found of suppression (not even partial) by tones lower in frequency than a fiber's BEF.

Most fibers required a somewhat larger relative intensity difference than that shown in Fig. 5-11 to achieve total suppression. Nevertheless they all showed a systematic increase in the strength of suppression as the relative intensity of the suppressor tone was increased. This graded level of suppression is illustrated in Fig. 5-12, which presents a plot of the firing rate of a fiber in response to a fixed BEF tone as the intensity of the suppressor tone was gradually increased. The firing rate to the 300-Hz BEF tone alone, presented at 10 dB above threshold, was 52 spikes/sec. For this unit the best suppression frequency was 700 Hz and, as the relative intensity of this second tone was increased, the spike rate to the BEF tone decreased monotonically. When the level of the suppressor tone was 25 dB greater than that of the excitatory tone, suppression was complete and the fiber fired at its average spontaneous rate (2 spikes/sec). Spontaneous activity was not suppressible.

As was the case with responses to difference tones, two-tone suppression was not present in all fibers nor could all those that showed some degree of suppression be totally inhibited (for relative intensity differences that were thought to be within the linearity range of the stimulus system). There are two related reasons for this. The first is that, in general, fibers with higher BEF's require a relatively more intense suppressor tone to produce the same degree of suppression than do fibers with lower BEF's. This

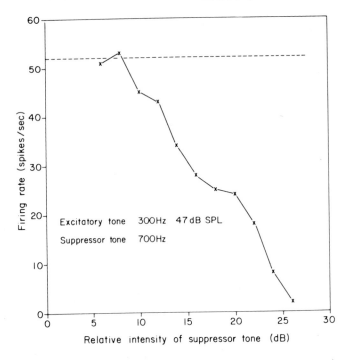

Figure 5-12. Spike-rate curve for a low-frequency-sensitive fiber in response to a tone at its BEF (300 Hz) with an intensity of 47 dB SPL (10 dB above excitatory threshold) in the presence of a suppressor tone (700 Hz) of increasing intensity. When the level of the suppressor tone reached 72 dB SPL, namely 25 dB greater than the excitatory tone, the firing rate was reduced to the level of spontaneous activity. The dashed line at the top represents the firing rate (52 spikes/sec) to the excitatory tone alone.

relationship is illustrated in Fig. 5-13 which is a plot of the relative intensity difference between BEF and BSF tones required to produce a 75% decrease in each fiber's response to its BEF tone presented at 10 dB above excitatory threshold. This percentage decrease was chosen for illustration, rather than total (100%) suppression, since most fibers could be suppressed by this amount, whereas several of the higher BEF fibers could not be totally suppressed. While there is some scatter in the plot in Fig. 5-13, nevertheless it is obvious that higher BEF fibers require a relatively more intense suppressor tone. Part of the scatter in this plot is probably due to the fact that there seems to be a relationship between threshold at BEF and relative intensity for equivalent suppression. For fibers having approximately the same best excitatory frequency, the more sensitive fibers tended to require a greater relative intensity of the suppressor tone than did a less sensitive fiber.

The other reason that only some of the amphibian papilla fibers can be suppressed is that as BEF increases so too does BSF. This is illustrated in Fig. 5-14, which is a plot of best excitatory frequency vs. best suppression frequency for 90 fibers from 8 animals. There is a clear correlation between the two frequencies. For fibers with best excitatory frequencies around 550 Hz, the required best suppression frequency

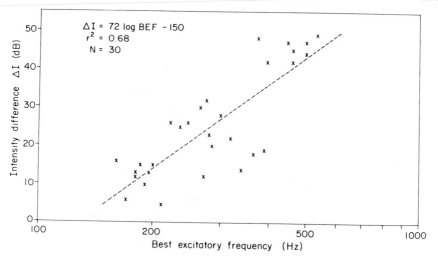

Figure 5-13. Intensity of best suppressor tone relative to intensity of best excitatory tone required to produce 75% suppression for 30 fibers of different best excitatory frequencies. Dotted line represents the linear regression $\Delta I = 72 \log BEF - 150$ with a correlation coefficient $r = 0.82$.

falls in the region 1 400 Hz to 1 500 Hz which approaches the upper cutoff frequency sensitivity of the amphibian papilla. This dependency, combined with the requirement of an increasing relative intensity difference between the suppressor and excitor tones as BEF increases, results in functional suppression only for those fibers whose best excitatory frequencies fall below about 550 Hz. It is interesting that only this same

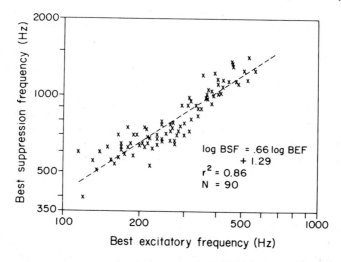

Figure 5-14. Best suppression frequency vs. best excitatory frequency for 90 auditory nerve fibers that exhibited two-tone suppression. Best fit line is represented by the equation $\log BSF = 0.66 \log BEF + 1.29$ with a correlation coefficient $r = 0.93$.

Stimulus

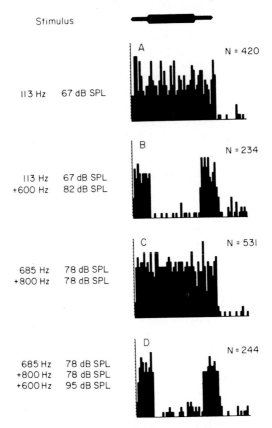

113 Hz	67 dB SPL

A   N = 420

113 Hz	67 dB SPL
+600 Hz	82 dB SPL

B   N = 234

685 Hz	78 dB SPL
+800 Hz	78 dB SPL

C   N = 531

685 Hz	78 dB SPL
+800 Hz	78 dB SPL
+600 Hz	95 dB SPL

D   N = 244

Figure 5-15. Sequence of histograms showing suppression of single-tone excitation and of difference-tone excitation for a representative low-frequency-sensitive fiber: (A) excitatory response to a single BEF tone at 10 dB above excitatory threshold; (B) suppression of the single tone response by adding a second tone; (C) excitation by a tone pair whose frequency difference is approximately equal to the fiber's BEF; (D) suppression of difference tone response by introducing a third tone during the presentation of the tone pair. The symbol at the top of the figure indicates the time pattern of the excitatory tones (300 msec duration) in which the larger amplitude suppressor tones (200 msec duration) were embedded in (B) and (D). To the left of each histogram are shown the corresponding tonal frequencies and intensities. N denotes the total number of spikes in each histogram (bin width = 5 msec).

population of low-frequency-sensitive fibers can be excited by difference tones. These results thus raise the possibility that difference-tone generation and two-tone suppression derive from a common nonlinearity.

In the studies of cancellation of difference-tone excitation, the quadratic distortion tone $f_2 - f_1$ behaved as if it were an actual tone delivered to the ear. It was therefore predicted that low-frequency difference tones may be suppressed by adding an appropriate third tone. Figure 5-15 gives the results of such a test. The top histogram (Fig.

5-15A) shows the response of a fiber to a 300-msec excitatory tone (113 Hz) at 10 dB above threshold. To this excitatory tone was added a 200-msec tone of 600 Hz with a relative intensity of 15 dB, resulting in typical two-tone suppression shown in Fig. 5-15B. The histogram in Fig. 5-15C shows the excitatory response to a 300-msec tone pair of f_1 = 685 Hz and f_2 = 800 Hz presented at 78 dB SPL; neither primary by itself had any effect. The frequency difference of these two tones was 115 Hz, which was very close to the fiber's BEF and clearly produced a vigorous response. By adding a 200-msec third tone of 600 Hz in the middle of this tone pair, almost total suppression of the difference-tone response was achieved, as shown in the bottom histogram (Fig. 5-15D). Similar results in other low-frequency-sensitive fibers were consistent: difference-tone excitation exhibited the same two-tone suppression characteristics as single tone excitation.

To identify the locus of two-tone suppression, in one animal the VIIIth nerve was cut close to its point of entry into the brain and records were taken from single fibers in the peripheral stump. It was found, in agreement with the earlier study by Frishkopf and Goldstein (1963), that two-tone suppression was still present, thus verifying that it is not due to those few efferent fibers that innervate the amphibian papilla (Robbins, Bauknight, and Honrubia 1967). Two-tone suppression was checked for in the animal whose eardrums and columella were destroyed to determine the source of difference-tone generation. It was found that its characteristics remained the same when the animal was stimulated by bone conducted sound. These two tests indicate that two-tone suppression, like difference-tone generation, is a nonlinear property of the inner ear itself.

4 Conclusion

Our study of the response properties of eighth nerve fibers have verified two very prominent nonlinearities in the anuran's peripheral auditory system: intermodulation distortion $f_2 - f_1$, and two-tone suppression. Every auditory fiber in the American toad's eighth nerve with a best excitatory frequency below about 550 Hz could be excited by the quadratic difference tone $f_2 - f_1$. Energy at this difference frequency was very prominent; it was maximum when the amplitudes of the two primary tones were approximately equal. For most of these fibers the sound pressure level required of the two primaries was less than 20 dB above the single tone threshold at best excitatory frequency; for several units this difference was as small as 4 dB. These relative values for detection of $f_2 - f_1$ are comparable to those reported by Hall (1972a, 1972b) in his psychoacoustic studies and by Pfeiffer et al. (1974) in their electrophysiological recordings in the cat's eighth nerve. Such low relative amplitudes of the primary tones for generation of this difference tone indicate that the underlying nonlinearity is not due to overloading. This is further supported by our finding that amplitude of the distortion product $f_2 - f_1$ decreases rapidly with increasing frequency separation of the two primaries. A similar relationship was noted by Hall (1972a) in his psychoacoustic studies. The overall characteristics of the $f_2 - f_1$ difference tone led him to conclude that it is not due to a classical, high-level square-law nonlinearity in the mammalian ear. We are led to this same conclusion, namely generation of $f_2 - f_1$ in the anuran's peripheral auditory system involves an essential nonlinearity. Our measurements using laser scattered light spectroscopy have shown that the anuran's middle ear is very linear (Moffat and Capranica 1978). Furthermore $f_2 - f_1$ is present when the inner ear is stimulated directly by bone-conducted sounds. Thus, the nonlinearity must reside in the anuran's inner ear.

We have demonstrated that auditory fibers in the anuran's eighth nerve exhibit excitatory tuning curves to quadratic difference tones that are very similar to tuning curves obtained with single tones. We have also shown that the responses to this difference tone can be canceled by adding a third tone of appropriate amplitude and phase at the difference frequency. And we have found that difference tones and single excitatory tones exhibit the same two-tone suppression effects. This leads us to conclude that the difference tone behaves as if it were an actual tone presented to the ear. A similar conclusion has been reached with regard to the cubic difference tone $2f_1 - f_2$ in the mammalian peripheral auditory system (e.g., Smoorenberg et al. 1976). In mammals it has been proposed that this cubic difference tone propagates as a mechanical disturbance along the basilar membrane (e.g., Kim and Molnar 1976, Siegel et al. 1977, Buunen and Rhode 1978, Kim et al. 1978). But this raises a contradiction. Direct measurements of basilar membrane motion in the cochlea fail to reveal a mechanical component at the cubic difference frequency (Wilson and Johnstone 1972, Rhode 1977). So how does this disturbance propagate? It leads to the conclusion that perhaps the disturbance propagates in some other way (Wilson and Johnstone 1972, Duifhuis 1974, Goldstein 1977, Wilson 1977).

As we pointed out in the introduction, the amphibian papilla in the anuran's inner ear doesn't possess a basilar membrane. The hair cells rest on stationary structures so that the primary mechanical coupling to the ciliary processes occur via the overlying tectorial membrane. Yet this organ still exhibits a place mechanism for frequency analysis (Capranica and Moffat 1977, Lewis and Leverenz 1979). And it exhibits difference-tone excitation whose properties are very similar to those in the cochlea. We're led to the conclusion that the tectorial membrane likely plays an important role in the generation and propagation of intermodulation distortion.

Low-frequency-sensitive fibers from the amphibian papilla show two-tone suppression. The characteristics of this suppression in the toad's ear are similar to those previously reported in other anurans species (e.g., the leopard frog *Rana pipiens* by Liff et al. 1968 and Liff and Goldstein 1970). Suppression occurs on the high-frequency side of the tuning curve, is graded, and can be complete for many fibers when the relative intensity of the suppression tone is only 4 dB to 10 dB above the level of the excitatory tone. Such effective suppression indicates that it is not due to overloading of the auditory apparatus. Two-tone suppression has also been seen in the microphonic potential within the anuran's ear (Capranica et al. 1966, Paton 1971). It also persists in the inner ear after transection of the eighth nerve. These results, as in two-tone suppression in the mammalian cochlea, indicate that it arises within the inner ear due to a nonlinearity.

An interesting finding in our study is that two-tone suppression only occurs in those fibers that are sensitive to difference tones. This result implies that they are related. Such a relationship has been found in models incorporating nonlinear characteristics (e.g., Engebretson and Eldridge 1968, Hall 1974, 1977b). Our results provide experimental demonstration that a common nonlinearity may very well underlie these two phenomena.

At this point we might wonder whether the so-called "second filter" (Evans 1973) that is responsible for neural sharpening is also related to this same common nonlinearity (Robertson 1976, Hall 1977a, Schmiedt 1977). Our observations indicate that they arise from different mechanisms. On induction of hypercapnia by CO_2 intake into the lungs of *Bufo americanus,* we found that the tuning curves of fibers from the amphibian papilla rapidly broaden and thresholds become elevated (Capranica and Moffat 1976). On restoration of air into the lungs, the tuning curves quickly return to their original sharpness and thresholds. These results have been interpreted by us as evidence that fibers from the amphibian papilla derive part of their tuning from a metabolically sensitive second filter. But during hypercapnia, two-tone suppression and $f_2 - f_1$ excitation remain unaltered. These results have convinced us that the second filter mechanism is separate from the nonlinearity underlying two-tone suppression and difference tone excitation.

In conclusion, our study has revealed that two-tone suppression in the anuran's peripheral auditory system only occurs in those fibers that respond to difference tones. We have consistently found that these two phenomena are related. Since their characteristics are quite similar to those in higher vertebrates, we believe that such a relationship may also exist in the cochlea. Since the anuran's amphibian papilla lacks a basilar membrane, we suggest that its tectorial membrane supports the common essential nonlinearity underlying both of these phenomena. Such a conclusion also accounts for Holton and Weiss' (1978) finding that two-tone suppression in. the alligator lizard's ear is only present in those fibers that innervate hair cells supplied by a tectorial membrane. Since all the hair cells in the cochlea are supplied by a tectorial membrane and since its auditory fibers show such homogeneous response properties, the functional role played by different structures in the inner ear is not so easy to identify. This is one of the principal advantages of a comparative approach to studies of the vertebrate peripheral auditory system.

Acknowledgments. This research was supported by the U. S. Public Health Service (N.I.H. Research Grant NS-09244) and the National Science Foundation (Grant BMS 77-06803). We thank Mary Vella for typing our manuscript.

References

Abbas, P. J., Sachs, M. B.: Two-tone suppression in auditory nerve fibers: extension of a stimulus-response relationship. J. acoust. Soc. Amer. 59, 112-122 (1976).

Arthur, R. M., Pfeiffer, R. R., Suga, N.: Properties of two-tone inhibition in primary auditory neurons. J. Physiol. 212, 593-609 (1971).

Buunen, T., Rhode, W. S.: Responses of fibers in the cat's auditory nerve to the cubic difference tone. J. acoust. Soc. Amer. 64, 772-781 (1978).

Capranica, R. R.: Morphology and physiology of the auditory system. In: Frog Neurobiology. Llinás, R., Precht, W. (eds.). Berlin-Heidelberg: Springer, 1976, pp. 551-575.

Capranica, R. R., Flock, Å., Frishkopf, L. S.: Microphonic response from the inner ear of the bullfrog. J. acoust. Soc. Amer. 40, 1262 (1966).

Capranica, R. R., Moffat, A. J. M.: Excitation, inhibition and "disinhibition" in the inner ear of the toad (Bufo). J. acoust. Soc. Amer. 55, 480 (1974a).

Capranica, R. R., Moffat, A. J. M.: Evidence for mechanical origin of peripheral inhibition in the anuran inner ear. J. acoust. Soc. Amer. 55, S85 (1974b).

Capranica, R. R., Moffat, A. J. M.: Auditory responses from the saccule: further evidence for the mechanical origin of inhibition. J. acoust. Soc. Amer. 56, S12 (1974c).

Capranica, R. R., Moffat, A. J. M.: Selectivity of the peripheral auditory system of spadefoot toads (Scaphiopus couchi) for sounds of biological significance. J. comp. Physiol. 100, 231-249 (1975).

Capranica, R. R., Moffat, A. J. M.: Effects of anoxia on the response properties of auditory nerve fibers in the American toad. J. acoust. Soc. Amer. 59, S46 (1976).

Capranica, R. R., Moffat, A. J. M.: Place mechanism underlying frequency analysis in the toad's inner ear. J. acoust. Soc. Amer. 62, S85 (1977).

Dallos, P., Cheatham, M. A., Ferraro, J.: Cochlear mechanics, nonlinearities and cochlear potentials. J. acoust. Soc. Amer. 55, 597-605 (1974).

Duifhuis, H.: An alternative approach to the second filter. In: Facts and Models in Hearing. Zwicker, E., Terhardt, E. (eds.). Berlin-Heidelberg-New York: Springer, 1974, p. 103.

Engebretson, A. M., Eldredge, D. H.: Model for the nonlinear characteristics of cochlear potentials. J. acoust. Soc. Amer. 44, 548-554 (1968).

Evans, E. F., Wilson, J. P.: The frequency selectivity of the cochlea. In: Basic Mechanisms of Hearing. Møller, A. R. (ed.). New York: Academic Press, 1973, pp. 519-551.

Feng, A. S., Narins, P. M., Capranica, R. R.: Three populations of primary auditory fibers in the bullfrog (Rana catesbeiana): their peripheral origins and frequency sensitivities. J. comp. Physiol. 100, 221-229 (1975).

Frishkopf, L. S.: Excitation and inhibition of primary auditory neurons in the little brown bat. J. acoust. Soc. Amer. 36, 1016 (1964).

Frishkopf, L. S., Goldstein, M. H., Jr.: Responses to acoustic stimuli from single units in the eighth nerve of the bullfrog. J. acoust. Soc. Amer. 35, 1219-1228 (1963).

Geisler, C. D., van Bergeijk, W. A., Frishkopf, L. S.: The inner ear of the bullfrog. J. Morph. 114, 43-58 (1964).

Goldstein, J. L.: Auditory nonlinearity. J. acoust. Soc. Amer. 41, 676-689 (1967).

Goldstein, J. L.: Comment on Rhode's paper. In: Psychophysics and Physiology of Hearing. Evans, E. F., Wilson, J. P. (eds.). London-New York: Academic Press, 1977, p. 41.

Goldstein, J. L., Kiang, N. Y.-S.: Neural correlates of the aural combination tone $2f_1 - f_2$. Proc. I.E.E.E. 56, 981-992 (1968).

Greenwood, D. P., Merzenich, M. M., Roth, G.: Some preliminary observations on interrelations between two-tone suppression and combination tone driving in anteroventral cochlear nucleus of the cat. J. acoust. Soc. Amer. 59, 607-633 (1976).

Gross, N. B., Anderson, D. J.: Single unit responses recorded from the first order neuron of the pigeon auditory system. Brain Res. 101, 209-222 (1976).

Guinan, J. J., Peake, W. T.: Middle ear characteristics of anesthetized cats. J. acoust. Soc. Amer. 41, 1237-1261 (1967).

Hall, J. L.: Auditory distortion products $f_2 - f_1$ and $2f_1 - f_2$. J. acoust. Soc. Amer. 51, 1863-1871 (1972a).

Hall, J. L.: Monaural phase effect: cancellation and reinforcement of distortion products $f_2 - f_1$ and $2f_1 - f_2$. J. acoust. Soc. Amer. 51, 1872-1881 (1972b).

Hall, J. L.: Two-tone distortion products in a nonlinear model of the basilar membrane. J. acoust. Soc. Amer. 56, 1818-1828 (1974).

Hall, J. L.: Spatial differentiation as an auditory "second filter": assessment on a nonlinear model of the basilar membrane. J. acoust. Soc. Amer. 61, 520-524 (1977a).

Hall, J. L.: Two-tone suppression in a nonlinear model of the basilar membrane. J. acoust. Soc. Amer. 61, 802-810 (1977b).

Holton, T., Weiss, T. F.: Two-tone rate suppression in lizard cochlear nerve fibers, relation to receptor organ morphology. Brain Res. 159, 219-222 (1978).

Javel, E., Geisler, C. D., Ravindran, A.: Two-tone suppression in auditory nerve of the cat: rate-intensity and temporal analyses. J. acoust. Soc. Amer. 63, 1093-1104 (1978).

Johnstone, B. M., Taylor, K.: Mechanical aspects of cochlear function. In: Frequency Analysis and Periodicity Detection in Hearing. Plomp, R., Smoorenburg, G. F. (eds.). Leiden: Sijthoff, 1970, pp. 81-90.

Johnstone, B. M., Yates, G. K.: Basilar membrane tuning curves in the guinea pig. J. acoust. Soc. Amer. 55, 584-587 (1974).

Kiang, N. Y.-S., Watanabe, T., Thomas, E. C., Clark, L. F.: Discharge Patterns of Single Fibers in the Cat's Auditory Nerve. Cambridge, Mass.: M.I.T. Press, 1965.

Kim, D. O., Molnar, C. E.: Spatiotemporal patterns of primary and distortion components of cochlear responses to phase-locked two tones. Program for Society for Neuroscience Sixth Annual Meeting, Toronto, Canada (1976).

Kim, D. O., Molnar, C. E., Pfeiffer, R. R.: A system of nonlinear differential equations modeling basilar membrane motion. J. acoust. Soc. Amer. 54, 1517-1529 (1973).

Kim, D. O., Siegel, J. H., Molnar, E. C.: Cochlear nonlinear phenomena in two-tone responses. Proceedings of the Symposium on Models of the Auditory System and Related Signal Processing Techniques, Münster, Germany, September 1978.

Kohllöffel, L. U. E.: A study of basilar membrane vibrations. II. The vibratory amplitude and phase pattern along the basilar membrane (post-mortem). Acustica 27, 68-81 (1972).

Legouix, J.-P., Remond, M.-C., Greenbaum, H. B.: Interference and two-tone inhibition. J. acoust. Soc. Amer. 53, 409-419 (1973).

Lewis, E. R., Leverenz, E. L.: Direct evidence for an auditory place mechanism in the frog amphibian papilla. Soc. Neurosci. Abstr. 5, 25 (1979).

Liff, H., Goldstein, M. H., Jr., Frishkopf, L. S., Geisler, C. D.: Best inhibitory frequencies of complex units in the eighth nerve of the bullfrog. J. acoust. Soc. Amer. 44, 635-636 (1968).

Liff, H., Goldstein, M. H., Jr.: Peripheral inhibition in auditory nerve fibers in the frog. J. acoust. Soc. Amer. 47, 1538-1547 (1970).

Moe, C. R.: An experimental study of subjective tones produced within the human ear. J. acoust. Soc. Amer. 14, 159-166 (1942).

Moffat, A. J. M., Capranica, R. R.: Effects of temperature on the response properties of auditory nerve fibers in the American toad. J. acoust. Soc. Amer. 60, S80 (1976).

Moffat, A. J. M., Capranica, R. R.: Middle ear sensitivity in anurans and reptiles measured by light scattering spectroscopy. J. comp. Physiol. 127, 97-107 (1978).

Moffat, A. J. M., Capranica, R. R.: Phase cancellation of auditory nerve fiber responses to combination tones, $f_2 - f_1$. J. acoust. Soc. Amer. 65, S82 (1979).

Nomoto, M., Suga, N., Katsuki, Y.: Discharge pattern and inhibition of primary auditory nerve fibers in the monkey. J. Neurophysiol. 27, 768-787 (1964).

Paton, J.: Microphonic potentials in the inner ear of the bullfrog. M. S. Thesis, Cornell University, Ithaca, New York, 1971.

Pfeiffer, R. R., Molnar, C. E.: Characteristics of the $(f_2 - f_1)$ component in response patterns of single cochlear nerve fibers. J. acoust. Soc. Amer. 56, S221 (1974).

Pfeiffer, R. R., Molnar, C. E., Cox, J. R., Jr.: The representation of tones and combination tones in spike discharge patterns of single cochlear nerve fibers. In: Facts and Models in Hearing. Zwicker, E., Terhardt, E. (eds.). New York-Heidelberg-Berlin: Springer, 1974, pp. 323-331.

Plomp, R.: Detectability threshold for combination tones. J. acoust. Soc. Amer. 37, 1110-1123 (1965).

Rhode, W. S.: Observations of the vibration of the basilar membrane in squirrel monkey using the Mössbauer technique. J. acoust. Soc. Amer. 49, 1218-1231 (1971).

Rhode, W. S.: Some observations on two-tone interaction measured with the Mössbauer effect. In: Psychophysics and Physiology of Hearing. Evans, E. F., Wilson, J. P. (eds.). London-New York: Academic Press, 1977, pp. 27-41.

Rhode, W. S.: Some observations on cochlear mechanics. J. acoust. Soc. Amer. 64, 158-176 (1978).

Robbins, R. G., Bauknight, R. S., Honrubia, V.: Anatomical distribution of efferent fibers in the VIIIth cranial nerve of the bullfrog, *Rana catesbeiana*. Acta otolaryng. (Stockh.) 64, 436-448 (1967).

Robertson, D.: Correspondence between sharp tuning and two-tone inhibition in primary auditory neurons. Nature 259, 477-478 (1976).

Sachs, M. B., Kiang, N. Y.-S.: Two-tone inhibition in auditory nerve fibers. J. acoust. Soc. Amer. 43, 1120-1128 (1968).

Sachs, M. B., Young, E. D., Lewis, R. H.: Discharge patterns of single fibers in the pigeon auditory nerve. Brain Res. 70, 431-447 (1974).

Schmiedt, R. A.: Single and two-tone effects in normal and abnormal cochleas: a study of cochlear microphonics and auditory-nerve units. Special Report, Institute for Sensory Research, Syracuse University, December, 1977.

Siegel, J. H., Kim, D. O., Molnar, C. E.: Cochlear distortion products: effects of altering the organ of Corti. J. acoust. Soc. Amer. 61, S2 (1977).

Smoorenburg, G. F., Gibson, M. M., Kitzes, L. M., Rose, J. E., Hind, J. E.: Correlates of combination tones observed in the response of neurons in the anteroventral cochlear nucleus of the cat. J. acoust. Soc. Amer. 59, 945-962 (1976).

von Békésy, G.: Experiments in Hearing. New York: McGraw-Hill, 1960.

Wever, E. G.: The ear and hearing in the frog, *Rana pipiens*. J. Morph. 141, 461-478 (1973).

Wever, E. G., Bray, C. W., Lawrence, M.: The interference of tones in the cochlea. J. acoust. Soc. Amer. 12, 268-280 (1940).

Wilson, J. P.: Towards a model for cochlear frequency analysis. In: Psychophysics and Physiology of Hearing. Evans, E. F., Wilson, J. P. (eds.). London-New York: Academic Press, 1977, pp. 115-124.

Wilson, J. P., Johnstone, J. R.: Capacitive probe measures of basilar membrane vibration. In: Symposium on Hearing Theory. Eindhoven, The Netherlands: Institute Perception Research, 1972, pp. 172-181.
Wilson, J. P., Johnstone, J. R.: Basilar membrane and middle-ear vibration in guinea pig measured by capacitive probe. J. acoust. Soc. Amer. 57, 705-723 (1975).
Zwicker, E.: Der ungewöhnliche Amplitudengang der Nichtlinearen Verzerrungen des Ohres. Acustica 5, 67-74 (1955).

PART THREE

Reptiles

The variation in the structure of the inner ear among reptiles is extraordinary. Miller (Chapter 6) illustrates the various dimensions of this variation and shows clear correlations with phylogenetic relationships among the different groups. These types of variation create valuable opportunities to unravel some of the basic and persistent questions of vertebrate auditory function. While we are still waiting hopefully for a successful psychophysical approach to these questions, Turner (Chapter 7) reviews the recent growth in studies of peripheral neurophysiology and makes clear the great promise that study of the reptilian auditory systems holds for general understanding of structure-function relations in vertebrate audition.

Chapter 6

The Reptilian Cochlear Duct

MALCOLM R. MILLER*

1 Introduction: Historical Overview

Before the 1950s only two major studies of the reptilian cochlea had been reported. They were the now classical works of Retzius (1884) and de Burlet (1934). In these studies some important gross and some microscopic features of a few representative species were described and figured.

In the 1950s the study of reptilian cochleae gained considerable momentum and numerous contributions have been made in the past quarter century. Shute and Bellairs (1953) were probably the first to recognize the phylogenetic significance of the differences and similarities of lizard cochlear duct structure. Schmidt (1964) later confirmed and extended these observations. Basic anatomical studies of the cochlear duct and the surrounding periotic spaces were reported by Baird (1960) and Hamilton (1960, 1964). In 1966a Miller reported the results of detailed studies of the cochleae of 205 species of lizards representing 131 genera and 18 families, and in 1968 he reported detailed studies of the cochleae of 189 species of snakes representing 160 genera and 12 families (all known living families). In 1970 Baird synthesized the then known information concerning the reptilian ear and later published further summaries (1974a, 1974b).

Probably the most extensive study of the reptilian ear is that by Wever (1978), which has taken place over the past 25 years. In these studies 247 species of reptiles were investigated, including 186 species of lizards belonging to 16 families, 19 species of snakes, 14 species of amphisbaenids, 24 species of turtles, 3 species of crocodilians, and the single living species of rhynchocephalians.

Wever's studies were both anatomical and physiological. The anatomical studies were, in large part, carried out by the preparation and study of serial celloidin sections, and the physiological correlates, by use of the cochlear potential method. The results of Wever's vast effort have appeared in numerous publications since the early 1950s,

*Department of Anatomy, University of California, San Francisco, California 94143.

and have been presented completely and beautifully in a book published late in 1978 (Wever 1978).

In the 1970s the detailed surface structure of the papilla basilaris of most types of reptiles has been studied by scanning electron microscopy (Baird 1974a, 1974b, Miller 1978a, 1978b, Bagger-Sjöbäck and Wersäll 1973, von Düring 1974, Karduck and Richter 1974, Leake 1977). Some, but not a great deal, of transmission electron microscopic detail has been reported by Mulroy (1968, 1974), Baird (1970, 1974b), and Bagger-Sjöbäck (1976). Concomitant with the anatomical study of the reptilian ear have been an increasing number of physiological investigations. The functional aspects of the reptilian ear and cochlea is the subject of another chapter in this volume (see Turner, Chapter 7).

As a result of extensive studies over the past 25 to 30 years, probably more is known concerning the comparative morphology and physiology of the reptilian ear than that of any other vertebrate class. This statement applies to the knowledge of the gross, histological, and ultrastructural details of a wide range of reptiles. Studies of mammalian species, on the other hand, while not as comprehensive from the standpoint of gross and histological comparative anatomy are more intensive in relation to the depth of study of a few laboratory animals such as the cat, monkey, and guinea pig.

In view of the relatively large amount of recently published information concerning the reptilian ear, the present account will not attempt to cover external and middle ear anatomy (see Wever 1978), but rather, will outline the major structural features of the cochlear duct. Particular attention will be given to structural variations of the papillae basilares as these are closely related to functional capacity and phylogenetic relationships.

2 Origin of the Reptilian Cochlear Duct

The cochlear duct of reptiles may either have arisen de novo in this class of vertebrates or it may have been derived from an outpocketing found on the posterior saccular wall of amphibians in which both a macula lagenae and a papilla basilaris are found (Lombard, Chapter 4). The major differences between the amphibian and reptilian papillae basilares are the lack of a basilar membrane and the lack of direct exposure to perilymphatic sound pressure changes in the amphibian papilla basilaris. Present evidence is insufficient to decide whether the reptilian cochlear duct is a new or a derived structure.

3 Anatomy of the Reptilian Cochlear Duct

The general anatomical features of the reptilian cochlear duct are described and illustrated in Figs. 6-1 to 6-5. Although the lizard cochlear duct is used to describe general reptilian cochlear duct structure, other more primitive or more specialized ducts are easily understood from study of the lizard cochlear duct.

The cochlear duct (Figs. 6-1 to 6-5) is the most inferior portion of the membranous labyrinth and is connected with the sacculus by the sacculo-cochlear duct. The latter

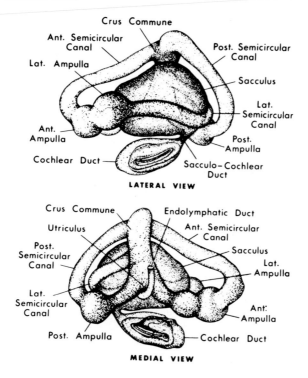

Figure 6-1. Lateral and medial views[1] of the left membranous labyrinth of a lizard, *Xantusia vigilis*. (From Miller 1966a).

usually arises from the posteroinferior aspect of the sacculus, but may be more infero-medial in location in some families (Hamilton 1964).

The cochlear duct in reptiles as exemplified in *Crotaphytus wislizeni* (Fig. 6-2) characteristically contains two sensory areas. The macula lagenae usually occupies the anterior and anteroinferior portion of the duct but frequently extends both antero-laterally and anteromedially. The macula lagenae is always covered by an otolithic membrane.

The papilla basilaris is a sensory area usually associated with an overlying tectorial membrane. (Outline sketches showing the lateral and medial aspects of the cochlear duct of *Crotaphytus* are presented in Fig. 6-2. Figure 6-3 is a cross section of the cochlear duct of *Gekko gecko*.)

The vestibular membrane makes up the lateral wall of the cochlear duct. The medial

[1]*Note on terminology*: Since these earlier papers (Miller 1966a, 1973a) were published, I have been using a different terminology to describe the orientation of the cochlear duct (see Fig. 6-4). The dorsal or posterior and superior end of the cochlear duct is the *basal* end, and the ventral or anteroinferior end is the *apical* extremity. The side of the papilla basilaris toward the neural limbus is the neural side, while the opposite side or direction, originally termed triangular, is now termed abneural.

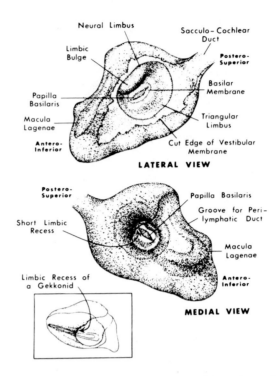

Figure 6-2. Lateral and medial views of the left cochlear duct of the lizard, *Crotaphytus wislizeni.* (From Miller 1966a).

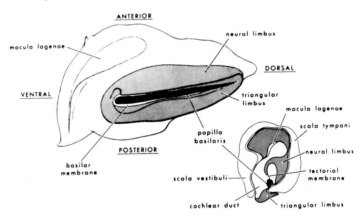

Figure 6-3. Upper left: Lateral view of the left cochlear duct of the lizard, *Gekko gecko.* Lower right: Cross sectional view of cochlear duct. (From Miller 1973a).

wall is more complex and in good part is formed of a modified connective tissue containing abundant intercellular substance that imparts a flexible quality to this portion of the duct. The supporting tissue is thickened where it surrounds the basilar membrane and is termed the limbus. The anterosuperior portion of the limbus is thicker and larger than the posteroinferior limb and is variously modified and sculptured on

KINOCILIAL ORIENTATION PATTERNS
IN REPTILIAN PAPILLAE BASILARES

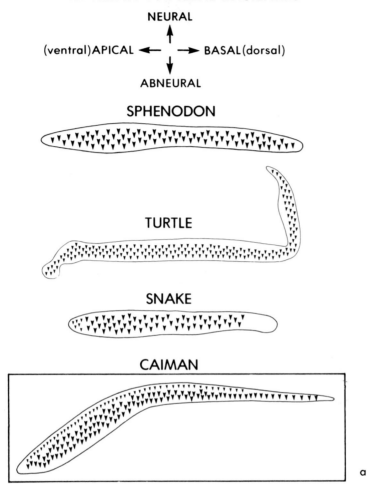

Figure 6-4. (a) Graphic representation of the general shape of the auditory papilla and the orientation of the kinocilia in *Sphenodon*, a turtle (*Kinosternon sp.*), a snake (*Epicrates cenchris*), and a caiman (*Caiman crocodilus*). The above papillae are not represented strictly to scale as the fixed and dried lengths are approximately the following: *Sphenodon*, 1.0 mm (estimated); *Kinosternon sp.*, 1.5 mm; *Epicrates cenchris*, 0.65 mm; and *Caiman crocodilus*, 4.0 mm. The caiman papilla is outlined in a box because of its greater size.

both medial and lateral surfaces. Since the auditory nerve is closely apposed to the medial side of the anterosuperior limbus, this portion of the limbus is designated the neural limbus. The lateral surface of the neural limbus may be merely a thin sculptured plate or it may be thickened or give rise to a liplike projection. A tectorial membrane usually arises from the lateral aspect of the neural limbus and overlies the cellular surface of the papilla basilaris.

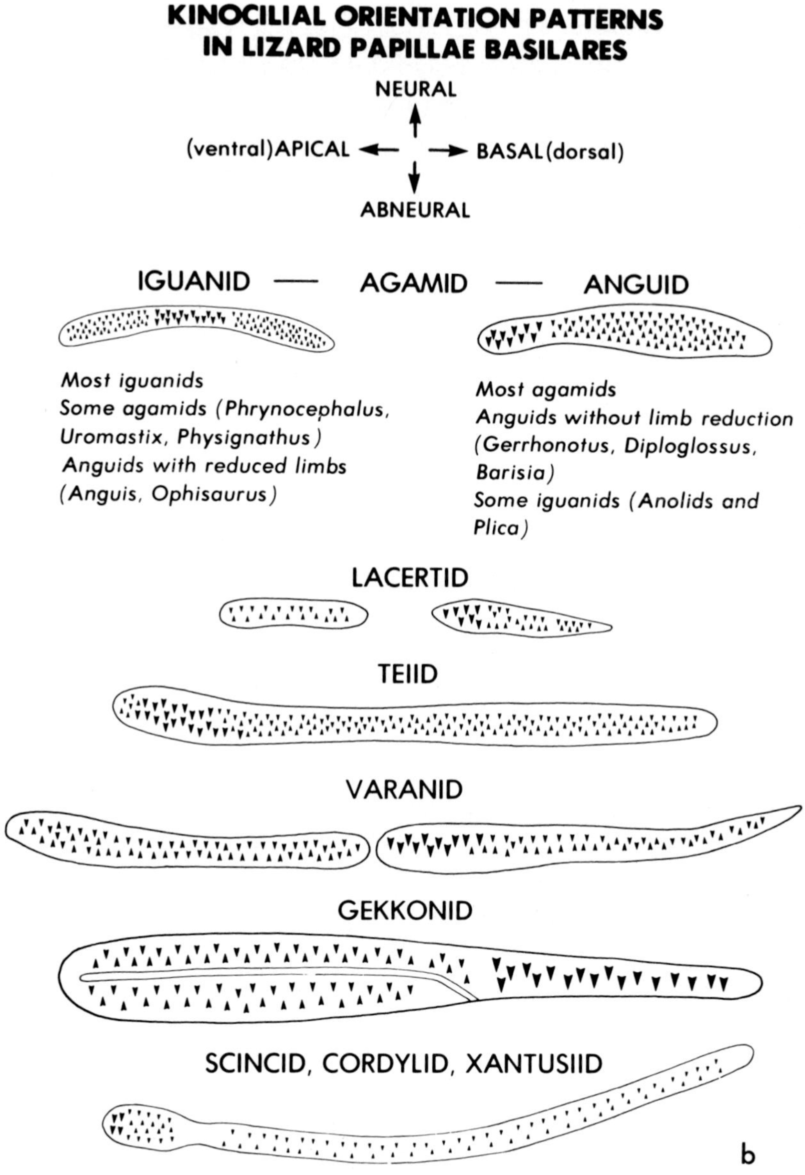

KINOCILIAL ORIENTATION PATTERNS
IN LIZARD PAPILLAE BASILARES

NEURAL

(ventral)APICAL ◄━━ ━━► BASAL(dorsal)

ABNEURAL

IGUANID —— AGAMID —— ANGUID

Most iguanids
Some agamids (*Phrynocephalus,*
Uromastix, Physignathus)
Anguids with reduced limbs
(*Anguis, Ophisaurus*)

Most agamids
Anguids without limb reduction
(*Gerrhonotus, Diploglossus,*
Barisia)
Some iguanids (*Anolids and*
Plica)

LACERTID

TEIID

VARANID

GEKKONID

SCINCID, CORDYLID, XANTUSIID

b

Figure 6-4. (b) Graphic representation of the general shape and approximate size of the auditory papillae and the orientation of the kinocilia in a variety of lizard families. In most lizard families, the length of the papilla varies with species size, and larger species have larger and longer papillae (Miller 1966a). The papillae as represented here are only approximately to scale. Average iguanid, agamid, anguid, and lacertid papillae are 0.3 to 0.5 mm long; teiids, ca. 0.8 to 0.9 mm long; scincids, ca. 1.1 to 1.2 mm long; gekkonids, ca. 1.2 mm; and varanids, ca. 1.6 to 1.7 mm long. (Miller 1966a).

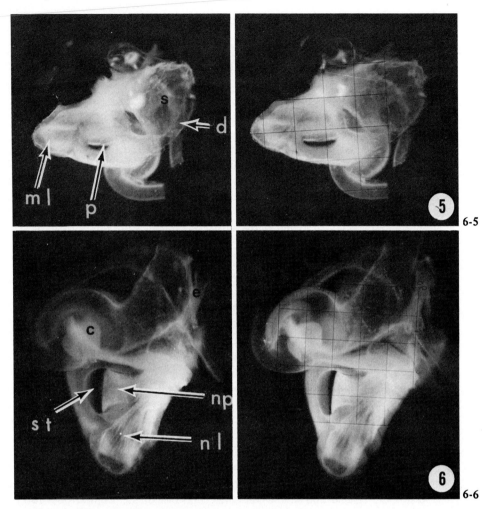

Figure 6-5. Stereophotograph[2] of a lateral view of the left cochlear duct of an iguanid lizard, *Iguana iguana*. The basal end of the cochlear duct is attached by the sacculo-cochlear duct (d) to the posterior and inferior aspect of the sacculus (s). The arrow tip (p) is on the midportion of the papilla basilaris, which in this species can be seen even at this low magnification to consist of shorter hair cells in the central region. In the apical region of the duct is the macula lagenae (ml). The ampulla of the posterior semicircular canal lies behind the cochlear duct. (X 18).

Figure 6-6. Stereophotographs of a medial view of the cochlear duct of *Iguana iguana*. Abbreviations: c-crista of posterior ampulla; e-endolymphatic duct; np-nerve to papilla basilaris; st-scala tympani overlying basilar membrane; nl-nerve to macula lagenae. (X 18).

[2] *Note on viewing stereophotographs*: These stereophotographic pairs are mounted for direct, not for cross-eyed, viewing. For fusion of the stereopairs, the viewer should have good and equal illumination of the photographs. A piece of stiff paper or cardboard placed vertically between photographic pairs is of help. It is necessary to try to focus one's eyes at a distance when first looking at the photopairs, and then they will rapidly form a stereoscopic image.

The medial side of the limbus is differently sculptured, in large part, to accommo-
date the perilymphatic duct and sac that conduct sound-pressure waves through the
perilymphatic fluid from the perilymphatic cistern to the medial aspect of the basilar
membrane. In some types of cochlear ducts, the perilymphatic sac is not enclosed by
any portion of the limbus along the greater part of the medial aspects of the basilar
membrane (Fig. 6-2 [note insert]). In other types of ducts, a considerable extent of
the basilar membrane on the medial side may be enclosed by connections between the
two portions of the limbus (Fig. 6-2 insert).

The posteroinferior part of the limbus is not as modified as the neural limb and is
referred to as the abneural part of the limbus (Fig. 6-4). It should be understood that
the limbus is one complete structure varying in shape from a saucer-like plate to an
elongated ovoid and has in its central portion an opening or hiatus across which is
stretched the basilar membrane.

Supported on the lateral aspect of the basilar membrane is the papilla basilaris. The
limbic hiatus and papilla basilaris vary in size and shape from small circular or ovoid to
large elongate structures. The papilla basilaris may be a simple continuous strip of cells,
or it may be divided by a limbic connective tissue bar. The papilla may also be evenly
contoured, or one end may be widened producing a fusiform-shaped structure.

That portion of the cochlear duct housing the limbus, basilar membrane, and papilla basilaris was termed the *pars basilaris* by Retzius (1884). Since the basilar membrane
is only a limited portion of this part of the cochlear duct, this author refers to it as the
limbic rather than basilar portion of the duct.

The anterosuperior part of the cochlear duct is supported along its anterior edge by
the same type of modified connective tissue that makes up the limbus. This heavy supporting tissue may also extend onto the anterolateral and anteromedial portions of the
duct. On the inner luminal surface of this portion of the cochlear duct is a strip of sensory epithelium, the macula lagenae, that is covered by an otolithic membrane. The
size and shape of the macula lagenae differs from one lizard family to another. The
portion of the cochlear duct housing the macula lagenae may be designated the lagena
or lagenar portion of the duct. Reference to Figs. 6-1 and 6-3 will make it clear that in
most lizards the lagenar and limbic-containing portions of the cochlear duct form a
united structure. Thus, while the location of the papilla basilaris and the macula lagenae
is definite, there is not a clear-cut distinction between the lagenar and limbic portions
of the duct.

4 Classification of Living Reptiles

Living reptiles constitute a class of vertebrate animals made up of four separate orders.
The following order of listing corresponds to my conception of the apparent state of
development of the cochlear duct and papilla basilaris. Thus, the order with the most
primitive cochlear duct and papilla is listed first and the most specialized, last.

Order Rhynchocephalia   (The Tuatara, *Sphenodon*)
Order Testudines   (turtles, tortoises, terrapins)

Order Squamata
    Suborder Serpentes   (snakes)
    Suborder Amphisbaenia   (Amphisbaenids)
    Suborder Lacertilia   (lizards)
Order Crocodylia   (crocodiles, alligators, gavials)

5 Order Rhynchocephalia

The Tuatara, *Sphenodon punctatus*, is the sole living species of an order of reptiles extending back some 155 to 190 million years. Along with lizards and snakes, *Sphenodon* probably stems from eosuchian reptiles that, while not the most primitive of reptilian stock, is not far removed from the stem reptiles. Since *Sphenodon* is of such ancient lineage, its structure assumes great importance.

A distinct cochlear duct is present in *Sphenodon* as well as in all other reptiles. This tube-like saccular appendage houses two separate and distinct regions of specialized neuroepithelium, namely, the macula lagenae and the papilla basilaris. The cochlear duct of *Sphenodon* (Figs. 6-6 to 6-9) arises by a relatively broad sacculo-cochlear duct (ductus reuniens) from the superior medial aspect of the sacculus. By contrast, the sacculo-cochlear duct of squamates (lizards and snakes) and crocodilians arises from the posterior inferior aspect of the sacculus. The origin of the sacculo-cochlear duct in turtles is posterior superior in location and is thus intermediate in its position of origin between *Sphenodon* and the squamates.

Because of the superior (high) origin of the cochlear duct in *Sphenodon*, most of the duct is medial to the sacculus. In other reptiles, the cochlear duct is essentially inferior to the sacculus.

The cochlear duct of *Sphenodon* is approximately 3.5 mm to 4.0 mm long. The apical (ventral or inferior) third of the duct consists of an almost spherical sacculation that contains the macula lagenae. The more dorsal or basal part of the duct is separated from the lagenar portion by a marked constriction and houses the papilla basilaris. Partial separation of the lagenar and basilar portions of the cochlear duct is characteristic of *Sphenodon*, turtles, and snakes. In lizards, the constriction between lagenar and basilar portions of the duct is much less pronounced and the duct has the appearance of a single united structure.

The basal (superior or dorsal) portion of the cochlear duct contains on its medial wall, the papilla basilaris, which rests on a fairly large basilar membrane. The medial wall of the duct in the region surrounding the basilar membrane is thickened and is termed the limbus.

In *Sphenodon*, the limbus is not much thickened and shows no particular modifications such as are seen in most lizards and in amphisbaenids. The medial aspect of the cochlear duct shows a moderate inpocketing over the basilar membrane. The edges of the limbus do not enclose this space (a part of the scala tympani), however.

Wever (1978) describes the basilar membrane as large and ovoid and approximately 725 μm long and 350 μm wide at its broadest point. The basement membrane is supported by mildly thickened periotic tissue, a thin "fundus" in Wever's terminology, or

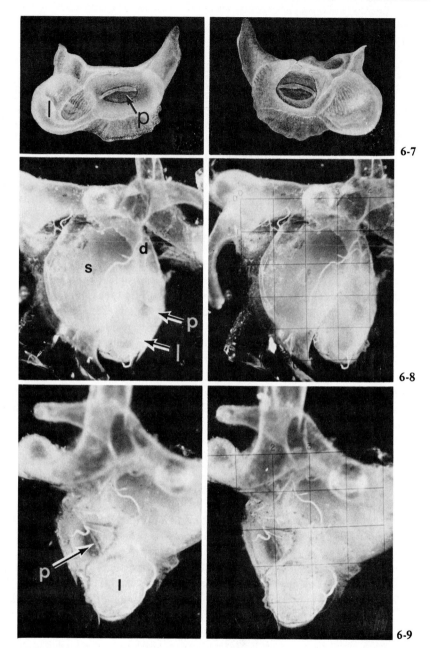

6-7

6-8

6-9

a "limbic bar" in this author's. Such support of the basement membrane is better developed in lizards but is not present in snakes or turtles.

According to Wever (1978), a tectorial membrane arises from the limbus, and the thickened distal portion of the membrane (tectorial plate according to Wever's terminology) covers nearly all the hair cells. The papilla basilaris is approximately 725 μm long and consists of four to eight longitudinal rows of hair cells. The total number of hair cells is approximately 225. In two specimens of *Sphenodon* studied, the papilla basilaris measured 1.2 mm in length (Miller 1966a). Hair cell orientation is, in large part, abneural (posterior in Wever's terminology), but one hair cell was seen to have an opposite (neural or anterior) orientation. Also of note is the presence of a few hair cells resting directly on the limbic tissue. The latter is a regular and distinctive feature of the turtle papilla basilaris.

Neither scanning nor transmission electron microscopic studies of the *Sphenodon* papilla have been described, but in all probability its ultrastructure is similar to that of turtles. It is likely that the papillae basilares of *Sphenodon* and turtles are fairly primitive (cf. Section 6) and resemble the ancestral reptilian papilla more than that of any other living reptile.

6 Order Testudines

The turtles are representatives of an ancient group of reptiles probably derived from stem reptilian stock, the cotylosaurs.

The cochlear duct (Figs. 6-10 to 6-13) arises postero-superiorly from the sacculus and lies mainly posterior to the sacculus. The lagenar and basilar parts of the cochlear duct are not as much separated one from the other as in *Sphenodon* and snakes but are less united than in the cochlear duct of lizards (Miller 1966a).

A distinctive feature of the turtle duct is the presence of numerous strands (periotic reticulum) connecting the duct to the surrounding walls of the cochlear recess. The

Figure 6-7. (A) Drawing of a lateral view of the left cochlear duct of *Sphenodon punctatus*. The apical end of the duct is to the reader's left and consists of a saclike compartment, the pars lagenae (l), which houses the macula lagenae. On the right side is the basilar portion of the dict. The papilla basilaris (p) is seen resting on the thin basilar membrane. The edges of the papilla are irregular because this drawing is of a poorly preserved museum specimen. Note that the medial face of the ring of tissue (limbus) surrounding the basilar membrane bears no special elevation. (B) Drawing of a medial view of the left cochlear duct of *Sphenodon*. The basal end of the duct is to the left. (About X 10) (From Miller 1966a).

Figure 6-8. Stereophotographs of a lateral view of the left cochlear duct of *Sphenodon*. The broad sacculo-cochlear duct (d) takes origin from the superior and mid-posterior aspect of the sacculus (s). The pars basilaris (p) and pars lagenae (l) are not especially clear because of poor specimen preservation. (About X 10).

Figure 6-9. Stereophotographs of a medial view of the left cochlear duct of *Sphenodon*. Abbreviations: p - papilla basilaris; l - lagenar portion of the cochlear duct. Apical end of the duct is directed toward bottom of the page. (About X 10).

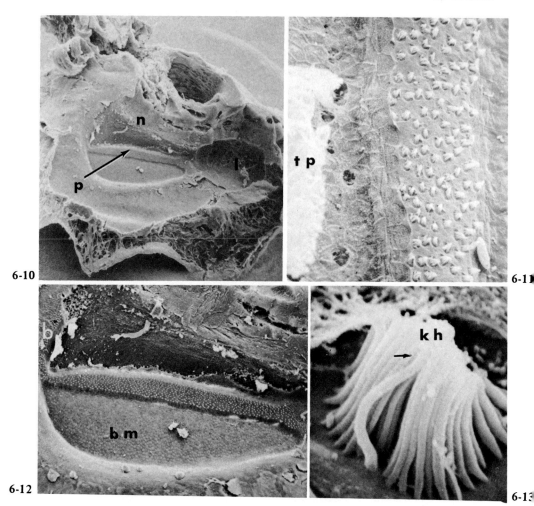

Figure 6-10. Entire right cochlear duct of the Mud Turtle, *Kinosternon sp.* Abbreviations: l-pars lagenae; p-papilla basilaris; n-neural limbus. (X 37) (From Miller 1978a).

Figure 6-11. Higher power view of a portion of the papilla basilaris of *Kinosternon sp.* showing widely spaced hair cells. A portion of the heavy tectorial plate (tp) has been displaced to the left. All hair cells are abneurally oriented (to the right). (X 580).

Figure 6-12. Papilla basilaris of *Kinosternon sp.* Note hair cells extending onto basal end (b) of the limbus as well as onto the apical end. Note that most of the papilla rests on the basilar membrane (bm). (X 63).

Figure 6-13. *Kinosternon sp.* View of the cilia of one hair cell. Abbreviations: kh-kinocilial bulb or head. Arrow indicates points at which the kinocilial shaft is attached to the tallest stereocilia. (X 14 000).

limbus is not particularly thick and the supporting periotic connective tissue is confluent with the contiguous lagenar and saccular walls. The medial limbic wall has no particular thickening nor sculpturing. The basilar membrane is fairly large, ovoid in shape
and thin, and is not thickened under the papilla as in lizards. The papilla basilaris is
unusual in that a considerable number of hair cells extend onto the limbic walls both
apically and basally (Fig. 6-11). The greater part of the papilla basilaris rests on the
free surface of the basilar membrane close to the edge of the neural limbus. A relatively large expanse of non-papilla-covered basilar membrane extends in the abneural direction (Fig. 6-11).

A limbus-attached tectorial membrane terminates in a thick plate-like mass that
overlies the hair cells that are supported by the basilar membrane. Those hair cells
resting on the limbic tissue are covered by a thin layer of tectorial tissue that is in continuity with the tectorial plate but has no direct limbic attachment. The tectorial tissue is attached to the enlarged heads of the kinocilia that in turn are attached to the
tallest stereocilia. This is the usual mode of attachment of the tectorial membrane to
the hair cells in the Reptilia (Figs. 6-39, 6-42).

In addition to an overlying tectorial membrane, turtles and many other reptilian
species possess a mat-like network of fibrillar material, similar to the tectorial membrane, that overlies the papillar supporting cells and serves to delimit the groups of
cilia arising from each hair cell. Delicate fibrillar strands connect this underlying papillar mat with the overlying tectorial membrane proper.

Approximately 30 species of turtles representing most families have been studied
by either Wever (1978) or Miller (1978a). The length of the papilla basilaris ranges
from approximately 0.5 to 1.5 mm and the number of hair cells, from about 500 to
1 500; and exceptionally, in a large marine turtle, *Chelonia mydas,* the Green Turtle,
it ranges to about 2 500.

The hair cells of the turtle papilla basilaris are widely spaced because of the relatively large surface area and the number of supporting cells (Fig. 6-12). This author has
calculated the density of the hair cells on the turtle papilla basilaris and found it to be
150/10 000 μm^2 in *Kinosternon* (Mud Turtle), 245 in *Chelydra serpentina* (Snapping
Turtle), and 247 in *Pseudemys scripta* (Red-eared Turtle). By contrast, in lizards,
where supporting cells are fewer and the surfaces are smaller, hair cell density was calculated to be 320/10 000 μm^2 in *Gekko gecko* (Tokay Gecko), 320 in *Varanus bengalensis* (Monitor), and 400 in *Mabuya carinata* (Skink).

Every hair cell has approximately 75 to 90 stereocilia that are 6 to 8 μm in length
and an orienting kinocilium attached to the five tallest stereocilia. Almost all kinocilia
are abneurally oriented (Fig. 6-4), although an occasional hair cell may be oriented at
an odd angle to the usual abneural direction.

The turtle papilla basilaris is probably not much different from that possessed by
reptilian stem stock, and some of its features may be considered primitive, namely,
low hair cell density, unidirectional kinocilial orientation, synapses with both afferent
and efferent nerve terminals, lack of special cytological modifications of the hair cell,
and a large number of stereocilia (Baird 1974a, Wersäll and Flock 1965). Although a
considerable variety of turtles have been studied, no significant variations in cochlear
duct or papillar structures have been noted.

7 Order Squamata

7.1 Suborder Serpentes

The sacculo-cochlear duct originates approximately mid-dorsally from the posterior saccular wall. A more dorsal origin of the sacculo-cochlear duct is found in *Sphenodon*, a few lizards, and most snakes, and is a primitive feature (Hamilton 1964).

The cochlear duct of snakes (Figs. 6-14 to 6-17) is a sac-like structure composed of two distinct portions, the limbic (or basilar) and the lagenar. The limbic region of the duct is a rounded oval to elongated oval tube, while the medial wall is composed of a ring of moderately dense periotic connective tissue. The basement membrane is stretched across the limbic hiatus and supports an elongated basilar membrane. There is no condensation of periotic connective tissue forming a papillar bar (fundus) as is found in lizards. There are no elevations or specializations of the lateral face of the neural limbus.

All the cochlear duct structural features are similar to those found in *Sphenodon* and in the turtles, and probably represent a primitive condition.

As will be seen, the snake papilla basilaris, while primitive in some characteristics, is somewhat advanced in others. The papilla basilaris of snakes varies greatly in size and length, which is primarily correlated more with habitat or life mode than with taxonomic relationships (Figs. 6-14 to 6-17). In general, the larger and longer papillae are found in burrowing species (Figs. 6-14, 6-15), small to moderate-sized papillae in terrestrial species (Figs. 6-16, 6-17), and small papillae in arboreal species (Miller 1968).

The snake papilla basilaris is covered by a tectorial membrane that may be a highly fenestrated or plate-like structure (Figs. 6-18, 6-19). Often the terminal ends of the papilla are not covered by tectorial membranes (Fig. 6-18).

As in turtles, the papillar hair cells are arranged in irregular rows. A study of approximately 200 species of snakes (Miller 1968) showed the papilla to range in length from 0.1 mm to 1.5 mm with the greater number of species having a papilla length of 0.3 mm to 0.8 mm. The number of hair cells varies from approximately 50 to 1 500 (Wever 1978).

The large majority of hair cells are abneurally oriented and only an occasional hair cell is oppositely polarized. As in the great majority of reptilian species, the tectorial membrane is attached to an enlarged kinocilial head that, in turn, is attached to the tallest stereocilia.

As seen in both transmission (Baird 1970) and scanning electron microscopy (Miller 1978a), the hair cells of the snake papilla are less densely packed than in lizards. Baird's (1970) studies of several snake species showed that the papillar supporting cells form a larger part of the papillar surface than in lizards. However, in some snake species, the number and distribution of cytoskeletal microtubules in the supporting cells approaches that found in lizards, and is greater than that found in turtles and amphisbaenids. Thus, in this regard the snake papilla is intermediate between the more primitive condition found in turtles and amphisbaenids and the more advanced state of development found in lizards.

As far as innervation of the snake papillary hair cells is concerned, Mulroy (1968) studied the ultrastructure of two species and found only a few efferent terminals. More

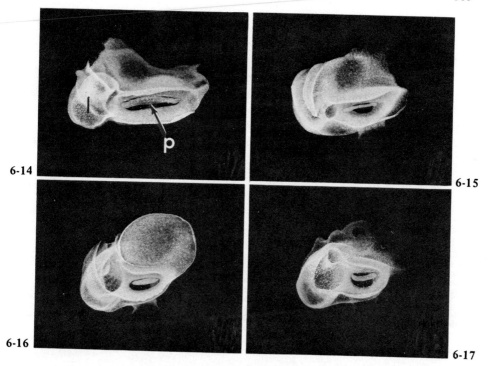

Figure 6-14. *Xenopeltis unicolor* (Sunbeam Snake, a burrowing species). The lagenar sac (l) lies to the left (apical) and the elongated papilla basilaris (p) is situated in the limbic part of the duct. (X 21) (From Miller 1966c).

Figure 6-15. *Eryx johni* (Sand Boa, a burrowing species). Same orientation as in Figure 14. (X 14) (From Miller 1966c).

Figure 6-16. *Naja nigricollis* (Spitting or Black-necked Cobra). Same orientation as in Figure 14. (X 11) (From Miller 1966c).

Figure 6-17. *Ancistrodon contortrix* (Copperhead Snake). Same orientation as in Figure 14. (X 20) (From Miller 1966c).

specific information is lacking regarding the innervation of the snake papilla basilaris, but like the turtle and other primitive papillae, it is probably supplied with both afferent and efferent nerve terminals (Bullock, Chapter 16).

While some degree of advancement over the turtle cochlear duct and papilla may be observed, an overall assessment indicates that the snake cochlea and papilla basilaris is relatively unspecialized and is not far removed from a primitive state.

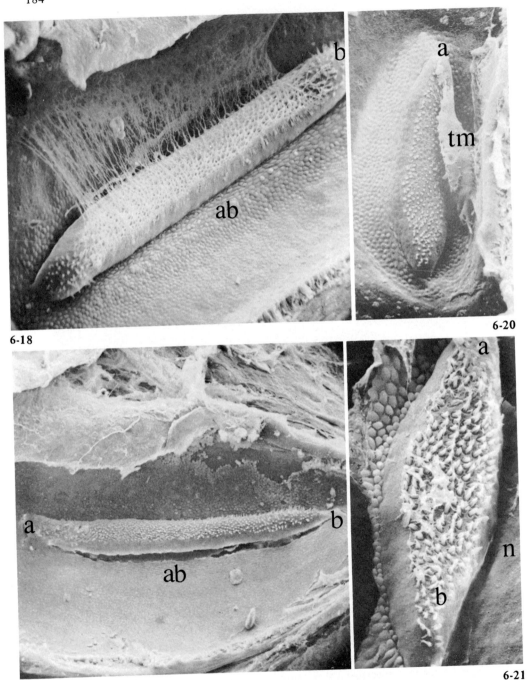

6-18

6-20

6-19

6-21

Figure 6-18. Left papilla basilaris of the boid snake, *Constrictor constrictor* (Boa Constrictor). Note the highly fenestrated nature of the tectorial attachment to the neural limbus as well as the tectorial plate overlying the hair cells. Both apical and basal extremities of the papilla are not covered by tectorial membrane. Abbreviations: tm - tectorial membrane; a - apical; b - basal; ab - abneural. (X 230) (From Miller 1978a).

7.2 Suborder Amphisbaenia

This group of reptiles are undoubtedly squamates, but their relationship to either lizards or snakes is uncertain. Probably the amphisbaenians, snakes, and lizards share common ancestry, but the three subgroups took separate lines of development at an early time and have long been separated.

On the basis of cochlear anatomy, the amphisbaenids are interesting in that they exhibit features that are, on the one hand, possibly primitive and, on the other hand, are regressed from a once more advanced condition. It should be noted that this group of reptiles has developed a specialized mechanism for the transmission of sound information to the internal ear (Wever 1978). In the overall balance, it can be assumed that the amphisbaenids were derived from lacertilian stock sometime after the ophidians had separated from the common squamatan line.

The cochlear duct arises from the sacculus wall anterior to its posterior limit and the duct lies essentially parallel to the inferior border of the sacculus. The cochlear duct is somewhat square to slightly elongate in shape (Figs. 6-22 to 6-27) and is only superficially divided by a slight constriction between the pars lagenae and pars basilaris. There is a tendency for increase in cochlear size with increase in species size. The pars lagenae is much reduced in size and is represented by the short, blunt inferior extremity of the cochlear duct (Figs. 6-22 to 6-27).

The limbus is a thin, almost circular concave plate surrounding a moderate sized, central, oval-shaped basilar membrane. The superior edge of the neural limbus projects laterally forming an overhanging lip (Figs. 6-22, 6-24, 6-26).

Wever's (1978) studies of the amphisbaenid papilla basilaris showed that a tectorial membrane is always present and that the thickened tectorial edge (plate) covers all but a few terminal hair cells. This author's studies of several amphisbaenid species (Miller unpublished) revealed an ovoid to ovoid-elongate papilla ranging from 0.15 mm to 0.3 mm in length. Further observations by Wever (1978) show the total hair cell number to be 38 to 154. Wever does not state the direction of kinocilial orientation but his figure 21 to 39 of the papilla of *Amphisbaena alba,* based on light microscopy, shows bidirectional kinocilial orientation. Baird (1974b), on the other hand, says that preliminary observations by transmission electron microscopy of *Bipes biporus* and *Amphisbaena caeca* show unidirectional orientation.

Baird (1974b) states that when examined by light microscopy, the organization of the papilla of amphisbaenids resembles that of *Chameleo* (true Chameleon).

Figure 6-19. Left papilla basilaris of *Constrictor constrictor* with the tectorial membrane removed to show shape of papilla and the irregular arrangement of the rows of hair cells. Abbreviations: a - apical; b - basal; ab - abneural. (× 127) (From Miller 1978a).
Figure 6-20. Left papilla basilaris of the colubrid snake, *Boaedon lineatum* (African House Snake). The tectorial membrane (tm) is rather dense but is shrunken and pulled away from the papillar surface. Many cilia have been pulled away with the tectorial membrane, leaving the hair cell surfaces denuded. Abbreviations: a - apical; b - basal. (× 200) (From Miller 1978a).
Figure 6-21. Left papilla basilaris of *Natrix sp.* The papilla in this genus tends to be canoe-shaped. The hair cell rows are irregular and kinocilial orientation is predominantly abneural. Abbreviations: a - apical; b - basal; n - neural. (× 333) (From Miller 1978a).

Further, the sensory cells of *Bipes biporus* and *Amphisbaena caeca* are more loosely ordered than are those of any lizard yet examined by electron microscopy, and supporting cells form a significant part of the papillar surface. Also, the supporting cells are less highly specialized than those of lizards. Baird (1974b) and Wever (1978) also state that a thin papillar bar (fundus) underlies the papilla basilaris.

The amphisbaenid cochlear duct appears to possess some lizard-like characteristics in the union of its parts and in the possibility of possessing kinocilial bidirectionality of the papillar hair cells. On the other hand, Baird's (1974b) observations indicate a good deal of primitiveness in the papillary structure (kinocilial unidirectionality, loose cellular organization, and the lack of cytological specializations).

Further study of the amphisbaenid papilla would be of great interest as it might help resolve the difficult interpretation of whether the amphisbaenid papilla is basically primitive or is a regressed form of a once more advanced condition.

7.3 Suborder Lacertilia

The lizard cochlear duct is by far the most interesting of reptilian cochlear ducts because of the diversity of its many features. Every family and, in some cases, even genera of lizards have "experimented" with certain features of duct morphology and in so doing have created a wide range of variations. As a consequence of these "natural experiments," both the taxonomist and the neurophysiologist is challenged to surmise the significance of these amazing variations.

The sacculo-cochlear duct arises from the postero-inferior portion of the sacculus (Figs. 6-1, 6-5) and gives rise to a somewhat triangular, square or elongated structure. Unlike the condition in *Sphenodon* and snakes, the apical (ventral) lagenar, and the basal (dorsal) limbic or basilar portions of the cochlear duct are conjoined to form a unified structure (Figs. 6-2, 6-3, 6-5).

Studies of well over 200 species of lizards (Miller 1966a, 1974, 1978b), representing all living lizard families but one (Shinosauridae), reveal differences in the cochlear duct in

1. overall size and configuration
2. relative proportions of the limbic and lagenar regions
3. size and shape of the limbus
4. degree of development of the lateral limbic lip
5. size and shape of the papilla basilaris
6. development of the papillar bar
7. development of the limbic recesses
8. tectorial membrane specializations
9. number and types of hair cells
10. innervation of the hair cells

Details of these differences have been thoroughly documented in the papers listed in the introductory section of this chapter. What follows is essentially a brief summary of what is known about lizard cochleae.

Perhaps the most important observation regarding the lizard cochlear duct is that, for the most part, the cochlear duct of each lizard family is different from that of any other lizard family. Thus, cochlear duct anatomy is distinctive at the family level of

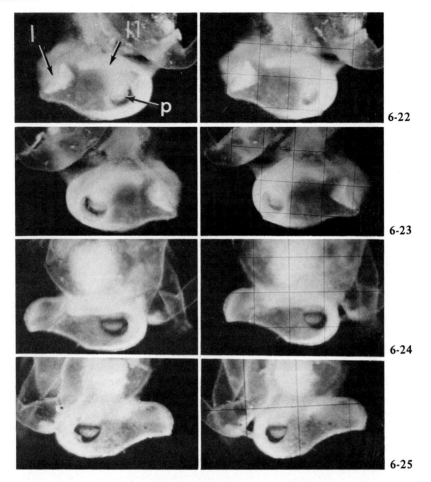

Figure 6-22. Lateral view; *Amphisbaena fuliginosa*. The apical lagenar portion of the cochlear duct is foreshortened and contains a macula lagenae (l) which is overlaid by an otolithic membrane. An overhanging limbic lip (ll) is characteristic of many amphisbaenid cochlear ducts. The papilla basilaris (p) is circular to ovoid in shape. (X 22).

Figure 6-23. Medial view; *Amphisbaena fuliginosa* (Cf. Fig. 6-22). (X 22).

Figure 6-24. Lateral view; *Bipes biporus* (Cf. Fig. 6-22). (X 22).

Figure 6-25. Medial view; *Bipes biporus* (Cf. Fig. 6-22). (X 22).

classification. Further detailed studies of the papilla in some families show that generic and even specific differences are present. While there may be some differences in cochlear duct and papilla basilaris anatomy between families, in some cases, related families show similarities of structure. Thus, cochlear anatomy is of considerable value in defining phylogenetic relationships within the lacertilian suborder.

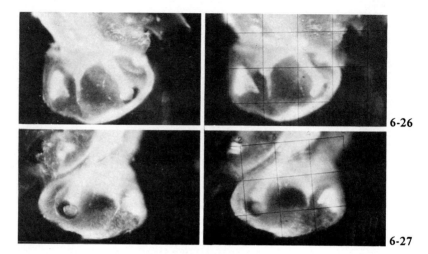

6-26

6-27

Figure 6-26. Lateral view; *Anops kingi* (Cf. Fig. 6-22). (X 22).

Figure 6-27. Medial view; *Anops kingi* (Cf. Fig. 6-22). (X 22).

While marked differences in features such as the size, shape, and extent of the macula lagenae, neural limbic modifications and limbic recesses are present, not much emphasis is given to these features. Rather the most attention has been given to differences in tectorial membrane modifications and to the detailed structural variations in the papilla basilaris. Since the latter features are probably the most important elements related to hearing, and apparently also to phylogenetic relationships, these are properly the most deserving of attention.

Detailed studies of the lizard papilla basilaris have been limited primarily to gross, light microscopic and scanning electron microscope observations (Miller 1978b, Wever 1978). Some transmission electron microscopic observations have been reported (Baird 1970, 1974b, Mulroy 1968, Bagger-Sjöbäck 1976), but such studies are not extensive. On the basis of a study of approximately 400 species of lizards (Miller 1966a, 1978b, Wever 1978) representing all the known families but one (Shinosauridae), several variations occur.

7.3.1 Iguanidae, Agamidae, and Anguidae

The papilla basilaris of species of these three families differ from those of other families in that there are two distinct types of hair cells. The first type is a short-ciliated hair cell that is abneurally, unidirectionally oriented and covered by a thickened tectorial mass (tectorial plate) attached to the limbus. The kinocilial heads of the short-ciliated hair cells (Fig. 6-42) are firmly attached to the tops of the tallest stereocilia and, in this way, are similar to the hair cells found in the basilar papillae of most lizard families. The second type is a long-ciliated hair cell that is bidirectionally oriented (kinocilial orientation toward the midpapillary axis) and is covered only by a loose type of tectorial material that has no limbic attachment. The kinocilia of the long-ciliated hair cells differ from those of other lizards or other types of reptile in that they are shorter than the tallest stereocilia, have small heads, and are not firmly attached to the stereocilia (Fig. 6-38).

In the Iguanidae, the short-ciliated unidirectional hair cells usually are confined to the midportion of the papilla and the long-ciliated hair cells, to the two extremities (Fig. 6-28). The anolids and *Plica* (Wever 1978) are exceptions in that the unidirectional hair cells are located at the apical end of the papilla and are thus like most agamids and some anguids (Fig. 6-4).

Studies of several genera and species of iguanid lizards (Miller unpublished) show that while the general pattern described is the case, some genera, and even species, reveal variations within this pattern that are sufficiently different to characterize a genus or species of lizard.

In the Agamidae, the short-ciliated, unidirectional, tectorial plate-covered hair cells are usually situated at the apical end of the papilla, and the tall, bidirectional, and lightly covered or uncovered hair cells occupy the basal portion of the papilla (Fig. 6-29). In three agamid genera, however, the short-ciliated hair cells are located in the midpapillary region as they are in most iguanids. These exceptions are species of the genera *Phrynocephalus, Uromastix,* and *Physignathus* (Wever 1978, Miller 1974). An interesting correlation is that the species of all three of these variant genera are burrowing animals.

A markedly variant agamid (*Leiolepis belliana*) is described by Wever (1978) in which all the hair cells are apparently short-ciliated and covered by a tectorial plate in the approximate apical three-fourths of the papilla and by sallets in the basal quarter. Interestingly, although this species is terrestrial, it lives in burrows (Schmidt and Inger 1957). Wever (1978) also notes that *Leiolepis* has a very superior auditory sensitivity as compared with other agamids.

In the family Anguidae, short-ciliated, unidirectional, tectorial plate-covered hair cells are found in an apical location in species that do not have reduced limbs, such as in species of the genera *Gerrhonotus* (Fig. 6-35), *Diploglossus,* and *Barisia.* In species with reduced limbs, however, such as in the genera *Anguis* and *Ophisaurus,* the short-ciliated, unidirectional, tectorial plate-covered cells are located in the midpapillary region similar to the common iguanid pattern.

The significance of the structural similarities and differences found in the Iguanidae, Agamidae, and Anguidae is conjectural. Probably species of these families acquired their basic papillar anatomy from a common ancestor. Close relationship of the iguanids and agamids is not surprising, but the relationship of the anguids here is problematical, as consideration of other anatomical systems does not show the anguids to be closely related to the iguanids and agamids. For the present, one must conclude that while the anguids were probably separated early and differ in other anatomical systems from the iguanids and agamids, the ear structure in these three families has remained remarkably similar.

In the past there has been a tendency to think of the iguanid-agamid-anguid type of ear structure as "primitive." However, when compared with the papillary structure of the turtle and *Sphenodon,* which are probably much more primitive, the iguanid-

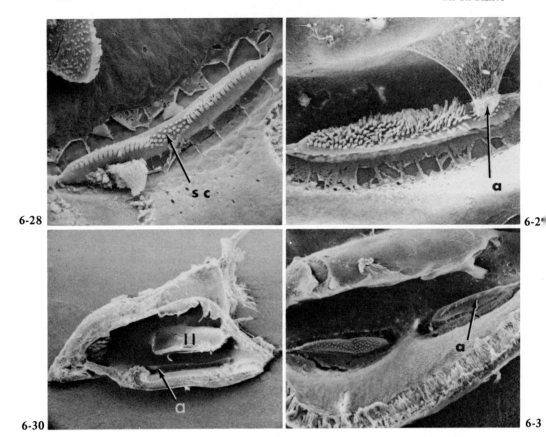

6-28

6-2

6-30

6-3

Figure 6-28. Left papilla basilaris of the iguanid lizard, *Sceloporus magister*. Note central short-ciliated hair cell segment (sc) and the long-ciliated hair cells located on each end of the papilla. The tectorial cap has been removed from the central segment. (X 242)..

Figure 6-29. Right papilla basilaris of the agamid lizard, *Agama agama*. Note that the apical (a) (to the right) short-ciliated hair cells are covered by a tectorial cap. Long-ciliated hair cells occupy the remainder of the papilla and are not covered by more than a few loose strands of tectorial tissue. (X 153) (From Miller 1978b).

Figure 6-30. Entire left cochlear duct of the teiid lizard, *Ameiva ameiva*. Note the pronounced limbic lip (ll). The papilla has a small constriction near the apical end (a). (X 25) (From Miller 1973c).

Figure 6-31. Entire right cochlear duct of the lacertid lizard, *Lacerta galloti*. The papilla is divided into two completely separate segments. The apical segment (a) is covered by a tectorial plate. The hair cells of the basal segment are not covered in this specimen, but according to Wever (1978), they are covered by sallets (Cf. Figs. 6-36, 6-37). (X 93).

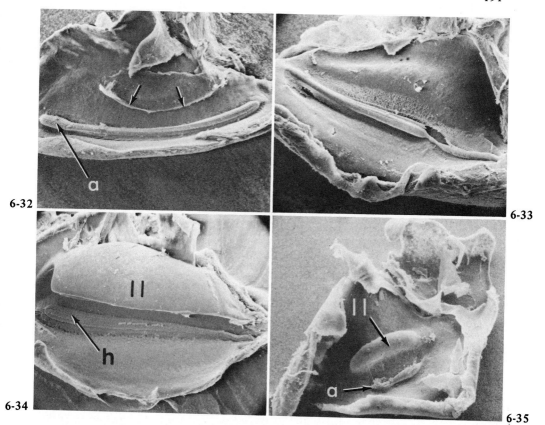

Figure 6-32. Left cochlear duct of the scincid lizard, *Mabuya carinata*. The arrows point to the site of the apparently vestigial remnants of the limbic portion of the tectorial membrane. In the adult animal, there is no connection between the limbus and the modified tectorial material covering the hair cells. The apical swelling (a) is covered by a large tectorial cap that has been lost in this specimen. The basal hair cells are covered by tectorial modifications termed "sallets." (Cf. Figs. 6-36, 6-37). (X 49) (From Miller 1973b).

Figure 6-33. Right cochlear duct of the varanid lizard, *Varanus exanthematicus*. A heavy tectorial plate covers all the hair cells and is attached throughout its length to the limbus. All hair cells are short-ciliated and are mostly bidirectionally oriented. (X 33) (From Miller 1974).

Figure 6-34. The left cochlear duct of the gekkonid lizard, *Gekko gecko*. The mid-axial hiatus (h) is seen at the arrow tip. Overhanging the papilla is an awning-like limbic lip (ll). Tectorial membrane detail is seen in Fig. 6-37. (X 42) (From Miller 1973a).

Figure 6-35. A left cochlear duct of an anguid lizard, *Gerrhonotus multicarinatus*. The limbic lip (ll) is quite pronounced. The apical papillar tip (a) is made up of short, unidirectionally oriented hair cells covered by a tectorial cap. (X 38).

agamid-anguid papilla is considerably changed and shows a number of modifications
that are not obviously primitive. While it is true that these papillae are relatively short
(0.1 mm to 0.7 mm) and contain a relatively small number of hair cells (50 to 300) as
compared with species of teiids, skinks, and gekkos (up to 2 500 hair cells), the
iguanid-agamid-anguid papillae are not truly primitive and in some cases exhibit remarkably good functional capacity.

7.3.2 Xenosauridae and Anniellidae

The Xenosauridae and Anniellidae (represented by only five or six species) have been
placed here since studies of the cochlear duct and papilla basilaris (Miller 1968, Wever
1978) of *Xenosaurus grandis* and *Anniella pulchra* (Limbless Lizard) reveal what may
be regressed structures, but are probably derived from the iguanid-agamid-anguid line
of auditory papilla. While the papilla is reduced to a small ovoid structure, it nevertheless possesses centrally located short-ciliated, largely abneural, unidirectionally
oriented hair cells overlaid by a tectorial plate, and basal and apical groups of long-
ciliated, bidirectional hair cells not covered by tectorial material.

   Both Underwood's (1971) summary of lizard phyletic relationships and Northcutt's
(1978) evaluation on the basis of brain structure place the xenosaurids close to the
anguids.

7.3.3 Chameleonidae (True Chameleons)

While the true Chameleons have been considered an early offshoot of the agamids
(Camp 1923, Broom 1924, Brock 1940), there has been marked reduction in both
the external and middle ear. As a possible consequence of the loss of a complete
external and middle ear, the papilla basilaris has probably been reduced in size and
has lost all resemblance to the agamid type of papilla.

   The papilla basilaris of chameleons is small and ovoid (approximately 0.10 mm to
0.15 mm in its greatest dimension) and contains approximately 40 to 60 hair cells.
According to Wever (1978), the tectorial membrane terminates in strand-like fila-
ments, each of which is attached to a hair cell. Wever terms this a "dendritic" type
of tectorial membrane attachment. Wever reports the hair cells to be largely, but
not exclusively, abneurally oriented. Baird (1974a) reported the hair cells in *Chameleo
zeglonicus* to be opposingly oriented in pairs or groups.

7.3.4 Lacertidae, Teiidae, and Varanidae

The papilla basilaris of lacertid lizards (Fig. 6-31) is remarkable in that it is completely
divided into two separate papillae. Both the apical and basal papillae are fairly short
(0.1 mm to 0.2 mm) and the total number of hair cells varies from 90 to 150.

   In the species of lacertids studied by scanning electron microscopy (Miller 1974,
1978b), the apical papilla is covered by a heavy tectorial membrane. The tectorial
structures were always lost in specimen preparation, but it was assumed that some

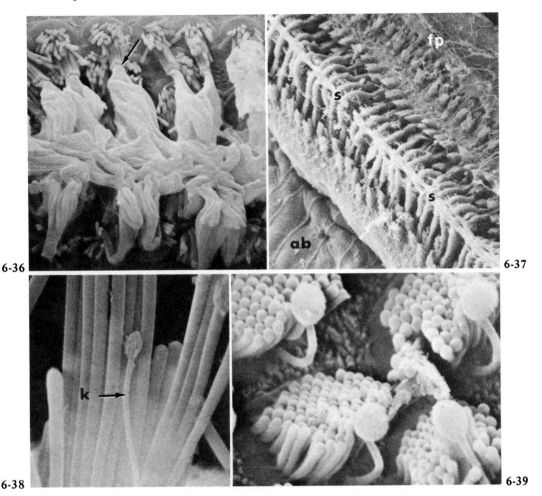

6-36

6-37

6-38

6-39

Figure 6-36. Part of the papilla basilaris of the scincid lizard, *Mabuya quinquelineata* demonstrating tectorial membrane modifications termed sallets. Sallets, like any other reptilian tectorial membrane modification, are attached to the kinocilial heads of the hair cells (indicated by an arrow). (X 2 040) (From Miller 1973b).

Figure 6-37. Part of the papilla basilaris of the gekkonid lizard, *Gekko gecko*, showing sallets (s) covering the abneural longitudinal hair cell strip, and another flattened and limbus-attached tectorial membrane modification (termed finger processes, Wever 1978) (fp) covering the neural longitudinal hair cell strip. Abbreviation: ab - abneural. (X 1 250) (From Miller 1973a).

Figure 6-38. Hair cell of the agamid lizard, *Uromastix sp.* A short kinocilium (k) only loosely attached to the stereocilia is characteristic of the long-ciliated hair cells found in the papillae basilares of iguanid, agamid, and anguid lizards. (X 7 670) (From Miller 1974).

Figure 6-39. Four hair cells from the papilla basilaris of the teiid lizard, *Ameiva ameiva*. Note the larger terminal kinocilial head attached to the five tallest stereocilia. The kinocilial stalk is always elongated and bent. The reason for this is not known. (X 12 000).

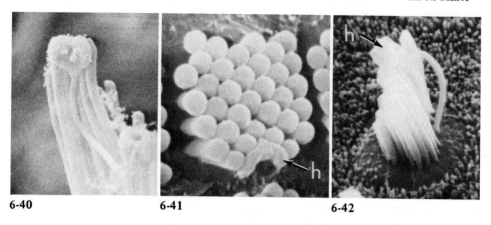

6-40   **6-41**   **6-42**

Figure 6-40. Hair cell of the scincid lizard, *Mabuya carinata*. View of the leaf- or heart-shaped kinocilial head characteristic of the basal bidirectional hair cells of scincids, xantusiids, and cordylids. By contrast, the apical kinocilial heads are spheroidal in shape. (X 7 488) (From Miller 1973b).

Figure 6-41. Hair cell of the scincid lizard, *Eumeces agilis*. Looking down on the hair cell surface, the kinocilial heads (h) appear to be arrow-shaped. (About X 10 000).

Figure 6-42. Detail of a short-ciliated hair cell from the central segment of the iguanid lizard, *Crotaphytus collaris*. Note the elongated kinocilial stalk and ellipsoidal kinocilial head (h) attached to the taller stereocilia. Compare with kinocilia of long-ciliated hair cells (Fig. 6-38). (X 7 980) (From Miller 1973c).

sort of tectorial structure covered these cells. Wever (1978) reports on several other lacertid species and describes a combination of tectorial plates and sallets (cf. Figs. 6-36, 6-37) covering the hair cells. Whether covered by tectorial plate or sallet, the hair cells are all short-ciliated, and except for a small number of unidirectionally oriented hair cells at the apical end of the basal segment, all other hair cells are bidirectionally oriented. Both uni- and bidirectionally oriented hair cells are alike in the lacertids, teiids, and varanids.

In the respect that both the teiids and lacertid cochlear ducts have well-developed neural limbic elevations or lips and a tendency for the teiid papilla to show constriction or near separation into two parts, there may be some relationship between these two families. On the other hand, the presence of sallets in the lacertids, together with some degree of papillar division in the skinks that are well endowed with sallets, suggests a possible relationship between the lacertids and skinks. On the basis of brain structure (Northcutt 1978), the lacertids may be closely related to the skinks, and thus the apparent resemblance between the cochlear duct structure of lacertids and teiids may not be significant.

The papilla basilaris of the teiid lizards (Fig. 6-30) is moderately long (0.5 mm to 1.3 mm) and consists of a relatively large number of hair cells (250 to 1 400). For the most part, the hair cells are covered by a moderately thick tectorial plate that in certain regions may break up into longitudinally running sallet-like formations (Miller unpublished). Wever (1978) states that sallets may be present at both extremities and a few hair cells may be uncovered.

Apparently much variation exists in kinocilial orientation in teiids. In *Ameiva ameiva*, 600 of 691 hair cells are directed abneurally. In *Tupinambis teguixin*, kinocilial orientation is very complex with change in orientation occurring as much as four times across the width of the papilla in some regions (Miller 1974, 1978b).

In both *Tupinambis* and *Ameiva*, approximately one-fourth the distance from the apical to the basal end of the papilla, there is a region of purely unidirectionally oriented hair cells. The location of the unidirectional hair cells, the presence of a heavy tectorial plate, a tendency for papillar constriction or separation into two parts, and a relatively large number of hair cells are features found in both teiids and varanid lizards and indicate a close relationship between these families at least as far as cochlear morphology is concerned. Northcutt's (1978) observations on brain structure corroborates the concept of close relationship between these two families.

The teiid lizards are a large and diverse group of lizards and, on the whole, their ear structure is little known. Further study of this group would be greatly rewarding as the central auditory structures of these lizards are better developed than those of most other lizards (Miller 1975).

The varanids have an elongated papilla basilaris (1.0 mm to 2.4 mm) (Fig. 6-33) containing approximately 500 to 1 800 hair cells. The papilla may be nearly or completely divided into two segments at a point approximately one-third the distance from the apical to the basal end. The entire papilla is covered by a rather heavy tectorial plate. In *Varanus bengalensis*, most hair cells are bidirectionally oriented; the apical quarter of the basal segment of the papilla, however, is comprised of unidirectional hair cells. The strong relationships between the teiid and varanid papillae are noted above.

7.3.5 Helodermatidae and Lanthanotidae

The gross (Miller 1966a, 1966b) and microscopic (Wever 1978) structure of the papillae of these small families suggest a possible relationship to the varanids that is in keeping with the phyletic concepts of Underwood (1971) and Northcutt (1978). In Heloderma (Beaded Lizard or Gila Monster), the papilla tends to be divided into two segments as in some varanid species. The basal region is larger than the apical, and most of the papilla is covered by a tectorial plate. The total number of hair cells approximates 300 (Wever 1978).

7.3.6 Scincidae, Xantusiidae, and Cordylidae

Species of all three of these families have elongated, relatively narrow papillae basilares (0.4 mm to 1.8 mm in length) and a relatively large number of hair cells (300 to 700 in xantusiids and cordylids, 400 to 1 400 in skinks) (Fig. 6-32). All have an ovoid apical papillar segment that is completely (xantusiids) or nearly separated (scincids and cordylids) from the remainder of the papilla. The apical segment is covered by a heavy tectorial cap that is attached to the limbus in some xantusiids (*Xantusia*) but not in others (*Lepidophyma*). That part of the tectorial membrane attached to the limbus is vestigial in both scincids and cordylids and does not reach the adult papilla.

The hair cells under the apical cap are bidirectional except for the apical-most hair cells (two or three horizontal rows) in skinks, but are mostly unidirectional in xantusiids, and are all unidirectional in cordylids. The kinocilial head of the apical hair cells is elongate-spheroid in shape.

The much longer basal segment of the papilla is covered by horizontal rows of sallets (Fig. 6-36), the basal end of the papilla is concave instead of convex, the kinocilia are fairly strictly oriented toward the midpapillary axis (bidirectional), and the kinocilial heads are uniquely leaf- or heart-shaped (Figs. 6-40, 6-41). The presence of different types of kinocilial heads on the apical and basal hair cells in the skinks, xantusiids, and cordylids indicates a certain degree of dissimilarity in the kinocilia of these lizard types. The degree of differentiation of hair cell types is greatest in the iguanids, agamids, and anguids; present but not as marked in the skinks, xantusiids, and cordylids; and least evident (if present at all) in the lacertids, teiids, and varanids as well as in the gekkonid-pygopodid group.

On the basis of the many remarkable similarities, the scincids, xantusiids, and cordylids are probably closely related. This corroborates Camp's (1923) ideas but is at variance with the concepts of Underwood (1971) and Northcutt (1978). It is still possible, if not probable, that ear anatomy may have developed independently of other systems, and thus present day relationships may be evaluated only by analysis of all anatomical systems.

7.3.7 Gekkonidae and Pygopodidae

The gekkonids are a large group of lizards that have developed a relatively complex auditory papilla exhibiting several unique features. The papilla varies in length from 0.6 mm to 2.0 mm and contains 400 to 2 000 or more hair cells. The apical half to two-thirds of the papilla is divided into two separate longitudinally running strips by a mid-axial papillar region devoid of hair cells (Fig. 6-34). Each of these longitudinally running strips contains bidirectionally (mid-axial) oriented hair cells. The basal third or half of the papilla is narrower than the apical region and the hair cells are not divided into separate longitudinal regions. In the greater number of species, the basal hair cells are unidirectionally oriented. In several genera, *Teratolepis, Phelsuma,* and *Aristelliger,* the basal papillar region has only bidirectionally oriented hair cells (Wever 1978), and thus, these species have no region of unidirectionally oriented hair cells.

At least three tectorial modifications are usually present on the same papilla in gekkonids (Fig. 6-37). The abneural side of the apical longitudinal strip is covered with sallets; the neural strip, by a limbus-attached, moderately thick tectorial material; and the basal hair cells, by a rather delicate tectorial membrane.

Thus, the unique and complex gekkonid cochlear duct and papilla basilaris is probably highly derived and far removed from ancestral stock. It is significant that Paull,

Williams, and Hall (1976) state that on the basis of karyotype as well as other morphological evidence, the Gekkonidae may be far separated from plausible basal stock.

The cochlear duct and papilla basilaris of pygopodids are remarkably similar to those of the gekkonids in almost all details. In those species so far studied by light microscopy (Wever 1978), the basal hair cells are bidirectionally oriented such as those found in three genera of gekkos. On the whole, there is no doubt of the close relationship of the gekkonids and pygopodids.

7.3.8 Dibamidae and Anelytropsidae

Since only the gross structure of the cochlear duct of these small families is known, little can be said about their relationship to other families.

8 General Summary of the Significance of Cochlear Duct Structure in Lizards

In Section 7, on the structure of the lizard cochlear duct and papilla basilaris, similarities and differences have been described and discussed in relation to their phylogenetic importance. In the following outline, lizard families are grouped according to the structural similarities of the cochlear duct and papilla basilaris.

I. Iguanidae
Agamidae
Anguidae
Xenosauridae
Anniellidae
II. Chameleonidae (relationships not clear)
III. Lacertidae (uncertain if related more to skinks or to teiids)
IV. Teiidae (possibly related to varanids)
V. Varanidae
VI. Helodermatidae ⎫
VII. Lanthanotidae ⎭ possibly related to varanids
VIII. Scincidae, Cordylidae, Xantusiidae (very close similarities in a large number of duct and papillar structural details)
IX. Dibamidae and Anelytropsidae (very close relationship)
X. Gekkonidae and Pygopodidae (very close affinities)

The above scheme is quite similar to that proposed by Wever (1978) with the exception that this author is less certain of the relationships of the lacertids, teiids, and varanids than is Wever. For the present, the relationships of the Lacertidae, Teiidae, and Varanidae have been left open, while Wever lists a superfamily Varanoidea consisting of varanids, helodermatids, and *Lanthanotus,* and a superfamily Lacertoidea embracing the Teiidae and Lacertidae.

The phylogenetic relationships of lizards based on cochlear duct structure for the most part is in agreement with the ideas concerning lizard relationships derived from the study of other anatomical systems (Underwood 1971, Northcutt 1978). Two major differences concern the relationships of the families Xantusiidae and Anguidae. According to Camp (1923), the xantusiids are a derivative of the scincomorphic line and this theory is supported by ear anatomy. Other anatomical relationships place the xantusiids closer to the gekkonids (Underwood 1971, Northcutt 1978). However, Northcutt's scheme (Northcutt 1978, p. 57) shows xantusiids being derived from the same base stock as the scincid lizards. Thus, it is possible that the xantusiids retained ear anatomy similar to that exhibited by many scincid types, but other systems evolved in directions more similar to that of the gekkonids.

The position of the Anguidae is more difficult to reconcile, however, for in none of the projected phyletic schemes do the iguanoids (Iguanidae and Agamidae) and the Anguoidea (Anguidae, Xenosauridae) share recent common ancestry. At this point, one can only surmise that the anguid cochlear duct was carried through a long period of evolution without major changes in the ear. That the anguoids and iguanoids are only related at the very early base or stem level but have each retained a very similar cochlear duct anatomy is suggestive that the type of ear found in these groups is a fairly early or a possibly "primitive" type of lizard ear. While it may be that the iguanoid-anguinoid ear is primitive, it is only so in relation to the more derived types of ears found in other lizard families. It must be kept in mind that the iguanoid-anguinoid ear is considerably advanced over that of snakes, amphisbaenids, *Sphenodon,* and turtles.

9 Order Crocodilia

The cochlear duct of alligators and crocodiles (Fig. 6-43) is considerably longer than that of other reptiles. The papilla basilaris is about 4 mm in length and is thus about twice the length of the best developed lizard papillae. In addition to greater length, the papilla is relatively broad and shows a marked expansion from a narrower basal part to a broader apical end. A heavy-appearing tectorial membrane covers all the hair cells and, according to Wever (1978), is dendritic in type with every hair cell connected to a tectorial fiber.

Figure 6-43. Montage of a right papilla basilaris of the crocodilian, *Caiman crocodilus.* The upper part is the basal half, and the lower, the apical half of the papilla. The neural side is toward the top of the page. Note the gradual increase in papillar width from the basal to the apical end (a). The papilla is about 4 mm long. (✕ 55) (From Leake 1977).

Figure 6-44. Enlarged view of a portion of the papilla basilaris of *Caiman crocodilus.* The hair cells above the arrows (toward page top) are the columnar inner hair cells. Below the arrows are the lenticular outer hair cells. (✕ 264) (From Leake 1977).

Figure 6-45. Outer hair cells of *Caiman crocodilus* showing the large cuticular plate, eccentric disposition of cilia and the kinocilia (k) as seen from behind (the kinocilia are facing abneurally). (✕ 6 286) (From Leake 1977).

Figure 6-46. Outer hair cells of *Caiman crocodilus* near the basal end of the papilla. Note eccentric disposition of stereocilia, large cuticular plate, and narrowed edges of supporting cells bearing numerous microvilli. (✕ 10 000) (From Leake 1977).

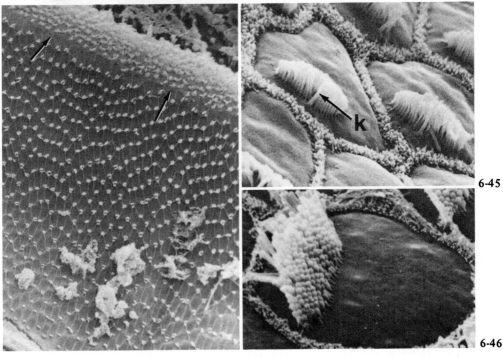

Table 6-1. Summary of major structural characteristics of the papillae basilares in the different orders of reptiles.

	Length in mm		Tectorial membrane	Types of hair cells
Rhynchocephalia	0.725	(Wever 1978)	Limbic-attached tectorial	one
	1.0 to 1.2	(Miller 1966)	plate covers all hair cells	
Testudines (Turtles, Tortoises, Terrapins)	0.5 to 1.5		Limbic-attached tectorial plate covers hair cells resting on basilar membrane; thinner non-limbic-attached material covers hair cells resting on limbus at both ends	one
Squamata Serpentes (Snakes)	0.5 to 1.5		Thick or fenestrated limbic-attached tectorial plate covers most hair cells. Hair cells at papillar ends are usually not covered by tectorial material.	one
Amphisbaenia (Amphisbaenids)	0.15 to 0.3		Limbic-attached tectorial plate covers most hair cells; papillar ends not covered	one
Lacertilia (Lizards)	0.2 to 2.5		Great variation from limbic-attached tectorial plate covering all or but a few cells in varying locations, to tectorial modifications in the form of free or attached masses of differing shapes (sallets, culmens, fine loose material), to apparently totally unencumbered hair cells	one or two
Crocodilia (Crocodiles, Alligators)	ca. 4.0		Limbic-attached tectorial plate	two

Table 6-1 (con't.)

Disposition of hair cells (kinocilial orientation)	Total number of hair cells	Shape of kinocilial head	Innervation
Largely unidirectional and abneural; rare hair cell is neurally oriented	ca. 225	Not known	Not known
Largely unidirectional and abneural; rare hair cell is neurally oriented	500 to to 1 500 (except in *Chelonia mydas*, 2 500)	Elongated spheroid	Afferent and efferent nerve fibers
Largely unidirectional and abneural; occasional neural orientation in boid snakes	50 to 1 500	Large elongated spheroid	Probably both afferent and efferent nerve fibers
(?) possibly either bidirectional or unidirectional varying with the species	38 to 154	Not known	Not known
Almost always specific regions of bidirectionally oriented hair cells	50 to 2 500	Spheroidal, elongate spheroidal or leaf-shaped	In regions of bidirectional hair cells, afferent nerve fibers only; in regions of unidirectional hair cells usually both afferent and efferent nerve fibers
Unidirectional and abneural	9 000 to 13 000	Somewhat thickened upper part of kinocilium	Afferent and efferent nerve fibers

The hair cells rest on a basilar membrane that is somewhat thickened by periotic connective tissue, but there is no dense papillar bar as there is in lizards. The number of hair cells is about 11 000 in *Alligator mississippiensis*, 11 500 in *Caiman crocodilus*, and 13 700 in *Crocodylus acutus* (Wever 1978, Leake 1977).

Hair cells are of two types: a smaller number of cells (3 000 of the 11 500 in *Caiman crocodilus*, Leake 1977) closer to the neural limbus are columnar in shape, while the larger number of hair cells (8 500 in *Caiman crococilus*) are lenticular in shape'and more abneurally placed and occupy the greater extent of the basilar membrane (Fig. 6-44). The columnar hair cells have been likened to the mammalian inner hair cells and the lenticular, to the outer hair cells (Leake 1976).

Stereocilia are laterally disposed on the cuticular plate in the outer hair cells (Figs. 6-45, 6-46) but are more centrally placed on the inner hair cell cuticular plates. Kinocilia are present, but the kinocilial heads are smaller than those found in other reptilian papillae. All hair cells are abneurally oriented.

The supporting cells are highly specialized and are structurally similar to mammalian pillar cells in that the cytoskeleton of the crocodilian supporting cells is made up of clusters of microtubules. Supporting cells throughout the Reptilia have a microtubular apparatus, but its development is much greater in the Crocodilia than in other reptiles.

In all anatomical details the papilla basilaris of the crocodylids is well advanced over that of other reptiles. The major structural variations found in the papillae basilares of reptiles are summarized in Table 6-1.

10 Innervation of Hair Cells

One final note in relation to the innervation of the reptilian hair cells should be made. On the basis of present information, the unidirectional hair cells of the turtle, snake, and crocodilian papillae are innervated by both afferent and efferent nerve terminals. Bidirectionally oriented hair cells, which comprise the greater number of hair cells in lizards, are lacking in efferent nerve terminals. The unidirectionally oriented hair cell regions of lizard papillae may vary; Nadol has found efferent nerve endings in *Gerrhonotus multicarinatus* (see Weiss et al. 1976) as have Baird and Marovitz (1971) in *Iguana iguana*, but an extensive study of *Calotes versicolor* by Bagger-Sjöbäck (1976) revealed no efferent terminals in the unidirectional region as well as in the bidirectional part of the papilla. Considerably more study of the innervation of the reptilian papilla basilaris is needed. The significance of the lack of efferent innervation in portions of the lizard papilla is not known.

Acknowledgments. The author wishes to express gratitude to Ms. Michiko Kasahara for constant expert help in all phases of this work, to Mr. Dave Akers for skilled photographic assistance, and to Mr. Wayne Emery and Mr. David Factor for the drawings. This work was supported by a USPHS Grant No. R01 NS 11838.

References

Bagger-Sjöbäck, D.: The cellular organization and nervous supply of the basilar papilla in the lizard, *Calotes versicolor*. Cell Tiss. Res. 165, 141-156 (1976).

Bagger-Sjöbäck, D., Wersäll, J.: The sensory hairs and tectorial membrane of the basilar papilla in the lizard *Calotes versicolor*. J. Neurocytol. 2, 329-350 (1973).

Baird, I. L.: A survey of the periotic labyrinth in some representative recent reptiles. Univ. Kansas Sci. Bull. 41, 891-981 (1960).

Baird, I. L.: The anatomy of the reptilian ear. In: Biology of the Reptilia. Gans, C., Parsons, T. S. (eds.). Vol. 2, pp. 193-275. London-New York: Academic Press 1970.

Baird, I. L.: Anatomical features of the inner ear in submammalian vertebrates. In: Handbook of Sensory Physiology. Keidel, W. D., Neff, W. D. (eds.). Vol. V., pt. 1, pp. 159-212. Berlin-Heidelberg-New York: Springer, 1974a.

Baird, I. L.: Some aspects of the comparative anatomy and evolution of the inner ear in submammalian vertebrates. Brain, Behav. Evol. 10, 11-36 (1974b).

Baird, I. L., Marovitz, W. F.: Some findings of scanning and transmission microscopy of the basilar papilla of the lizard, *Iguana iguana*. Anat. Rec. 169, 270 (1971).

Brock, G. T.: The skull of the chameleon *Lophosaura ventralis* (Gray); some developmental stages. Proc. Zool. Soc. London, Ser. B, 110, 219-241 (1940).

Broom, R.: On the classification of the reptiles. Bull. Amer. Mus. Nat. Hist. 51, 39-65 (1924).

Camp, C. L.: Classification of the lizards. Bull. Amer. Mus. Nat. Hist. 48, 289-481 (1923).

de Burlet, H. M.: Vergleichende Anatomie des statoakustischen Organs. a) Die innere Ohrsphäre; b) Die mittlere Ohrsphäre. In: Handbuch der vergleichende Anatomie der Wirbeltiere, 2, 2. Hälfte. Berlin und Wien: Urban und Schwartzenberg, 1934.

Hamilton, D. W.: Observations on the morphology of the inner ear in certain gekonoid lizards. Univ. Kansas Sci. Bull. 41, 983-1024 (1960).

Hamilton, D. W.: The inner ear of lizards. I. Gross structure. J. Morp. 115, 255-271 (1964).

Leake, P. A.: Scanning electron microscopy of labyrinthine sensory organs in *Caiman crocodilus*. In: Scanning Electron Microscopy, Proceedings of the Workshop on Advances in Biomedical Applications of the SEM, IIT Research Institute. Chicago, Vol. II, pp. 277-284 (1976).

Leake, P. A.: SEM observations of the cochlear duct in *Caiman crocodilus*. In: Scanning Electron Microscopy, Proc. of the Workshop on Biomedical Applications— SEM Studies of Sensory Organs, IIT Research Institute. Chicago, Vol. II, pp. 437-444 (1977).

Miller, M. R.: The cochlear duct of lizards. Proc. Calif. Acad. Sci. 33, 255-359 (1966a).

Miller, M. R.: The cochlear ducts of *Lanthanotus* and *Anelytropsis* with remarks on the familial relationship between *Anelytropsis* and *Dibamus*. Occ. Papers Calif. Acad. Sci., no. 60, pp. 1-15 (1966b).

Miller, M. R.: The cochlear duct of lizards and snakes. Am. Zool. 6, 421-429 (1966c).

Miller, M. R.: The cochlear duct of snakes. Proc. Calif. Acad. Sci. 35, 425-475 (1968).

Miller, M. R.: A scanning electron microscope study of the papilla basilaris of *Gekko gecko*. Z. Zellf. 136, 307-328 (1973a).

Miller, M. R.: Scanning electron microscope studies of some skink papillae basilares. Cell Tiss. Res. 150, 125-141 (1973b).

Miller, M. R.: Scanning electron microscope studies of some lizard basilar papillae. Am. J. Anat. 138, 301-330 (1973c).

Miller, M. R.: Scanning electron microscopy of the lizard papilla basilaris. Brain Behav. Evol. 10, 95-112 (1974).

Miller, M. R.: The cochlear nuclei of lizards. J. Comp. Neurol. 159, 375-406 (1975).

Miller, M. R.: Scanning electron microscope studies of the papilla basilaris of some turtles and snakes. Am. J. Anat. 151, 409-435 (1978a).

Miller, M. R.: Further scanning electron microscope studies of lizard auditory papillae. J. Morph. 156, 381-418 (1978b).

Mulroy, M. J.: Ultrastructure of the Basilar Papilla of Reptiles. Doctoral Dissertation, University of California, San Francisco, Ca., 1968.

Mulroy, M. J.: Cochlear anatomy of the alligator lizard. Brain Behav. Evol. 10, 69-87 (1974).

Northcutt, R. G.: Forebrain and midbrain organization in lizards and its phylogenetic significance. In: Behavior and Neurology of Lizards. Greenberg, N., MacLean P. D. (eds.). pp. 11-64, Nat. Insts. Mental Health, U. S. Dept. HEW, 1978.

Paull, D., Williams, E. E., Hall, W. P.: Lizard karyotypes from the Galapagos Islands: Chromosomes in Phylogeny and Evolution. Breviora, Mus. Comp. Zool. 441, 1-31 (1976).

Retzius, G.: Das Gehörorgan der Wirbeltiere. II. Das Gehörorgan der Reptilien, der Vögel und der Säugetiere. Stockholm: Samson und Wallin, 1884.

Schmidt, R.: Phylogenetic significance of the lizard cochlea. Copeia, pp. 542-549 (1964).

Schmidt, K. P., Inger, R. F.: Living Reptiles of the World, Garden City, New York: Hanover House, 1957.

Shute, C. D. D., Bellairs, A. D'A.: The cochlear apparatus of Gekkonidae and Pygopodidae and its bearing on the affinities of these groups of lizards. Proc. Zool. Soc. London 123, 695-709 (1953).

Underwood, G. L.: A modern appreciation of Camp's "Classification of the lizards". In: Camp's Classification of the Lizards. A facsimile reprint published by the Society for the Study of Amphibians and Reptiles, 1971.

von Düring, M., Karduck, A., Richter, H.-G.: The fine structure of the inner ear in *Caiman crocodilus*. Anat. Entwickl.-Gesch. 145, 41-65 (1974).

Weiss, T. F., Mulroy, M. J., Turner, R. J., Pike, C. L.: Tuning of single fibers in the cochlear nerve of the alligator lizard: relation to receptor morphology. Brain Res. 115, 71-90 (1976).

Wersäll, J., Flock, Å.: Functional anatomy of the vestibular and lateral line organs. In: Contributions to Sensory Physiology, vol. I. New York-London: Academic Press, 1965.

Wever, E. G.: The Reptile Ear. Princeton, New Jersey: Princeton University Press, 1978.

Chapter 7

Physiology and Bioacoustics in Reptiles

ROBERT G. TURNER*

1 Introduction

Recent anatomical studies have revealed a unique and important characteristic of the reptilian auditory system. Auditory anatomy is extremely diverse among reptiles, particularly in the cochlea where morphology can vary significantly across taxonomic families (see Miller, Chapter 6). This anatomical diversity has stimulated interest in the physiology of hearing in reptiles by providing an excellent opportunity to investigate the fundamental relation between anatomical structure and physiological response.

A better understanding of the reptilian ear should result in a better understanding of the mammalian ear. Wever (1978, p. 981) states: "The ears of birds and mammals followed the pattern of the reptilian ear. This seems to be true despite the fact that avian and mammalian branches of the vertebrate line have arisen separately from early reptiles and have remained apart for many millions of years." Research has shown that there are anatomical structures and physiological patterns that are common to the ears of reptiles and mammals, suggesting certain common mechanisms for processing acoustic information.

This chapter provides an overview of hearing in reptiles with emphasis on the relation between anatomy and physiology. It is assumed that the reader is familiar with basic mammalian anatomy and physiology; however, simplified descriptions of reptilian anatomy have been included for the reader's convenience. See Chapter 6 (Miller) in this text for a more detailed discussion of reptilian auditory anatomy.

Almost all available physiological data have been reviewed in this chapter except the "cochlear potential" data that have recently been presented in Wever's book *The Reptile Ear* (1978). Unfortunately, the published data for reptiles are insufficient to provide a comprehensive description of auditory function. Not only is the quantity of data relatively small, but it is difficult to generalize results to other reptiles because of the great variations in reptilian anatomy. Even with these limitations, this chapter

*Otological Research Laboratory, Department of Otolaryngology, Henry Ford Hospital, 2799 West Grand Blvd., Detroit, Michigan 48202.

should provide the reader with an adequate introduction to auditory physiology and bioacoustics in reptiles.

2 External and Middle Ear

2.1 Anatomy

In some lizards and all crocodilians, the tympanic membrane is recessed from the surface of the head, resulting in a small external auditory meatus. For other lizards, as as turtles, snakes, amphisbaenians, and *Sphenodon,* the external ear is absent. Either there is no tympanic membrane, or the tympanic membrane is flush with the surface of the head.

The middle ear of turtles, crocodilians, and most lizards consists of a tympanic membrane that communicates with the cochlea via two ossicles, the columella (stapes) and extracolumella (extrastapes). The extracolumella attaches to the medial surface of the tympanic membrane and to the columella (Fig. 7-1). The columella is a slender, rod-like structure. Its medial end expands to form a footplate that occupies the oval

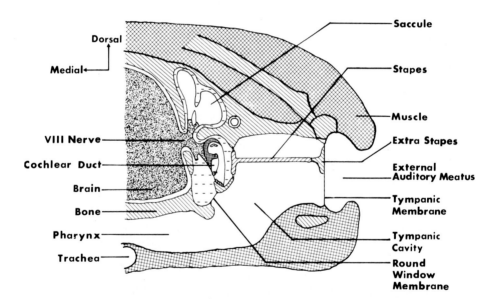

Figure 7-1. "Typical" lizard ear. This diagram represents a transverse section through the head and is representative for most lizards. The ear of crocodilians is similar, containing a tympanic membrane and a round window membrane. Other reptiles differ in middle ear and/or cochlear anatomy. Some lizards as well as snakes, amphisbaenians, and *Sphenodon* have no tympanic membrane. These reptiles, like the turtles, employ a reentrant fluid circuit instead of a round window membrane for fluid pressure relief in the cochlea. In this figure, the columella and extracolumella are labeled stapes and extrastapes, respectively. (Adapted from Weiss, Mulroy, and Altmann 1974).

window. When this type of middle ear is combined with an inner ear containing a round window membrane, as in most lizards and crocodilians, the ear is designated a "typical" lizard ear (Fig. 7-1).

The middle ear of snakes, amphisbaenians, *Sphenodon,* and some lizards differs from the middle ear described above in that the tympanic membrane is absent and the extracolumella is absent or modified. The columella attaches directly, or via the extracolumella, to bones of the skull, such as the quadrate, or to tissues in the side of the head.

2.2 Tympanic Membrane Vibration

Using the Mossbauer technique, Manley (1972b) measured the amplitude of vibration of the tympanic membrane in the lizard *Gekko gecko* and concluded that the membrane did not vibrate as a stiff plate. At frequencies below 2 000 Hz, the membrane vibrates in phase and is well coupled to the extracolumella. At higher frequencies, the vibratory pattern becomes complex, reducing the "effective" area of the tympanic membrane and the impedance matching capability of the middle ear.

2.3 Middle Ear Transfer Function

One measure of the performance of the reptilian middle ear is the middle ear transfer function, a plot of columella amplitude versus frequency for constant sound pressure at the tympanic membrane. The middle ear transfer function has been measured in the lizards *Gekko gecko* (Manley 1972a, 1972b), *Amphibolures reticulatus,* and *Phyllurus millii* (Saunders and Johnstone 1972). In addition, the middle ear transfer function can be estimated from other measures of middle ear motion in the lizards *Gerhonotus multicarinatus* (Weiss, Peake, Ling, and Holton 1978) and *Gehyra variegata* (Manley 1972c). All of these lizards have a "typical" lizard ear.

To evaluate the performance of the lizard middle ear, it is useful to compare the middle ear transfer function for these lizards and the cat (Guinan and Peake 1967). As in the cat, the middle ear transfer function for these lizards has the form of a low-pass filter (Fig. 7-2). Columella displacement is relatively constant in amplitude for frequencies below the middle ear resonant frequency. Resonant frequency corresponds to maximum columella velocity and a slope of the transfer function of -20dB/decade. Above resonant frequency, columella displacement decreases with increasing frequency. Resonant frequencies and ossicular displacements at low frequencies are grossly similar for the lizards and the cat. However, the slope of the transfer function above resonant frequency is consistently greater (more negative) for the lizards (-60dB/decade to -80dB/decade) than for the cat (-40dB/decade). Even though the high frequency slope is significantly greater for the lizard, the difference in displacement amplitude for the lizards studied and for the cat is less than about 20dB for frequencies below 10 000 Hz.

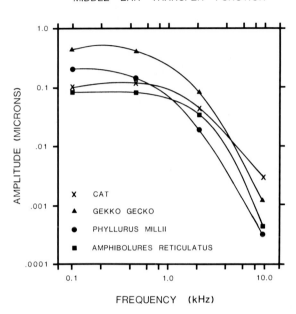

Figure 7-2. Middle ear transfer functions for certain lizards and domestic cat. These curves are a plot of columella or stapes displacement (zero-to-peak) as a function of frequency for a constant intensity of 100 dB SPL. These curves were adapted from published data for the cat (Guinan and Peake 1967) and the lizards *Gekko gecko* (Manley 1972b), *Phyllurus millii,* and *Amphibolures reticulatus* (Saunders and Johnstone 1972).

2.4 Impedance Matching

One function of the mammalian middle ear is to match the impedance of the external acoustic medium to that of the cochlea. Two mechanisms that contribute to impedance matching are the lever ratio and the area ratio. Wever and Werner (1970) proposed that the columella and extracolumella in the "typical" lizard ear can act as a lever with the extracolumella rotating about a point at the tip of the superior process (Fig. 7-3). The displacements of the inferior process and the columella have been compared in the lizards *Amphibolures reticulatus, Phyllurus millii* (Saunders and Johnstone 1972), *Gekko gecko,* and *Gehyra variegata* (Manley 1972a, 1972b). At low frequencies, the displacement of the inferior process is 10dB to 14dB greater than the displacement of the columella. For most lizards, this difference in displacement increased at higher frequencies. Manley (1972c) concluded that above 4 000 Hz the inferior process can flex, causing a relative reduction in columella displacement and a loss of transmission through the middle ear. These data indicate the presence of a lever action in the middle ear. At low frequencies, the displacement of the tip of the inferior process is about four times that of the columella; however, the lever ratio is probably less than four since the force exerted by the tympanic membrane is distributed along

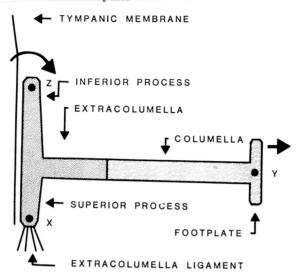

Figure 7-3. The lever action of the middle ear ossicles in the lizard. The extracolumella rotates about a point (X) in the superior process so that the tip of the inferior process (Z) has greater displacement than the footplate (Y). In this situation the ossicles act as a lever to increase the force at the footplate.

the extracolumella and is not concentrated at the tip of the inferior process. Even so, the lever ratio for these lizards is probably greater than the 1.3 lever ratio of man (Yost and Nielsen 1977, p. 38).

The area ratio is the ratio of the "effective" area of the tympanic membrane to the area of the footplate. Wever (1978, pp. 168, 179, 187, 578) has calculated area ratios for several species of lizards. The area ratio varies from 13 in *Mabuya carianata* to 40 in *Gekko gecko,* compared to an area ratio of 17 for man (Yost and Nielsen 1977, p. 38). Wever (1978, p. 851) calculated a combined lever and area ratio of 11.7 for the turtle *Chrysemys scripta.* He suggests that the lower ratio in this turtle represents a compromise between aerial and aquatic reception of sound.

2.5 Performance of the Reptilian Middle Ear

Measurements in reptiles of middle ear performance have been limited to lizards with a "typical" ear. Since the crocodilian ear is anatomically similar to the "typical" lizard ear, the performance of the crocodilian middle ear is probably similar to that of "typical" lizard middle ear. Very little is known about middle ear performance in those reptiles with ears lacking tympanic membranes and/or round window membranes. This would include turtles, snakes, amphisbaenians, *Sphenodon,* and some lizards.

The performance of the "typical" lizard middle ear appears comparable to the middle ear of mammals for frequencies below 10 000 Hz. Impedance matching capability, as reflected in measurement of lever and area ratio, is as good in the lizard as in man.

The transmission of high frequencies is poorer in the lizard than in the cat because of the greater slope of the lizard transfer function above resonant frequency; however, the actual difference in transmission is less than 20dB below 10 000 Hz.

3 Inner Ear

3.1 Anatomy

There are two basic types of cochleae found in reptiles. The first type, found in crocodilians and most lizards, is similar to the cochleae of birds and mammals, employing a round window membrane for the release of fluid pressure (Fig. 7-4). The second type of cochlea, found in turtles, snakes, amphisbaenians, *Sphenodon,* and a small number of lizards, employs a reentrant fluid circuit for pressure relief. Wever (1978) states:

> In these animals the round window is absent or (as in certain lizards of the genus *Phyrnosoma*) its outer surface is covered with fluid and rendered relatively immobile. These ears contain a fluid passage that leads from the

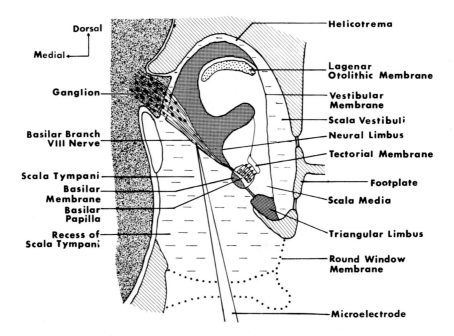

Figure 7-4. The cochlea of the lizard *Gerrhonotus multicarinatus.* Like other reptiles, the cochlea of *Gerrhonotus* contains many structures that are found in the mammalian cochlea; however, the scalae are not coiled but are chambers of irregular shape. The dotted lines indicate the bone and round window membrane removed to expose the auditory nerve. (Adapted from Weiss, Mulroy, and Altmann 1974).

inner boundary of the cochlea by a circuitous path outward to the lateral face of the columella footplate . . . Therefore an inward movement of the footplate produces a fluid displacement that involves not only the cochlear pathway but the complete circuit back to the footplate. (p. 91)

Both types of cochleae contain the three scala—vestibuli, tympani, and media. These scalae are not coiled, as in mammals, but are chambers or irregular shape.

The basilar papilla is the auditory receptor organ in the reptilian cochlea. Two features of the morphology of the papilla, hair cell orientation and tectorial structures, are particularly important to the study of auditory physiology in reptiles. Three basic patterns of hair cell orientation are found in reptiles: unidirectional, segregated-bidirectional, and integrated-bidirectional. In general, an individual hair cell is oriented either toward (neural) or away from (abneural) the auditory nerve. In a unidirectional population, all hair cells are oriented abneurally (see the Anguid family, Fig. 7-5). In the segregated-bidirectional population, hair cells are oriented toward a midpapillary axis. Hair cells with neural orientation are located on the abneural side of the mid-papillary axis; hair cells with abneural orientation are located on the neural side of the

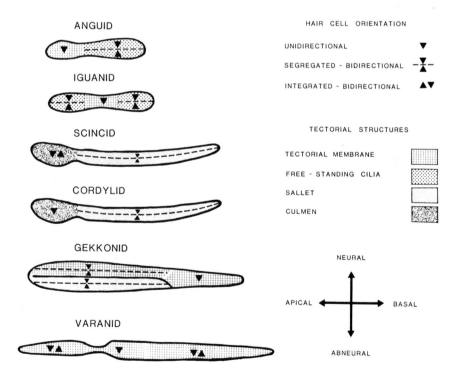

Figure 7-5. Patterns of hair cell orientation and tectorial structures in seven families of lizards. These patterns represent the most typical pattern observed for that family. For some species within a family there can be variations from what is pictured. Neural means toward the auditory nerve. This figure is derived primarily from the work of Miller.

axis. Therefore, the hair cells are separated into two groups with all hair cells within the group having the same orientation (see the Anguid family, Fig. 7-5). In the integrated-bidirectional population, neurally and abneurally oriented hair cells are intermixed within a region of the papilla (see the Varanid family, Fig. 7-5).

Four basic tectorial structures are found in the reptilian cochlea: the tectorial membrane, the sallet, the culmen, and free-standing cilia. The tectorial membrane is a thin membrane that attaches at one end to the neural limbus and at the other end to the cilia of hair cells. Wever (1967, 1978) has identified several variations of this structure in reptiles. The sallet is a large mass lying free in the cochlear fluid except for contact with the cilia of one or more hair cells. The culmen, usually found in association with sallets, is a large mass that contacts the cilia of all the hair cells of a population. Free-standing cilia are characterized by the absence of any type of additional tectorial structure. The hair cell cilia stand free in the cochlear fluid.

For all reptiles except lizards, the morphology of the basilar papilla is quite similar among species of the same taxonomic order or suborder; however, this morphology does differ between orders. For example, the papillae of different species of snakes are similar, but the papillae of snakes and turtles are different. Even though the same basic patterns of hair cell orientation and tectorial structure (unidirectional orientation/tectorial membrane) are found in snakes, turtles, *Sphenodon,* and crocodilians, the papillae differ in other morphological features. Amphisbaenians have a tectorial membrane, but hair cell orientation is not yet known. The lizard is unique in that papilla morphology is similar within a family but differs among families (Fig. 7-5), and there can even be variations on the basic pattern within a family. In lizards, seven combinations of hair cell orientations and tectorial structures have, thus far, been identified. Tectorial membranes are found with all three patterns of hair cell orientation; free-standing cilia and sallets are found with segregated-bidirectional hair cell orientations; and culmen are found with either unidirectional or integrated-bidirectional orientation.

3.2 Basilar Membrane Mechanics

In mammals, different points along the basilar membrane are tuned to different frequencies that increase systematically from the apical to basal end of the membrane. In fish and amphibians, the receptor organs are situated on substantial structures that are unlikely to vibrate or to be tuned as in the mammal. In reptiles and birds, the auditory receptor organs are located on basilar membranes that may, therefore, be tuned as in mammals.

The tonotopic organization of the mammalian basilar membrane results, in part, from the systematic change in the mechanical properties of the membrane. In crocodilians and some lizards, the basilar membrane, like the mammalian membrane, is long and narrow and increases in width in the apical direction. In turtles, snakes, amphisbaenians, *Sphenodon,* and many lizards, the basilar mambrane assumes a variety of shapes quite different from the mammalian basilar membrane.

Using the Mossbauer technique, Weiss, Peake, Ling, and Holton (1978) measured basilar membrane motion in the lizard *Gerrhonotus multicarinatus*. The basilar membrane is oval in shape and the papilla is larger at each end than in the middle. Weiss, Peake, Ling, and Holton determined that the membrane is tuned, but all points on the membrane are tuned to approximately the same frequency, which is correlated with the resonant frequency of the middle ear. The resonant frequency varies across individuals of the species from about 1 500 Hz to 4 000 Hz.

3.3 Endolymphatic and Intracellular Potentials

The endolymphatic potential (EP) has consistently been found to be smaller in reptiles than in mammals. In six species of lizards (Schmidt and Fernandez 1962, Weiss, Altmann, and Mulroy 1978, Trincker, Khan, and Mueller-Arnecke 1978), the maximum EP has varied from 15mV to 30mV. Schmidt and Fernandez (1962) recorded a maximum EP of 15mV in one species of snake and a smaller EP (less than 6mV) in three species of turtle and crocodilian. However, Trincker et al. found a maximum EP of 20mV in the crocodilian *Caiman crocodilus*.

Weiss, Altmann, and Mulroy (1978) recorded intracellular resting potentials from the cochlea of *Gerrhonotus multicarinatus* and found that the mean resting potentials of hyaline epithelial cells (-113mV) and supporting cells (-93mV) were more negative than those of hair cells (-73mV). They were uncertain whether these differences represented real differences in the resting potentials of different types of cells or resulted from the insertion of the microelectrode into the cell. In the crocodilian *Caiman crocodilus* and the lizard *Gekko gecko*, hair cell resting potentials varied from about -60mV to -80mV (Trincker et al. 1978), values consistent with the mean values in *Gerrhonotus*. Potentials recorded from various other cells ranged from -30mV to -90mV. In general, these potentials were smaller (less negative) than similar potentials recorded in *Gerrhonotus*.

Weiss, Mulroy, and Altmann (1974) recorded intracellular responses to click stimuli from both hair cells and supporting cells in the cochlea of *Gerrhonotus multicarinatus*. They could distinguish the responses of hair cells from supporting cells and could correlate the initial polarity of the intracellular response of hair cells with hair cell orientation. For a rarefaction click, there is an initial depolarization of hair cells with abneural orientation and an initial hyperpolarization of hair cells with neural orientation. By calculating the Fourier transform of the intracellular responses to clicks, it was possible to obtain a best frequency for cells. The best frequencies form a low frequency group (350 Hz to 580 Hz) and a high frequency group (1 300 Hz to 2 600 Hz). In general, hair cells in the low frequency group were located in the apical hair cell population and hair cells in the high frequency group were located in the basal population.

Weiss et al. (1974) suggested that the responses recorded in supporting cells result from direct electrical coupling of supporting cells to hair cells. Nadol, Mulroy, Goodenough, and Weiss (1976) observed gap junctions between hair cells and supporting cells in *Gerrhonotus*. The gap junction provides a mechanism for direct electrical coupling between cells.

Mulroy, Altmann, Weiss, and Peak (1974) recorded intracellular responses to tones and tone bursts from hair cells and supporting cells in *Gerrhonotus*. The tone burst response consisted of two components that resembled the cochlear microphonic and the summating potential recorded extracellularly in mammals. The responses to tones and the microphonic-like responses to tone bursts were sinusoidal with the fundamental frequency equal to stimulus frequency. Similar responses have been recorded intracellularly from hair cells in the turtle *Chrysemys scripta* for tone bursts with frequencies below 1 000 Hz (Crawford and Fettiplace 1978).

3.4 Gross Response

Electrical responses to acoustic stimuli can be recorded in reptiles using a wire electrode placed on or near the round window. These electrical potentials, appropriately called the gross response by Weiss et al. (1974), consist, in theory, of many components, including the cochlear microphonic, the summating potential, the whole-nerve response, the activity of higher-order auditory structures, biological noise, and electrical artifact. There has been little systematic investigation of the gross response to determine the magnitude of the various components. In the lizards *Gerrhonotus multicarinatus* (Weiss et al. 1974, Turner 1975), *Gekko gecko* (Hepp-Reymond and Palin 1968), and *Coleonyx variegatus* (Campbell 1969), the whole-nerve response is the largest component of the gross response for click stimuli and can reach 40mV peak-to-peak.

Several investigators have recorded the gross response in lizards for tones or tone bursts and found that the microphonic-like response was less than about 20mV peak-to-peak. Johnstone and Johnstone (1969a) examined several species but studied only the lizard *Trachysaurus rugosus* in detail. They concluded that they could record the cochlear microphonic, the whole-nerve response, and the summating potential. The cochlear microphonic is, in general, smaller than the whole-nerve response or the summating potential. Hepp-Reymond and Palin (1968) concluded that for tones, the gross response in *Gekko gecko* is primarily cochlear microphonic except at low frequencies where there is some neural "contamination." Turner (1975) studied the gross response in the *Gerrhonotus multicarinatus* (Fig. 7-6). At low frequencies, the gross response contains distinct diphasic components, most likely the whole-nerve response, that occur with the same frequency as the stimulus. As the frequency is increased, the diphasic components begin to overlap and appear sinusoidal. Therefore, in *Gerrhonotus* the gross response to low frequency tones and tone bursts contains a large neural component.

Schmidt and Fernandez (1962) reported a large cochlear microphonic, approximately 150mV peak-to-peak, in the *Caiman* for 3 000 Hz tone bursts. The size of the microphonic may be correlated with the number of hair cells, which is one to two orders of magnitude greater in crocodilians than in lizards. Also, the basilar papillae of most species of lizards contain at least one bidirectional hair cell population. The contributions to the microphonic by oppositely oriented hair cells may cancel. The size of the microphonic may also be affected by the technique used to record it.

Some investigators have referred to the gross response as the "cochlear potentials," implying that these potentials are primarily of cochlear origin. Wever (1978) presents an extensive review of "cochlear potential" data for reptiles. These data are of interest

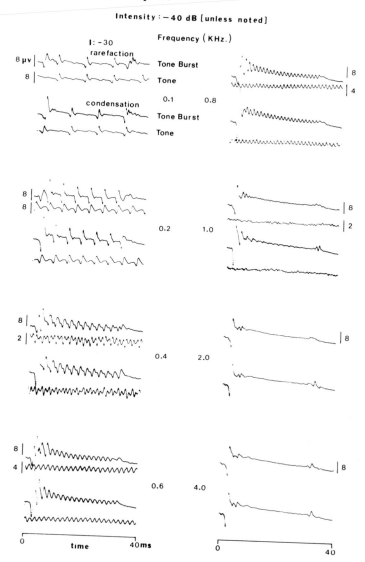

Figure 7-6. Averaged gross responses to tones and tone bursts for *Gerrhonotus multi-carinatus*. The gross responses were recorded using a wire electrode placed on the bone near the round window. The diphasic components evident at low frequencies are most likely the whole-nerve response. At higher frequencies, these diphasic components overlap and appear sinusoidal. These curves indicate that in *Gerrhonotus* the gross responses to low frequency tones and tone bursts contain a large neural component. One hundred responses were averaged to produce each curve. A stimulus intensity of −40 dB equals 70 dB SPL to 80 dB SPL, depending on frequency. The vertical bar indicates amplitude in μV. (From Turner 1975).

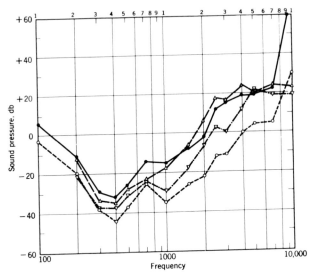

Figure 7-7. Sensitivity curves for four specimens of *Gerrhonotus multicarinatus*. These curves represent the stimulus intensity necessary to maintain the fundamental component of the gross response to continuous tones at 0.1 μV. These sensitivity curves are "contaminated" by neural responses at low frequencies. Sound pressure is relative to 1 dyne/sq. cm. (From Crowley 1964).

because they represent the only physiological data available for many species of reptiles. When the stimulus is a tone, the "cochlear potentials" have typically been analyzed in terms of intensity functions and sensitivity functions. The intensity function is the relation between stimulus intensity and the magnitude of the potentials. The sensitivity function represents the stimulus intensity, as a function of frequency, necessary to maintain the fundamental of the "cochlear potentials" at a designated level, usually 0.1 μV. Considering the results of Turner (1975), the sensitivity function of *Gerrhonotus* (Fig. 7-7) is probably determined in the low frequencies by the whole-nerve response, not by potentials of cochlear origin. In other reptiles, the gross response to low frequency tones may also contain a significant neural component. If intensity and sensitivity functions are to be used to investigate peripheral physiology in reptiles, then the various components of the gross response must be carefully identified in each species.

4 Auditory Nerve

4.1 Anatomy

The eighth nerve in reptiles is divided into an anterior and a posterior branch. Fibers of the anterior branch innervate the crista of the anterior and lateral ampulla and the utricular macula. The posterior branch sends fibers to the saccular macula, the crista of the posterior ampulla, the macula neglecta, the lagena macula, and the basilar papilla (Wever 1978, p. 69).

The branch to the basilar papilla is called the auditory or cochlear nerve. In the lizard *Gerrhonotus multicarinatus* the auditory nerve contains about 600 bipolar neurons. The cell bodies of these neurons are located primarily in the internal auditory canal (see Fig. 7-4). As the fibers approach the basilar papilla, they form a wide, thin sheet and become unmyelinated as they pass through the basilar membrane. Fiber diameter, excluding the myelin sheath, is about 1 μm (Weiss, Mulroy, Turner, and Pike 1976). This basic description for *Gerrhonotus* is applicable to other reptiles, although the number of auditory nerve fibers may vary significantly in relation to basilar papilla size.

4.2 Overview of Physiology

Two techniques can be used to record from primary nerve fibers in reptiles. When the round window membrane and surrounding bone are removed, the primary fibers are accessible via scale tympani (see Fig. 7-4). It is possible to record from fibers peripheral to the ganglion and close to the papilla. Since the nerve forms a sheet close to the papilla, this technique facilitates measurements of the tonotopic organization of the nerve. The second technique utilizes a central approach to record from fibers as they emerge from the internal auditory canal.

Single-unit recordings from auditory nerve fibers have been obtained from five species of lizards and one species of crocodilian. While this represents a small sample of the existing species of reptiles, these data do provide significant insight into auditory function in reptiles. Before reviewing the data for individual species, it is worthwhile to provide a brief overview.

All fibers examined have been tuned with characteristic frequencies (CF) below about 4 500 Hz. The tuning curves are "V"-shaped, although in some species tuning curves may be "complex," meaning that the tuning curves contain additional peaks at frequencies above and/or below CF. Both symmetrical and asymmetrical tuning curves have been reported. When the tuning curve is asymmetrical, the high frequency side generally has the greater slope. Tuning curve "sharpness" is indicated by the measure Q_N, which equals the CF divided by the bandwidth of the tuning curve at "N" dB re threshold at CF. Values of Q_N vary significantly with CF for fibers in the same species, and the relation between Q_N and CF is different for different species. It is difficult, however, to compare tuning curve properties for different species because the tuning curves were obtained using different experimental procedures.

Separate fiber populations have been identified in several species of lizards on the basis of differences in tuning curve properties and/or other measures of the response of the fibers, and the different fiber populations have been correlated with morphologically different hair cell populations. Fibers have been found in most species studied demonstrating a physiology much like that of mammalian primary fibers, and at least some fibers in every reptile have spontaneous activity with rates comparable to those in mammals. The most sensitive fiber thresholds at CF varied slightly with species over a range of 5 dB SPL to 20 dB SPL.

4.3 Anguid Lizards

The papilla of *Gerrhonotus multicarinatus* is typical of the Anguid family, containing two hair cell populations (see Fig. 7-5). The apical population has unidirectional orientation and a tectorial membrane. The basal population has segregated-bidirectional orientation and free-standing cilia. The papilla is about 400 μm long and contains 150 hair cells, about one-third of which are in the apical population.

Weiss et al. (1976) recorded from auditory nerve fibers in *Gerrhonotus* and found that, on the basis of tuning curve properties, the fibers can be divided into a low CF population (200 Hz to 800 Hz) and a high CF population (900 Hz to 4 000 Hz). Low CF fiber tuning curves are asymmetrical and are sharper than the symmetrical high CF fiber tuning curves. These differences are evident in measurements of "Q" and the slopes of the sides of the tuning curves (Figs. 7-8, 7-9, 7-10). Low CF fiber tuning curves resemble cat tuning curves of the same CF. Q_{10} for the low CF fibers varies from 0.3 to 6.0; this is comparable to Q_{10} for cat fibers with CF below 1 000 Hz (Kiang, Watanabe, Thomas, and Clark 1965).

Weiss et al. (1976) used a dye-marking technique to determine the tonotopic organization of the auditory nerve close to the papilla. Since the nerve forms a sheet as it enters the papilla, its tonotopic organization is probably a good representation of the tonotopic organization of the papilla. Their results indicate that the low CF fibers enter the apical region of the papilla and most likely synapse to hair cells with unidirectional orientation and a tectorial membrane. The high CF fibers enter the basal region of the papilla and apparently synapse to hair cells with free-standing cilia in the basal bidirectional population (Fig. 7-11).

Measures of fiber response other than tuning curves can also be used to distinguish

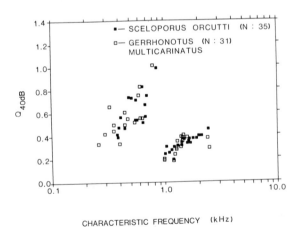

Figure 7-8. Comparison of Q_{40} for *Sceloporus orcutti* and *Gerrhonotus multicarinatus*. Q_{40} is defined as CF/bandwidth of tuning curve at 40 dB above threshold at CF. Measurements of Q_{40} clearly indicate two distinct fiber populations in *Sceloporus* and *Gerrhonotus*. Note also the quantitative agreement of Q_{40} for the two species. N = number of fibers. (From Turner and Nielsen 1979).

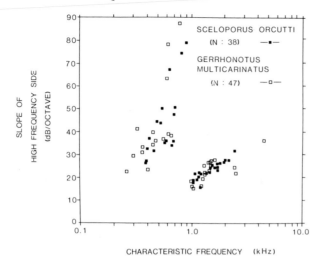

Figure 7-9. Comparison of the high frequency slope of tuning curves for *Sceloporus orcutti* and *Gerrhonotus multicarinatus*. High frequency slope is the slope of a line from CF to a point on the high frequency side of the tuning curve that is 40 dB in intensity above CF. Note the two fiber populations and the quantitative agreement of high frequency slope for the two species. N = number of fibers. (From Turner and Nielsen 1979).

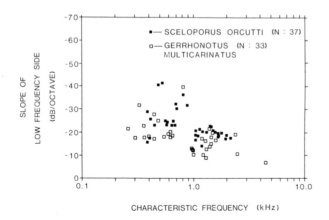

Figure 7-10. Comparison of the low frequency slope of tuning curves for *Sceloporus orcutti* and *Gerrhonotus multicarinatus*. Low frequency slope is the slope of a line from CF to a point on the low frequency side of the tuning curve that is 40 dB in intensity above CF. Note the two populations and the quantitative agreement of low frequency slope for the two species. N = number of fibers. (From Turner and Nielsen 1979).

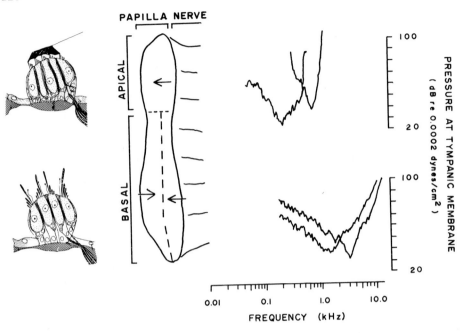

Figure 7-11. The relation of papilla morphology to tuning curve properties in *Gerrhonotus multicarinatus*. The hair cells in the apical population have unidirectional orientation and a tectorial membrane. Auditory nerve fibers that enter this region of the papilla have "sharp" tuning curves and CFs below 900 Hz. The hair cells in the basal population have bidirectional orientation and free-standing cilia. Fibers that enter this region of the papilla have "broad" tuning curves and CFs greater than 900 Hz. The tuning curves are from Weiss et al. 1976; the anatomical cross sections of the papilla are from Mulroy 1974.

two fiber populations in *Gerrhonotus* (Turner 1975, Turner, Nielsen, and Teas 1976). The physiology of low CF fibers is much like that of cat fibers of the same CF. For click stimuli, the post-stimulus-time (PST) histograms are characterized by multiple peaks that interleave when stimulus polarity is reversed (Fig. 7-12). The interpeak interval is a function of fiber CF; however, as CF increases, the interval becomes greater than predicted by 1/CF (Fig. 7-13). The fibers demonstrate once-per-period, phase-locked response to tone bursts with frequencies below 1 500 Hz (Fig. 7-14). Once-per-period, phase-locked response means that there is a time during one period of the sinusoidal stimulus for which the fiber has a maximum probability of firing. This is indicated in the PST histograms by peaks that occur once during each period of the tone burst. The noise burst response is designated "primary-like" because it is similar to the response of primary fibers in the cat (Fig. 7-15). Two-tone rate suppression (TTRS) is present in the low CF fibers (Holton 1977, Holton and Weiss 1978).

As with tuning curves, the discharge patterns of the high CF fibers are much different from those of the low CF fibers and mammalian fibers. The click PST histogram

COMPOUND PST HISTOGRAMS
STIMULUS. CLICK

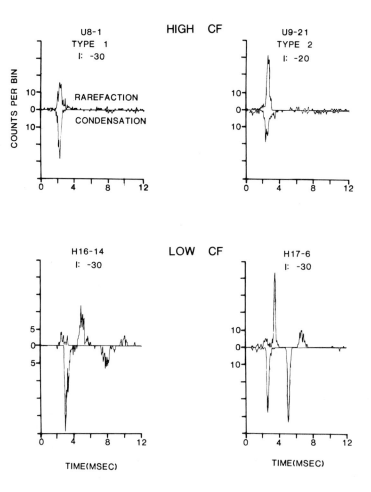

Figure 7-12. Compound PST histograms of the response of auditory nerve fibers to clicks in *Gerrhonotus multicarinatus*. Note that for high CF fibers a single peak is present in the histogram for either click polarity, whereas for low CF fibers multiple peaks are present for either click polarity. An explanation of Type 1 and Type 2 fibers appears later in the text. Intensity (I) is expressed in dB relative to a peak pressure of 170 dynes/sq. cm. The top histogram in each compound histogram is for a rarefaction click, the bottom histogram is for condensation. There were 256 stimulus presentations for each histogram. Zero time corresponds to the onset of the stimulus. (From Turner et al. 1979).

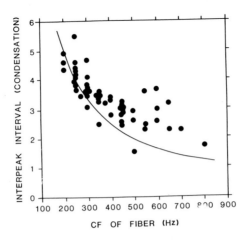

Figure 7-13. The relation between interpeak interval and fiber CF for low CF fibers in *Gerrhonotus multicarinatus*. Interpeak interval is the time between peaks in the PST histogram for condensation clicks. The solid line is a plot of I/CF. As CF increases, the interpeak interval tends to be longer than predicted by the relation I/CF.

has a single peak for either click polarity (see Fig. 7-12). The responses to high frequency tone bursts and to noise bursts are similar but are different from "primary-like" (see Fig. 7-15). The PST histograms are characterized by a large peak at the onset of the stimulus, followed by a relatively constant discharge for the duration of the stimulus. This response pattern has been designated "peaky" by Manley (1977). The high CF fibers also demonstrate once-per-period, phase-locked response to low frequency tones (see Fig. 7-14). Two-tone rate suppression has not been observed for high CF fibers (Holton and Weiss 1978).

Both the low CF and the high CF fibers in *Gerrhonotus* differ from mammalian fibers in that the latency to the first peak in the click PST histogram does not decrease with increasing CF, but is approximately the same for all fibers (Turner et al. 1976). This indicates the absence of a traveling wave, a result consistent with the type of basilar membrane tuning in *Gerrhonotus* (Weiss, Peake, Ling, and Holton 1978).

Hair cell orientation in *Gerrhonotus* determines the relation between fiber response and stimulus polarity (Turner, Nielsen, and Teas 1975, Turner, Teas, and Nielsen 1979). The high CF fibers can be classified as type 1 or type 2. The response of a type 1 fiber to a stimulus with one polarity (e.g., condensation) is similar to the response of a type 2 fiber to the same stimulus with opposite polarity (e.g., rarefaction). Also, high CF fibers can be divided into two groups on the basis of the phase of their response to low frequency tones. These two groups differ in phase by approximately 180° (Fig. 7-16). The low CF fibers cannot be divided into two or more groups using the same criteria employed for the high CF fibers. These results are consistent with the unidirectional and bidirectional orientation of the two hair cell populations and suggest that an individual high CF fiber synapses to hair cells of the same orientation in the basal bidirectional population.

COMPOUND   PST   HISTOGRAMS

STIMULUS: TONE BURST (0.6KHZ)

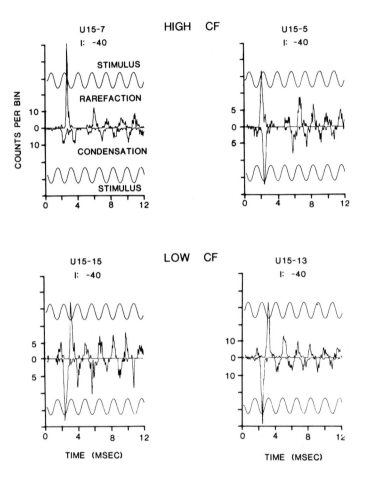

Figure 7-14. Compound PST histograms of the responses of high CF and low CF fibers in *Gerrhonotus multicarinatus* to 600 Hz tone bursts. Both the high CF and the low CF fibers demonstrate once-per-period, phase-locked response to sinusoidal stimuli. Tone burst intensity was 64 dB SPL. The waveform labeled "stimulus" depicts the electrical signal to the earphone. Each histogram is for 256 stimulus presentations. (From Turner et al. 1979).

The whole-nerve response has been analyzed in detail only in *Gerrhonotus multicarinatus* (Turner 1975), although there are limited data from other lizards (Campbell 1969a, Hepp-Reymond and Palin 1968, Johnstone and Johnstone 1969a). In *Gerrhonotus,* the high CF fiber population contributes a component to the whole-nerve response that is independent of stimulus polarity and most likely results from the bidirectional orientation of the basal hair cell population. The contribution of low CF

TONE BURST (f = 2.0 kHz)
AND NOISE BURST
PST HISTOGRAMS

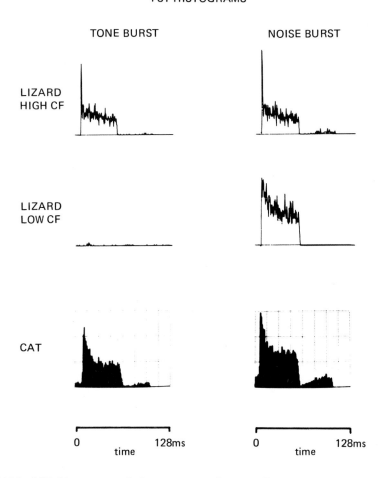

Figure 7-15. PST histograms of the response of high CF and low CF fibers in *Gerrhonotus multicarinatus* to tone bursts and noise bursts. High CF fibers have "peaky" tone burst and noise burst response patterns that are characterized by a sharp peak at the onset of the response. The low CF fibers show very little response to high frequency tone bursts, but a noise burst response that is "primary-like." For comparison, data from the cat (Kiang et al. 1965) are shown.

fiber population changes with stimulus polarity, consistent with the unidirectional orientation of the apical hair cell population. The averaged gross response to click stimuli contains a large, diphasic N_1 - P_1 that is the whole-nerve response (Fig. 7-17). For a condensation click, the initial discharges of the high and low CF fibers occur with approximately the same latency producing the large N_1 - P_1. When click polarity is reversed, the low CF fibers no longer discharge synchronously with the high CF fibers resulting in a smaller N_1 - P_1 and the broadening of P_1. The waves N_2 - P_2 and

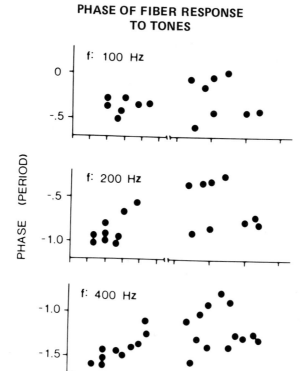

PHASE OF FIBER RESPONSE TO TONES

Figure 7-16. Phase of fiber response to continuous tones. High CF and low CF fibers in *Gerrhonotus multicarinatus* demonstrate phase-locked response to continuous low frequency tones. The frequency (f) of the tone is indicated on each plot. The phase of the phase-locked response was measured relative to the electrical signal to the earphone. Any phase measurement could be adjusted by integer multiples of a period. The phase data for the high CF fibers form two populations approximately one-half period (180°) apart. The phases of the low CF fibers vary with fiber CF but do not form two distinct populations. (From Turner et al. 1979).

N_3 - P_3 are probably generated by higher-order auditory structures. No large population of primary fibers discharges synchronously with a latency appropriate to generate N_2 - P_2 or N_3 - P_3.

4.4 Iguanid Lizards

The basilar papillae of Iguanids contain the same two types of hair cell populations found in the Anguid: unidirectional orientation/tectorial membrane and segregated-bidirectional/free-standing cilia (see Fig. 7-5). Miller (1973, p. 310) states: "In many

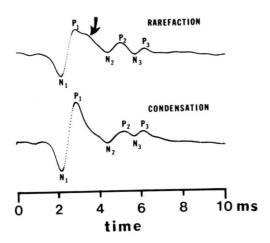

Figure 7-17. Averaged gross response to click stimuli for *Gerrhonotus multicarinatus*. The gross response was recorded using a wire electrode placed on the bone near the round window. The diphasic N_1 - P_1 is the whole-nerve response. The amplitude of N_1 - P_1 is greater for condensation clicks. The arrow indicates the broadening of P_1 for a rarefaction click. One hundred responses were averaged to produce each curve; positive is up. Zero time corresponds to the onset of the stimulus. (From Turner 1975).

details, the papilla basilaris of *Gerrhonotus multicarinatus* is more similar to that of the Iguanid than any other so far described group of lizards." The papillae differ in one important aspect: the papilla of the Iguanid contains two segregated-bidirectional/free-standing cilia populations. A bidirectional population is located at each end of the papilla with the unidirectional population in the middle.

Because of the similarity in anatomy, a comparison of single-unit response in the Anguids and Iguanids is of interest. In the Iguanid, *Sceloporus orcutti*, the fibers can be divided into a low CF (200 Hz to 900 Hz) population and a high CF (900 Hz to 3 500 Hz) population (Turner 1978, Turner and Nielsen 1979) (Fig. 7-18). Two fiber populations are evident from measurements of Q_{40} (see Fig. 7-8) and the slopes of the sides of the tuning curves (see Figs. 7-9, 7-10). The low CF fibers are more sharply tuned than the high CF fibers. A comparison of tuning curve properties for *Sceloporus* and *Gerrhonotus* reveals an impressive quantitative agreement in CF ranges, Q40, and slopes (Figs. 7-8, 7-9, 7-10).

The two fiber populations in *Sceloporus* can be distinguished on the basis of their responses to click, tone burst and noise burst acoustic stimuli. The responses of high CF and low CF fibers in *Sceloporus* to these stimuli are basically the same as described previously for *Gerrhonotus* (see Section 4.3).

Both *Gerrhonotus* and mammals have an apical to basal tonotopic organization of low to high frequency. If this were true in *Sceloporus*, then the apical bidirectional and the basal bidirectional hair cell populations would correspond to different frequencies, resulting in three nerve fiber populations. However, only two populations are evident and these are almost physiologically identical to the two fiber populations

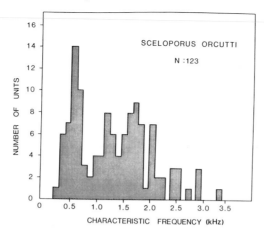

Figure 7-18. Histogram of characteristic frequency (CF) of auditory nerve fibers in *Sceloporus orcutti*. The fibers form a low CF (200 Hz to 900 Hz) and a high CF (900 Hz to 3 500 Hz) population. N = total number of fibers (units) in histogram. (From Turner and Nielsen 1979).

in *Gerrhonotus*, implying that both bidirectional populations are sensitive to high frequencies. The tonotopic organization of the nerve has been grossly measured by recording fiber CF as the microelectrode is "stepped" across the nerve. The results indicate a tonotopic organization of high CF-low CF-high CF.

4.5 Varanid Lizards

The papillae of Varanid lizards are constricted or completely divided into two segments (see Fig. 7-5). The entire papilla is covered by a tectorial membrane, and hair cell orientation is integrated-bidirectional except for the apical end of the basal segment where orientation is unidirectional.

Manley (1977) recorded from auditory nerve fibers in *Varanus bengalensis* and found that the fibers form a low CF (250 Hz to 1 100 Hz) and a high CF (1 300 Hz to 3 000 Hz) population. Fibers with CF less than 550 Hz have multiple peaks in their click PST histograms that interleave when polarity is reversed, and an interpeak interval of approximately I/CF. Their tone burst response is "primary-like" and similar to the response of the low CF fibers in *Gerrhonotus multicarinatus* and *Sceloporus orcutti*. The low CF fibers with CF greater than 550 Hz enter the region of the basal papilla with bidirectional hair cell orientation. The PST histograms of these fibers have multiple peaks in response to clicks, but the peaks do not change latency when polarity is reversed. The tone burst response is "intermediate" between "primary-like" and "peaky." Close to the papilla, the low CF fibers have a tonotopic organization that increases in the basal direction.

The high CF fibers enter the apical segment of the papilla with CF increasing in the apical direction. The multiple peaks in the histograms do not change latency when click polarity is reversed, and the interpeak interval is not a function of fiber CF. The tone burst response is "peaky" and is similar to the response of high CF fibers in *Gerrhonotus* and *Sceloporus*.

4.6 Other Lizards

Johnstone and Johnstone (1969b) recorded from a small number of fibers in *Trachysaurus rugosus* (Family Scincidae). The fibers have relatively "broad" tuning curves with CF as great as 3 000 Hz. Eatock and Manley (1976) examined the effect of temperature on tuning curves in *Gekko gecko* (Family Gekkonidae) and found that an increase in temperature increases fiber CF and frequently improves sensitivity at CF. Manley (1974) recorded from five auditory nerve fibers in *Gekko gecko*. For tone bursts three fibers had a "peaky" response, one a "primary-like" response, and one an "on" response.

4.7 Crocodilians

The crocodilians have the largest basilar papilla of any reptile. In *Caiman crocodilus* (Leake 1977) the papilla can reach about 4 mm in length and contain more than 10 000 hair cells. A tectorial membrane overlays the hair cells, all of which have unidirectional orientation. Two distinct hair cell populations, designated inner and outer hair cells, are evident, although it is unknown if they are homologous to the inner and outer hair cells in mammals. The two types of hair cells are distributed along the papilla with as many as 10 inner and 19 outer rows of hair cells situated across the papilla. The papilla and basilar membrane increase in width in the apical direction.

Klinke and Pause (1977) recorded from primary neurons in the *Caiman* (the particular species was not indicated in the article). The fibers are sharply tuned with Q_{10} increasing with CF from 1.0 to 6. CF was as great as 2 800 Hz; however, it was not possible to separate the fibers into two or more distinct populations on the basis of tuning curve properties even though 40% of the fibers contact the outer hair cells. The physiology of these fibers is much like that of mammals and the low CF fibers in *Gerrhonotus multicarinatus*, *Sceloporus orcutti*, and *Varanus bengalensis*. Multiple peaks are evident in the click PST histograms for fibers with CF below 1 200 Hz and the interpeak interval is approximately I/CF. At higher CFs, only a single peak may be present. Fibers demonstrate once-per-period, phase-locked response to tones with frequency below 1 500 Hz and two-tone rate suppression. The fibers in *Caiman* differ from the low CF fibers in lizards; in *Caiman* the latency to the first peak in the click PST histograms decreases as much as 4 msec with increasing CF. This indicates a traveling wave and the possibility of basilar membrane tuning like that in mammals.

4.8 Transduction Processes in the Reptilian Cochlea

Direct measurement has shown that the basilar membrane in the lizard *Gerrhonotus multicarinatus* is not tuned like the basilar membrane in mammals. Is "mammalian tuning" present in any reptilian cochlea? The answer to this question must be inferred from available anatomical and physiological data since basilar membrane motion has not been measured in any other reptile. The crocodilians are the best candidate for "mammalian tuning" because among reptiles the anatomy of the crocodilian cochlea is most like that of mammals. Single-unit data from *Caiman* indicate a traveling wave, a physiological property consistent with "mammalian tuning." The best candidate for "mammalian tuning" among lizards is the family Gekkonidae, which has a highly developed and anatomically distinct cochlea. Its papilla is one of the longest among lizards, and papilla and basilar membrane dimensions systematically decrease from the apical to the basal end (Wever 1978, pp. 482-537). Unfortunately, no physiological data are available relevant to the issue of basilar membrane tuning. The Varanid lizards also have a large papilla; however, the tonotopic organization of the auditory nerve in *Varanus bengalensis* indicates a tontopic organization of the papilla that is quite different from the tonotopic organization of the mammalian cochlea (Manley 1977). In addition, single-unit data from *Varanus* are inconclusive with regard to this issue. Very little can be said about basilar membrane tuning in snakes, turtles, amphisbaenians, and *Sphenodon* except that cochlear anatomy in these reptiles does not suggest the systematic variations in mechanical properties associated with "mammalian tuning."

In *Gerrhonotus* the tuning of the basilar membrane contributes little to the tuning of auditory nerve fibers. The fibers are tuned over a wide range of frequencies and, therefore, other mechanisms must be responsible. Furthermore, since intracellular recordings in *Gerrhonotus* indicate that the hair cells are tuned with a tonotopic organization much like that of the primary fibers, the mechanisms responsible must operate at or near the level of the hair cell. Wever (1971) suggested that the tectorial structures are involved in the mechanics of hair cell stimulation. Weiss, Peake, Ling, and Holton (1978) proposed that the tectorial structures are involved in the tuning of fibers and, in particular, that systematic variations in basal hair cell cilia length determine the tonotopic organization of the high CF fibers. Single-unit data from *Gerrhonotus multicarinatus* and *Sceloporus orcutti* indicate the importance of the tectorial structures for determining the response of primary fibers. Even in lizards belonging to diferent families, fibers that innervate morphologically similar hair cell populations have a remarkable similarity in physiology. Thus, it is reasonable to conclude that the tectorial structures are involved in hair cell stimulation and the coding of acoustic information.

Wever (1971) also concluded that the different tectorial structures stimulate hair cells by different mechanisms. Single-unit data from *Gerrhonotus* and *Sceloporus* support this statement; fibers innervating hair cell populations with different tectorial structures display different physiological patterns. Can particular patterns of fiber response be associated with particular tectorial structures? In all reptiles studied, primary fibers with "sharp" tuning curves apparently innervate hair cell populations with tectorial membranes. In *Gerrhonotus* and *Sceloporus* fibers entering the region of the papilla with free-standing cilia have "broad" tuning curves. The limited data on

Trachysaurus rugosus (Family Scincidae) also indicate "broad" tuning curves. These fibers innervate hair cell populations with sallets or culmen. These results suggest that "sharp" tuning curves are associated with tectorial membranes and "broad" tuning curves with other tectorial structures. In *Varanus bengalensis,* however, the only tectorial structure is the tectorial membrane, and the high CF fibers have "broad" tuning curves. The data are too limited, particularly for fibers innervating hair cell populations with sallets and culmens, to resolve this fundamental question concerning the relation of tuning curve properties to tectorial structure.

The relation between fiber CF and papilla morphology is unclear. Maximum fiber CF is not correlated with papilla size. The *Caiman* papilla is about ten times as long as the papilla in *Gerrhonotus,* but the reported maximum fiber CF is greater in *Gerrhonotus* than in *Caiman* (4 000 Hz versus 2 800 Hz). Also, there is no obvious correlation between CF range and tectorial structure.

With the exception of one "on-unit" reported by Manley (1974) for *Gekko gecko,* only "primary-like" and "peaky" tone burst patterns have been found. In *Gerrhonotus, Sceloporus,* and *Varanus,* where it is possible to correlate a fiber with a hair cell population, fibers with "primary-like" tone burst responses are associated with unidirectional/tectorial membrane hair cell populations. In addition, neurons in the cochlear nucleus of *Caiman crocodilus* display a "primary-like" tone burst response (see Section 5.3). The entire *Caiman* papilla has unidirectional hair cell orientation and a tectorial membrane. These results suggest a correlation between a "primary-like" tone burst response and unidirectional/tectorial membrane hair cell populations. In *Gerrhonotus* and *Sceloporus,* fibers with a "peaky" response are associated with segregated-bidirectional/free-standing cilia hair cell populations, and in *Varanus* with integrated-bidirectional/tectorial membrane hair cell populations. For these fibers the only apparent correlation is between a "peaky" tone burst response and bidirectional hair cell orientation.

The fibers in *Gerrhonotus, Sceloporus, Varanus,* and *Caiman* with "primary-like" tone burst response also have many other response properties similar to mammalian auditory nerve fibers. Like mammalian fibers, these fibers apparently innervate hair cells with unidirectional orientation and a tectorial membrane. This common anatomy and physiology suggest that there may be common transduction mechanisms operating in the mammalian and reptilian cochlea.

5 Central Nervous System

5.1 Anatomy

Auditory nerve fibers from the basilar papilla terminate in as many as four nuclei in the cochlear nucleus, including nucleus angularis, nucleus magnocellularis medialis, nucleus magnocellularis lateralis, and nucleus laminaris (Miller 1975). Second-order fibers project bilaterally to the superior olivary complex and contribute fibers primarily to the contralateral lateral lemniscus and its nucleus (Leake 1976). Fibers of the lateral lemniscus terminate in the midbrain in the torus semicircularis, the homologue of the mammalian inferior colliculus. Fibers from the torus semicircularis may project to nucleus Z and nucleus reuniens, a midline nucleus of the diencephalon. Ultimately, ascending fibers terminate in the dorsolateral area of the telencephalon (Campbell and Boord 1974).

5.2 Lizards

Recordings from the cochlear nucleus of the Iguanid lizards *Crotophytus wislizenii* and *Sceloporus cyanogenys* (Manley 1970a, 1974) indicate that neurons are tuned with CFs as high as 1 600 Hz and 3 000 Hz, respectively. In both species, the average Q_{10} is smaller for neurons with CF greater than 1 000 Hz than for neurons with CF below 1 000 Hz, suggesting two populations of fibers. Almost all units recorded from *Crotophytus* had a "primary-like" tone burst response and a CF less than 1 000 Hz. These results are consistent with measurements of Q_{40} and the tone burst responses of the high CF and the low CF auditory nerve fibers in *Sceloporus orcutti* (see Section 4.4).

In the cochlear nucleus of the Gekkonid lizards *Gekko gecko* (Manley 1972a, 1974) and *Coleonyx variegatus* (Manley 1970a, Suga and Campbell 1967) neurons have been found with CFs as great as 5 000 Hz and 4 500 Hz, respectively. In both species, Q_{10} increases with CF to a maximum between 5 and 6. In *Gekko*, the tone burst response is "gecko-type," which is essentially the same as "peaky." The one neuron reported with CF below 1 000 Hz had a "primary-like" tone burst response. In *Coleonyx*, responses to auditory stimuli could be recorded from neurons in the region of the torus semicircularis (Kennedy 1974).

In the Varanid lizard *Varanus bengalensis,* the physiology of neurons in the cochlear nucleus is very similar to the physiology of auditory nerve fibers (Manley 1976, 1977). There is good agreement in CF range, tone burst response, and the description of a low CF and a high CF fiber population (see Section 4.5). Fibers in the auditory nerve and the cochlear nucleus differ primarily in the sharpness of the high CF fibers, which have an average Q_{10} significantly less than the average Q_{10} of neurons in the cochlear nucleus. Manley (1977) suggests that this difference results from surgical damage to the auditory nerve; however, no evidence is presented to support this.

5.3 Crocodilians

Neurons in the cochlear nucleus of *Caiman crocodilus* are tonotopically organized with CFs as great as 2 900 Hz. The Q_{10} appears to increase with CF up to 4.5 and the tone burst response is "primary-like" (Manley 1970b). Midbrain (torus semicircularis) auditory neurons in *Caiman crocodilus* have CFs as great as 1 850 Hz and demonstrate a tonotopic organization within the nucleus (Manley 1971). Two or more physiologically distinct populations are not evident for neurons in the cochlear nucleus or the torus semicircularis.

5.4 Turtles

Neurons in the cochlear nucleus of turtle *Terrapene carolina major* have CFs from 100 Hz to 500 Hz (Manley 1970a). This upper limit of CF is much lower than has been found for lizards and crocodilians. The relation of Q_{10} to CF suggests two populations with CFs near 140 Hz and 400 Hz.

5.5 Snakes

Recordings of gross electrical potentials from the midbrains of Crotalid, Boid, and Colubrid snakes indicate that their auditory systems respond to aerial sound and substrate vibration in the frequency range 150 Hz to 600 Hz with best sensitivity at 250 Hz to 300 Hz (Hartline and Campbell 1969, Hartline 1971a, 1971b). For aerial sound, responses were obtained at intensities as low as 50 dB SPL.

A second sensory system, discovered by Hartline and Campbell and named the somatic system, responds to aerial sound and substrate vibration over a greater frequency range (50 Hz to 1 000 Hz) than the auditory system; however, the auditory system is as much as 20 dB more sensitive for frequencies near 300 Hz. The response of the somatic system is relatively constant across frequencies and probably uses skin mechanoreceptors that transmit information to the central nervous system via the spinal cord.

Because the snake lacks an external ear and overt behavioral responses to sound, it is commonly believed that snakes are "deaf" to aerial sounds and hear via substrate vibration. However, the data of Hartline and Campbell plus additional physiological data of Wever and Vernon (1960) suggest that snakes are surprisingly sensitive to aerial sounds.

5.6 Peripheral versus Central Physiology

Manley (1977, p. 259) states that "the cochlear nuclei in lizards do little more than act as a relay station to higher brain centers." Auditory and cochlear nucleus neurons are physiologically more similar in reptiles than in mammals. There is good agreement in CF range and fiber thresholds and the same tone burst response patterns, "primary-like" and "peaky," are present at the level of the auditory nerve and the cochlear nucleus. Other patterns, such as "choppers" and "pausers," which have been recorded in the mammalian cochlear nucleus, have not been found in the cochlear nucleus of reptiles.

A detailed comparison of auditory nerve and cochlear nucleus physiology is possible only in *Varanus bengalensis*. This is the only species of reptile for which there exist sufficient data from both auditory and cochlear nucleus neurons. In *Varanus* there is good agreement between auditory nerve and cochlear nucleus physiology except for the sharpness of the high CF tuning curves (see Section 5.2). Considering this difference in *Varanus* and the small number of reptile species that have been studied, it is inappropriate to assume a one-to-one correspondence between peripheral and central physiology.

6 Behavioral Measurements

Behavioral measurements of auditory sensitivity in reptiles have been reported only for the turtle *Chrysemys scripta* (Patterson 1966). The auditory sensitivity curve is U-shaped with thresholds of about 85 dB SPL at 40 Hz and 1 000 Hz. Maximum sensitivity is about 40 dB SPL to 50 dB SPL in the frequency range of 200 Hz to 600 Hz (Fig. 7-19).

The behavioral data for *Chrysemys* can be compared to the recordings from data for cochlear nucleus neurons in the *Terrapene carolina major* (see Section 5.4), although it is important to remember that these data were obtained from two different species of turtles. The most sensitive fibers in *Terrapene* have thresholds of about 35 dB SPL to 45 dB SPL. These thresholds agree well with the behavioral sensitivity in *Chrysemys*. Behavioral sensitivity in *Chrysemys* decreases rapidly with frequency for frequencies about 600 Hz, correlating nicely with the upper CF limit of 500 Hz in *Terrapene*.

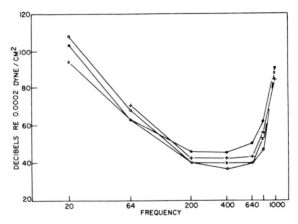

Figure 7-19. Behavioral auditory thresholds for four specimens of the turtle *Chrysemys scripta*. (From Patterson 1966).

7 Hearing in Reptiles

What are the auditory capabilities of reptiles relative to mammals? Although behavioral measurements of auditory performance are necessary to answer this question, available physiological data suggest that the reptile and the mammal differ primarily in their high frequency auditory sensitivity. The behavioral auditory sensitivity curve for the turtle indicates poor sensitivity for frequencies greater than about 800 Hz. Measurements of the performance of the "typical" lizard middle ear demonstrate a moderate inferiority to the mammalian middle ear in the ability to transmit high frequencies. The most crucial factor limiting high frequency performance in reptiles appears to be the lack of fibers with high CFs. No fibers have been found in any species of reptile with CF greater than 5 000 Hz, and in many species maximum CF is significantly lower.

For low frequencies, the performance of the reptilian ear is more like that of the mammalian ear. The "typical" lizard middle ear appears equal to the mammalian middle ear in footplate displacement and impedance matching. In most reptiles studied, there are low CF auditory nerve fibers with physiology much like mammalian auditory nerve fibers. In some reptiles, auditory capability may be limited by the small number of auditory nerve fibers. Even if individual fibers have thresholds and tuning comparable to mammalian fibers, a small fiber population could adversely affect certain dimensions of auditory performance such as frequency and intensity discrimination.

It is difficult to compare the hearing of different reptiles because the quantity of available physiological data is so small. Most physiological measurements have been made on reptiles with a "typical" lizard middle ear and a round window system for pressure relief in the cochlea. The data for the turtle and snake represent the only information for "divergent" forms of middle ear and cochlear anatomy. The turtle employs a reentrant fluid circuit for pressure relief; the snake also has a reentrant fluid circuit and lacks a tympanic membrane. The data indicate that hearing in turtles and snakes is limited to low frequencies. The effect of the reentrant fluid circuit on auditory performance is unknown, although Wever (1978, p. 91) suggests that this type of system may limit high frequency sensitivity. There has been no measurement of middle ear transfer function for a middle ear without a tympanic membrane. This type of middle ear is probably less sensitive to aerial sound than the "typical" lizard middle ear but more sensitive than has been previously assumed. If hearing is judged on the basis of sensitivity and high frequency capability for aerial sound, then the crocodilians and lizards probably have the best hearing among the reptiles.

Acknowledgments. The author would like to thank Dr. D. Nielsen, Dr. P. Cornett, and Ms. D. Hopkins for their assistance in the preparation of this manuscript. The author's research that appeared in this text was supported, in part, by a Henry Ford Hospital Institutional Grant from the Ford Foundation and by National Institutes of Health grants NS13462, NS06459, NS05475, and NS07287.

References

Campbell, C., Boord, R.: Central auditory pathways of nonmammalian vertebrates. In: Handbook of Sensory Physiology, Vol. V/1. Kerdel, W., Neff, D. (eds.). New York: Springer-Verlag, 1974, pp. 337-362.

Campbell, H. W.: The effects of temperature on the auditory sensitivity of lizards. Physiol. Zool. 42, 183-210 (1969).

Crawford, A., Fettiplace, R.: Ringing responses in cochlear hair cells of the turtle. J. Physiol. 284, 120-122 (1978).

Crowley, D.: Auditory responses in the alligator lizard. J. Auditory Research 4, 135-143 (1964).

Eatock, R. A., Manley, G. A.: Temperature effects on single auditory nerve fiber responses. J. Acoust. Soc. Amer. 60, Supp. 1, 580 (1976).

Guinan, J. J., Peake, W. T.: Middle-ear characteristics of anesthetized cats. J. Acoust. Soc. Am. 41, 1237-1261 (1967).

Hartline, P.: Physiological basis for detection of sound and vibration in snakes. J. Exp. Biol. 54, 349-371 (1971a).

Hartline, P.: Midbrain responses of the auditory and somatic vibration systems in snakes. J. Exp. Biol. 54, 373-390 (1971b).

Hartline, P., Campbell, H. W.: Auditory and vibratory responses in the midbrains of snakes. Science 163, 1221-1223 (1969).

Hepp-Reymond, M., Palin, J.: Patterns of cochlear potentials of the Tokay Gecko (Gekko gecko). Acta Otolaryng. 65, 270-292 (1968).

Holton, T.: Two-tone rate suppression in lizard cochlear nerve fibers. M. S. Thesis, Massachusetts Institute of Technology, Cambridge, Ma., 1977.

Holton, T., Weiss, T.: Two-tone rate suppression in lizard cochlear nerve fibers: relation to receptor organ morphology. Brain Research 159, 219-222 (1978).

Johnstone, J. R., Johnstone, B. M.: Electrophyiology of the lizard cochlea. Experimental Neurology 24, 99-109 (1969a).

Johnstone, J. R., Johnstone, B. M.: Unit responses from the lizard auditory nerve. Experimental Neurology 24, 528-537 (1969b).

Kennedy, M. C.: Auditory multiple-unit activity in the midbrain of the Tokay Gecko (Gekko gecko, L.) Brain Behav. Evol. 10, 257-264 (1974).

Kiang, N., Watanabe, T., Thomas, E., Clark, L.: Discharge patterns of single fibers in the cat's auditory nerve. Cambridge: M.I.T. Press, 1965.

Klinke, R., Pause, M.: The performance of a primitive hearing organ of the cochlea type: Primary fiber studies in the caiman. In: Psychophysics and Physiology of Hearing. Evans, E., Wilson, J. (eds.). New York: Academic Press, 1977, pp. 101-114.

Leake, P. A.: Scanning electron microscopic observations of labyrinthine sense organs and fiber degeneration studies of secondary vestibular and auditory pathways in Caiman crocodilus. Dissertation Abstracts International Vol. XXXVII, No. 2 (1976).

Leake, P. A.: SEM observations of the cochlear duct in Caiman crocodilus. Scanning Electron Microscopy Vol. II, 437-444 (1977).

Manley, G.: Comparative studies of auditory physiology in reptiles. Z. vergl. Physiol. 67, 363-381 (1970a).

Manley, G.: Frequency sensitivity of auditory neurons in the Caiman cochlear nucleus. Z. vergl. Physiol. 60, 251-256 (1970b).

Manley, G.: Frequency response of the ear of the Tokay Gecko. J. Exp. Zool. 181, 159-168 (1972a).

Manley, G.: The middle ear of the Tokay Gecko. J. Comp. Physiol. 81, 239-250 (1972b).

Manley, G.: Frequency response of the middle ear of Geckos. J. Comp. Physiol. 81, 251-258 (1972c).

Manley, G. A.: Activity patterns of neurons in the peripheral auditory system of some reptiles. Brain Behav. Evol. 10, 244-256 (1974).

Manley, G. A.: Auditory responses from the medulla of the monitor lizard Varanus bengalensis. Brain Research 102, 329-334 (1976).

Manley, G. A.: Response patterns and peripheral origin of auditory nerve fibers in the monitor lizard Varanus bengalensis. J. Comp. Physiol. 118, 249-260 (1977).

Manley, J.: Single unit studies in the midbrain auditory area of Caiman. Z. vergl. Physiol. 71, 255-261 (1971).

Miller, M. R.: Scanning electron microscope studies of some lizard basilar papillae. Amer. J. Anat. 138, 301-330 (1973).

Miller, M. R.: The cochlear nuclei of lizards. J. Comp. Neuro. 159, 375-406 (1975).

Mulroy, M. J.: Cochlear anatomy of the alligator lizard. Brain Behav. Evol. 10, 69-87 (1974).

Mulroy, M. J., Altmann, D. W., Weiss, T. F., Peake, W. T.: Intracellular electric responses to sound in a vertebrate cochlea. Nature 249, 482-485 (1974).

Nadol, J. B., Mulroy, M. J., Goodenough, D., Weiss, T. F.: Tight and gap junctions in a vertebrate inner ear. Amer. J. Anat. 147, 281-302 (1976).

Patterson, W. C.: Hearing in the turtle. J. of Auditory Research 6, 453-464 (1966).

Saunders, J. C., Johnstone, B. M.: A comparative analysis of middle ear function in non-mammalian vertebrates. Acta Otolaryng. 73, 353-361 (1972).

Schmidt, R. S., Fernandez, C.: Labyrinthine DC potentials in representative vertebrates. J. Cell Comp. Physiol. 59, 311-322 (1962).

Suga, N., Campbell, H. W.: Frequency sensitivity of single auditory neurons in the Gecko Coleonyx variations. Science 157, 88-90 (1967).

Trincker, D., Khan, N., Mueller-Arnecke, H.: Comparative studies on DC resting potentials of the mammalian and reptilian labyrinth organs. IRCS Medical Science 6, 211 (1978).

Turner, R. G.: The relation between whole-nerve and unit response of the auditory nerve (alligator lizard). Ph.D. dissertation, University of Florida, Gainesville, Florida, 1975.

Turner, R.: Physiology and bioacoustics in reptiles. J. Acoust. Soc. Amer. 64, Supp. 1, S3 (1978).

Turner, R., Nielsen, D.: Auditory nerve fiber activity in an iguanid lizard. In preparation (1979).

Turner, R. G., Nielsen, D. W., Teas, D. C.: Discharges of auditory nerve fibers in the alligator lizard in relation to basilar papilla morphology. J. Acoust. Soc. Amer. 58, Supp. 1, S103 (1975).

Turner, R. G., Nielsen, D. W., Teas, D. C.: Comparison of the response of the two fiber populations in the auditory nerve of the alligator lizard. J. Acoust. Soc. Amer. 60, Supp. 1, 81 (1976).

Turner, R. G., Teas, D. C., Nielsen, D. W.: The relation of hair cell orientation to auditory nerve fiber response in the alligator lizard. Submitted to the J. Acoust. Soc. Amer. (1979).

Weiss, T. F., Altmann, D. W., Mulroy, M. J.: Endolymphatic and intracellular resting potential in the alligator lizard cochlea. Pflugers Arch. 373, 77-84 (1978).

Weiss, T. F., Mulroy, M. J., Altmann, D. W.: Intracellular responses to acoustic clicks in the inner ear of the alligator lizard. J. Acoust. Soc. Amer. 55, 606-619 (1974).

Weiss, T., Mulroy, M., Turner, R., Pike, C.: Tuning of single fibers in the cochlear nerve of the alligator lizard: Relation to receptor morphology. Brain Research 115, 71-90 (1976).

Weiss, T. F., Peak, W. T., Ling, A., Holton, T.: Which structures determine frequency selectivity and tonotopic organization of vertebrate cochlear nerve fibers? Evidence from the alligator lizard. In: Evoked Electrical Activity in the Auditory Nervous System. Naunton, R. F. (ed.). New York: Academic Press, 1978, pp. 91-112.

Wever, E. G.: The tectorial membrane of the lizard ear: Types of structure. J. Morphol. 122, 307-319 (1967).

Wever, E. G.: The mechanics of hair cell stimulation. Ann. Otol. 80, 786-804 (1971).

Wever, E.: The Reptile Ear. Princeton: Princeton University Press, 1978.

Wever, E., Vernon, J.: The problems of hearing in snakes. J. Auditory Research 1, 77-83 (1960).

Wever, E. G., Werner, Y.: The function of the middle ear in lizards: *Crotaphytus collaris* (Iguanidae). J. Exp. Zool. 175, 327-342 (1970).

Yost, W., Nielsen, D.: Fundamentals of Hearing. New York: Holt, Rinehart and Winston, 1977.

PART FOUR

Birds

Birds are ideal subjects for auditory studies since they use sound extensively under natural conditions and are easily studied using behavioral and physiological approaches. The degree of peripheral structural variation is less than that of the fishes and reptiles, and the auditory system of birds appears to be fairly readily understood in terms of analogies with the mammalian auditory system (Saito, Chapter 8). Comparative behavioral data on auditory function shows sensitivity and discrimination capacities that rival those of mammals (Dooling, Chapter 9), in spite of birds and mammals having separately diverged from reptiles long before the development of "modern" auditory systems. In a treatment of the central neural mechanisms possibly responsible for the processing of complex signals such as species specific vocalization, Sachs, Woolf, and Sinnott (Chapter 11) illustrate that the response properties of auditory neurons in birds also show certain mammal-like characteristics. Despite much speculation, little was known about avian sound localization until recently. Knudsen (Chapter 10) reviews the recent and exciting work on localization, and its central neural correlates, which is likely to become a model experimental approach for use with other species.

Chapter 8

Structure and Function of the Avian Ear

NOZOMU SAITO*

1 Introduction

There is much behavioral and neurophysiological data on the auditory system of members of the class Ave but considerably less data regarding the structure and function of their receptor organs. The auditory discrimination capacities of avian species and their responses to "biologically relevant" sounds have been worked out in considerable detail. The audibility curves of the passerines and nonpasserines fall close to those of man (Dooling, Chapter 9; Dooling 1975b), while pigeons are now known to be sensitive to infrasound (Yodlowski, Kreithen, and Keeton 1977). The vocal frequency range of song birds tends to exceed the highest best frequency response of auditory neurons (Konishi 1969, Sachs and Simmott 1978). The response of the pigeon's auditory neuron does not appear to be qualitatively different from those of the mammal's (Sachs, Lewis, and Young 1974, Sachs, Woolf, and Sinnott, Chapter 11). Song birds are particularly interesting since they tend to respond to "biologically relevant" sounds (Dooling 1978, Leppelsack 1978, Scheich 1977) and also share with man an aptitude for vocal learning (Bullock 1977, Nottebohm, Konishi, Hillyard, and Marler 1972, Karten 1968, Konishi 1963).

Until recently, there has been little data regarding structural features of the ear of birds that might account for their exceptionally good hearing. However, two general structural characteristics are evident: (1) the middle ear of birds shows less variation than does that of reptiles (Henson 1974, Smith and Takasaka 1971) and (2) the avian basilar papillae have different kinds of hair cells and a well-organized pattern of nerve fibers that, in some ways, resemble the mammalian organ of Corti (Takasaka and Smith 1971). Von Békésy (1944) emphasized that the cochlear partitions in several mammals and in chickens have approximately the same elasticity, suggesting similar hearing sensitivity even though chickens have a shorter cochlea than mammals. The evidence of structural and functional features now provides some insight into the

*Department of Physiology, Dokkyo University School of Medicine, Mibu, Tochigi 321-02, Japan.

general principles of the avian auditory mechanisms underlying high sensitivity and perception of such complex sounds as species-specific vocalization.

2 External Ear

The external ear of birds consists of specialized feathers that surround the opening of the auditory canal. The feathers in front of the ear opening are adapted to minimize turbulence in flight. Since these feathers are not compact and the skin is thin, the entrance of sound waves is not obstructed. In species that dive into water, the opening is reduced in diameter, and the auditory canal can be closed using muscles (Kartashev and Ilyichev 1964). An enlarged ear funnel has developed in song birds and parrots (Schwartzkopff 1973).

The most extensive adaptive variation in the ear funnel is found among owls. Many species of owl have developed not only very large posterior ear flaps but also erectable anterior ear flaps. The skull bones that constitute the auditory canals are asymmetrical in these species, and the size and shape are different on each side, with an enlargement of the ear canal on the left. The axes of the eardrums and the ear apertures are also bi-laterally asymmetrical (Norberg 1968). In the barn owls (*Tyto alba*), the skin of the inside curving wall of the ear openings is specially developed; the skin is thickened into ridges, and this increases the surface available for the attachment of thick, densely packed feathers. These adaptions suggest that feathers may be operating as sound re-flectors, in this case (Payne 1971). The functional significance of these features seems to be in sound localization (see Knudsen, Chapter 10).

3 Middle Ear

Several cranial bones make up the middle ear cavity. The bones constitute an articular crest that transverses the middle ear cavity and communicates with the air filled spaces of the skull (Stellbogen 1930). The bilateral tympanic cavities are not isolated from one another, and changes of air pressure, particularly those of low frequency, are trans-ferred directly from one eardrum to the other (Wada 1924, Schwartzkopff 1952). The interaural changes in the phase of pure tones with different frequencies were recently measured for the chicken using the cochlear microphonic responses. These delays are much greater than would be expected from the interaural distances involved; the change in the phase was $180°$ at frequencies between 125 Hz and 500 Hz and $800°$ at 6 300 Hz (Rosowski and Saunders 1977).

Unlike in most lizards, the tympanic membrane in birds is located in a relatively deep external auditory meatus. A firmly fixed supportive framework enables the tympanic membrane to become thinner and more delicate, and thus allows the ratio of the area of the tympanic membrane to the stapedial footplate to increase. The highest ratio (40:1) being in owls. These features may contribute to greater high frequency sensitivity (Henson 1974).

The tympanic membrane makes up the ventral wall of the auditory canal. The epidermis of the canal continues upon the tympanic membrane, but the epidermis is only loosely attached to the underlying elastic membrane (membrana propria). The tympanic membrane of birds, in contrast to that of mammals, protrudes outward and is mainly supported by the extracolumellar cartilage. The shape of the columella and footplate is species-specific, and it is more delicate in those birds with good hearing abilities. The total mass of the middle ear system is derived from a combination of the tympanic membrane; the structures of the columella and extracolumella; the general suspensory system consisting of supporting ligaments, middle ear muscles, and footplate in the oval windows; and, finally, the fluid in the cochlea.

The characteristics of the transfer function from the tympanic membrane to the footplate were obtained by using the Mössbauer technique (Saunders and Johnstone 1972). The displacement amplitudes of the tympanic membrane and columella was found to decrease with increasing frequency up to 3 000 Hz. The lever ratio, expressed in dB, was calculated by dividing the amplitude of the tympanic membrane by that of the columella and was found to be constant within 10 dB up to 1 000 Hz (Saunders and Johnstone 1972) (Fig. 8-1B). This ratio is 1/2 that of a cat and 1/4 that of man (Guinan and Peake 1967, von Békésy 1941). It should also be recognized that the present methods for determining motion are only sensitive to piston-like movements of the tympanic membrane and footplate and that other more complex movements might contribute significantly to the small lever ratio in birds. In some owls the extracolumella in the tympanic membrane is arranged to cause a rocking movement during sound conduction. The footplate is sometimes rounded on its internal surface, which may result in less turbulent vibration within the perilymphatic fluid (Schwartzkopff 1973, Pumphrey 1961). The changes of footplate velocity with frequency are illustrated for representatives of mammals, reptiles, amphibians, and birds (Fig. 8-1A), and they all exhibit maximal values over a similar frequency range near 1 000 Hz and distribute within 5 dB below 3 000 Hz, except for amphibians (Saunders and Johnstone 1972). Finally, it is interesting to note that the response of the columella ear falls within 6 dB of the mammalian ear over the range of "communication" frequencies below 3 000 Hz.

4 General Features of Inner Ear

The avian osseous cochlea forms a tube that bends slightly medially. Considered bilaterally, the ends of the two cochleas approach each other at the base of the cranium. The cochlea is divided by a longitudinal cartilaginous frame over which the basilar membrane lies (see Fig. 8-7). The wall of the cochlear duct is covered with differentiated epithelial cells. The thick wall of the tegmentum vasculosum (see Figs. 8-6, 8-7 TV) separates the scala media from the scala vestibuli, the latter being very compressed beneath the roof cartilage. The tegmentum vasculosum provides electrical and chemical insulation for the basilar papilla as does Reissner's membrane in mammals. The sensory patch, or basilar papilla, divides the cochlear duct and is shaped somewhat like a spatula—wider distally than proximally. The basilar papilla contains supporting and sensory cells. The basal ends of the supporting cells rest on the basilar membrane, and their apical ends reach to the free surface of the papilla between sensory cells (see

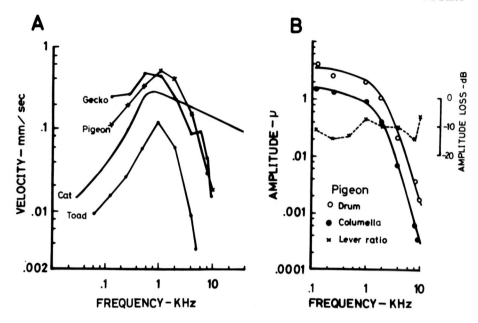

Figure 8-1. Transfer function of avian middle ear. (A) Velocity-frequency response curves in the amphibian, reptile, bird, and mammal. (B) Middle ear frequency response for the pigeon, *Streptopelia risoria*. The sound intensities are 100 db SPL for (A) and (B). (From Saunders and Johnstone 1972).

Fig. 8-8 MV). The sensory hair cells are closely packed among the interspersed supporting cells and do not touch the basilar membrane. The terms "anterior" and "posterior" in reference to hair cells on the basilar papilla are more appropriate than the mammalian "inner" and "outer" because of the location of the straight cochlea. The anterior hair cells are comparable to the inner side of the mammalian basilar membrane from the standpoint of histological organization. The anterior hair cells are elongated, while the hair cells oriented toward the posterior cartilage are shorter and thicker.

4.1 Cochlear Partition

The basilar membrane of birds is short in comparison to that of mammals, but the width of the membrane is greater. The total length of the inner ear (3 mm to 5 mm) does not increase substantially with an increase in body size, although small song birds do have relatively longer cochleas than nonsong birds. Only in the nocturnal owls does the shape and size of the basilar papilla approach that of the organ of Corti in mammals (Schwartzkopff 1973). The scala vestibuli has not developed in birds. The perilymphatic scala tympani widens basally near the round window while its blind end extends under the apical part of the basilar membrane to the lagena, where a long, narrow duct sometimes leads perilymph to the scala vestibuli. In the chicken, according to von Békésy (1944), there is no connection between the two perilymphatic canals that corresponds to the mammalian helicotrema. Thus, pressure in one canal will produce a sustained bulging of the partition, and such a pressure difference between two canals could cause tearing of the cochlear partition. While such incidental damage would not occur if a helicotrema were present, a helicotrema would not be advantageous in birds for other reasons. Because of the limited length of its cochlear canals and the relatively large size of the oval and round windows, an opening between the two canals would greatly reduce the mechanical displacement of the cochlear partition resulting in a loss of sensitivity.

4.2 Mechanical Properties

A schematic diagram of the chicken basilar membrane (Fig. 8-2) indicates the stapedial location (dotted lines). The lower graph of Fig. 8-2 represents the relation between sound stimulus frequency and the position of maximal displacement along the basilar membrane from the stapedial footplate (von Békésy 1947). Below 100 Hz, there is no longer any mechanical frequency analysis in the chicken, because then the partition vibrates as a whole, and this form of vibration does not change as the frequency is lowered. The upper limit of the frequency range can be extrapolated from the figure. Worthy of note is the density with which frequencies are approached within a very short span of the basilar papilla. Thus, the chicken's partition must be perceived to vary greatly in elasticity along its length.

The volume elasticity of the cochlear partition is compared using various animals in Fig. 8-3. The body sizes of the different species cover a wide range and the length of the various cochleas extends from 5 mm in the chicken to as much as 50 mm in the elephant. It is interesting that the partitions of all animals exhibit the same range of elasticity from 10^{-4} to 10^{-8} cm^3/1 cm water, regardless of the different partition lengths. The steepest relationship between displacement and length is for the chicken, whereas the elephant has the lowest slope. For a better comparison of the various animals, a circle has been added to each curve to indicate the volume elasticity at the position of the excitation peak for 1 000 Hz. This value is constant only for very small animals. The changes in elasticity for the chicken extend over a wide range along the shortest cochlear duct among the animals examined. The distal end of the cochlear partition in birds seems very elastic and is comparable to the longer cochleas of certain mammals; however, the basal area close to the stapes shows comparable stiffness. Von Békésy (1944) described this correlation by another parameter—mechanical separation of frequency resolution. If $\Delta\ell$ is the shift of resonance position along the cochlear partition for a percentage change of frequency, $\Delta n/n$ (n: frequency), the mechanical resolution of frequency, A, is defined as $A = \Delta\ell/(\Delta n/n)$, as shown in Fig. 8-4. The chicken has the least developed cochlea, however, its limited resolution would be improved by a steeper change in elasticity along the length of the basilar membrane.

The mammalian tectorial membrane, Reissner's membrane, and the basilar membrane vibrated completely in phase, so that it is proper to consider the entire cochlear

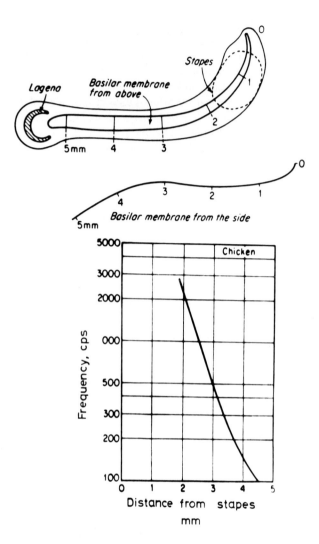

Figure 8-2. Positions of maximal stimulation along the cochlear partition of the chicken for various frequencies. Upper illustration is the top and side view of the basilar membrane showing distances from the stapes. (From von Békésy 1944).

partition as a single structure (von Békésy 1947). The measurements of volume elasticity showed that the overall elasticity is determined not by the tectorial membrane but by the basilar membrane, which exhibits a continuous change in volume elasticity along the partition (Fig. 8-5A). The tectorial membrane is noteworthy for the ease with which it may be moved in a transverse direction, perpendicular to the basilar membrane. This is seen as a concave depression elongated transversely on the tectorial membrane when a point pressure is applied (Fig. 8-5B) (von Békésy 1947). In the longitudinal direction, the tectorial membrane possesses a substantial stiffness. It may

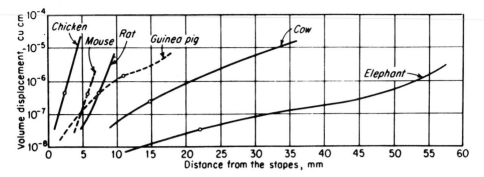

Figure 8-3. Comparison of different animals as regards the elasticity of the cochlear partition, measured as the volume displacement produced in a 1-mm segment by a pressure of 1 cm of water. (From von Békésy 1944).

be assumed that this longitudinal stiffness of the tectorial membrane produces a pressure drop along the cochlear duct which, in turn, increases the degree of mechanical separation of frequency resolution ($A = \Delta\ell/\Delta n/n$). For the avian basilar papilla, the cilia of hair cells are inserted into niches in the tectorial membrane, apparently more firmly than in mammals (Tanaka and Smith 1975). This permits frequency analysis to be much more precise and efficient along the shorter avian cochlear duct than along the mammal's.

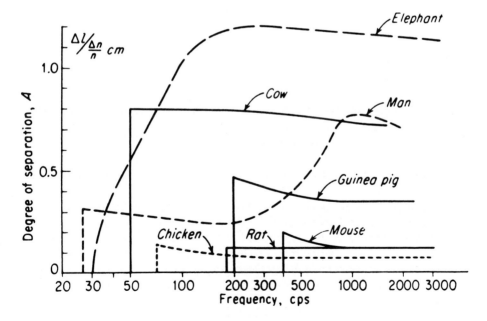

Figure 8-4. Comparison of different animals as regards mechanical separation of frequencies. (From von Békésy 1944).

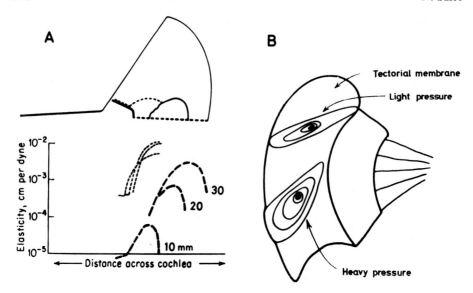

Figure 8-5. Elastic properties of the cochlear partition. (A) Elastic changes across the cochlear partition at three different cross sectional planes superimposed. Above the graph is a schematic outline of the structure. Dashed lines on both the graph and the scheme indicate the basilar membrane; dotted lines indicate the tectorial membrane. The figures on the curve indicate spiral or longitudinal distances from the stapes. (B) Elongated depressions of the tectorial membrane produced by fine test hair seen from above. (From von Békésy 1947).

4.3 Tegmentum Vasculosum

The tegmental structure of scala media, shown in Fig. 8-6B for a longitudinal section of the chicken cochlea is characteristic for birds. The thick foldings protrude and traverse the roof of scala media opposite the basilar papilla. One of the puzzling features of the avian cochlear partition is that rich blood vessels within the tegmentum vasculosum should disturb hearing by the slightest dilatations (von Békésy 1953).

Tissue of the tegmentum vasculosum is richly supplied with blood vessels and differentiates into various cell types. The light stained cells are assumed to secrete endolymphatic fluid, while the dark, bottle-shaped cells probably resorb sodium (Fig. 8-6A) (Kuijpers, Houben, and Bonting 1970). $Na^+ - K^+ -$ ATP-ase activities appear to be predominantly located in the tegmentum vasculosum, and its absolute activity is comparable to the mammalian stria vascularis (Kuijpers et al. 1970). The tegmentum may have a role in maintaining the cochlear cation gradient so that sodium is resorbed and potassium is secreted and the endocochlear potential maintained as in the stria vascularis.

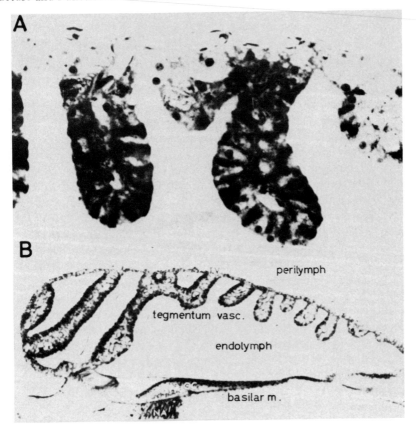

Figure 8-6. The structures of the tegmentum vasculosum. (A) Detail of the tegmentum vasculosum showing dark and light cells. (B) Longitudinal section of cochlea of a one-day-old chicken showing transverse structures of thick folds. (From Kuijpers et al. 1970).

4.4 Cochlear Potentials

The cochlear duct generates a positive DC endolymphatic potential (EP). Although it measures considerably less than in mammals (Schmidt and Fernandez 1962), the EP in birds seems to be caused by an active transport mechanism of the tegmentum vasculosum, where potassium- and sodium-sensitive ATP-ase is highly localized. This enzyme is considered to actively exchange Na^+ and K^+ at the inner side of the tegmentum vasculosum (Kuijpers et al. 1970) and thereby control the ionic composition of the endolymph. The endolymphatic fluid of high K^+ and low Na^+ concentration supposedly contributes to hair cell function. If this is true, the EP might be only viewed as an epiphenomenon or a secondary by-product of the ionic constitution. It has been variously assumed that the endocochlear potential of mammals represents an important part of the hearing mechanism (von Békésy 1951, Davis 1957) or that it is nothing more than an incidental by-product of some other factor such as fluid transport (Dohlman 1960). Unfortunately, the data on the avian EP does not provide very much insight into this question. The avian EP is of lower amplitude, less than 20 mV, although the value fluctuates in different determinations. The difference in size of the mammalian and avian EP cannot be explained by the degree of Na^+- K^+- ATP-ase, which contributes to the generation of EP, since the ATP-ase is practically equal in both classes (Kuijpers et al. 1970). The structural difference between the stria and the tegmentum may offer an explanation for this; the former is constructed of multi-layered epithelium in which only the external layer is enzyme-active. However, the tegmentum is a single-layered, although richly folded, epithelium (Fig. 8-6A) (Schwartzkopff 1973). Necker (1970) describes the mean amplitude of EP as 15 ± 4 mV for 28 song birds, including the blackbird, starling, and sparrow. The avian EP responds dramatically to anoxia, as does the EP in the mammals, although the former is only 25% of the latter in amplitude. The EP drops rapidly to a negative value during anoxia (Necker 1970, Schmidt and Fernandez 1962), at the conclusion of which it is immediately restored to its original value. Amphibians, turtles, lizards, snakes, and crocodiles failed to show any clear response to anoxia (Schmidt and Fernandez 1962). Therefore, the results can be divided into two groups, according to the response to anoxia, and will be referred to as "anoxia sensitive" and "anoxia insensitive" (Schmidt and Fernandez 1962). Because all amniotes seem to have a very definite stria vascularis, the phylogenetic distribution of this response is not correlated with the presence of a stria. Another possible correlation—with body temperature—was examined and also found to be insignificant (Schmidt and Fernandez 1962). Accordingly, the EP of poikilotherms is likely to result from a biochemical mechanisms of their stria being different from that of homeotherms.

The cochlear microphonic potential (CM) and summating potential (SP) of birds, except for some details, behave similarly to those of mammals during anoxia. The negative component of the CM is exclusively sensitive to anoxia while the CM+ is not. Further, the SP does not change polarity during anoxia in contrast to its behavior in mammals (Necker 1970). The differences between birds and mammals are considered not to have much functional significance. The electrogenic origin of the CM and SP has been satisfactorily interpreted by an equivalent circuit composed of two or three parallel electromotive forces, each with a series conductance, located within the hair cells and the stria. According to this scheme, the different behavior of the avian CM- and SP during anoxia, in comparison with those of mammals, derives from either a change in the ratio of electromotive force of the hair cells to that of the stria or to a change in the ratio of their series conductances.

4.5 Hair Cell

According to Takasaka and Smith (1971), tall, short, and intermediate types of hair cells can be recognized in the pigeon (*Columba livia*) basilar papilla. In a cross section of the basilar papilla, 20 to 40 closely packed hair cells can be seen in one row, unlike the mammalian basilar papilla, which has an outer and inner row divided by the pillar and covered by the reticular membrane (Fig. 8-7). To reduce intrinsic distortion in the

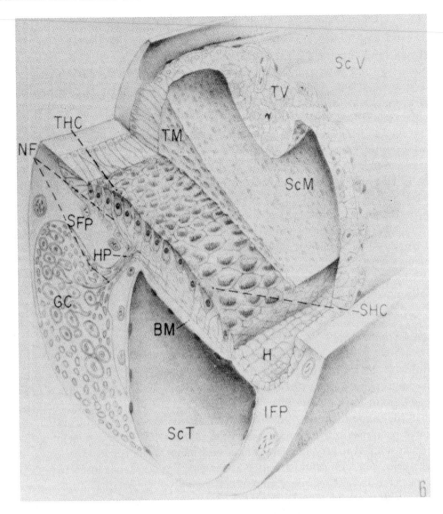

Figure 8-7. Schematic drawing of transverse section of a pigeon's cochlear duct. Short (SHC) and tall (THC) hair cells rest on basilar membrane (BM) over superior (anterior) fibrocartilaginous plate (SFP). Over the inferior (posterior) plate (IFP), hyaline cells (H) are attached to the tegmentum vasculosum (TV), which separates scala media (ScM) and scala vestibuli (ScV). Peripheral nerve fibers (NF) run from cochlear ganglion cells (GC) through the habenular perforation (HP). Also note scala tympani (ScT) and the tectorial membrane (TM). (From Takasaka and Smith 1971).

organ of Corti, mammalian hair cells are rigidly supported by the frame of the pillar that contains Deiter's cells and the reticular membrane. In contrast, the avian cells rest loosely on the basilar membrane but are connected firmly with the tectorial membrane, which is much thicker than in mammals. The avian tall hair cells are located over the superior cartilaginous plate (Fig. 8-7) and have small efferent and large afferent nerve endings. The hair cells are covered by the thicker part of the tectorial membrane. Fifty

to seventy stereocilia and a single kinocilium are located in the center of the pentagonal top surface of the sensory cells with a deep cuticular plate on the root (Fig. 8-8A)
(Takasaka and Smith 1971). The shorter hair cells are only found on the free basilar
membrane and are covered by a more attenuated part of the tectorial membrane. The
short cells have large efferent and small afferent nerve endings. The stereocilia are
twice as numerous as in the taller cells and are embedded in a large diameter of the
cuticular plate (Fig. 8-8B).

The chicken has two types of well-differentiated hair cells and nerve endings. The
majority of the ultrastructural findings are similar to those described for pigeons
(Tanaka and Smith 1978, Jahnke, Lundquist, and Wersäll 1969). However, Tanaka and
Smith (1978) emphasized the difference of the structures of the hair cells between
chickens and pigeons. They report that most of the short cells and many of the tall
cells located on the central part of the chicken papilla are without kinocilia (Tanaka
and Smith 1978, Kurokawa 1978). The cochlear microphonic responses of the chicken
were compared with those of the pigeon (Gates, Perry, and Coles 1975), and no clear
differences were found between the two species. Observations of the parakeets and a
species of passerine also reveal the same distribution of the two types of hair cells
(Tanaka unpublished).

4.6 Nerve Innervation

The afferent fibers of the pigeons run from the habenular perforation on the anterior
cartilaginous plate, as indicated by a silver stained preparation (Figs. 8-9A, C) (Takasaka and Smith 1971). These fibers lose their myelin sheath and continue their course
through the supporting cells, directly transversing the distal part of the papilla. In the
proximal part, they run obliquely toward the proximal direction as seen in Fig. 8-9C.
It is interesting that very little branching of afferent fibers occurs between the supporting cells, and terminal ramification as revealed by silver stains seems to be limited
to a restricted area. The mammalian afferents extensively innervate multiple outer hair
cells, but this does not occur in the pigeon. Multiple innervation by the afferent fibers
could not be found in all parts of the papilla. However, Fig. 8-9B shows many fewer
efferent fibers when acetylcholine esterase staining was used to selectively reveal the
efferent fibers. The arrangement of the efferent fibers within the papilla is quite different from that of the afferent fibers. The efferent fibers ramify extensively and form
networks through the hyaline cell layer beneath the papilla (Fig. 8-10) where an unusual relationship is found between the hyaline cell and the efferent fibers (Takasaka
and Smith 1971). Numerous slender branches of the efferent fibers tightly connect the
hyaline cells, and, thus, the connecting junctions seem to mimic the synaptic structures. However, no synaptic organelles were visible in either side of the hyaline cells
or the efferent fibers, and the physiological significance of the finding is still unknown.
No axodendritic synapses between afferent and efferent fibers are found in any part
of the papilla in the pigeon, in contrast to the mammalian efferent system (Takasaka
and Smith 1971).

The sensory pattern of the pigeon cochlea resembles that of mammals (Takasaka
and Smith 1971). For example, mammalian inner hair cells and the tall hair cells of

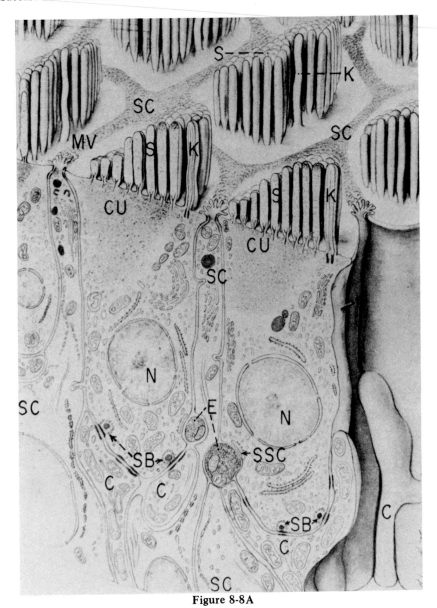

Figure 8-8A

Figure 8-8. Schematic three-dimensional view of tall (Fig. 8-8A) and short (Fig. 8-8B) (see page 254) hair cells and supporting cells. Microvillous tufts (MV) on the supporting cell surface (SC) form pentagonal borders surrounding each hair cell. Cochlear nerve endings (C) are large, finger-like (Fig. 8-8A) or small (Fig. 8-8B), with synaptic ball (SB) and subsynaptic cisterna (SSC). Efferent terminals (E) are small (Fig. 8-8A) and large (Fig. 8-8B). Stereocilia (S) rest on a cuticular plate (CU). K = kinocilium. (From Takasaka and Smith 1971).

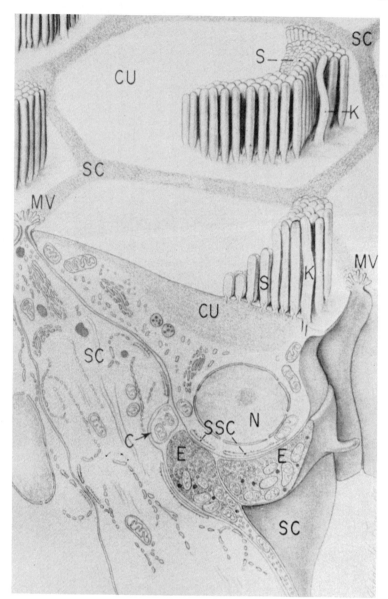

Figure 8-8B

pigeons are located on the less flexible part of the papilla (superior edge, upper side of the diagram of Fig. 8-11), while the mammalian outer cells and short cells of the pigeon are located on the flexible part (inferior edge, lower side). Further, the large efferent terminals are present on the more flexible part of the papilla in birds and mammals, while they decline distally (dashed line in the lower graph of Fig. 8-11). However, the afferent nerve count increases (solid line) there.

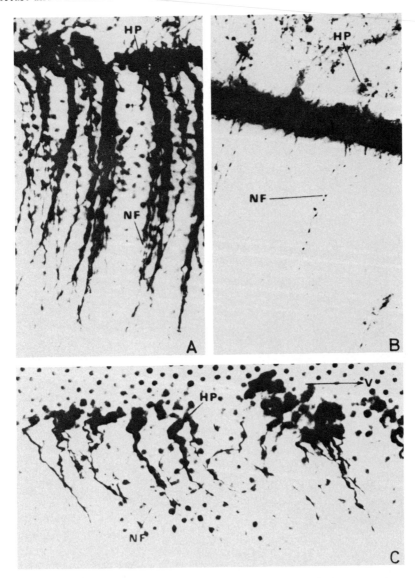

Figure 8-9. Horizontal section of pigeon's basilar papilla showing afferent fibers stained by silver impregnation (Fig. 8-9A and Fig. 8-9C) and by acetylcholine-esterase method (Fig. 8-9B). (A) Afferent fibers (NF) run through the habenular perforation (HP) across the basilar papilla (distal one-third), straight to posterior edge. Some others (*) turn in the opposite direction. (B) Specimens from same region as in Fig. 8-9A, but stained by the AChE method that selectively reveals efferent fibers. The thick shadowed structure is a blood vessel. (C) Proximal one-third of a basilar papilla showing the oblique course of nerve fibers (NF). Arrow indicates the direction of the vestibuli. (From Takasaka and Smith 1971).

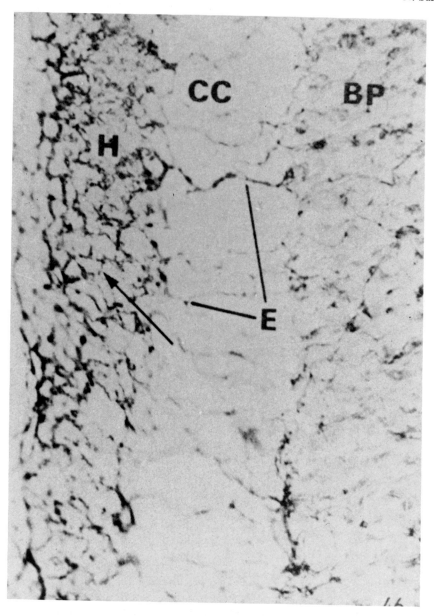

Figure 8-10. Networks of efferent nerve fibers of a pigeon's basilar papilla. Acetyl-choline-esterase-positive fibers (E) in the basilar papilla (BP) cross cuboidal cells (CC) and form dense networks (arrow) between hyaline cells (H). (From Takasaka and Smith 1971).

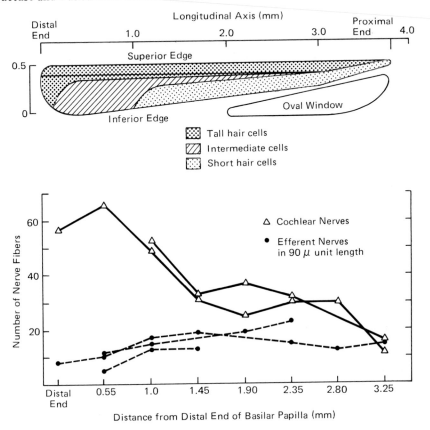

Figure 8-11. Dimensions of a pigeon's basilar papilla and distribution of hair cell types (upper diagram) and nerve types (lower graph). Analyses in upper diagram were from cross sections of an Araldite embedded ear. Comparative populations of cochlear nerve fibers (open triangles) and efferent fibers (filled circles) were counted within 90μ intervals along the papilla. (From Takasaka and Smith 1971).

5 Conclusion

There is a wide divergence of adaptations in the external ear of birds. The ear openings of birds that are adapted for flight or water diving involve delicate feathers in front of the ear openings while adaptations for sound localization in owls involves thick feathers located in the inside of the ear canal.

The middle ears are located within a deep meatus. The tympanic membrane and the columella are delicate and thus provide good transfer characteristics for either amplitude or velocity over the range of the "communication" frequencies. The tympanic cavity is not isolated, and interaural connections provide great variation in the phase of the sound at the two ends.

The mechanical properties of the cochlear partition are characterized by a high elasticity of the basilar membrane. This elasticity along with the stiffness of the tectorial membrane in a longitudinal direction are factors that tend to increase the frequency resolution of the relatively short cochlea of the chicken. The amplitude of the avian EP is lower than that of mammals and may be due to the structure of the tegmentum vasculosum rather than the ATP-activity. The pigeon and the chickens have two types of well-differentiated hair-cell/nerve-ending units, although the lack of kinocilia may be unique in chickens.

Acknowledgments. I thank Dr. J. E. Hinde, University of Wisconsin, for kindly reading the manuscript and Drs. Y. Katsuki, National Center for Biological Sciences, and I. Taniguchi, Dokkyo University, for helpful comments of an earlier draft.

References

Bullock, T. H. (ed.): Life Science Report 5, Dahlem Konferenzen, Berlin (1977).

Davis, H.: Biophysics and physiology of the inner ear. Physiol. Rev. 37, 1-49 (1957).

Dohlman, G. F.: Histochemical studies of vestibular mechanisms. In: Neural Mechanisms of the Auditory and Vestibular Systems. Rasmussen, G. L. et al. (eds.). Thomas (1960).

Dooling, R. J., Saunders, J. C.: Hearing in the parakeet (*Melopsittacus undulatus*): Absolute thresholds, critical ratios, frequency difference limens, and vocalizations. J. Comp. Physiol. Psych. 88, 1-20 (1975a).

Dooling, R. J., Saunders, J. C.: Auditory intensity discrimination in the parakeet (*Melopsittacus undulatus*). J. Acoust. Soc. Am. 58, 1308-1310 (1975b).

Dooling, R. J., Zoloth, S., Baylis, J.: Auditory sensitivity, equal loudness, temporal resolving power and vocalizations in the house finch (*Carpodacus mexicanus*). J. Comp. Physiol. Psych. 92, 867-876 (1978).

Gates, G. R., Perry, D. R., Coles, R. B.: Cochlear microphonics in the adult domestic fowl (*Gallus domesticus*). Comp. Biochem. Physiol. 51A, 251-252 (1975).

Guinan, J. J., Jr., Peake, W. T.: Middle-ear characteristics of anaesthetized cats. J. Acoust. Soc. Am. 41, 1237-1261 (1967).

Hensen, O. W.: Comparative anatomy of the middle ear. In: Handbook of Sensory Physiology, Vol. 1. Keidel, W. D., Neff, W. D. (eds.). Springer-Verlag, pp. 39-110 (1974).

Jahnke, P. G. L., Lundquist, P.-G., Wersäll, J.: Some morphological aspects of sound perception in birds. Acta Oto-laryng. 67, 583-601 (1969).

Kartashev, N. N., Ilyichev, V. D.: Über das Gehörorgan der Alkenvögel. J. Ornithol. 105, 113-136 (1964).

Karten, H. J.: The ascending auditory pathway in the pigeon (*Columba livia*). II, Telecephalic projections of the nucleus ovoidalis thalami. Brain Res. 11, 134-153 (1968).

Konishi, M.: The role of auditory feedback in the vocal behavior of the domestic fowl. Zeitschrf. Tierphsychol. 20, 349-367 (1963).

Konishi, M.: Hearing, single unit analysis, and vocalizations in songbirds. Science 166, 1178-1181 (1969).

Kuijpers, W., Houben, N. M. D., Bonting, S. L.: Distribution and properties of ATP-ase activities in the cochlea of the chicken. Comp. Biochem. Physiol. 36, 669-676 (1970).

Kurokawa, N.: The ultrastructure of the basilar papilla of the chick. J. Comp. Neur. 181, 361-364 (1978).

Leppelsack, H. J.: Unit responses to species-specific sounds in the auditory forebrain center of birds. In: Auditory Processing and Animal Sound Communication (symp. 1977). Proc. Fed. 37, No. 10, 2336-2341 (1978).

Necker, R.: Zur Entstehung der Cochlea-potentiale von Vögeln: Verhalten bei O_2-Mangel, Cyanidvergiftung und Unterkühlung sowie Beobachtungen über die räumliche Verteilung. Z. vergl. Physiol. 69, 367-425 (1970).

Norberg, A.: Physical factors in directional hearing in *Aegolius funereus* (Linne) (Strigiformes), with special reference to the significance of the asymmetry of the external ears. Ark. Zool. 22, 181-204 (1968).

Nottebohm, F., Konishi, M., Hillyard, S., Marler, P.: Ontogeny of acoustic behavior. In: Auditory Processing of Biologically Significant Sounds. Worden, F. G. et al. (eds.). Neurosciences Res. Prog. Bull. 10, 31-49 (1972).

Payne, R. S.: Acoustic localization of prey by barn owls (*Tyto alba*). J. Exp. Biol. 54, 535-573 (1971).

Pumphrey, R. J.: Sensory organs. In: Biology and Comparative Physiology of Birds. Marshall, A. J. (ed.). 2, 69-86 (1961).

Rosowski, J. J.: Acoustic properties of the interaural pathway in the chicken. J. Acoust. Soc. Am. 61, S3 (A) (1977).

Rosowski, J. J., Saunders, J. C.: Phase interactions resulting from the avian interaural pathway. J. Acoust. Soc. Am. 62, Suppl. 1, 586 (1977).

Sachs, M. B., Lewis, R. H., Young, E. D.: Discharge patterns of single fibers in the pigeon auditory nerve. Brain Res. 70, 431-447 (1974).

Sachs, M. B., Sinnott, J. M.: Responses to tones of single cells in nucleus magnocellularis and nucleus angularis of the redwing blackbird (*Agelaius phoeniceus*). J. Comp. Physiol. 126, 347-361 (1978).

Saunders, J. C., Johnstone, B. M.: A comparative analysis of middle-ear function in non-mammalian vertebrates. Acta Otolaryng. 73, 353-361 (1972).

Scheich, H.: Central processing of complex sounds and feature analysis. In: Life Science Report 5, Dahlem Konferenzen, Berlin, 161-182 (1977).

Schmidt, R. S., Fernández, C.: Labyrinthine DC potentials in representative vertebrates. J. Cell. Comp. Physiol. 59, 331-322 (1962).

Schwartzkopff, J.: Unterschungen über die Arbeitsweise des Mittelohres und Richtungsgehören der Singvögel unter Verwendung von Cochlea–Potentialen. Z. vergl. Physiol. 34, 46-68 (1952).

Schwartzkopff, J.: Mechanoreception. In: Avian Biology 3. Farner, D. S. et al. (eds.). Academic Press, 417-477 (1973).

Schwartzkopff, J.: Inner ear potentials in lower vertebrates: dependence on metabolism. In: Basic Mechanism in Hearing. Møller, A., Boston, P. (eds.). Academic Press, 423-452 (1973).

Smith, C. A., Takasaka, T.: Auditory receptor organs of reptiles, birds and mammals. In: Contribution to Sensory Physiology (V). Neff, W. D. (ed.). Academic Press, 129-178 (1971).

Stellbogen, E.: Über das äussere und mittelere Ohr des Waldkauzes. Z. Morphol. Oekol. Tiere. 19, 686-731 (1930).

Takasaka, T., Smith, C. A.: The structure and innervation of the pigeon's basilar papilla. J. Ultrastruct. Res. 35, 20-65 (1971).

Tanaka, K., Smith, C. A.: Structure of avian tectorial membrane. Anal. Oto-Rhin-Laryng. 84, 287-296 (1975).

Tanaka, K., Smith, C. A.: Structure of the chicken's inner ear: SEM and TEM study. Am. J. Anat. 153, 251-272 (1978).

von Békésy, G.: Über die Messung der Schwingungsamplitude der Gehörknöchelchen mittels einer kapazitiven Sonde. Akust. Zeits. 6, 1-16 (1941).

von Békésy, G.: Über die mechanishe Frequenzanalyse in der Schnecke verschiedener Tiere. Akust. Zeits. 9, 3-11 (1944).

von Békésy, G.: The variation of phase along the basilar membrane with sinusoidal vibrations. J. Acoust. Soc. Amer. 19, 452-460 (1947).

von Békésy, G.: DC potential and energy balance of the cochlear partition. J. Acoust. Soc. Am. 23, 576-582 (1951).

von Békésy, G.: Description of some mechanical properties of the organ of Corti. J. Acoust. Soc. Am. 25, 770-785 (1953).

Wada, Y.: Beiträge zur vergleichenden Physiologie des Gehörorgans. Pflügers Arch. Gesamte Physiol. Menschen Tiere. 202, 46-69 (1924).

Yodlowski, M. L., Kreithen, M. L., Keeton, W. T.: Detection of atmospheric infrasound by homing pigeons. Nature 265, 725-726 (1977).

Chapter 9

Behavior and Psychophysics of Hearing in Birds

ROBERT J. DOOLING*

1 Introduction

It is generally agreed that birds and mammals share a common ancestry in the class Reptilia dating back about 250 million years (Brodkorb 1971). In considering the evolution of vertebrate auditory systems, it is therefore not uncommon to find birds placed between reptiles and mammals, particularly on the basis of anatomical criteria. For instance, the basilar membrane is generally short in reptiles, longer in birds, and longest in mammals, with some degree of overlap (Manley 1971, 1973).

While birds as a class are quite similar morphologically, they have undergone an extensive and complex adaptive radiation (Brodkorb 1971, Nottebohm 1972). Thus it is no more appropriate to speak of *the* bird than to speak of *the* mammal or *the* fish in regard to auditory function. There are about 8 600 living species of the birds that have been grouped into 29 orders by most taxonomists (Van Tyne and Berger 1976, Storer 1971). The most recently evolved order, the Passeriformes, contains about 5 100 of these 8 600 species (Brodkorb 1971). Fortunately, behavioral studies of hearing in birds are almost equally divided between passerines and nonpasserines, affording a reasonable comparison of auditory capability of the "evolutionarily" old versus the new and of song birds versus nonsong birds. Song birds are particularly interesting since they tend to have acoustically-complex vocal signals and also share with man the characteristic of vocal learning (Nottebohm 1972).

Behavioral observations have long suggested that many members of the class Aves have a requirement for good absolute auditory sensitivity and excellent frequency and time perception (see, for example, Thorpe 1961, Falls 1963, Konishi 1973). Until recently, however, neither anatomical nor psychophysical evidence was adequate to support the notion of exceptionally good hearing in birds. Within the last decade, the application of rigorous psychophysical techniques to the measurement of hearing in birds has brought our knowledge of auditory capabilities in this group in closer agreement with ethological observations. There are now at least partial answers to some of

*The Rockefeller University, Field Research Station, Tyrrel Road, Millbrook, New York 12545.

the important questions: How do birds compare with other vertebrate groups in terms of basic auditory capabilities? Are there differences between song birds and nonsong birds? Do auditory discrimination data allow any reasonable inferences as to the peripheral auditory system function in birds? How well do auditory discrimination data agree with the precision and stereotypy of vocal output characteristic of the song birds? Finally, are there any parallels between the role that hearing plays in the development of a young bird's song and the crucial role that hearing plays in the development of speech in human infants?

2 Absolute Auditory Sensitivity

To date, behavioral audibility curves are available for 16 species of birds. To facilitate a comparison of audibility functions among these 16 species, six arbitrary descriptive parameters similar to those used by Masterton, Heffner, and Ravizza (1969) for mammals were selected. The comparison shown in Table 9-1 is based on these parameters defined as follows: bandwidth (in octaves) is the frequency range in octaves at an intensity level 40 dB above the most sensitive point in the audibility function; best intensity (in decibels) is the lowest point on the audibility function; low frequency sensitivity (in decibels) is the threshold at 500 Hertz; high frequency cutoff (in kilohertz) is the highest frequency a bird can hear at a sound pressure level of 60 dB (SPL); low frequency slope (in decibels/octave) is the rate at which sensitivity declines for frequencies below the most sensitive region of the audibility function; and high frequency slope (in decibels/octave) is the rate at which sensitivity declines for frequencies above the most sensitive region of the audibility function.

2.1 Comparison of Audibility Functions

There is a striking similarity among the absolute threshold curves of the 16 avian species studied. The lowest thresholds occur between 2 kHz and 4 kHz, the region of maximum sensitivity extends over a relatively narrow range (1 to 2 octaves), and sensitivity to high frequencies (> 10 kHz) is extremely poor. A typical avian audibility curve can be constructed by plotting the median value of the 16 thresholds available at each test frequency. This composite audibility curve for the class Aves is shown in Fig. 9-1. For comparison, absolute threshold curves are also shown for two mammals commonly used in auditory research—*Felis catus* (cat), and *Chinchilla lanigera* (chinchilla)—and for the only member of the class Reptilia that has been tested behaviorally—*Pseudemys scripta* (common turtle).

At least for birds and mammals, a comparison among these audibility functions offers a reasonable representation of the relative capabilities of each vertebrate class. Since the behavioral data from the class Reptilia are so limited, not much can be said. Both mammalian species have audibility curves that span a considerably greater range of frequencies than does the median avian audibility curve including substantially better sensitivity to frequencies above 10 kHz for the two mammals. In this comparison the turtle shows the poorest absolute sensitivity, but these data are for sound in air.

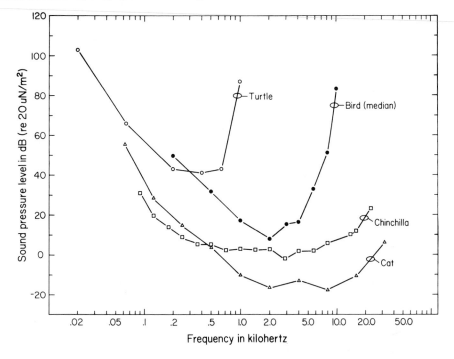

Figure 9-1. The median avian audibility function is compared to behaviorally obtained absolute threshold curves for the turtle (Patterson 1966), the cat (Miller, Watson, and Covell 1963), and the chinchilla (Miller 1970).

Underwater thresholds have not yet been measured and may very well be different. There is a marked similarity in shape and relative bandwidth between the typical avian audibility curve and that of the turtle. In fact, if these four threshold curves were plotted relative to their best frequency and best intensity, the only glaring difference remaining would be the greater range in octaves above the most sensitive point in the audibility curve for the two mammals. Finally, it should be mentioned that the cat, as a nocturnal predator, has excellent absolute sensitivity and that the two nocturnal predators from the class Aves—*Tyto alba* (barn owl) and *Bubo virginianus* (great horned owl)—do as well (see Table 9-1, Best intensity).

It goes without saying that, in all the animal kingdom, the psychoacoustic thresholds for man are clearly the most rigorously obtained and precisely defined. For this reason alone, a comparison with human data is often useful and such a comparison is shown in Fig. 9-2. The variability among birds in terms of absolute thresholds can be described by the interquartile range of the 16 avian threshold curves. As can be seen, the hearing sensitivity of birds as a class falls well within the area of audibility for man, particularly near the upper and lower frequency limits of avian hearing. Within the class Aves, a comparison between thresholds of passerines and nonpasserines tested reveals that passerines tend to have better high frequency sensitivity and poorer low frequency sensitivity than nonpasserines.

Table 9-1

Bird and Reference	40 dB bandwidth (octave)	Best intensity (dB)	Low frequency sensitivity (dB)	High frequency cutoff (kHz)	Low frequency slope (dB/oct)	High frequency slope (dB/oct)
Song Birds						
blue jay (Cohen, Stebbins and Moody 1978)	6.02	11.5	38.0	7.8	12.0	38.0
bullfinch (Schwartzkopff 1949)	4.61	-4.5	33.7	12.0	18.5	25.0
canary (Dooling, Mulligan and Miller 1971)	4.08	8.0	49.0	9.6	17.0	35.0
cowbird (Hienz, Sinnott, and Sachs 1977)	5.24	13.0	37.6	9.7	20.0	125.2
crow (Trainer 1946)	3.81	-16.0	12.0	7.0	29.0	50.0
field sparrow (Dooling, Peters, and Searcy 1979)	5.19	6.8	31.3	11.0	12.8	38.0
house finch (Dooling, Zoloth, and Baylis 1978)	4.69	8.2	38.2	7.2	15.8	33.0
red-winged blackbird (Hienz, Sinnott, and Sachs 1977)	5.52	13.1	33.2	9.6	15.0	70.0
starling (Trainer 1946)	4.05	8.0	44.0	8.6	25.0	31.0
Nonsong Birds						
barn owl (Konishi 1973)	5.70	-18.0	4.0	12.5	13.5	103.0
great-horned owl (Trainer 1946)	4.45	-16.0	-1.0	7.0	16.0	46.0
mallard (Trainer 1946)	5.42	14.5	35.0	6.3	10.5	49.0
parakeet (Dooling and Saunders 1975a)	5.08	0.2	18.1	8.0	16.7	63.7

pigeon (Trainer 1946)	5.85	20.7	32.7	5.8	8.8	40.0
pigeon (Heise 1953)	5.72	17.0	22.0	5.5	10.0	70.0
pigeon (Stebbins 1970)	5.12	15.0	30.0	5.6	14.0	28.5
pigeon (Harrison and Furumoto 1971)	4.67	1.0	12.0	7.3	24.0	60.0
pigeon (Hienz, Sinnott, and Sachs 1977)	5.85	11.0	16.5	5.6	8.0	75.0
sparrow hawk (Trainer 1946)	4.88	6.0	25.0	7.4	17.0	60.0
turkey (Maiorana and Schleidt 1972)	5.43	17.5	32.0	6.6	17.0	43.0

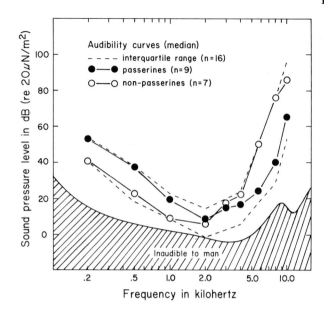

Figure 9-2. Interquartile range of the 16 avian audibility curves (dashed lines) is shown in relation to the auditory space of man (Robinson and Dadson 1956). Median thresholds of 9 species of passerines (closed symbols) are compared to median thresholds of 7 species of nonpasserines (open symbols).

2.2 Factors Affecting the Shape of the Audibility Function

Some of the factors that contribute to the shape of the audibility curves for members of the class Aves are now becoming known. Part, but certainly not all, of the difference in high frequency sensitivity between birds and mammals may be the performance of the middle ear. There is a 15 to 20 dB/octave difference between a columella and a three-bone ossicular chain in the rates of roll-off in sensitivity above 6 kHz to 7 kHz (Saunders and Johnstone 1972). Since this difference is not enough to account entirely for the 70 to 100 dB/octave roll-off in sensitivity observed in many birds (Sachs, Sinnot, and Hienz 1978), the search must continue for other factors (i.e., inner ear impedance).

Limitations on low frequency sensitivity must be related to a number of the factors known to affect middle ear impedance, including small eardrums with small air spaces behind them (Schwartzkopff and Winter 1960, Dooling, Mulligan, and Miller 1971, Webster and Webster 1972, 1975). The function of the lagena in the avian cochlea and whether or not it contributes to low frequency sound perception is not clear. Definitive anatomical data on the existence of a structure(s) in birds comparable to the helicotrema in the mammalian cochlea is also lacking. Békésy (1960) reported that *Gallus domesticus* (chicken) does not show a connection between scala tympani and scala vestibuli so that a pressure in one canal produces a persistent bulging of the basilar papilla. This suggests a potential pressure sensitivity that could function as a highly sensitive barometer (Kreithen and Keeton 1974). Schwartzkopff and Winter (1960), on the other hand, report that song birds show not one but two connections between scala vestibuli and scala tympani. It has been suggested that in mammals, the helicotrema functions as a high pass filter (Dallos 1970, Ehret 1977). A small helicotrema improves reception of low frequency sound by allowing traveling wave amplitude to remain high nearer the apex. Thus, for birds to show sensitivity to low frequency sounds, a small or nonexistent helicotrema would seem to be indicated.

This issue of what limits low frequency sensitivity in birds acquires new significance in view of recent experiments demonstrating that *Columba livia* (domestic pigeon) are surprisingly sensitive to very low frequency sound (Yodlowski, Kreithen, and Keeton 1977, Quine 1978, Kreithen and Quine 1979). Over the years, behavioral audibility curves have been reported for the pigeon in five different investigations (see Table 9-1). These results, shown in Fig. 9-3, reveal an admirable consistency, considering that different procedures and different laboratories were involved. The exciting comparison, however, involves the most recent auditory thresholds reported for the pigeon spanning the frequencies from 200 Hz down to .05 Hz. In the region of 1 to 10 Hz, the pigeon is approximately 50 dB more sensitive than man (Yeowart and Evans 1974, Yeowart, Bryan, and Tempest 1967). It seems certain that sensitivity to very low frequency sounds in pigeons involves the peripheral auditory system in that columellar destruction raises thresholds about 40 dB while cochlear destruction completely abolishes the phenomenon (Yodlowski et al. 1977, Quine 1978). Whether this unexpected sensitivity to low frequency sounds is unique to pigeons is unknown, since at the present time the only other vertebrate that has been tested at such low frequencies is man.

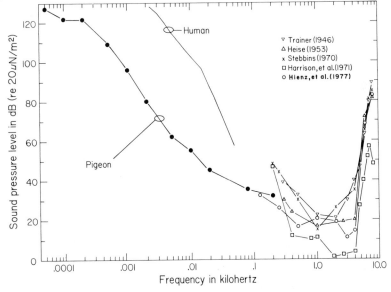

Figure 9-3. Behaviorally determined audibility thresholds for the pigeon are shown as reported in five different investigations. For comparison, low frequency thresholds are shown for the pigeon (Quine 1978) and man (Yeowart and Evans 1974).

Finally, ambient noise can affect the shape of the audibility curve. For organisms with small critical ratios (i.e., good signal-to-noise ratios) at low frequencies (i.e., below 200 Hz where acoustic insulation in testing booths is less effective), it is always difficult to rule out the possibility that absolute thresholds may in fact be masked thresholds (Miller 1970). Proving that ambient noise is not the limiting factor in absolute sensitivity requires a demonstration that the background noise level in the testing environment is more than one critical ratio below the subject's absolute threshold. Fig. 9-4 shows the audibility curve of *Melopsittacus undulatus* (parakeet), the audibility curve minus the critical ratio, and the upper bound of the spectrum level of ambient noise in the testing booth. At no frequency is the spectrum level of ambient noise greater than one critical ratio below the absolute threshold curve. This is an important demonstration in that it eliminates ambient noise level as a factor in determining the shape of the audibility curve for the parakeet. Thus, biological factors must be responsible.

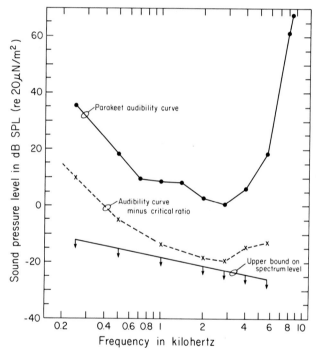

Figure 9-4. The relation between absolute auditory sensitivity for the parakeet (solid symbols), the absolute threshold minus the critical ratio for the parakeet (dotted line), and the upper bound on the spectrum level of noise in the testing chamber (solid line and arrows).

3 Auditory Discrimination

Studies of auditory discrimination in birds, nonexistent 10 years ago, now provide an array of behavioral data rivaling that available for man. Some of these results indicate differences between birds and mammals in peripheral auditory system processing. The traditional experiments of intensity, frequency, and temporal resolving power have been conducted with birds.

3.1 Intensity Resolving Power

Five species of birds have been tested on an intensity discrimination task involving pulsed sinusoids in a repeating background procedure. Fig. 9-5 shows the results from these five species in comparison to human data collected under similar stimulus conditions. The results are plotted in terms of the relative difference limen in decibels. All species were tested at or near the frequency to which they are most sensitive. Where the human shows a relative difference limen in dB of about 1.0 dB at a sensation level of 60 dB (Jesteadt, Wier, and Green 1977), birds cover a fairly wide range from 1.5 dB to about 3.0 dB. Most vertebrates that have been tested, including the cat (Raab and Ades 1946), *Mus musculus* (house mouse) (Ehret 1975a), *Rattus norvegicus* (white rat) (Hack 1971), and *Carassius auratus* (goldfish) (Jacobs and Tavolga 1967), show similar levels of sensitivity. Thus, in comparison to other vertebrates, birds do not appear to be unusual with regard to the ability to discriminate intensive differences in pulsed sinusoids.

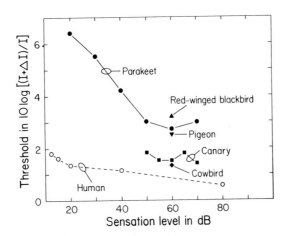

Figure 9-5. Relative intensity difference limen in decibels for five species of birds compared to similar measures for man (Jesteadt, Wier, and Green 1977). The data for the parakeet are from Dooling and Saunders (1975b); for the cowbird, the red-winged blackbird, and the pigeon from Sinnott, Sachs, and Hienz (1976); and for the canary from Dooling (unpublished).

3.2 Temporal Resolving Power

A similar picture emerges when reviewing the data on temporal resolving power. In spite of arguments to the contrary from song learning data (see, for example, Pumphrey 1961, Greenewalt 1968), from cochlear anatomy (Schwartzkopff 1968, 1973), and from single unit recordings (Konishi 1969), birds do not appear to be capable of an unusual degree of temporal resolving power. To be sure, there are many ways to vary the temporal characteristics of an acoustic signal, and some of these may be more salient to the avian ear than others. However, on the basis of gap detection, duration discrimination, and temporal summation, birds appear similar to other vertebrates that have been tested.

Gap detection thresholds have been measured for *Carpodacus mexicanus* (house finch) (Dooling, Zoloth, and Baylis 1978), and the two-click threshold has been measured for the pigeon as well as *Pyrrhula pyrrhula* (bullfinch) and *Carduelis chloris* (greenfinch) (Wilkinson and Howse 1975). The avian data all show levels of sensitivity in the 3 msec to 5 msec range, placing them in close agreement with a large body of literature suggesting 2.0 msec as the limit of temporal resolution for humans (Green 1971).

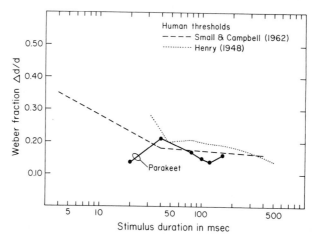

Figure 9-6. The Weber fraction $\Delta d/d$ for the parakeet is compared to similar measures from man. The data for the parakeet are replotted from Dooling and Haskell (1978).

Discrimination of auditory duration by the parakeet also agrees well with the levels of sensitivity reported for man (Dooling and Haskell 1978, Henry 1948, Small and Campbell 1962). These results are shown in Fig. 9-6 and indicate a fairly good agreement between parakeet and man over the range of 20 msec to 200 msec. To date, the only other vertebrates tested on duration discrimination have been the pigeon (Kinchla 1970) and the bottlenosed dolphin (Yunker and Herman 1974). The parakeet's sensitivity to changes in the duration of an acoustic signal is greater than that reported for the pigeon and less than that reported for the bottlenosed dolphin.

Temporal summation data are available for both *Spizella pusilla* (field sparrow) and parakeet (Dooling 1979). Data for both of these avian species are surprisingly similar to that reported for other vertebrates including man (Watson and Gengel 1969), house mouse (Ehret 1976a), chinchilla (Henderson 1969), *Tursiops truncatus* (bottlenosed dolphin) (Johnson 1968), and *Macaca mulatta* (monkey) (Clack 1966). The results for the parakeet tested at three different frequencies are shown in Fig. 9-7. The functions

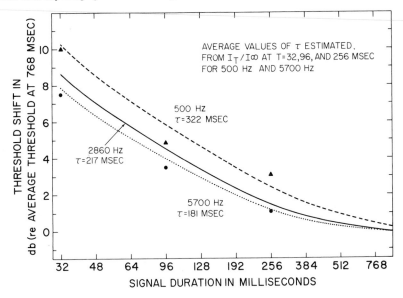

Figure 9-7. Temporal integration functions for the parakeet are shown at three test frequencies. The data for 2 860 Hz are based on threshold measurements at ten tone durations between 32 msec and 768 msec (Dooling 1979). The line of best fit is, according to Plomp and Bouman (1959), $I_t/I_\infty = 1(1 - e^{-t/\tau})$.

relating absolute threshold and tone duration are generated using the procedures of Plomp and Bouman (1959). The function for 2 860 Hz is taken from Dooling (1979). As has been described for man (Watson and Gengel 1969) and the house mouse (Ehret 1976a), there also appears to be a frequency effect of temporal integration for the parakeet with estimates of τ for 500, 2 860, and 5 700 Hertz being 322, 217, and 181 msec respectively. These data stand in marked contrast to another vertebrate, the goldfish, which shows no temporal summation (Popper 1972).

3.3 Frequency Resolving Power

While there seem to be no great differences between bird and mammal in intensity or temporal resolving power, there do appear to be differences between these two vertebrate groups in frequency resolving power. Fig. 9-8 compares frequency resolving power for *Molothrus ater* (cowbird), *Agelaius phoeniceus* (red-winged blackbird), and parakeet along with cat, chinchilla, and man in a plot of log ΔF versus frequency. Differences among these species are evident. Below the region of 500 to 1 000 Hz, ΔF for the mammals remains constant or improves slightly. For the three birds, ΔF increases as frequency decreases. These results are puzzling in view of electrophysiological data from the avian auditory system such as tuning curve width and temporal synchrony, which suggest that birds appear to code stimulus frequency as well as mammals (Sachs, Young, and Lewis 1974, Konishi 1969, Sachs et al. 1978, Sachs, Woolf, and Sinnott, Chapter 11).

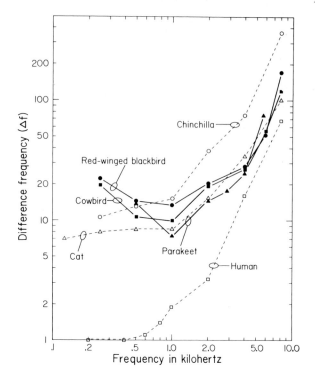

Figure 9-8. Log frequency discrimination threshold (Δf) plot for three birds—parakeet, cowbird, and red-winged blackbird—compared to similar measures for three mammals —the cat (Elliot, Stein, and Harrison 1960), the chinchilla (Nelson and Kiester 1978), and man (Wier, Jesteadt, and Green 1977). The data for the parakeet are from Dooling and Saunders (1975a); for the cowbird and the red-winged blackbird, data are from Sinnott, Sachs, and Hienz (1976).

4 Auditory Masking and Threshold Shift

Of all the psychophysical data available for birds, the results of broad band and narrow band masking and noise exposure studies are perhaps the most telling in terms of peripheral auditory system function. The relation among frequency difference limens, critical bands and critical ratios has served as a unifying principle in investigations of human auditory function (Scharf 1970). It is generally accepted that the place of maximum displacement is organized logarithmically along the basilar membrane (Békésy 1960) and, as a consequence, critical bands are viewed as representing constant distances along the basilar membrane for any one species (Watson 1963, Ehret 1975b, 1976b). Thus there is a pervasive notion that for most vertebrates, basilar membranes and associated cochlear events are perfect scale models of one another (Greenwood 1961a, 1961b, 1962). Results from recent masking experiments in the parakeet and *Rhinolophus ferrumequinum* (horseshoe bat) are at variance with this hypothesis (Dooling and Saunders 1975a, Long 1977).

4.1 Results of Broad Band Masking

Critical ratio data for a number of vertebrate species are shown in Fig. 9-9. The general pattern observed is a 3 dB increase in critical ratio for each doubling of frequency. This pattern prevails for all species except the parakeet and the horseshoe bat. It is well established, at least in the domestic chicken and the pigeon (and most probably in other birds), that frequency is organized logarithmically along the basilar papilla (Békésy 1960). If this is true, then the shape of the critical ratio function for the parakeet rules out the possibility that critical bands represent equal distances along the basilar papilla (Dooling and Saunders 1975a). In the case of the horseshoe bat, the departure from the expected critical ratio function corresponds to a thickening of the basilar membrane at a point 4.5 mm from the base (Bruns 1976a, 1976b).

The peculiar shape of the parakeet critical ratio function has invited retesting of the validity of these results. A recent experiment by Saunders, Denny, and Bock (1978) provided one such test by measuring the critical bandwidth directly using band-narrowing procedures. These investigators found that the relation between directly measured critical bandwidths and critical ratios for parakeets is nearly identical to that observed for man (Scharf 1970). Critical bandwidths are about 2.5 times greater than bandwidths inferred from critical ratio data.

The fundamental discrepancy between directly measured critical bandwidths and bandwidths inferred from broad band masking data has recently been resolved by Bilger (1976). According to Bilger, the assumption that at masked threshold the power in a masked tone is equal to the power in a critical band of frequencies surrounding the tone is incorrect. Rather, the task involved in a critical ratio experiment is to detect the intensity increment between a critical band of noise alone and a critical band of noise plus signal. Thus critical bands and critical ratios are related by way of the Weber fraction for intensity. In other words, multiplying the critical ratio by the reciprocal of the Weber fraction for intensity will result in the correct estimate of critical bandwidth. Dooling and Searcy (1979) reasoned that intensity difference limens using amplitude modulated pure tones may be closer to a simultaneous masking task (i.e., like that used to collect critical ratio and critical band data) than pulsed sinusoids presented in a repeating background (Dooling and Saunders 1975b). Fig. 9-10 shows the relation between critical ratios, critical bands, and intensity difference limens measured in this way for the parakeet. In spite of the unusual masking pattern observed in the parakeet, the relation between critical ratio, critical band, and intensity difference limen is essentially the same as that observed for man (Bilger 1976).

While critical band data are intimately tied to the mechanisms of peripheral auditory system function, they carry another implication as well. Auditory mechanisms do not evolve in the near absolute quiet of a testing booth. On the contrary, an organism's detection of biologically meaningful sounds takes place in the presence of continuous environmental noise. Thus, in some ways, a more revealing comparison among vertebrates in terms of basic auditory capabilities might involve critical ratio functions rather than the more traditional absolute threshold curves. For instance, the sonar signals of the horseshoe bat coincide with the lowest point in the critical ratio function. Furthermore, not only does the parakeet show the lowest critical ratio at 2.86 kHz of all other vertebrates tested, but this frequency is also the approximate center frequen-

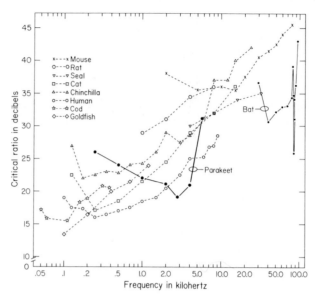

Figure 9-9. Critical ratio functions are shown for nine vertebrates: mouse (Ehret 1975b), white rat (Gourevitch 1965), *Pusa hispida* (ringed seal) (Terhune and Ronald 1975), cat (Watson 1963), chinchilla (Miller 1964), man (Hawkins and Stevens 1950), *Gadus morhua* (cod) (Hawkins and Chapman 1975), goldfish (Fay 1974), horseshoe bat (Long 1977), and parakeet (Dooling and Saunders 1975a).

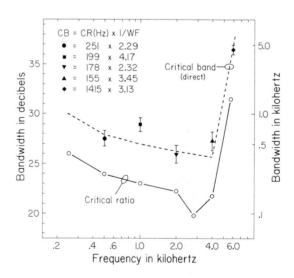

Figure 9-10. The relation between the Weber fraction for intensity ($\Delta I/I$) for the parakeet and both indirect (Dooling and Saunders 1975a) and direct (Saunders, Denny, and Bock 1978) measures of the critical bandwidth are shown. Solid symbols are the critical ratio in hertz multiplied by the reciprocal of the Weber fraction for intensity at that test frequency. These data are replotted from Dooling and Searcy (1979).

cy of the species' vocalizations. This means essentially that a parakeet can detect its conspecific vocalization in the presence of environmental noise at a considerably greater distance from the source than any of these other species. This fact is not at all apparent when comparing absolute threshold curves among these same species.

4.2 Results of Noise Exposure Experiments

Another line of evidence suggesting differences in peripheral auditory system processing between birds and mammals comes from experiments involving threshold shifts from noise exposure in parakeets. Dooling and Saunders (1974) exposed parakeets for 72 hours to four levels of a 1/3 octave band of noise centered at 2.0 kHz. The general pattern of threshold shift observed in man and other mammals (Mills, Gengel, Watson, and Miller 1970, Mills 1973) was also observed in the parakeet. This pattern consists of a growth of threshold shift that reaches a plateau or asymptote within 12 to 24 hours independent of the intensity of the noise, a spread of threshold shift that is greater for frequencies above the noise band than for those below, and a decay of threshold shift after removal from the noise, the duration of which depends on the conditions of the noise exposure (Saunders and Dooling 1974).

There were, however, several intriguing differences between the pattern of threshold shift in parakeets and that observed in mammals. First, the parakeets showed much less threshold shift than would be observed for a mammalian ear, supporting Pumphrey's (1961) claim that birds are relatively immune to acoustic trauma from loud noises. Second, and perhaps more importantly, the maximum threshold shift occurs at the center of frequency of the noise (rather than 1/2 to 1 octave above), and there is relatively little spread of threshold shift into the higher frequencies. These results are shown in Fig. 9-11 (lower). Both the reduced amount and the somewhat more symmetrical pattern of threshold shift near the exposure frequency prompted an anatomical investigation of the basilar papilla of parakeets exposed to this same noise (Bohne and Dooling 1974). While no sensory cell loss was observed even at the highest level of exposure, there were morphological changes in the sensory cells over portions of the basilar papilla. These ranges are indicated schematically in Fig. 9-11 (upper) and show a particularly interesting relation to the hearing loss audiograms as measured behaviorally. These data, along with the observations of Békésy (1960) and the review by Saito (see Saito, Chapter 8) support the notion of a place analysis of frequency occurring along the basilar papilla of birds.

4.3 Narrow Band and Pure Tone Masking

The symmetrical pattern of threshold shift near the exposure frequency observed in parakeets suggests that the shape of the traveling wave along the basilar papilla may also be sharper and more symmetrical than in mammals. Recent experiments on narrow band masking in parakeets support this notion showing threshold shifts that closely follow the spectral characteristics of the masking noise even at high intensity levels

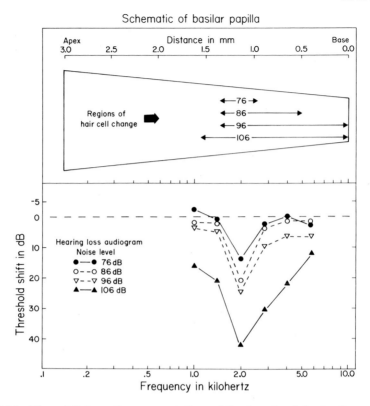

Figure 9-11. (Upper) Schematic representation of the length of the basilar papilla of the parakeet based on measurements of Bohne and Dooling (1974). Width measurements are not to scale. Frequency mapping along the papilla is estimated from Békésy's (1960) measurements of the chicken and the known hearing range of the parakeet (Dooling and Saunders 1975a). Areas of lesions shown by arrows. (Lower) Hearing loss audiograms for the parakeet as measured behaviorally (Dooling and Saunders 1974, Saunders and Dooling 1974).

(Saunders, Bock, and Fahrbach 1978). Results from studies employing one pure tone to mask another also show a reduced spread of masking to higher frequencies in the parakeet compared to that found in man (Saunders, Else, and Bock 1978, Vogten 1974, Egan and Hake 1950, Wegel and Lane 1924). Psychophysical tuning curves generated from pure tone masking data indicate that the low frequency arm of the tuning curve is about as steep as the high frequency arm for the parakeet but not for the two mammals that have been tested to date. These data are shown in Fig. 9-12 normalized along both axes to allow a simple comparison on the basis of the shape of the masking pattern for these three species. Notice that the low frequency "tails" of these curves are more pronounced for the two mammalian species than for the parakeet. This is essentially a demonstration of less spread of masking to frequencies higher than the masker frequency for the parakeet. A comparison of 8th nerve single unit tuning curves from the pigeon with similar data from the cat reveals the same pattern (Sachs, Young, and Lewis 1974, Sachs et al., Chapter 11).

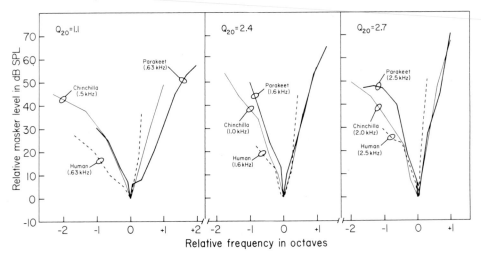

Figure 9-12. Psychophysical tuning curves (masker level versus masker frequency) are shown for the parakeet (Saunders, Else, and Bock 1978), the chinchilla (McGee, Ryan, and Dallos 1976), and man (Saunders and Rintelmann 1978). For a measure of tuning curve width, the Q_{20} values are shown for the parakeet. These values are increasing (more narrow tuning) as frequency is increased. Both masker intensity and frequency are in relative units to facilitate a comparison of psychophysical tuning curves obtained in the low, medium, and high frequency regions of each of these species.

5 Correspondence between Hearing and Vocalizations

One of the more fascinating aspects of hearing in birds is the relation between auditory perception and vocal output. It has been known for some time that individual and species recognition can be conveyed by subtle acoustic cues (Thorpe 1961, Falls 1963, Brooks and Falls 1975a, 1975b). Furthermore, song learning experiments and deafening experiments have demonstrated that hearing is critically involved in the development of normal vocal behavior in many species (Konishi 1964, 1965, Marler and Mundinger 1971).

5.1 Relations between Auditory Sensitivity and Vocal Characteristics

There have been a number of reports documenting a relation between hearing capability and some characteristic of vocal output. For instance, it is generally true that the peak in the long term average power spectrum of the species song coincides with or is slightly above the most sensitive point in the audibility curve (Dooling et al. 1971, Konishi 1970, Dooling and Saunders 1975a, Hienz, Sinnott, and Sachs 1977). While this notion seems to hold for songs, caution must be exercised in assuming the same relation for calls. There are numerous examples of alarm calls, for instance, with a considerable amount of energy in the frequency range of 4 kHz to 8 kHz (Marler 1955, Thorpe 1961). The adaptive significance of placing alarm vocalizations in the frequency region of diminished auditory sensitivity is not clear. Needless to say, selective pressures are involved in the design of these vocal signals other than those serving to maximize transmission distance.

Processes of auditory species recognition, mate selection, and territorial defense all involve sophisticated discrimination between complex acoustic patterns. This has prompted an attempt in the literature to consider auditory discrimination in birds from the standpoint of song learning and the analysis of the fine structure of vocalizations (Greenewalt 1968). Essentially, it is argued that birds must be able to discriminate the complex acoustic pattern of a song in order to produce a similar song by imitation. Very precise measurements of the acoustic fine structure of avian vocalizations have supported the notion that the ability to maintain such a remarkable degree of stereotypy may be tied in some ways to the resolving power of the avian ear (Dooling and Saunders 1975a).

It might also be argued that a bird embarking on song learning has an auditory memory or template for a certain vocal pattern (i.e., a syllable) and that each production of this vocal pattern represents an attempt to match this template precisely. If attempts to attain a perfect match involve the ear, then one might suppose that the degree of variability that occurs from model to imitation or across repeated occurrences of the same vocalization (i.e., as defined by the coefficient of variation) is directly related to the auditory system's ability to make acoustic discriminations (i.e., as defined by the Weber fraction). Both discrimination and vocal production data are available for canary and parakeet. Comparisons for the dimension of frequency, intensity, and duration are given in Table 9-2. There is quite a good match between resolving power measured psychophysically and the coefficient of variation in vocal production.

5.2 Relation of Hearing to Vocal Ontogeny

There is considerable evidence that human infants are predisposed to be responsive to particular aspects of speech sounds before speaking (Eilers and Minifie 1975, Streeter 1975, Kuhl 1976). For many species of birds, normal hearing is clearly a requirement for the development of a normal vocal behavior. But, how intrusive are the influences of auditory perceptual processes in the development of the species song? Comparative studies of vocal learning in birds have shown that early auditory experience (10 days to 50 days) often has a profound effect on subsequent vocal performance (Konishi and Nottebohm 1969, Marler and Mundinger 1971, Konishi 1978). In many song birds, exposure to songs of conspecifics during this sensitive period is necessary for normal

Table 9-2. Coefficient of variation in note production compared to Weber fraction for discrimination

Acoustic dimension	Species	Coefficient of variation (SE/$\bar{X}$)	Weber fraction $\Delta X/X$
Frequency	parakeet	.008 – .020	.007 – .009
Intensity	canary	.20 – .80	.35 – .50
Duration	parakeet	.10	.13 – .20

vocal development to occur, and young males of some species will selectively imitate songs of conspecifics that are presented together with songs of other species (Thorpe 1961, Marler 1970). The problem with these studies stems from the difficulty in separating sensory and motor components of song learning. Whether or not a particular song pattern has been learned can only be determined later when the bird sings it. Thus, it has always been an open question whether the phenomenon of selective vocal learning is due to early perceptual preferences or whether later motor constraints play a role. Recent studies showing differences in vocal learning between two closely related species of sparrow highlight the issue. When raised in the laboratory under controlled conditions and tutored with both normal and synthetic songs of both species, *Melospiza georgiana* (swamp sparrow) proved to be highly discriminating, learning only their own syllable types (Marler and Peters 1977). When *Melospiza melodia* (song sparrow) were tutored with both species' song, they proved less discriminating and readily learned both swamp and song sparrow syllables (Marler and Peters, in press).

Dooling and Searcy (in press) have found evidence based on the cardiac response (Fig. 9-13, upper), for an early perceptual selectivity in 5-day-old swamp sparrows that is evident at initial exposure to conspecific song and thus may contribute to selective vocal learning in this species. Cardiac responses were obtained to the presentation of a number of vocalizations of the two species (Fig. 9-13, lower). Swamp sparrows showed significantly greater deceleration to conspecific song than to song sparrow song even though they were hearing both sets of sounds for the first time. Song sparrows, on the other hand, failed to show a significant difference in response to the two sets of sounds. These data, from an auditory perceptual standpoint, are in good accord with known differences in song learning between these two species. Thus, species-specific perceptual predispositions are at least partly responsible for the differences in song learning between these two species.

6 Conclusion

A review of the currently available psychoacoustic data for members of the class Aves reveals interesting and sometimes surprising comparisons with hearing data from other vertebrates. Many questions remain to be answered. Absolute threshold curves for birds are narrowly tuned to the region of about 1 kHz to 5 kHz and show generally poor high frequency sensitivity. There are differences between passerine and nonpasserine audibility curves but the reasons for these differences are not at all clear. A simple explanation based on the fact that passerines tend to be smaller and have higher-pitched vocalizations than nonpasserines is inadequate since there are several glaring exceptions. Compare, for instance, cowbirds and parakeets on the basis of body size, hearing sensitivity, and vocalization spectra.

On the basis of the present review, birds as a class show a remarkably consistent pattern of absolute threshold sensitivity, clearly showing much less variability in this regard than other vertebrate classes. It would be interesting to test the hearing of very small birds such as members of the family *Trochilidae* (hummingbirds) and very large birds such as *Struthio camelus* (ostrich) to see if the consistency remains. There is also evidence for echolocation in the class Aves (Griffin 1954, Griffin and Suthers

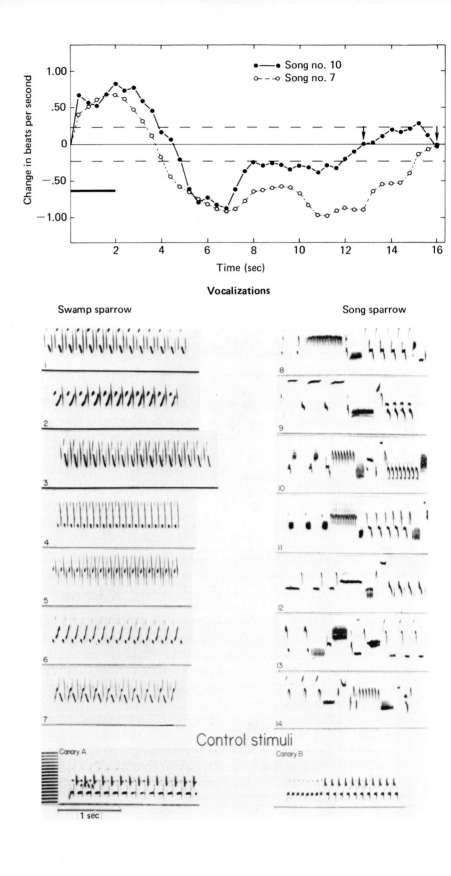

Vocalizations

Swamp sparrow

Song sparrow

Control stimuli

Canary A

Canary B

1 sec

1970). Do birds that echolocate show an unusual pattern of absolute sensitivity as a result of this special adaptation?

The low frequency sensitivity of the pigeon is surprising and poses a number of questions as to whether or not the pigeon is truly lacking a structure comparable to a helicotrema, whether or not other birds are as sensitive to low frequency sounds as the pigeon, and whether or not other vertebrates are as insensitive to these same low frequency sounds as man appears to be.

Results of a variety of experiments on auditory discrimination in birds show clearly that birds do as well or better than most mammals and approach the levels of sensitivity commonly reported for man. Earlier suggestions (Schwartzkopff 1973, Thorpe 1963, Greenewalt 1968, Pumphrey 1961) that birds may be as much as ten times more sensitive than man to changes in the frequency or temporal parameters of an acoustic signal are not yet supported by data.

The notion that a common place-dependent filter mechanism may be operating in the cochleas of birds and mammals is supported by frequency discrimination data for birds above 1.0 kHz, by Békésy's (1960) observations of a traveling wave along the avian basilar papilla and by narrowband noise exposure data for the parakeet resulting in restricted lesions near the basal end of the papilla. Critical band and critical ratio data for the parakeet, however, do not sit well with this hypothesis. A solution to this problem is not obvious. Perhaps critical bands represent equal distances along the basilar papilla of the parakeet over only a very narrow range of frequencies (i.e., 2.8 kHz to 4.0 kHz) where a 3 dB/octave increase in signal-to-noise occurs. Then one would argue that frequencies outside of this range do not map on the papilla in a logarithmic fashion. Alternatively, of course, it may be that the mechanism that accounts for critical band effects in the parakeet is simply not as peripheral (i.e., mechanical) as its counterpart in mammals. It is interesting in this regard that the relation between critical bands, critical ratios, and ΔI holds in spite of the unusual shape of the masking functions.

Psychophysical tuning curve data, narrow band noise exposure and masking data, and electrophysiological tuning curve data all support the notion of a sharper and perhaps more symmetrical traveling wave envelope along the avian basilar papilla. This may be part of the mechanism by which birds attain excellent frequent resolving power in spite of a short receptor surface.

The remarkable precision and stereotypy so characteristic of song bird vocalizations has, in the past, served as a foundation for assertions about the discriminatory abilities of birds. Several examples have been cited of a correspondence between resolving power of the auditory system and the variability in repeated productions of the same vocal signal. Thus, the relation between hearing and song learning might be viewed in

Figure 9-13. (Opposite) (Upper) Heart rate responses of a swamp sparrow to the presentation of a conspecific song (# 7) and to the song of a song sparrow (# 10). Each datum point is the mean heart rate over a 400 msec time interval. Stimulus presentation is indicated by a dark line. Dashed lines indicate the 95% confidence intervals based on a prestimulus baseline. (Lower) Sonographic examples of sixteen vocalizations presented to young swamp sparrows and song sparrows. Frequency and time markers are shown at the bottom.

the following way. Consider a hypothetical note defined as a pure tone of 2 000 Hz having a duration of 20 msec and an intensity of 60 dB SPL. The present review of hearing in the class Aves suggests that a young bird attempting to imitate this note to the best of its auditory ability should come within ± 14 Hz in frequency ($\Delta f/f$ = .007), ± 2 msec in duration ($\Delta d/d$ = .1), and ± 1 dB in intensity ($\Delta I/I$ = .40). One must realize that these measured levels of sensitivity for the avian ear are equal to the reported levels of precision observed in song learning experiments and more than adequate to ensure identical tracings on a sonagraph.

There are already a number of parallels between song learning in birds and the development of speech in man (Marler 1970, 1975). The heart rate recordings from the swamp sparrow suggest yet another. Young swamp sparrows faced with the task of learning their own species song early in life, bring to the task a perceptual predisposition to guide the vocal learning process and reduce the risk of learning the wrong song.

As should be clear, the class Aves represents a tremendously wide and varied array of auditory issues. The present review has considered only behavioral comparisons, and several general comments are in order. On the one hand, ethologists are often quick to point out that psychophysical training and testing procedures are usually far removed from the conditions found in the animal's natural environment. Thus while reports of absolute and differential thresholds (i.e., Dooling and Saunders 1975a) are certainly valid, animal psychophysicists must always be prepared to learn that an animal is even more sensitive when tested with some other procedure. On the other hand, the present review indicates there is remarkable consistency in the psychoacoustic data among various species of birds. In spite of widely different training and testing methods, similar threshold curves and psychoacoustic relations recur. This consistency can be taken as evidence that the basic characteristics of the different auditory systems are what is being measured and compared rather than the response proclivities peculiar to each species. Lastly, the similarities in the hearing data of birds and mammals are quite remarkable and deserve comment. In spite of substantial differences in the anatomy of the middle ear, inner ear, and the acoustic pathway of birds and mammals, many psychoacoustic relations are the same. This means, of course, that some of the basic mechanisms underlying detection and discrimination of acoustic signals in mammals may be independent of: 1) outer, middle, and inner ear specializations; 2) two distinct types of hair cells; and 3) the complex innervation scheme seen in the cochlea of mammals.

Acknowledgments. Preparation of this manuscript was supported by Public Health Service Awards MH 14651 and MH 31165. I thank G. Ehret, R. Larkin, P. Marler, D. Quine, and M. Yodlowski for comments on an earlier draft. M. Searcy patiently prepared all of the illustrations.

References

Békésy, G. V.: Experiments in Hearing. New York: McGraw-Hill, 1960.

Bilger, R. C.: A revised critical band hypothesis. In: Hearing and Davis: Essays Honoring Hallowell Davis. Hirsh, S. K., Eldredge, D. H., Hirsh, I. J., Silverman, S. R. (eds.). St. Louis: Washington Univ. Press, 1976, pp. 141-148.

Bohne, B. A., Dooling, R. J.: Morphological changes in the ears of noise exposed parakeets. J. Acoust. Soc. Am. 55, S77(A) (1974).

Brodkorb, P.: Origin and evolution of birds. In: Avian Biology I. Farner, D. S., King, J. R., Parkes, K. C. (eds.). New York-London: Academic Press, 1971, pp. 19-55.

Brooks, R. J., Falls, J. B.: Individual recognition by song in white-throated sparrows. I. Discrimination of songs by neighbors and strangers. Can. J. of Zool. 53, 879-888 (1975a).

Brooks, R. J., Falls, J. B.: Individual recognition by song in white-throated sparrows. III. Song features used in individual recognition. Can. J. of Zool. 53, 1749-1761 (1975b).

Bruns, V.: Peripheral auditory tuning for fine frequency analysis by the CF-FM bat, *Rhinolophus ferrumequinum*. I. Mechanical specializations of the cochlea. J. Comp. Physiol. 106, 77-86 (1976a).

Bruns, V.: Peripheral auditory tuning for fine frequency analysis by the CF-FM bat, *Rhinolophus ferrumequinum*. II. Frequency mapping in the cochlea. J. Comp. Physiol. 106, 87-97 (1976b).

Clack, T. D.: Effect of signal duration on the auditory sensitivity of humans and monkeys (*Macaca mulatta*). J. Acoust. Soc. Am. 40, 1140-1146 (1966).

Cohen, S. M., Stebbins, W. C., Moody, D. B.: Audibility thresholds of the blue jay. Auk. 95, 563-568 (1978).

Dallos, P.: Low frequency auditory characteristics: species dependence. J. Acoust. Soc. Am. 48, 489-499 (1970).

Dooling, R. J.: Temporal summation of pure tones in birds. J. Acoust. Soc. Am. 65, 1058-1060 (1979).

Dooling, R. J., Haskell, R. J.: Auditory duration discrimination in the parakeet (*Melopsittacus undulatus*). J. Acoust. Soc. Am. 63, 1640-1642 (1978).

Dooling, R. J., Mulligan, J. A., Miller, J. D.: Auditory sensitivity and song spectrum of the common canary (*Serinus canarius*). J. Acoust. Soc. Am. 50, 700-709 (1971).

Dooling, R. J., Peters, S., Searcy, M. H.: Auditory sensitivity and vocalizations of the field sparrow (*Spizella pusilla*). Bull. Psychonomic Soc. 14, 106-108 (1979).

Dooling, R. J., Saunders, J. C.: Threshold shift produced by continuous noise exposure in the parakeet (*Melopsittacus undulatus*). J. Acoust. Soc. Am. 55, S77(A) (1974).

Dooling, R. J., Saunders, J. C.: Hearing in the parakeet (*Melopsittacus undulatus*): Absolute thresholds, critical ratios, frequency difference limens, and vocalizations. J. Comp. Physiol. Psych. 88, 1-20 (1975a).

Dooling, R. J., Saunders, J. C.: Auditory intensity discrimination in the parakeet (*Melopsittacus undulatus*). J. Acoust. Soc. Am. 58, 1308-1310 (1975b).

Dooling, R. J., Searcy, M. H.: The relation among critical ratios, critical bands, and intensity difference limens in the parakeet (*Melopsittacus undulatus*). Bull. Psychonomic Soc. 13, 300-302 (1979).

Dooling, R. J., Searcy, M. H.: Early perceptual selectivity in the swamp sparrow. Dev. Psychobiol. (in press).

Dooling, R. J., Zoloth, S. R., Baylis, J. R.: Auditory sensitivity, equal loudness, temporal resolving power and vocalizations in the house finch (*Carpodacus mexicanus*). J. Comp. Physiol. Psych. 92, 867-876 (1978).

Egan, J. P., Hake, H. W.: On the masking patterns of a simple auditory stimulus. J. Acoust. Soc. Am. 22, 622-630 (1950).

Ehret, G.: Frequency and intensity difference limens and non linearities in the ear of the house mouse (*Mus musculus*). J. Comp. Physiol. 102, 321-336 (1975a).

Ehret, G.: Masked auditory thresholds, critical ratios, and scales of the basilar membrane of the house mouse (*Mus musculus*). J. Comp. Physiol. 103, 329-341 (1975b).

Ehret, G.: Temporal auditory summation for pure tones and white noise in the house mouse (*Mus musculus*). J. Acoust. Soc. Am. 59, 1421-1427 (1976a).

Ehret, G.: Critical bands and filter characteristics in the ear of the house mouse (*Mus musculus*). Biol. Cybernet. 24, 35-42 (1976b).

Ehret, G.: Comparative psychoacoustics: Perspectives of peripheral sound analysis in mammals. Naturwissen. 64, 461-470 (1977).

Eilers, R. E., Minifie, F. D.: Fricative discrimination in early infancy. J. Speech Hear. Res. 18, 158-167 (1975).

Elliott, D. N., Stein, L., Harrison, M. J.: Determination of absolute-intensity thresholds and frequency difference thresholds in cats. J. Acoust. Soc. Am. 32, 380-384 (1960).

Falls, J. B.: Properties of territorial song eliciting responses from territorial males. Proc. 13th Intern. Ornithol. Congr. 259-271 (1963).

Fay, R. R.: Masking of tones by noise for the goldfish (*Carassius auratus*). J. Comp. Physiol. Psych. 87, 708-716 (1974).

Gourevitch, G.: Auditory masking in the rat. J. Acoust. Soc. Am. 37, 439-443 (1965).

Green, D. M.: Temporal auditory acuity. Psychol. Rev. 78, 540-551 (1971).

Greenewalt, C. H.: Bird song: Acoustics and physiology. Washington, D. C.: Smithsonian Institution Press, 1968.

Greenwood, D. D.: Auditory masking and the critical band. J. Acoust. Soc. Am. 33, 484-502 (1961a).

Greenwood, D. D.: Critical bandwidth and the frequency coordinates of the basilar membrane. J. Acoust. Soc. Am. 33, 1344-1356 (1961b).

Greenwood, D. D.: Approximate calculation of the dimensions of traveling-wave envelopes in four species. J. Acoust. Soc. Am. 34, 1364-1369 (1962).

Griffin, D. R.: Acoustic orientation in the oil bird, *Steatornis*. Proc. Nat. Acad. Sci. 39, 885-893 (1954).

Griffin, D. R., Suthers, R. A.: Sensitivity of echolocation in cave swiftlets. Biol. Bull. 139, 495-501 (1970).

Hack, M.: Auditory intensity discrimination in the rat. J. Comp. Physiol. Psychol. 74, 315-318 (1971).

Hawkins, A. D., Chapman, C. J.: Masked auditory thresholds in the cod, *Gadus morhua* L. J. Comp. Physiol. 103, 209-226 (1975).

Hawkins, J. E., Jr., Stevens, S. S.: The masking of pure tones and of speech by white noise. J. Acoust. Soc. Am. 22, 6-13 (1950).

Harrison, J. B., Furumoto, L.: Pigeon audiograms: comparison of evoked potentials and behavioral thresholds in individual birds. J. Aud. Res. 11, 33-42 (1971).

Hienz, R. D., Sinnott, J. M., Sachs, M. B.: Auditory sensitivity of the red-winged blackbird (*Agelaius phoeniceus*) and brown-headed cowbird (*Molothrus ater*). J. Comp. Physiol. Psych. 91, 1365-1376 (1977).

Heise, G. A.: Auditory thresholds in the pigeon. Am. J. Psychol. 66, 1-19 (1953).

Henderson, D.: Temporal summation of acoustic signals by the chinchilla. J. Acoust. Soc. Am. 46, 474-475 (1969).

Henry, F. M.: Discrimination of the duration of a sound. J. Exp. Psychol. 38, 734-743 (1948).

Jacobs, D., Tavolga, W.: Acoustic intensity limens in the goldfish. Anim. Behav. 15, 324-335 (1967).

Jesteadt, W., Wier, C. C., Green, D. M.: Intensity discrimination as a function of frequency and sensation level. J. Acoust. Soc. Am. 61, 169-177 (1977).

Johnson, C. S.: Relation between absolute thresholds and duration-of-tone in the bottlenosed porpoise. J. Acoust. Soc. Am. 43, 757-763 (1968).

Kinchla, J.: Discrimination of two auditory durations by pigeons. Per. Psychophys. 8, 299-307 (1970).

Konishi, M.: Effects of deafening on song development in two species of juncos. Condor. 66, 85-102 (1964).

Konishi, M.: The role of auditory feedback in the control of vocalization in the white-crowned sparrow. Z. Tierpsychol. 22, 770-783 (1965).

Konishi, M.: Time resolution by single auditory neurons in birds. Nature. 222, 566-567 (1969).

Konishi, M.: Comparative neurophysiological studies of hearing and vocalizations in song birds. Z. vergl. Physiol. 66, 257-272 (1970).

Konishi, M.: How the barn owl tracks its prey. Am. Sci. 61, 414-424 (1973).

Konishi, M.: Auditory environment and vocal development in birds. In: Perception and Experience. Walk, R. D., Pick, H. L. (eds.). New York: Plenum Press, 1978, pp. 105-118.

Konishi, M., Nottebohm, F.: Experimental studies in the ontogeny of avian vocalizations. In: Bird Vocalizations. Hinde, R. A. (ed.). London: Cambridge Univ. Press, 1969, pp. 29-48.

Kreithen, M. L., Keeton, W. T.: Detection of changes in atmospheric pressure by the homing pigeon (Columba livia). J. Comp. Physiol. 89, 73-82 (1974).

Kreithen, M. L., Quine, D. B.: Infrasound detection by the homing pigeon: A behavioral audiogram. J. Comp. Physiol. (1979).

Kuhl, P. K.: Speech perception in early infancy: the acquisition of speech-sound categories. In: Hearing and Davis: Essays honoring Hallowell Davis. Hirsh, S. K., Eldredge, D. H., Hirsh, I. J., Silverman, S. R. (eds.). St. Louis: Washington Univ. Press, 1976, pp. 265-280.

Long, G. R.: Masked auditory thresholds from the bat, Rhinolophus ferrumequinum. J. Comp. Physiol. 116, 247-255 (1977).

Maiorana, V. A., Schleidt, W. M.: The auditory sensitivity of the turkey. J. Aud. Res. 12, 203-207 (1972).

Manley, G. A.: Some aspects of the evolution of hearing in vertebrates. Nature. 230, 506-509 (1971).

Manley, G. A.: A review of some current concepts of the functional evolution of the ear in terrestrial vertebrates. Evolution. 26, 608-621 (1973).

Marler, P.: Characteristics of some animal calls. Nature. 176, 6-7 (1955).

Marler, P.: A comparative approach to vocal learning: Song development in white-crowned sparrows. J. Comp. Physiol. Psych. 71, 1-25 (1970).

Marler, P.: Bird song and speech development: Could there be parallels? Am. Sci. 58, 669-673 (1970).

286

Marler, P.: On the origin of speech from animal sounds. In: The Role of Speech in Language. Kavanagh, J. F., Cutting, J. E. (eds.). Cambridge, Mass.: MIT Press, 1975.

Marler, P., Mundinger, P.: Vocal learning in birds. In: Ontogeny of Vertebrate Behavior. Moltz, H. (ed.). New York: Academic Press, 1971, pp. 389-449.

Marler, P., Peters, S.: Selective vocal learning in a sparrow. Science. 198, 519-522 (1977).

Marler, P., Peters, S.: Bird song and speech: Evidence for special processing. In: Perspectives on the Study of Speech. E.mas, P., Miller, J. (eds.). Hillsdale, N. J.: Erlbaum (in press).

Masterton, B., Heffner, H., Ravizza, R.: The evolution of human hearing. J. Acoust. Soc. Am. 45, 966-985 (1969).

McGee, T., Ryan, A., Dallos, P.: Psychophysical tuning curves of chinchillas. J. Acoust. Soc. Am. 60, 1146-1150 (1976).

Miller, J. D.: Auditory sensitivity of the chinchilla in quiet and in noise. J. Acoust. Soc. Am. 36(A), 2010 (1964).

Miller, J. D.: Audibility curve of the chinchilla. J. Acoust. Soc. Am. 48, 513-523 (1970).

Miller, J. D., Watson, C. S., Covell, W. P.: Deafening effects of noise on the cat. Acta Otolaryngol. Suppl. 176, 1-81 (1963).

Mills, J. H.: Temporary and permanent threshold shifts produced by nine-day exposures to noise. J. Speech. Hear. Res. 15, 426-438 (1973).

Mills, J. H., Gengel, R. W., Watson, C. S., Miller, J. D.: Temporary changes of the auditory system due to exposure to noise for one or two days. J. Acoust. Soc. Am. 48, 524-530 (1970).

Nelson, D. A., Kiester, T. E.: Frequency discrimination in the chinchilla. J. Acoust. Soc. Am. 64, 114-126 (1978).

Nottebohm, F.: The origins of vocal learning. Am. Nat. 106, 116-140 (1972).

Patterson, W. C.: Hearing in the turtle. J. Aud. Res. 6, 453-464 (1966).

Plomp, R., Bouman, A.: Relation between hearing threshold and duration for tone pulses. J. Acoust. Soc. Am. 31, 749-758 (1959).

Popper, A. N.: Auditory threshold in the goldfish (Carassius auratus) as a function of signal duration. J. Acoust. Soc. Am. 52, 596-602 (1972).

Pumphrey, R. J.: Sensory organs; hearing. In: Biology and Comparative Anatomy of Birds. Marshall, A. J. (ed.). New York: Academic Press, 1961.

Quine, D. B.: Infrasound detection and ultra low frequency discrimination in the homing pigeon (Columba livia). J. Acoust. Soc. Am. 63, S75 (1978).

Raab, D., Ades, H.: Cortical and midbrain mediation of a conditional discrimination of acoustic intensities. Am. J. Psychol. 59, 59-83 (1946).

Robinson, D. W., Dadson, R. S.: A re-determination of the equal loudness relations of pure tones. Brit. J. Appl. Phys. 7, 166-181 (1956).

Sachs, M. B., Sinnott, J. M., Hienz, R. D.: Behavioral and physiological studies of hearing in birds. Fed. Proc. 37, 2329-2335 (1978).

Sachs, M. B., Young, E. D., Lewis, R. H.: Discharge patterns of single fibers in the pigeon auditory nerve. Brain Res. 70, 431-447 (1974).

Saunders, J. C., Bock, G. R., Fahrbach, S. E.: Frequency selectivity in the parakeet (Melopsittacus undulatus) studied with narrow-band noise masking. Sens. Proc. 2, 80-89 (1978).

Saunders, J. C., Denny, R. M., Bock, G. R.: Critical bands in the parakeet (Melopsittacus undulatus). J. Comp. Physiol. 125, 359-365 (1978).

Saunders, J. C., Dooling, R. J.: Noise-induced threshold shift in the parakeet (*Melopsittacus undulatus*). Proc. Nat. Acad. Sci. 71, 1962-1965 (1974).

Saunders, J. C., Else, D. V., Bock, G. R.: Frequency selectivity in the parakeet (*Melopsittacus undulatus*) studied with psychophysical tuning curves. J. Comp. Physiol. Psych. 92, 406-415 (1978).

Saunders, J. C., Johnstone, B. M.: A comparative analysis of middle ear function in non-mammalian vertebrates. Acta Otolaryngol. 73, 353-361 (1972).

Saunders, J. C., Rintelmann, W. F.: Frequency selectivity in man: The relation between critical band, critical ratio, and psychophysical tuning curves. Paper presented at Mid-winter meeting of the Association for Research in Otolaryngology, St. Petersburg, Fla., Jan. 29-Feb. 1, 1978.

Scharf, B.: Critical bands. In: Foundations of Modern Auditory Theory. Tobias, J. V. (ed.). New York-London: Academic Press, 1970, pp. 157-202.

Schwartzkopff, J.: Über Sitz und Leistung von Gehör und Vibrationssinn bei Vögeln. Z. vergl. Physiol. 31, 527-603 (1949).

Schwartzkopff, J.: Structure and function of the ear and the auditory brain areas in birds. In: Hearing Mechanisms in Vertebrates. DeReuck, A. V. S., Knight, J. (eds.). Boston: Little, Brown, 1968, pp. 41-59.

Schwartzkopff, J.: Mechanoreception. In: Avian Biology III. Farner, D. S., King, J. R., Parkes, K. C. (eds.). New York: Academic Press, 1973, pp. 417-477.

Schwartzkopff, J., Winter, P.: Zur Anatomie der Vogel-Cochlea unter natürlichen Bedingungen. Biol. Zentralblatt. 79, 602-625 (1960).

Sinnott, J. M., Sachs, M. B., Hienz, R. D.: Differential sensitivity to frequency and intensity in songbirds. J. Acoust. Soc. Am. 60, S87 (1976).

Small, A. M., Jr., Campbell, R. A.: Temporal differential sensitivity for auditory stimuli. Am. J. Psychol. 75, 401-410 (1962).

Stebbins, W. C.: Studies of hearing and hearing loss in the monkey. In: Animal Psychophysics: The Design and Conduct of Sensory Experiments. Stebbins, W. C. (ed.). New York: Appleton, Century, Crofts, 1970, pp. 41-66.

Storer, R. W.: Classification of birds. In: Avian Biology I. Farner, D. S., King, J. F., Parkes, K. C. (eds.). New York-London: Academic Press, 1971, pp. 1-18.

Streeter, L. A.: Language perception of 2-month-old infants shows effects of both innate mechanisms and experience. Nature. 259, 39-41 (1975).

Terhune, J. M., Ronald, K.: Masked hearing thresholds of ringed seals. J. Acoust. Soc Am. 58, 515-516 (1975).

Thorpe, W. H.: Bird-song: The biology of vocal communication and expression in birds. Cambridge Monographs in Experimental Biology, No. 12, Cambridge Univ. Press, 1961.

Thorpe, W. H.: Antiphonal singing in birds as evidence for avian auditory reaction time. Nature. 197, 774-776 (1963).

Trainer, J. E.: The auditory acuity of certain birds. Unpublished doctoral dissertation, Cornell University (1946).

Van Tyne, J., Berger, J.: Fundamentals of Ornithology. New York: John Wiley, 1976.

Vogten, L. L. M.: Pure-tone masking: A new result from a new method. In: Facts and Models in Hearing. Zwicker, E., Terhardt, E. (eds.). New York: Springer-Verlag, 1974, pp. 142-155.

Watson, C. S.: Masking of tones by noise for the cat. J. Acoust. Soc. Am. 35, 167-172 (1963).

Watson, C. S., Gengel, R. W.: Signal duration and signal frequency in relation to auditory sensitivity. J. Acoust. Soc. Am. 46, 989-997 (1969).

Webster, D., Webster, M.: Kangaroo rat auditory thresholds before and after middle ear reduction. Brain, Behav. and Evol. 5, 41-53 (1972).

Webster, D., Webster, M.: Auditory systems of heteromyidae: functional morphology and evolution of the middle ear. J. Morph. 146, 343-376 (1975).

Wegel, R. L., Lane, C. E.: The auditory masking of one pure tone by another and its probable relation to the dynamics of the inner ear. Phys. Rev. 23, 266-285 (1924).

Wier, C. C., Jesteadt, W., Green, D. M.: Frequency discrimination as a function of frequency and sensation level. J. Acoust. Soc. Am. 61, 177-184 (1977).

Wilkinson, R., Howse, P. E.: Time resolution of acoustic signals by birds. Nature. 258, 320-321 (1975).

Yeowart, N. S., Bryan, M. E., Tempest, W.: The monaural M.A.P. threshold of hearing at frequencies from 1.5 to 100 c/s. J. Sound Vib. 6, 335-342 (1967).

Yeowart, N. S., Evans, M. J.: Thresholds of audibility for very low-frequency pure tones. J. Acoust. Soc. Am. 55, 814-818 (1974).

Yodlowski, M. L., Kreithen, M. L., Keeton, W. T.: Detection of atmospheric infrasound by homing pigeons. Nature. 265, 725-726 (1977).

Yunker, M. P., Herman, L. M.: Discrimination of auditory temporal differences of the bottlenosed dolphin and by the human. J. Acoust. Soc. Am. 56, 1870-1875 (1974).

Chapter 10

Sound Localization in Birds

Eric I. Knudsen*

1 Introduction

Birds perform a wide variety of acoustically guided behaviors that place a demand on their abilities to localize sound sources in space. Consider, for example, a male songbird foraging on the forest floor. Should it hear the song of a conspecific male up in the canopy, it will localize the song and fly to the intruder to defend its territory (Weeden and Falls 1959, Falls 1963, Emlen 1971, Krebs 1976). Or consider the barn owl that silently flies over meadows at night in search of food—it hears the rustle of an unsuspecting field mouse, localizes the source, and dives for its prey (Payne and Drury 1958, Payne 1962). Notice that for birds the task of sound localization is complicated by their need to localize accurately in two dimensions: azimuth and elevation.

At first glance birds seem ill equipped for such a formidable task, since their heads are small (affording little sound shadow), and their inconspicuous ears are closely set (rendering little interaural time difference). Furthermore, behavioral studies show that birds typically hear only a narrow band of relatively low sound frequencies (up to 10 000 Hz, Dooling, Chapter 9), compared with the average mammal (up to 50 000 Hz, Masterton, Heffner, and Ravizza 1969).

Nevertheless, birds succeed in extracting accurate spatial information from sound stimuli. How do they do it? In general, analytical data on sound localization by birds are sparse and only tentative conclusions can be drawn from bioacoustical considerations. The one exception is in the case of the barn owl (*Tyto alba*), where sufficiently complete behavioral and neurophysiological data have accrued to justify a theory of sound localization.

*Department of Neurobiology, Stanford University School of Medicine, Stanford, California 94305.

2 Behavioral Evidence of Sound Localization

Common birds with unspecialized auditory systems have well documented sound localization abilities (Gatehouse and Shelton 1978, Shalter 1978, Jenkins and Masterton 1979). Ducklings localize and follow their mother's call without visual cues (Gottlieb 1971). A domestic hen acoustically locates, and comes to the aid of, her chicks when they give a distress call (Brückner 1933). Engelmann (1928) estimated the angular acuity of chickens in localizing calls to be approximately 4°.

Songbirds, which have well developed auditory systems, have evolved lifestyles in which sound localization plays a crucial role (Emlen 1971, Krebs 1976). These birds employ various vocalizations in territorial defense, mate attraction, filial interactions, and flocking, for which the location of the calling bird is an integral part of the message. The importance of "locatability" to the function of these vocalizations is attested to by their physical structure (Marler 1955, 1959): they consist of short, interrupted sounds and contain a wide range of frequencies. As we shall see, these properties optimize locatability. By contrast, certain alarm calls, given to warn of an approaching aerial predator, are tonal with gradual onsets. Such sounds are comparatively difficult to localize (Knudsen and Konishi 1979) and can, therefore, warn other birds of danger without revealing the location of the calling bird. These basic spectral and temporal properties are similar in a wide variety of species (Marler 1955, 1959). Thus, the very structure of the songs and calls of songbirds suggests that locatability has exerted a selective pressure on their evolution. Unfortunately the accuracy of songbirds in localizing such vocalizations has yet to be tested rigorously.

Behavioral studies on owls provide the most impressive evidence of sound localization by birds. Many owl genera include nocturnal raptors capable of catching prey at night when the light level is often too low for visual detection (Dice 1945, Curtis 1952, Payne 1962). As an alternative to vision, other sensory modalities such as infrared sensitivity (Hecht and Pirenne 1940, Payne 1962) and olfaction (Payne 1962, 1971) have been suggested and tested. However, only audition has been confirmed as an alternate modality that provides spatial information to the owl. The evidence comes from behavioral studies on the barn owl by Payne and Drury (1958), Payne (1962, 1971), and Konishi (1973b). Under controlled laboratory conditions, barn owls were able to capture prey in total darkness and were readily trained to localize and attack hidden speakers that emitted noise. Based on the accuracy of the owl's performance, which involved flying from its perch and landing on the sound target, Payne (1962, 1971) estimated the angular acuity of the barn owl to be about 1° in azimuth and elevation. Further behavioral experiments that measured the accuracy with which the barn owl orients its head to sound targets, have determined the owl's angular acuity to be just under 2° in both dimensions (Knudsen, Blasdel, and Konishi 1979).

3 Auditory Cues for Sound Localization

Which parameters of sound might a bird use to localize a sound source? There are two basic categories of localization cues: binaural and monaural. Binaural cues derive from interaural comparisons of a sound's arrival time, phase, intensity, and spectrum. Interaural arrival time (ΔT) refers to the delay in the onset of a sound at the two ears. Interaural phase ($\Delta\phi$) pertains to the relative phase shift in the ongoing waveforms at the two ears. Both ΔT and $\Delta\phi$ depend on the difference in the path lengths that sound must travel to reach each ear. Interaural intensity (ΔI) refers to the difference in the amount of sound energy reaching each ear. Interaural spectrum (ΔS) is the difference in the distribution of sound energy as a function of frequency (power spectrum) in each ear. Direction-dependent ΔI and ΔS cues are caused by changes in the sensitivity of each ear as a function of sound direction. The advantages of binaural cues are that they can be obtained quickly from even a brief sound, and they depend only on intrinsic properties of the binaural receiver.

Monaural localization cues include monaural spectrum and intensity scanning using head movement, neither of which relies on binaural comparison. A monaural spectrum cue is available if the ear's directionality (sensitivity as a function of direction) varies with sound frequency. If this is the case, a bird could identify a sound's direction by a characteristic spectrum. Man, for example, localizes sound in elevation by recognizing certain spectral notches that result from elevation-dependent filtering by the pinna and shoulders (Butler 1969, Blauert 1969, Hebrank and Wright 1974, Wright, Hebrank, and Wilson 1974). This cue requires a wide-band signal of familiar (predictable) spectral composition.

Intensity scanning involves moving the ear in the sound field. If the ear is directional, comparing the intensity of the sound at different head orientations could enable the bird to determine the direction of the source. The limiting requirement of this cue is that the sound must persist for a sufficient period of time to allow the bird to make sequential intensity judgments.

3.1 Auditory Morphology and Hearing Range

The value of these localization cues depends greatly on the dimensions of the bird's head and external ears relative to the dimensions of the sound wavelengths that the bird can detect. Thus, a consideration of the morphology of the avian auditory periphery and the frequency range of hearing helps in ascertaining the potential contribution of each of these cues to sound localization by birds.

The basic layout of the avian external and middle ears can be modeled as shown in Fig. 10-1. The average bird's head is small. Consequently the distance between the ear openings (interaural distance) is rarely more than a few centimeters. Furthermore, birds do not possess a homologue to the mammalian pinna, although many owls have elaborated an analogous reflective structure called a facial ruff (see Section 4.1). For the most part, the ear canals of birds are simple tubes, devoid of conspicuous sound-gathering structures. Each external meatus runs along the base of the skull from an opening behind the jaw articulation to its termination in the exoccipital bone. A tympanic membrane forms the inner wall of the meatus near its termination. Behind the tympanic membrane lies a large middle ear cavity that comprises a number of interconnecting air spaces. The middle ear cavities on each side communicate extensively with each other by two major air passages: the anterior air space, which is a single cavity lying ventral to the brain case that connects by large bony canals to both middle ears; and the large, patent eustachian tubes that originate in the middle ears

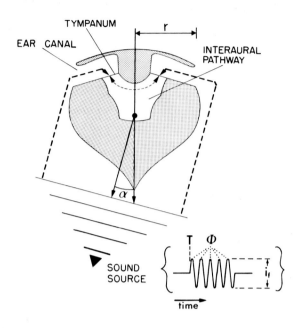

Figure 10-1. A schematic model of the head of a typical bird in a sound field. The external and middle ear cavities are illustrated in horizontal section with the bird facing downward. The path that sound would travel from the sound source to each ear is indicated by the dashed lines. Notice that in birds, sound has access to both sides of the tympanum through the ear canal and the interaural pathway. To apply this schema to owls, concave reflecting structures should be positioned behind each ear to represent the facial ruff. The parameters of sound in time are shown in brackets. Abbreviations: r-the "radius" of the bird's head or half its interaural distance; α-the azimuthal angle of the sound source with respect to the bird's median sagittal plane; T-onset of the sound; ϕ-phase of the sound; i_f-intensity of the sound at frequency f.

and fuse at the midline before they exit into the nasopharynx (Wada 1923, Stellbogen 1930, Payne 1971, Henson 1974, Norberg 1978). Thus, in birds as in other nonmammalian vertebrates (Wever and Vernon 1957, Strother 1959, Henson 1974, Pettigrew, Chung, and Anson 1978), sound pressure has access to both sides of the tympanic membrane: through both the external meatus and the interaural air passages.

To judge the capability of such an acoustic receiver to derive spatial information from sound, we must define the frequency range over which it must operate (Fig. 10-2). The audible range of birds has been determined behaviorally by measuring conditioned responses (for review see Dooling, Chapter 9) and neurophysiologically using single unit thresholds or cochlear microphonics (Wever and Bray 1936, Schwartzkopff 1955, van Dijk 1973, Coles 1977, Konishi 1970, Knudsen and Konishi 1978b), for a large number of species representing six different orders: Anseriformes, Falconiformes, Galliformes, Caprimulgiformes, Passeriformes, and Strigiformes. From these studies it can be concluded that the acoustic sensitivity of the average bird is maximal between 1 kHz and 5 kHz, and rarely extends much above 8 kHz (Dooling, Chapter 9).

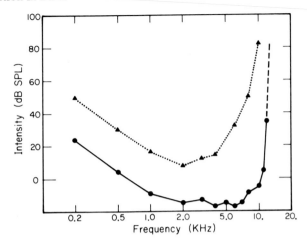

Figure 10-2. The hearing range of birds. The dotted curve represents the audibility function for the "average" bird, derived from the median absolute thresholds for 16 different species (data from Dooling, Chapter 9). The solid curve indicates the audible range of the barn owl (data from Konishi 1973b).

Even the oil bird (*Steatornis caripensis*), once thought to be capable of unusually high frequency hearing because of its ability to echolocate in dark caves (Griffin 1953), has been found to be most sensitive to 2 kHz tones, and insensitive to frequencies above 8 kHz, which is the normal frequency range for birds (Konishi and Knudsen 1979). The only birds with audible ranges that often exceed 8 kHz are certain species of songbirds (Konishi 1970, Dooling, Chapter 9) and owls (van Dijk 1973, Konishi 1973b). Of these, the barn owl hears the highest frequencies and its audibility curve declines sharply above 12 kHz (Konishi 1973b). Hence the typical bird must localize sound using frequencies of less than 8 kHz, which corresponds to wavelengths greater than 4.2 cm.

3.2 Interaural Arrival Time (ΔT)

Definition: the time delay between the arrival of a distinct acoustic event in the left and right ears.

When a sound originates from a location not on the median sagittal plane, the difference in the distance that the sound must travel to reach each ear results in a small interaural temporal disparity in the detected signal. Two manifestations of this interaural disparity are commonly distinguished: (1) interaural arrival time (ΔT), which is the delay in the onset of the sound detected by the two ears—this is the cue that gives rise to the "precedence effect" in man (Wallach, Newman, and Rosenzweig 1949); and (2) phase delay (Δφ), which is caused by the temporal shift in the ongoing waveforms at the two ears (see Section 3.3).

Because the ears lie in the horizontal plane (except in some owls, see Section 4.1), ΔT will vary systematically with the azimuthal angle of the sound source (α, Fig. 10-1).

When the source is located in the bird's median sagittal plane ($\alpha = 0°$), ΔT will be zero; when the source is directly to the side ($\alpha = 90°$), ΔT is maximum. By modeling the bird's head as a sphere with ears at opposite poles, ΔT can be evaluated approximately by the following equation:

$$\Delta T = \frac{r}{c} (\alpha + \sin \alpha),$$

where r is the radius of the bird's head, and c is the speed of sound (33.4 cm/msec). This equation, originally formulated by Woodworth (1962) to account for observed ΔT values in man, points out three important aspects of the ΔT cue. First, since ΔT is a property of sound onset, it is independent of wavelength. Second, at any given angle α, ΔT increases linearly with interaural distance. For a "typical" bird, with a head radius (r) of 1 cm, the maximum possible ΔT ($\alpha = 90°$) is approximately 60 μsec. Third, since ΔT is a sine function of α, the maximum change in ΔT per degree of change in α (angular resolution) is greatest at $\alpha = 0°$ and decreases sinusoidally as α approaches $90°$. Because of the short interaural distance of the average bird, a $10°$ change in the location of a sound source, even in the region of greatest angular resolution, would cause a change in ΔT of only about 10 μsec. This value represents the threshold of man's ability to lateralize dichotically presented, impulsive sounds (Harris 1960, Yost, Wightman, and Green 1971). Thus, for the typical bird to determine the azimuth of a sound within $\pm 5°$ using ΔT under optimal conditions, its auditory system must be capable of an interaural time resolution comparable to that of man. Unfortunately no experimental evidence is available that bears directly on the interaural time resolution of birds. However, the anatomy of the bird's inner ear (Takasaka and Smith 1971) suggests that the ΔT resolution of birds may in fact be superior to that of man. The avian basilar membrane is typically short and wide (4 x 0.5 mm in pigeon, versus 35 x 0.05 in man), and may contain greater than 50 hair cells in transverse section; in man there are only 4 to 5 hair cells across the basilar membrane. Consequently, in the avian ear, about 10 times as many hair cells register the occurrence of a stimulus event at any point along the basilar membrane. By integrating the inputs of the 50 or so parallel receptors, the bird's central auditory system could determine the precise timing of a given stimulus event with high reliability; much higher, perhaps, than can man with only 4 parallel receptors. Such high fidelity encoding of the timing of an acoustic event in each ear is just what a bird would need to effectively exploit the small ΔT cue afforded it for sound localization.

The question of whether birds use ΔT to localize sound is unsettled. In a behavioral study on sound localization by the bullfinch, Schwartzkopff (1950) tested the ability of these birds to localize either tone bursts or continuous tones and found that the minimum audible angle was the same in both cases ($20°$ to $25°$). He cites this as evidence against ΔT as a sound localization cue. However, recent behavioral studies on the barn owl (Knudsen and Konishi 1979), indicate that this bird does use ΔT in sound localization (see Section 5.2). Also, Marler (1955, 1959) has argued on ethological grounds for the importance of transients in the songs of songbirds for their locatability. The premise of this argument is that birds exploit the transients to gain a ΔT cue for sound localization.

3.3 Interaural Phase (ϕ)

Definition: the relative phase shifts in the Fourier components of a sound at the left and right ears.

Interaural phase, although similar to ΔT in that it depends on the difference in the path lengths to each ear, is frequency-dependent. For the bird to use $\Delta\phi$ for localizing sounds, not only must $\Delta\phi$ be sufficiently large, but the bird's auditory system must be capable of following the waveform of the sound with high reliability.

The magnitude of $\Delta\phi$ depends on sound wavelength (λ) as follows:

$$\Delta\phi = \frac{2\pi r\,(\alpha + \sin\alpha)}{\lambda}.$$

The shorter the wavelength, the greater will be $\Delta\phi$ for any given difference in path length. This means that the angular resolution of the $\Delta\phi$ cue should improve with frequency. However, a limit is reached when $\Delta\phi$ equals and exceeds $180°$ ($\Delta\phi \geqslant \pi$) or $\lambda \leqslant 4r$. Beyond this point, the leading and lagging ears cannot be distinguished and the laterality of the sound source becomes ambiguous. For mammals this ambiguity restricts the potential usefulness of $\Delta\phi$ to the lowest frequencies of their typically high audible ranges. Most birds, on the other hand, would not experience this ambiguity, since their small heads and low frequency hearing assures that λ is always greater than $4r$ (twice the interaural distance).

However, birds are subject to different limitations. Due to their closely set ears, $\Delta\phi$ becomes extremely small at low frequencies. As a result, sound phase will change little as a function of sound direction, and the spatial resolution of $\Delta\phi$ at low frequencies will deteriorate accordingly. At the high frequency end of a bird's audible range (4 kHz to 6 kHz), since λ is still larger than $4r$, the limiting factor becomes the ability of the bird's auditory system to temporally encode high frequency waveforms. In this regard, the bird seems to be well adapted. Single unit studies of auditory nerve fibers in various songbirds report phase-locking discharges to tone frequencies up to 4 kHz (Konishi 1969, Sachs, Young, and Lewis 1974, Sachs, Woolf, and Sinnott, Chapter 11).

Because $\Delta\phi$ varies with the difference in pathlength to the two ears, $\Delta\phi$ like ΔT, provides potential information for localizing sounds in azimuth. Assuming a constant threshold for $\Delta\phi$ detection in birds that equals the $2.5°$ figure determined for man at 250 Hz to 500 Hz (Zwislocki and Feldman 1956), then the maximum spatial resolution that a typical bird might achieve using the $\Delta\phi$ cue would range from $14°$ at 500 Hz up to almost $1°$ at 6 kHz.

At present there are no behavioral data suggesting that birds do, in fact, exploit the $\Delta\phi$ cue for sound localization. To the contrary, evidence from the barn owl, discussed in Section 5.2, argues against the use of $\Delta\phi$, at least by this species.

3.4 Interaural Intensity (ΔI)

Definition: the difference in the total sound intensity detected by the two ears. In mathematical terms:

$$\Delta I = [i_{f_1} + i_{f_2} + \ldots i_{fn}]\, \text{Left} - [i_{f_1} + i_{f_2} + \ldots i_{fn}]\, \text{Right}$$

where i_{f_1} equals the intensity of sound at a frequency, f_1 in one ear.

An object becomes an effective obstacle to sound propagation only when its dimensions approach the wavelength of the sound. Longer wavelengths propagate around the object, whereas shorter wavelengths are reflected and diffracted, resulting in the creation of a sound shadow or decrease in sound intensity on the side of the object further from the sound source. Mammals take advantage of the intensity difference caused by the sound shadow cast by their heads to gain information about the direction of the sound (Masterton et al. 1969). If an animal's head has a radius r, the magnitude of its head shadow varies as r/λ. Mammals with small heads (small r) maintain a strong head shadow by sensing extremely high frequencies (small λ). In fact, there exists an inverse correlation among mammals between head size and the high frequency limit of their hearing (Masterton et al. 1969, Masterton 1974) (Fig. 10-3).

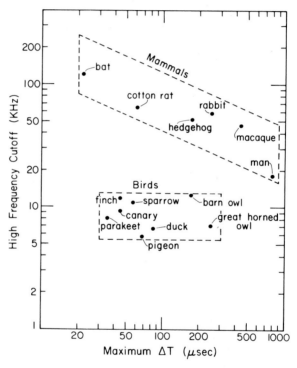

Figure 10-3. Upper limit of hearing versus maximum interaural time cue for birds and mammals. The strong inverse relationship demonstrated by mammalian species suggests a trade-off between a large interaural time cue and a large interaural intensity cue due to head shadow at high frequencies (data from Masterton 1974). Bird species have small ΔT ranges and they hear only low frequencies, which argues against their use of head shadow for sound localization. Maximum ΔT for each animal was calculated from interaural distance using Woodworth's (1962) equation. High frequency cutoff for the mammals was taken at 70 dB SPL and for birds at 60 dB SPL (data from Dooling, Chapter 9).

This inverse correlation is interpreted as demonstrating the importance of head shadow to sound localization.

Birds, on the other hand, have small heads ($r \approx 1$ cm), and they hear only low frequencies ($\lambda > 4.2$ cm). Thus, they do not show the inverse correlation that is characteristic of mammals (Fig. 10-3). Measurements of sound intensity at the external meatus of the quail, *Coturnix japonica*, ($r = 1.5$ cm) show a maximum head shadow of only about 6 dB at 8.0 kHz (Coles, Lewis, Hill, Hutchings, and Gower, in press).

Because head shadow is so weak, birds must either (1) not use ΔI in sound localization, (2) be far more sensitive to small values of ΔI than are mammals, or (3) enhance ΔI by a mechanism that is not available to mammals. There is now compelling evidence that the latter is true. First, evidence comes from the difference between the directionality of the ear measured in the external meatus, compared to the directionality of the cochlear microphonic (CM), which reflects the evoked auditory activity. CM directionality has been determined for a number of birds (Schwartzkopff 1952, 1962, Payne 1962, Knudsen and Konishi 1978b, Coles et al., in press). Even at low frequencies where directionality due to head shadow is negligible, the threshold of the CM can vary by as much as 25 dB as a function of sound direction. This enhancement of ear directionality over that due to head shadow appears to be a consequence of the interaural air passages that connect the two middle ears (Hill, Lewis, Hutchings, and Coles, in press). Coles et al. (in press) compared directionality at the entrance to the external meatus with the CM in the quail. They found for example at 315 Hz ($\lambda = 106$ cm), where no detectable change in sound pressure with direction occurred at the external meatus, that the CM sensitivity changed by as much as 24 dB (Fig. 10-4A)! Their CM directionality plot for 315 Hz contains a broad maximum on the ipsilateral side of the head and a sharp "null" on the contralateral side. Such a cardioid pattern is typical of an asymmetrical pressure gradient receiver. Also their results in Fig. 10-4B, show a profound loss in CM directionality following occlusion of the contralateral ear. Occlusion of the contralateral ear blocks the access of sound to the inside of the ipsilateral tympanic membrane (through the interaural pathway) and thereby reduces its normally subtractive influence. Thus, the pressure gradient operation of their tympanic membranes gives birds directional hearing that is independent of head shadow, as suggested by Fig. 10-3.

The CM directionality measurements by Schwartzkopff (1952) on the bullfinch stand in partial contradiction to the conclusions of Coles et al. Although the CM directionality (12 dB at 3.2 kHz) that Schwartzkopff measured was much larger than could be expected simply from head shadow, he found no change in CM amplitude or directionality following contralateral ear occlusion. Schwartzkopff did not discuss this paradox and concluded that each ear must operate independently.

Assuming that the directionality of one ear is the mirror image of the other with respect to the bird's median sagittal plane (Schwartzkopff 1952, Pumphrey 1961), then the spatial information that the bird would derive from ΔI would be azimuthal, as with ΔT and $\Delta \phi$. The angular resolution of this azimuthal information will vary with the sharpness of the ear's directionality and the ability of the bird to detect small differences in ΔI. Adequate data are not available on either ear directionality (CM) or ΔI sensitivity to make a prediction of the potential spatial resolution offered by ΔI. However, ear plugging experiments, which test the effect of altering sound intensity to

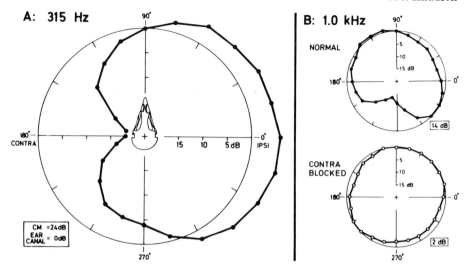

Figure 10-4. Evidence that the bird's ear functions as a pressure gradient detector. Directionality plots are shown for the right ear of the quail (*Coturnix coturnix japonica*). (A) The ear's sensitivity to a 315 Hz tone drops sharply when the sound source is located contralaterally. (B) Ear directionality to a 1 kHz tone before and after plugging the contralateral ear demonstrates the effect of blocking sound from the interaural pathway. (From Coles et al. in press).

one ear, have been performed on the chicken (Engelmann 1928) and the barn owl (Konishi 1973b, Knudsen and Konishi 1979). In both species, sound localization was severely disrupted. But ear plugging experiments simultaneously affect both ΔI and ΔS, and, as will be described in Section 5.2, additional experiments on the barn owl indicate that ΔS is in fact the important localization cue.

3.5 Interaural Spectrum (ΔS)

Definition: the difference in the intensity of the sound at each frequency in the two ears. In mathematical terms:

$$\Delta S = [i_{Lf_1} - i_{Rf_1}], [i_{Lf_2} - i_{Rf_2}], \dots [i_{Lfn} - i_{Rfn}]$$

where i_{Lf_1} equals the sound intensity at frequency f_1 in the left ear, and i_{Rf_1} the sound intensity at the same frequency f_1 in the right ear.

The frequency response of the bird's ear changes as a function of sound direction (Fig. 10-5). For any given direction, the two ears filter sound differently. Thus, a bird could identify the location of the source by comparing the difference in the power spectrum sensed by each ear. Such a ΔS cue is thought to be employed in elevational localization by man (Searle, Braida, Cuddy, and Davis 1975).

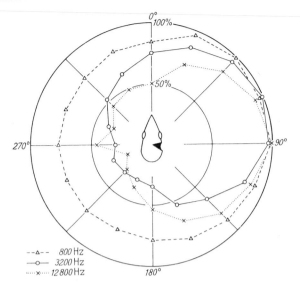

Figure 10-5. The frequency response of the bullfinch's (*Pyrrhula europaea*) ear changes as a function of sound direction. The sensitivity of the cochlear microphonic, measured in reference to the sound pressure necessary for a maximum response, is plotted for different sound directions. The data is for the right ear. (From Schwartzkopff 1962).

The salience of the ΔS cue depends on the variation in each ear's frequency response as a function of direction and on the difference between the directionalities of the two ears. Directionality plots for the bullfinch (Schwartzkopff 1952), quail (Coles et al. in press), and barn owl (Payne 1971, Knudsen and Konishi unpublished) show that the avian ear is more directional to high than to low frequency sounds (Figs. 10-5, 10-6). In a sense, the ear acts as a low-pass filter with a high frequency cutoff that is space dependent. To illustrate how ΔS might be used, consider the case of the bullfinch (Fig. 10-5). Its ear is most sensitive to all frequencies when the sound source is located directly to the ipsilateral side, and becomes less sensitive particularly to the higher frequencies as the source moves toward the contralateral side. Thus, if a wideband noise source were to move from the left of the finch to its right, its right ear would sense a distinct increase in the proportion of high frequency energy, while its left ear would sense a decrease in high frequency energy.

If the directionalities of the two ears are mirror images with respect to the median sagittal plane at all frequencies, the spatial information derived from ΔS will be the same as that derived from ΔI. This is probably the case for most birds because of the simplicity and symmetry of their ears. Therefore, for most birds, whether the ΔS or ΔI cue is used in sound localization, reduces to a question of how the auditory system processes intensity information. In this regard ΔS appears to be the correct term, since intensity information is not summed across all frequencies in the nervous system, but is maintained in discrete frequency specific channels that are organized systematically in each auditory nucleus in tonotopic distributions (see Sachs, Woolf, and Sinnott, Chapter 11). As an example, the cells in nucleus laminaris, the first site of binaural

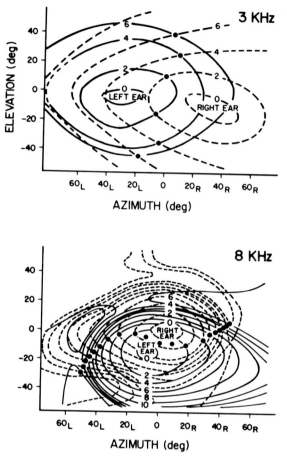

Figure 10-6. Ear directionality of the barn owl measured at 3 kHz and 8 kHz. Sound pressure level, measured with microphones placed in the left and right ear canals, is plotted as a function of sound source location. The measurements are in decibels of attenuation compared to the most sensitive reading for each ear. The contour lines, which represent directions to which the ear is equally sensitive, are in 2 dB increments. The dots designate the locations at which the tone causes equal intensities in both ears. Notice that at 3 kHz the plane of equal intensity is vertical, whereas at 8 kHz it is largely horizontal. The 3 kHz data are from Knudsen and Konishi, unpublished; the 8 kHz data are from Payne, 1962.

convergence in the avian auditory system, receive inputs from both cochlear nuclei: ipsilateral inputs to their dorsal dendrites and contralateral inputs to their ventral dendrites (Ramón y Cajal 1908, Boord 1968, Parks and Rubel 1975). The inputs represent acoustic stimulation at the same or similar frequencies in the two ears. Cells in the caudal portion of the nucleus respond selectively to low frequency stimulation; progressing rostrally, the cells respond to increasingly higher frequencies. To the extent that these cells perform a binaural comparison of sound intensity, the neuronal circuitry of nucleus laminaris is precisely that required for determining ΔS.

In certain genera of owls (Payne 1962, 1971, Norberg 1978) the directionalities of the two ears are not mirror symmetrical with respect to the median sagittal plane, moreover their plane of symmetry changes as a function of sound frequency (Fig. 10-6). For such birds, ΔI and ΔS offer different spatial information: whereas ΔI would define the location of the sound source along a single spatial plane, the multiple Δi_f terms of the ΔS cue, each defining a different spatial plane, would localize the sound source in both azimuth and elevation. Behavioral experiments on the barn owl—one of the owls with a frequency dependent plane of symmetry of ear directionality—suggest that this species does use ΔS for sound localization (Section 5.2). It is noteworthy that the nucleus laminaris in the barn owl is the largest of any bird studied to date (Winter and Schwartzkopff 1961, Winter 1963).

3.6 Monaural Cues and Two-Dimensional Sound Localization

Except for some owls, birds possess simple ears located symmetrically on either side of the head. Consequently, the binaural localization cues provide information only about the azimuth of the sound source. Yet there are numerous anecdotal examples of birds, especially songbirds, locating sounds in elevation as well as in azimuth (Marler 1959, Krebs 1976). This dilemma can be resolved if birds make use of monaural cues for sound localization in elevation.

The monaural spectrum cue requires that (1) the frequency response of the ear change as a function of direction, (2) the signal include a wide spectrum, and (3) the bird be able to anticipate the original power spectrum of the signal. For birds attempting to localize calls or songs of their conspecifics, each of these criteria appear to be satisfied. First, as described in the previous section, the frequency response of the avian ear is direction-dependent. Second, bird songs and calls (except for some alarm calls) characteristically contain a broad spectrum (Marler 1955). Third, the spectra of songs and calls are stereotyped for each species so that a bird can "expect" a certain power spectrum from its conspecifics. The difference between the power spectrum that the bird "expects" and the spectrum it actually hears would be due to the filter properties imposed by the ear, which in turn depend on the direction of the source. Thus, birds could determine a sound's azimuth by using binaural ΔT, $\Delta\phi$, and/or ΔS cues, and its elevation from their monaural spectrum cues. This is essentially the strategy used by man to localize sounds in azimuth and elevation (Searle, Braida, Davis, and Colburn 1976).

Birds could also use intensity scanning to localize sounds in two dimensions. The bird could simply move its head until the intensity of the sound is maximized in one or the other ear. The sound source would then be in a characteristic and identified direction with respect to the orientation of the bird's head.

A third possible, though unlikely, strategy for two-dimensional localization is for the bird to make sequential measurements of azimuth and elevation before and after tilting its head in the sound field. That is, the bird first could take a reading of the sound's azimuth using one or all of the binaural cues, then rotate its head so that the axis of its ears (and therefore of ΔT, $\Delta\phi$, and ΔS) contains a vertical component, and make a second binaural measurement to derive the sound's elevation.

All of the above strategies are of course speculative. Behavioral experiments on normal birds are needed to determine which of these, or any other, are actually employed for localizing sounds. However, for the highly specialized owls, our knowledge is considerably further advanced. Based on available bioacoustical, physiological, and behavioral data, particularly from the barn owl, we can resolve, to a fair extent, the strategy these birds use for two-dimensional sound localization.

4 The Auditory Systems of Owls

The auditory systems of owls exhibit a wide range of specialization the extent of which correlates with the ecological niche of the particular species (Winter and Schwartzkopff 1961, Winter 1963, Schwartzkopff 1968). In particular, the auditory systems of crepuscular and nocturnal species manifest various degrees of enlargement and elaboration. These auditory specializations increase the acoustic sensitivity of these owls (see Fig. 10-2) and their ability to localize sounds. Specializations that serve to increase the owl's sensitivity include expanded external and middle ear cavities, and a large eardrum to stapedial footplate area ratio (Schwartzkopff 1957, Norberg 1978). Specializations that also improve the owl's ability to localize sounds include acoustically reflective external structures or facial ruff (Fig. 10-7), asymmetrical ear openings and canals, enlarged interaural air passages connecting the middle ears, a long basilar membrane, disproportionately large auditory nuclei in the brain, and an audible range extending up to 10 kHz to 12 kHz. Due to space limitations, all of these specializations cannot be treated in detail. For further information refer to Schwartzkopff (1968), Payne (1971), Erulkar (1972), Norberg (1977, 1978). However, of direct importance to the behavioral studies that follow are the influences of the owl's facial ruff and asymmetrical ears on ear directionality.

4.1 Facial Ruff and Ear Asymmetry

A few owls, particularly in the genus *Asio* (Pycraft 1898), have developed a thick fold of skin (postaural fold) that runs behind and partially encircles their ear openings. In the majority of owls, the postaural fold itself is not so prominent but the specially modified feathers that it supports are (Norberg 1977). These unusually stiff, dense feathers stand in numerous tightly packed rows and form an acoustically reflective surface called the facial ruff that frames the owl's face and gives it its characteristically "owlish' appearance (although some of the diurnal owls virtually lack a facial ruff[1]). The facial ruff of the barn owl is particularly well developed and is typical of nocturnal owls (Fig. 10-7). Its rows of curved ruff feathers form two vertical troughs on the left and right sides of the face, each of which is about 1.5 cm to 2 cm wide and 6 cm to 8 cm high in an adult bird. When viewed from the side, these feathery troughs make paraboloid curves from the peak of the owl's forehead, behind the ears, then back out along the base of the lower mandible. The left and right troughs are

[1] Many bird species other than owls also have modified pre- or postaural feathers but none, except perhaps hawks in the genus *Circus*, is developed to the extent found among owls.

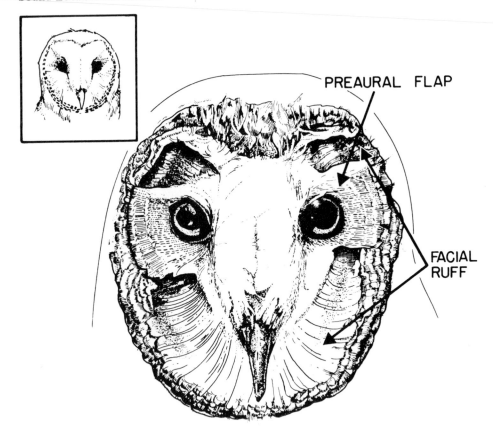

PREAURAL FLAP

FACIAL RUFF

Figure 10-7. The facial ruff of the barn owl (*Tyto alba*). This illustration depicts the face of the owl as it would appear if the owl's facial disc feathers were removed. The owl's normal appearance is shown in the upper left corner.

separated by a midline ridge of thick lore feathers above the beak, but communicate weakly below the beak. The ear openings themselves lie in vertical tunnels formed by the feathered trough behind, and a flap of skin (preaural flap) in front (Fig. 10-7).

The facial ruff collects sound energy over a large area and funnels it into the ears (Konishi 1973b). Theoretically the facial ruff should enhance the directionality (and sensitivity) of the ears to wavelengths of sound that approach or are smaller than its dimensions. In the case of the barn owl, whose ruff is 6 cm to 8 cm in length, this corresponds to sound frequencies of about 4 kHz and above. Ear directionality data show that this is precisely what happens.

Before discussing these data, however, one further aspect of the owl's auditory periphery must be mentioned: that of ear asymmetry. In most owl genera the left and right ears are symmetrical. However, in 9 out of the 29 genera of Strigiformes, either the soft tissues of the external ear or the ear canal in the skull itself are markedly asymmetrical (Norberg 1977). Asymmetry of the soft tissues takes several forms:

1. the ear openings are of different sizes (*Bubo bubo*, many species of *Ciccaba* and *Strix*)
2. the ear openings are at different vertical levels (*Tyto alba, Phodilus badius*)
3. the preaural ear flaps are at different vertical levels (*Tyto alba*) or have different shapes (some species of *Strix*)
4. complex diverticula in the ear canals have different geometries (*Rhinoptynx clamator, Pseudocops grammicus* and many species of *Asio*).

Asymmetries of the skull itself include ear canals of different shape (*Strix uralensis, S. nebulosa, Aegolius funereus*, and *A. acadicus*) or at different vertical levels (*Aegolius funereus* and *A. acadicus*). Based on the diversity of these various types of ear asymmetry, Norberg concludes that ear symmetry has evolved independently at least five times in the order *Strigiformes*. Since many of these species are known to hunt prey at night, the obvious inference is that ear asymmetry has evolved in response to a strong selective pressure for accurate sound localization.

4.2 Ear Directionality

The effect of ear asymmetry is to create a vertical disparity in the directionality of the two ears at frequencies to which the ruff is an effective reflector. This vertical disparity has been most carefully documented by ear canal measurements in the barn owl (*Tyto alba*) by Payne (1971) and Knudsen and Konishi (1978b, unpublished). The ear openings and preaural flaps of the barn owl are positioned differently in the vertical curves formed by the right and left troughs of the ruff: the opening and flap on the right are centered in the paraboloid curve, while those on the left are located high in the curve (Fig. 10-7). The cumulative effects of the ruff and the ear asymmetry on ear directionality are shown by the plots in Fig. 10-6. In general, both ears are maximally sensitive to areas of space directly in front of the owl; the owl's right ear tends to be most sensitive just to the right of the owl's median plane, its left ear is most sensitive just to the left of its median plane. The elevation of each ear's most sensitive region, however, is frequency dependent. At low frequencies both ears are most sensitive to a region just below the owl's horizontal plane (elevation = 0°). As the frequency of sound increases above 3 kHz, the region of space to which the owl's left ear is most sensitive drops, while that of the right ear rises. At 6 kHz the right ear is most sensitive to an area of space approximately 15° above that of the left ear. Also, above 5 kHz the sensitivity of each ear cuts off far more rapidly to the ipsilateral side than to the contralateral side. Thus, despite the fact that the left ear is maximally sensitive to an area slightly to the left, and the right ear slightly to the right, there exists a large frontal region, approximately 40° to either side of the median plane, within which both ears are equally sensitive to sound along the horizontal plane (see Fig. 10-6). Thus, at frequencies of 3 kHz and less, the directionalities of the owl's left and right ears are essentially mirror images with respect to its median sagittal plane—as was true of normal birds. As sound frequency increases above 3 kHz and the facial ruff becomes an effective reflector, this plane of symmetry rotates until at 6 kHz it becomes largely horizontal at least over the frontal 80° of space. This frequency-dependent rotation in the plane of symmetry of the ears' directionalities, provides the barn owl with the ΔS information that it apparently uses for two-dimensional sound localization (Section 5.2).

Norberg (1968) measured ear directionality in a modeled head of the Tengmalm's owl (*Aegolius funereus*), another nocturnal predator with grossly asymmetrical ears (Norberg 1978). He found that below 10 kHz there was no vertical asymmetry, while above 10 kHz there was strong vertical asymmetry. He proposed, therefore, that the owl might use low frequencies (4 kHz to 8 kHz) to localize sounds in azimuth, and high frequencies (12.5 kHz to 16 kHz) to localize in elevation. The problem, of course, is that the owl does not hear frequencies of 12.5 kHz to 16 kHz. In a recent paper, Norberg (1978) states (although he presents no data) that in fact significant vertical deviations of ear directionality appear already at 6.3 kHz and above, and previous measurements were in error due to imperfections in the head model. This later statement is in essential agreement with directionality plots made from the barn owl by Payne (1962, 1971) and Knudsen and Konishi (1978b, unpublished).

Schwartzkopff (1962) measured the directionality of CM responses in the left and right ears of the long-eared owl (*Asio otus*). Because he concentrated on low frequencies (mainly 3 kHz), he found no evidence of vertical disparity in ear directionality.

5 Sound Localization by the Barn Owl

Our most complete knowledge about the capacities and limitations of sound localization by any bird stems from behavioral studies on the barn owl (*Tyto alba*). The barn owl is a nocturnal predator capable of using acoustic cues to capture prey in the dark and, as might be expected, its auditory system manifests many of the specializations for sound detection and localization described in the previous sections. An impressive example of the barn owl's sound localization prowess is its ability to detect the direction of movement of a sound source, from which it infers the body orientation of its prey, so that it can orient its talons according to the prey's body axis and maximize its chance for a successful strike (Payne 1962).

The behavioral experiments that have been performed on the barn owl have assayed the accuracy of two different types of behavior: strike accuracy and head orientation accuracy. The strike accuracy paradigm (Payne 1962, 1971, Konishi 1973a, 1973b) required the owl to fly toward and land on a sound target. Data yielded by this approach are suited for evaluating the owl's relative accuracy under systematically varying conditions, but are not reliable for measures of absolute accuracy since flight errors are compounded with localization errors, and the location of the owl when it makes its final judgment and its land site can be only crudely approximated. The head orientation paradigm used by Knudsen et al. (1979) and Knudsen and Konishi (1979) is better suited for absolute measures of localization accuracy. This paradigm was based on the owl's natural head-orienting response to sound stimuli, and measured in degrees of error how precisely the owl aligned its head with a sound target. In these experiments the orientation of the owl's head was monitored by using a modification of the search-coil technique (Knudsen et al. 1979). The search coil was mounted on the owl's head, and the owl perched with its head at the center of an orthogonal array of magnetic fields. The complex electrical signal induced in the search coil by the fields was demodulated so that azimuthal and elevational components of the owl's head orientation could be measured separately.

The angular acuity of the barn owl, measured either in terms of strike accuracy (Payne 1971) or head orientation accuracy (Knudsen et al. 1979), is 1° to 2° in both azimuth and elevation. This acuity is better than that of any terrestrial vertebrate tested so far, including man (Mills 1958, Butler 1969, Gourevitch, Chapter 12) and is only slightly worse than that of the echolocating dolphin (Renaud and Popper 1975).

The effect of frequency on sound localization accuracy has been investigated by Payne (1962, 1971), Konishi (1973b), and Knudsen and Konishi (1979). Knudsen and Konishi found that the owl was most accurate when localizing tonal sounds of 4 kHz to 8 kHz (Fig. 10-8). However, even at the best frequency (6 kHz), the owl was significantly worse at localizing a tone than a noise. Data from Konishi (1973a, 1973b), who measured strike accuracy, are largely consistent with those of the above study.

Payne (1962, 1971) measured the influence of sound spectrum on localization accuracy by band filtering the "sounds of rustling leaves." He concluded that the "owl's acoustic orientation appears . . . to depend upon frequencies above 8.5 kHz." This conclusion is not in accord with those of the more recent studies and is not supported by the owl's audibility curve that cuts off steeply above 10 kHz (Fig. 10-2).

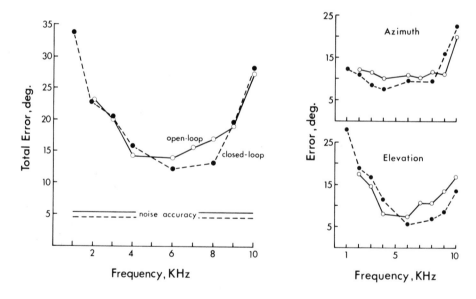

Figure 10-8. Localization of tones by the barn owl. Head orientation error to tonal targets was measured under open-loop (open circles) and closed-loop (closed circles) conditions. In open-loop trials the tone terminated before the owl began its head movement; in closed-loop trials the tone remained on throughout the owl's orientation response. The owl's mean total errors are plotted on the left; the azimuthal and elevational components of that error are plotted separately on the right. The sound targets were presented at various locations, 30° away from the owl's point of fixation. The owl's accuracy in localizing wideband noise at this angular distance is shown in the graph on the left. (From Knudsen and Konishi 1979).

The importance of signal bandwidth to sound localization accuracy was determined by Konishi (1973a). He used band-limited noise centered at 7.5 kHz and a 7 kHz pure tone. Surprisingly, his data show no difference in the owl's localization accuracy to a 1 kHz bandwidth noise versus the pure tone (Fig. 10-9). Even a 2 kHz bandwidth noise was not significantly better than the tone. It was not until the owl was presented with a 3 kHz bandwidth that its accuracy significantly improved, and it continued to improve up to a bandwidth of 4 kHz, which was the widest band that Konishi tested.

The effect of sound source location on localization accuracy was studied by Knudsen et al. (1979). This study measured the owl's head orientation accuracy to various target locations in the owl's horizontal and median planes under two stimulus conditions: open-loop and closed-loop. In open-loop trials the target sound (wide-band noise) was terminated before the owl could begin to move its head. Therefore the owl was forced to orient to a remembered target location without the help of subsequent acoustic input. In closed-loop trials the target sound persisted throughout the owl's orientation movement so that acoustic input was constantly available.

Under open-loop conditions the owl's accuracy deteriorated as the angular distance to the target increased, but even at target angles of 70° the owl's accuracy was still quite good (Fig. 10-10). Under closed-loop conditions, the owl's accuracy was independent of the target's location for angular distances greater than or equal to 30°. However, the owl improved significantly when the target angle was less than 30°. Within this frontal area ($\leqslant 30°$), the owl was equally accurate in open- and closed-loop trials.

The full implications of these results will be discussed later, but it should be emphasized that the ability of the owl to localize sounds in the open-loop condition means that the owl must determine both the azimuth and the elevation of the sound target "instantaneously," i.e., before it begins to move its head.

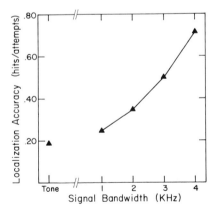

Figure 10-9. Strike accuracy of the barn owl improves with the bandwidth of the signal. Strike accuracy was quantified by dividing the number of direct hits on the target by the total number of attempts. The tone was 7 kHz; the noise bands were centered by 7.5 kHz. (Data from Konishi 1973a).

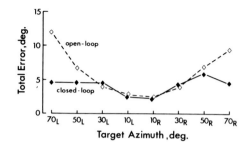

Figure 10-10. The effect of target location on sound localization accuracy of the barn owl. The owl's head orientation error was measured under open-loop (open circles) and closed-loop (closed circles) conditions at the designated speaker locations in the horizontal plane. The mean values of those errors are plotted as a function of the speaker's location. (From Knudsen and Konishi 1979).

5.1 Theories of Sound Localization

Any comprehensive theory of sound localization by the barn owl must take into account the following behavioral observations:
1. The owl's audible range extends from 0.1 kHz to 12 kHz (Konishi 1973b).
2. The owl can determine the azimuth and elevation of a sound source without moving its head (Knudsen et al. 1979).
3. The owl can localize pure tones in both azimuth and elevation, although more crudely than it does noise (Konishi 1973a, 1973b, Knudsen and Konishi 1979).

Basically two different theories of sound localization have been proposed: one by Payne (1962, 1971) and a second by Pumphrey (1948) that was restated by Norberg (1968, 1978). Payne's (1971) theory is the following:

> Because the owl will hear all frequencies in a complex sound at maximum intensity in both ears only when it directly faces a sound source, my theory demands just one thing from the owl, that it try to make *all* the frequencies audible to it in a complex sound as loud as possible in both ears. When this achieved, the owl will automatically be facing the sound. p. 568.

Thus, Payne's theory assumes that the owl must scan its head in the sound field during a prolonged or repetitive sound in order to localize the source. This is not true (observation 2). Also in discussing the elevational resolution that the owl would achieve by head scanning, Payne relies heavily on the ear's directionality to sound frequencies of 8.5 kHz and 13 kHz—frequencies that the owl does not localize well and, above 12 kHz, cannot even hear (observation 1).

The second theory, as originally formulated by Pumphrey (1948), proposed in essence that the owl localizes a sound source by using interaural spectrum (ΔS). The premises of Pumphrey's theory are:

> 1) The sound must be complex and the ears competent to resolve it into at least three bands of frequency in such a way that independent compari-

son of the signal arriving at the two ears is possible in each band. 2) The two ears must have a direction of maximum sensitivity which is different for each band and is different for the left and right ear for at least two of the bands. p. 193.

His rationale was as follows: a given Δi resulting from a frequency f_1 corresponds to a (curved) plane of points in space, the geometry of which depends on the directionality of the two ears to that frequency. If at some other frequency, f_2, the ears have a different directionality pattern, then the plane of points corresponding to a given Δi_2 will have a different geometry. If at a third frequency, f_3, the ears have yet a different directionality, a given Δi_3 will define a third plane of points. By making independent Δi measurements of the f_1, f_2, and f_3 components of a complex sound, the owl's auditory system could uniquely identify the location of the sound source by the intersection of the three planes. Pumphrey proposed his theory based simply on the morphological asymmetry of the owl's ears and without any data on the ear's directionality. Norberg (1968, 1978) later provided these data (see Section 4.3) and concurred with Pumphrey, although he rephrased the theory as follows: "Low frequencies (below 6 000 Hz) provide intensity cues to azimuth, high frequencies [provide] intensity cues to elevation."

The Pumphrey-Norberg theory is attractive because it predicts the ability of the barn owl to localize sounds without moving its head (observation 2); its weakness is that it cannot explain the barn owl's ability to localize tonal stimuli (observation 3). According to their theory, the owl should be able to localize a single frequency only to a plane in space. Yet even under open-loop conditions, the owl localizes a tonal source without giving any indication of a "plane of ambiguity" (Knudsen and Konishi 1979).

5.2 A New Theory of Sound Localization

A third theory of sound localization is here proposed for the barn owl: The owl assigns to each direction (D) in space, a set of interaural values or coordinates based on ΔT and ΔS cues, in the form

$$D = [\Delta T, \Delta i_{f_1}, \Delta i_{f_2} \dots \Delta i_{fn}],$$

that characterizes that direction (Fig. 10-11). Because the spatial planes defined by the individual coordinates differ from one another, the owl's accuracy of localization will increase with the number of coordinates that are derivable from the sound. This theory incorporates the Pumphrey-Norberg theory, and adds to it the term ΔT as an important parameter for sound localization. Both Pumphrey (1961) and Norberg (1978) clearly recognized the potential of the ΔT cue, but neither of them included it in their final formulations.

The necessity of including ΔT rests on several lines of behavioral evidence. First, the owl is capable of localizing tonal targets of 4 kHz to 8 kHz under open-loop conditions in both azimuth and elevation (Knudsen and Konishi 1979). Since only one

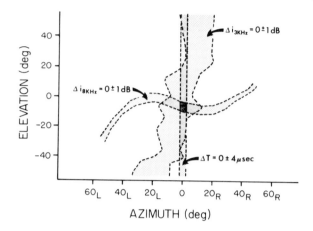

Figure 10-11. A graphic demonstration of the ΔT, ΔS coordinate model for sound localization. For simplicity, the hypothetical sound contains only two frequencies, 3 kHz and 8 kHz. In this example, the owl hears 3 kHz and 8 kHz equally in both ears ($\Delta i = 0$ dB), and the onset of the sound was simultaneous in both ears ($\Delta T = 0$ μsec). The three stippled bands in the figure represent the planes of space defined by each of these three binaural cues. The $\Delta i = 0$ dB planes for 3 kHz and 8 kHz were derived from the data in Fig. 10-6, and assume that the error of the owl's Δi measurement is $\pm$ 1 dB. The ΔT plane is assumed to be vertical and to have a measurement error of $\pm$ 4 μsec (see Section 5.2). The spatial information contained in the ΔS cue corresponds to the area of intersection of the $\Delta i_{3\ kHz}$ and $\Delta i_{8\ kHz}$ planes. The direction (D) of the sound source can be determined from the binaural coordinates D = [0 μsec, 0 dB$_{3\ kHz}$, 0 dB$_{8\ kHz}$], and is represented graphically by the area of intersection of all three planes.

frequency is available, Δi can only stipulate one dimension. To localize in two dimensions, the owl must use a second cue. However, two candidate cues must be considered: ΔT and $\Delta \phi$. The conclusion that ΔT is the one used is based on the frequency independence of the owl's azimuthal errors (Fig. 10-8). As discussed in Sections 3.2 and 3.3, the axis of spatial information provided by ΔT and $\Delta \phi$ is in azimuth. Thus, the contribution of ΔT or $\Delta \phi$ to sound localization will manifest itself in the azimuthal component of the owl's error. If the owl uses ΔT, its azimuthal error should remain constant across frequencies, since ΔT is a property of the wave envelope and not of the tone's frequency. On the other hand, if the owl uses $\Delta \phi$, its azimuthal error should be small at low frequencies, and should increase dramatically above 3 kHz, since the $\Delta \phi$ cue becomes ambiguous at wavelengths shorter than twice the interaural distance, which for the barn owl is 2 x 5 cm = 10 cm (see Section 3.3).[2] The experimental result was that the owl's azimuthal error was independent of frequency up to 9 kHz (Fig. 10-8), far above the upper frequency limit of an unambiguous $\Delta \phi$ cue.

Further evidence in support of ΔT and against $\Delta \phi$ comes from a comparison of the

---

[2] The contribution of the ΔS cue complicates these expectations slightly since below 4 Hz the axis of Δi_f becomes coincident with the axis of ΔT and $\Delta \phi$ (Fig. 10-11). However, at these low frequencies the ear's directionality is poor (Fig. 10-6). Thus the salience of Δi_f, and therefore its contribution to the owl's angular resolution, must be relatively small.

owl's accuracy in localizing tonal targets under open- and closed-loop conditions (Knudsen and Konishi 1979). In both conditions the ΔT cue is the same, since it depends only on the angular distance of the target from the bird's median plane (α) at the onset of the tone. In contrast, the $\Delta\phi$ cue differs in the two conditions. In the open-loop condition the owl can measure $\Delta\phi$ only once when the sound source is to its side (before it moves its head). In the closed-loop condition, however, the owl can increase the spatial resolution provided by $\Delta\phi$ as it orients toward the continuous sound source (see Section 3.2). Thus, if the owl uses ΔT, its azimuthal error should remain the same under open- and closed-loop conditions; but if it uses $\Delta\phi$, its azimuthal error should decrease in the closed-loop condition. The result was equal azimuthal accuracy in the two conditions (Fig. 10-8).

A third piece of evidence for ΔT as the localization cue is found in the owl's performance when localizing wide band noise under closed-loop conditions (Knudsen et al. 1979). For all azimuthal target angles of $30°$ or more, the accuracy of the owl's localization was fairly constant, since the continuing sound allowed it to correct major errors in its orientation. However, even with its correction, the owl could not localize the target as accurately as it could when the target was located at azimuthal angles of less than $30°$ (Fig. 10-10). Thus, the target had to be at a small azimuthal angle at the onset of the sound in order for the owl to achieve its maximum accuracy (this finding is in agreement with Payne's 1971 observation that the barn owl tends to wait, after orienting to its prey, until it hears a second sound before attacking). The only localization cue that is restricted to the onset of a sound, and is optimal when the sound comes from a small azimuthal angle, is ΔT. Implicit in this interpretation is that the owl requires a substantial discontinuity in the sound envelope to make its ΔT measurement, since it did not utilize transients in the ongoing noise burst to maximize its azimuthal accuracy to targets at large angles under closed-loop conditions (Fig. 10-10).

The foregoing arguments assumed that the owl determines the elevation of a sound source by using ΔS. The behavioral evidence in support of this conclusion is the following. It was Payne and Drury (1958) who first stated that an owl with one ear plugged could not strike prey. Ear occlusion does not significantly affect ΔT, but drastically disrupts both ΔI and ΔS (see Sections 3.4 and 3.5). Konishi (1973b) further noted that a trained owl with one ear plugged committed systematic localization errors: with its right ear plugged, it struck to the left and short of the target, with its left ear plugged, it struck to the right and beyond the target, Knudsen and Konishi (1979) quantified Konishi's original observations by using the head orientation paradigm. They demonstrated a dramatic localization error resulting from ear occlusion, the major component of which was elevational: a right ear plug caused the owl to orient to a point below the target, and a left ear plug caused it to orient above the target (Fig. 10-12). The tighter the plug the greater was the owl's error. It should be emphasized that the owl's error was never only elevational; the error resulting from an ear plug contained both an elevational and an azimuthal component, the relative magnitudes of each depended on the location of the sound target relative to the owl's head.

This result is expected from perturbation of the ΔS (or ΔI) cue, since the plane of symmetry for Δi_f at frequencies below 6 kHz is no longer horizontal, but begins to rotate, with decreasing frequency, toward the vertical plane (Figs. 10-6, 10-11) thereby

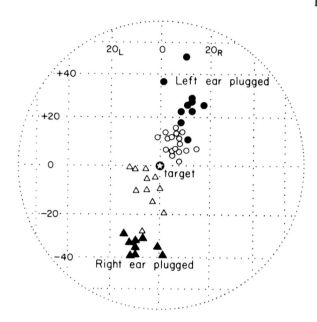

Figure 10-12. Effect of ear occlusion on sound localization by the barn owl. Head orientations (data points) to a single, wide-band noise target are plotted for four conditions of ear occlusion: (1) left ear, hard plug (closed circles); (2) left ear, light plug (open circles); (3) right ear, hard plug (closed triangles); and (4) right ear, light plug (open triangles). Left ear occlusion causes the owl to miss up and to the right; right ear occlusion, down and to the left. (From Knudsen and Konishi 1979).

increasing the azimuthal component of its information. The azimuthal error caused by ear occlusion probably results from the altered Δi_f values of these lower frequencies.

Evidence that specifically implicates ΔS as the localization cue comes from the effect of signal bandwidth on the owl's localization accuracy. According to Knudsen and Konishi (1979), even when localizing 6 kHz tones (the owl's best frequency for tonal localization) the owl's mean error was nearly three times greater than when it was localizing wide band noises (Fig. 10-8). Furthermore, Konishi's (1973a) data on the influence of noise bandwidth on strike accuracy show that an increase in bandwidth up to 1 kHz does not improve the owl's accuracy over its pure tone accuracy (Fig. 10-9). However, on widening the bandwidth beyond 1 kHz, the owl's accuracy rapidly improves and becomes maximal at a 4 kHz bandwidth. The requirement of a wide signal bandwidth for maximal localization accuracy is not expected if the owl is simply using ΔI. The ΔI cue is basically spectrum-independent (see Section 3.4); the accuracy of measuring ΔI may improve somewhat with bandwidth, but the improvement probably would not be on the order of the 270% reported by Knudsen and Konishi (1979). On the other hand, this tremendous improvement in accuracy with bandwidth is exactly what the ΔS cue would predict. When the target is a tone, ΔS offers only one Δi_f coordinate. Substantial improvement in localization accuracy should not occur until the signal bandwidth includes frequencies with significantly different planes of symmetry (Fig. 10-11). The wider the bandwidth, the greater will be the diversity in the orientations of Δi_f planes, and the smaller and more definite will be the area of space defined by the increasing number of Δi_f coordinates.

Knudsen and Konishi (1979) tested this "coordinate theory" for sound localization by measuring the localization accuracy of the barn owl after removing its facial ruff. They reasoned that ruff removal must greatly reduce the sharpness and vertical disparity of ear directionality at frequencies affected by the facial ruff, i.e., frequencies greater than 3 kHz (vertical disparity would not be totally eliminated since the owl still retains ear openings and preaural flaps that are asymmetrical in the thickened postaural fold). With this loss in high-frequency ear directionality, the owl should tend to hear these frequencies at almost equal intensity in both ears over a large range of target elevations. Under normal conditions (with ruff intact), equal intensities of high frequencies in both ears occur only when the sound source is located in the horizontal plane (Fig. 10-11). Thus, according to the theory, the owl without its ruff should make large elevational errors, and the direction of the errors should be toward the horizontal plane. On the other hand, ruff removal should not disturb ΔT and low frequency Δi_f (f < 4 kHz) cues that are not influenced by the facial ruff. Therefore, the unperturbed ΔT and low frequency Δi_f should allow the owl to localize sounds in azimuth with basically normal accuracy. The results of the experiment were as predicted above: the owls made small or no azimuthal errors while committing large elevational errors that were directed toward their horizontal planes (Fig. 10-13).

At low frequencies (less than 4 kHz), the directionality of the owl's ears is relatively poor, and, therefore, the spatial uncertainty associated with a Δi_f value will be large. However, as mentioned, the spatial plane defined by Δi_f at these low frequencies is the same as that defined by ΔT, i.e., primarily azimuthal (Fig. 10-11). Therefore, for sound sources at small azimuthal angles (α), where ΔT affords high spatial resolution (see Section 3.2), the ΔT value probably determines the owl's azimuthal acuity; at large azimuthal angles, where the spatial resolution of ΔT is poor, ΔS cues may become more important for localization in azimuth. If this argument is correct, the ΔT resolution of the barn owl can be estimated from the localization data for small angles published by Knudsen et al. (1979) using Woodworth's (1962) formula for ΔT (see Section 3.2). The assumption is that the standard deviation in the owl's azimuthal error when localizing sounds 10° to the right or left was due to the standard deviation in the owl's measurement of ΔT. At $\alpha = 10°$, the standard deviation of the owl's azimuthal errors ranged from $\pm 0.8°$ to $\pm 1.6°$ (Knudsen et al. 1979). Using the largest value, we can substitute $10° \pm 1.6°$ for α, and 2.5 cm for head radius (r) of the barn owl to obtain:

$$\Delta T = \frac{2.5 \text{ cm}}{.0334 \text{ cm}/\mu\text{sec}} \left[(11.6° - 8.4°) + (\sin 11.6° - \sin 8.4°) \right] \cong 8 \,\mu\text{sec}$$

or ΔT standard deviation $\cong \pm 4 \,\mu\text{sec}$.

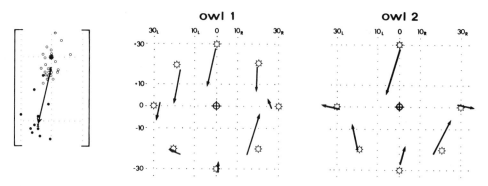

Figure 10-13. Effect of ruff removal on sound localization by the barn owl. The accuracy of two owls in localizing wide band noise bursts presented at various locations (open stars) was tested before and after their ruff feathers were removed. The data are presented in vector form. The vectors were derived as shown in brackets on the left. The tail of the vector indicates the median of the owl's fixation points under normal conditions (open circles). The head of the vector corresponds to the median of the fixation points following ruff removal (closed circles). Thus, each vector represents the magnitude and the direction of the effect of ruff removal for a given target location. Notice that azimuthal errors are relatively small following ruff removal, whereas elevational errors tend to be large and always directed toward the owl's horizontal plane. (From Knudsen and Konishi 1979).

6 Neural Correlates of Sound Localization

The behavioral experiments on the barn owl demonstrate that, at least for this species, sound localization is a binaural process. The effects of closed-field binaural stimulation on the responses of central auditory neurons has been rigorously examined only in the midbrain auditory nucleus of the chicken (Coles 1977). This nucleus, called mesencephalicus lateralis dorsalis or MLD, is the avian homologue or the mammalian inferior colliculus. Binaural units in MLD could be classified as bilateral excitatory (EE), excitatory-inhibitory (EI), and bilateral inhibitory (II); unit classes that are entirely analogous to those that have been exhaustively studied in various mammalian systems (Erulkar 1972). EI units, which predominated, were sensitive to changes in interaural intensity; EE units tended to prefer certain interaural time delays, although usually beyond the physiological range of ΔT. The role of such binaural units in encoding auditory space is still unclear in birds and mammals.

Greater insight into the manner in which the auditory system encodes space has been gained from a series of free-field studies on the midbrain (MLD) and forebrain auditory centers of the barn owl (Knudsen, Konishi, and Pettigrew 1977, Knudsen and Konishi 1978a, 1978b, 1978b). Knudsen et al. (1977) conducted their experiments in a large anechoic chamber, and studied directly, using a movable speaker, the influence of sound location on the response properties of these high order auditory neurons. This approach revealed a special class of auditory units that appear to be involved in encoding auditory space. These units, which occurred both in the midbrain and forebrain, were called limited field units (L-F units) and were distinguished from all other unit classes by their selective responses to sounds originating from only a restricted area of space, or receptive field (Fig. 10-14). The receptive fields of these units were largely unaffected by sound intensity and the type of sound presented. In the midbrain the receptive fields of L-F units were found to have a center-surrounded configuration, with locations outside a unit's receptive field behing inhibitory (Knudsen and Konishi 1978c). Although this receptive field property was not reported for the forebrain units, unpublished data show that it occurs in these units also. One difference between L-F units in the midbrain and forebrain was that midbrain units were selective primarily for the location of the sound, whereas many of the forebrain units were selective both for the type of sound and its location in space.

In the forebrain, L-F units tended to occur in clusters and often with overlapping receptive fields. In the midbrain the segregation and organization of L-F units was even more conspicuous. L-F units were found only in the lateral and anterior region of the MLD and were arranged systematically within this region according to the relative locations of their receptive fields so that they formed a map of auditory space (Fig. 10-15). Most of the map was devoted to contralateral auditory space, with azimuths from $0°$ to $20°$ contralateral receiving a disproportionately large representation in the map. Furthermore the map on each side of the brain extended $15°$ onto the ipsilateral side, thereby giving bilateral representation to this frontal region of space. This overrepresentation of frontal space in the neural map correlates well with increased spatial resolution that the owl demonstrates behaviorally when localizing sounds in this region (see Section 5.2).

Whether or not the receptive fields of L-F units depend on ΔT and ΔS, as predicted by the coordinate model of sound localization, has yet to be carefully explored. However, several preliminary findings and bits of evidence from midbrain L-F units suggest that they are. First, if after plotting the receptive field of an L-F unit, one plugs lightly the right ear of the owl, the unit's receptive field moves up and to the right; if one plugs the left ear, the field moves down and to the left (Knudsen and Konishi unpublished). This effect demonstrates that these receptive fields are crucially dependent on the balance of sound intensity in the two ears. Furthermore, the directions of these field shifts predict exactly the behavioral errors caused by ear plugging (Fig. 10-12), reinforcing the contention that L-F units are associated with sound localization. Second, the inhibitory surrounds of the unit receptive fields are derived from different and distinct bands of frequencies. In the example given by Knudsen and Konishi (1978c), the spectrum of frequencies that inhibited the unit on either side of its receptive field was 6 kHz to 8 kHz. However, the frequency spectrum that inhibited the unit from a location above the unit's field was 6 kHz to > 10 kHz, while the spectrum that inhibited the unit below its field was < 4 kHz to 9 kHz. Thus the Δi_f values are not simply summed and compared as ΔI, rather each frequency makes an individual contribution to the creation of the receptive field. Third, the responses of L-F units to an excitatory stimulus tend to be phasic, or have a predominant phasic component. The proportion of similarly phasic units in other regions of the midbrain or forebrain is far lower. This suggests that the onset of the stimulus, and therefore ΔT, is of unusual importance to the response of these units.

All of these neuronal properties are consistent with the behavioral sound localization results and support the proposed theory that the owl determines sound direc-

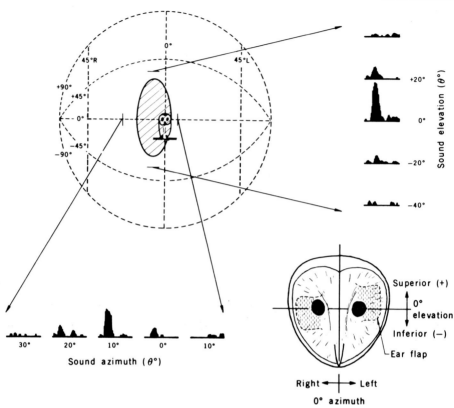

Figure 10-14. The receptive field of an auditory neuron in the forebrain of the barn owl is depicted from the observer's point of view. The owl is shown facing out from the center of the stimulus sphere (dashed globe), and the unit's receptive field ($25°$ in azimuth by $62°$ in elevation) is projected onto the sphere (diagonally lined area). The unit was located in the owl's left forebrain; its field dimensions were independent of stimulus intensity. Below and to the right are shown peristimulus-time histograms of the unit's responses to a sound stimulus presented at different locations within this receptive field. The stimulus was a 200-msec noise burst, 20 dB above threshold, delivered once per second. Each histogram is a 500-msec sample and represents 16 stimulus repetitions. Notice the increasing response vigor as the sound source approaches the center of the unit's receptive field. The owl head in the lower right corner illustrates the alignment of the owl in the stimulus sphere and defines the nomenclature used for describing auditory space. (From Knudsen et al. 1977).

tion from ΔS and ΔT cues. The extent to which L-F units actually underlie sound localization behavior awaits further investigation. However, because their space-dependent response properties and systematic representation of space must have been specifically generated through neuronal integration, it seems likely that these units are intimately involved in the process of sound localization.

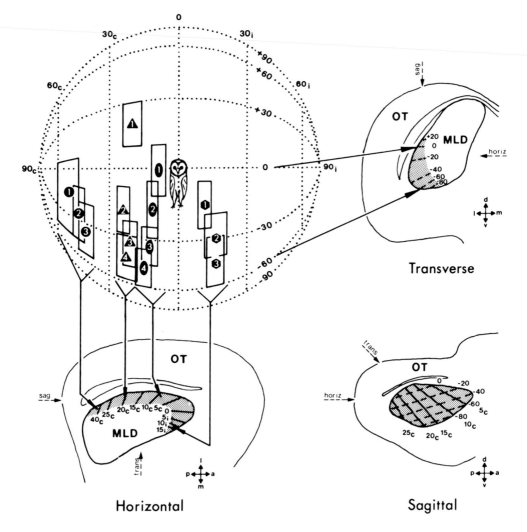

Figure 10-15. The representation of auditory space in the mídbrain (MLD) of the barn owl. In the upper left, the coordinates of auditory space are depicted as a dotted globe surrounding the owl. Projected onto the globe are the receptive fields (solid-lined rectangles) of 14 units that were recorded in four separate penetrations. The large numbers backed by similar symbols represent units from the same penetration; the numbers themselves signify the order in which the units were encountered and are placed at the centers of their receptive fields. The penetrations were made with the electrode oriented parallel to the transverse plane at the positions indicated in the horizontal section by the solid arrows. Below and to the right of the globe are illustrated three histological sections through the MLD in the horizontal, transverse, and sagittal planes. The stippled portion of the MLD corresponds to the region that contains neurons with small receptive fields. Isoazimuth contours, based on receptive field centers, are shown as solid lines in the horizontal and sagittal sections; isoelevation contours are represented by dashed lines in the transverse and sagittal sections. On each section dashed arrows indicate the planes of the other two sections. Solid, crossed arrows to the lower right of each section define the orientation of the section. Abbreviations: a-anterior, d-dorsal, l-lateral, m-medial, p-posterior, v-ventral. The length of the arrows corresponds to 600 μm. The optic tectum (OT) is labeled on each section. (From Knudsen and Konishi 1978a).

7 Conclusion

Birds live in a world where sound localization is an important capacity. Birds have evolved diverse sets of communicative vocalizations for which location of the calling bird is a primary component of the message. For certain owls, a keen ability to localize the sounds of prey is vital to their success as nocturnal predators.

The task of sound localization by birds is formidable, first because they must constantly localize sources both in azimuth and elevation, and second because their heads are small, rendering little interaural time difference or head shadow, and they hear only low sound frequencies (< 10 kHz). Nevertheless, behavioral experiments indicate that even birds with unspecialized auditory systems localize sounds quite well. The auditory cues that birds might exploit in sound localization include binaural comparisons of onset time (ΔT), phase ($\Delta \phi$), intensity (ΔI), and spectrum (ΔS), as well as monaural cues such as monaural spectrum and intensity scanning. A careful review of these cues reveals that all of them are potentially usable, despite the small head sizes of most birds.

Based on theoretical considerations, birds other than certain owls should only be able to derive the azimuthal coordinate of a sound source from binaural localization cues. It is therefore proposed that normal birds rely on monaural cues (either monaural spectrum or intensity scanning) to determine the elevation of a sound source.

Owls manifest tremendous auditory specializations that enhance their acoustic sensitivity and the directionality of their hearing. The genius of these specializations is that they also cause the spatial information offered by ΔT and ΔS cues to define different spatial planes and allow the owl to determine both azimuth and elevation from their binaural cues. Consequently, a theory for sound localization is proposed, the essence of which is that the owl assigns to each direction (D) in space, a set of interaural values or coordinates based on ΔT and ΔS, in the form

$$D = [\Delta T, \Delta i_{f_1}, \Delta i_{f_2} ... \Delta i_{f_n}],$$

which characterizes that direction. The results from numerous behavioral experiments on the barn owl (*Tyto alba*) support this theory.

A special class of auditory neurons has been found in the midbrain and forebrain of the barn owl that may well be involved in encoding auditory space. These neurons respond to sounds from only restricted areas of space or receptive fields. These receptive fields are largely unaffected by the intensity or spectral properties of the sound stimulus. Various properties of these neurons and their functional organization are consistent with expectations derived from behavioral localization experiments, and suggest that these specialized neurons contribute to the process of sound localization.

Acknowledgments. I thank M. Konishi, H. Leppelsack, D. Margoliash, D. Corey, R. Coles, and S. Guilman for helpful discussion and review of the manuscript, and Betty Hanson for her secretarial help. Preparation of this review was supported by National Institute of Health grant NS14617A.

References

Blauert, J.: Sound localization in the median plane. Acustica. 22, 205-213 (1969).

Boord, R. L.: Ascending projections of the primary cochlear nuclei and nucleus laminaris in the pigeon. J. Comp. Neurol. 133, 523-542 (1968).

Brückner, G. H.: Untersuchungen zur Tiersoziologie, insbesondere der Auflösung der Familie. Zs. Psychol. 128, 1-120 (1933).

Butler, R. A.: Monaural and binaural localization of noise bursts vertically in the median sagittal plane. J. Aud. Res. 3, 230-235 (1969).

Coles, R. B.: Physiological and anatomical studies of auditory units in midbrain areas of the domestic fowl (*Gallus gallus*). Thesis. Monash University, 1977.

Coles, R. B., Lewis, D. B., Hill, K. G., Hutchings, M. E., Gower, D. M.: Directional hearing in the Japanese quail (*Coturnix coturnix japonica*) II. Cochlear physiology. J. Exp. Biol., in press.

Curtis, W. E.: Quantitative studies of echolocation in bats (*Myotis l. lucifugus*); studies of vision of bats (*Myotis l. lucifugus* and *Eptesicus f. fuscus*) and quantitative studies of vision in owls (*Tyto alba pratincola*). Ph.D. Thesis, Cornell Univ., 1952.

Dice, L. R.: Minimum intensities of illumination under which owls can find dead prey by sight. Am. Nat. 79, 384-416 (1945).

Emlen, S. T.: An experimental analysis of the parameters of bird song eliciting species recognition. Behavior. 20, 130-171 (1971).

Engelmann, W.: Untersuchungen über die Schallokalisation bei Tieren. Z. Psychol. 105, 317-370 (1928).

Erulkar, S. D.: Comparative aspects of spatial localization of sound. Physiol. Rev. 52, 237-359 (1972).

Falls, J. B.: Properties of bird song eliciting responses from territorial males. Proc. Int. Ornith. Congr. 1, 259-271 (1963).

Gatehouse, R. W., Shelton, B. R.: Sound localization in bobwhite quail (*Colinus virginianus*). Behav. Biol. 22, 533-540 (1978).

Gottlieb, G.: Development of Species Identification in Birds. Chicago-London: Univ. of Chicago Press, 1971, pp. 44-52.

Griffin, D. R.: Acoustic orientation in the oil bird, *Steatornis*. Proc. Nat. Acad. Sci. U. S. 39, 884-893 (1953).

Harris, G. G.: Binaural interactions of impulsive stimuli and pure tones. J. Acoust. Soc. Am. 32, 685-692 (1960).

Hebrank, J., Wright, D.: Spectral cues used in the localization of sound sources on the median plane. J. Acoust. Soc. Am. 56, 1829-1834 (1974).

Hecht, S., Pirenne, M. H.: The sensibility of the nocturnal long-eared owl in the spectrum. J. Gen. Physiol. 23, 709-717 (1940).

Henson, O. W.: Comparative anatomy of the middle ear. In: Handbook of Sensory Physiology, Vol. V/1. Keidel and Neff (eds.). New York: Springer, 1974, pp. 39-110.

Hill, K. G., Lewis, D. B., Hutchings, M. E., Coles, R. B.: Directional hearing in the Japanese quail (*Coturnix coturnix japonica*) I. Acoustic properties of the auditory system. J. Exp. Biol., in press.

Jenkins, W. M., Masterton, R. B.: Sound localization in pigeon (*Columbia livia*). J. Comp. Physiol. Psychol. 93, 403-413 (1979).

Knudsen, E. I., Konishi, M.: A neural map of auditory space in the owl. Science. 200, 795-797 (1978a).

Knudsen, E. I., Konishi, M.: Space and frequency are represented separately in the auditory midbrain of the owl. J. Neurophysiol. 41, 870-884 (1978b).

Knudsen, E. I., Konishi, M.: Center-surround organization of auditory receptive fields in the owl. Science. 202, 778-780 (1978c).

Knudsen, E. I., Konishi, M.: Mechanisms of sound localization by the barn owl (*Tyto alba*). J. Comp. Physiol. 133, 13-21 (1979).

Knudsen, E. I., Blasdel, G. G., Konishi, M.: Sound localization by the barn owl measured with the search coil technique. J. Comp. Physiol. 133, 1-11 (1979).

Knudsen, E. I., Konishi, M., Pettigrew, J. D.: Receptive fields of auditory neurons in the owl. Science. 198, 1278-1280 (1977).

Konishi, M.: Time resolution by single auditory neurons in birds. Nature. 222, 566-567 (1969).

Konishi, M.: Comparative neurophysiological studies of hearing and vocalizations in song birds. Z. vergl. Physiol. 66, 257-272 (1970).

Konishi, M.: Locatable and nonlocatable acoustic signals for barn owls. Amer. Nat. 107, 775-785 (1973a).

Konishi, M.: How the owl tracks its prey. Amer. Sci. 61, 414-424 (1973b).

Konishi, M., Knudsen, E. I.: The oilbird: hearing and echolocation. Science. 204, 425-427 (1979).

Krebs, J.: Birdsong and territory defense. New Science. 70, 534-536 (1976).

Marler, P.: Characteristics of some animal calls. Nature. 176, 6-8 (1955).

Marler, P.: Developments in the study of animal communication. In: Darwin's biological work. Bell, P. R. (ed.). Cambridge Univ. Press, 1959, pp. 150-206.

Masterton, R. B.: Adaptation for sound localization in the ear and brainstem of mammals. Fed. Proc. 33, 1904-1910 (1974).

Masterton, R. B., Heffner, H. E., Ravizza, R. J.: Evolution of human hearing. J. Acoust. Soc. Am. 45, 966-985 (1969).

Mills, A. W.: On the minimum audible angle. J. Acoust. Soc. Am. 30, 237-246 (1958).

Norberg, R. Å.: Physical factors in directional hearing in *Aegolius funereus* (Linné) (Strigiformes), with special reference to the significance of the asymmetry of the external ears. Arkiv. Zool. 20, 181-204 (1968).

Norberg, R. Å.: Occurrence and independent evolution of bilateral ear asymmetry in owls and implications on owl taxonomy. Phil. Trans. Roy. Soc. Lond. B 280, 375-408 (1977).

Norberg, R. Å.: Skull asymmetry, ear structure and function, and auditory localization in Tengmalm's owl, *Aegolius funereus* (Linné). Phil. Trans. Lond. B 282, 325-410 (1978).

Parks, T. N., Rubel, E. W.: Organization and development of brainstem auditory nuclei of the chicken: organization of projections from nucleus magnocellularis to nucleus laminaris. J. Comp. Neurol. 164, 435-448 (1975).

Payne, R. S.: The acoustical location of prey by the barn owl (*Tyto alba*). Ph.D. Thesis, Cornell University, 1962.

Payne, R. S.: Acoustic location of prey by barn owls (*Tyto alba*). J. Exp. Biol. 54, 535-573 (1971).

Payne, R. S., Drury, W. H.: Marksman of the darkness. Nat. Hist. 67, 316-323 (1958).

Pettigrew, A., Chung, S.-H., Anson, M.: Neurophysiological basis of directional hearing in amphibia. Nature. 272, 138-142 (1978).

Pumphrey, R. J.: The sense organs of birds. Ibis. 90, 171-199 (1948).

Pumphrey, R. J.: Sensory organs: Hearing. In: Biology and Comparative Physiology of Birds, Vol. II. Marshall, A. J. (ed.). New York: Academic Press, 1961, pp. 69-86.

Pycraft, W. P.: A contribution towards our knowledge of the morphology of the owls. Part 1. Pterylography. Trans. Linn. Soc. Lond. Zool. 7, 223-275 (1898).

Ramón y Cajal, S.: Les ganglions terminaux du nerf acoustique des oiseaux. Trab. Inst. Cajal Invest. Biol. 6, 195-225 (1908).

Renaud, D. L., Popper, A. N.: Sound localization by the bottlenose porpoise *Tursiops truncatus*. J. Exp. Biol. 63, 569-585 (1975).

Sachs, M. B., Young, E. D., Lewis, R. H.: Discharge patterns of single fibers in the pigeon auditory nerve. Brain Res. 70, 431-447 (1974).

Schwartzkopff, J.: Beitrag zum Problem des Richtungshörens bei Vögeln. Z. vergl. Physiol. 32, 319-327 (1950).

Schwartzkopff, J.: Untersuchungen über die Arbeitsweise des Mittelohres und das Richtungshören der Singvögel unter Verwendung von Cochlea-Potentialen. Z. vergl. Physiol. 34, 46-68 (1952).

Schwartzkopff, J.: On the hearing of birds. Auk. 72, 340-347 (1955).

Schwartzkopff, J.: Die Grössenverhältnisse von Trommelfell, Columella-Fussplatte und Schnecke bei Vögeln verschiedenen Gewichts. Z. Morph. Ökol. Tiere. 45, 365-378 (1957).

Schwartzkopff, J.: Zur Frage des Richtungshörens von Eulen (*Striges*). Z. vergl. Physiol. 45, 570-580 (1962).

Schwartzkopff, J.: Structure and function of the ear and of the auditory brain areas in birds. In: Ciba Foundation Symposium "Hearing Mechanisms in Vertebrates" DeReuck, A. V. S., Knight, J. (eds.). London: Churchill, 1968, pp. 41-59.

Searle, C. L., Braida, L. D., Cuddy, D. R., Davis, M. F.: Binaural pinna disparity: Another auditory localization cue. J. Acoust. Soc. Am. 57, 448-455 (1975).

Searle, C. L., Braida, L. D., Davis, M. F., Colburn, H. S.: Model for auditory localization. J. Acoust. Soc. Am. 60, 1164-1175 (1976).

Shalter, M. D.: Localization of passerine seeet and mobbing calls by goshawks and pygmy owls. Z. Tierpsychol. 46, 260-267 (1978).

Stellbogen, E.: Über das ässere und mittlere Ohr des Waldkauzes (*Syrnium aluco* L.). Z. Morph. Ökol Tiere. 19, 686-731 (1930).

Strother, W. F.: The electrical responses of the auditory mechanisms in the bullfrog (*Rana catesbeiana*). J. Comp. Physiol. Psychol. 52, 157-162 (1959).

Takasaka, T., Smith, C. A.: Structure and innervation of the pigeon's basilar papilla. J. Ultrastruct. Res. 35, 20-65 (1971).

Van Dijk, T.: A comparative study of hearing in owls of the family Strigidae. Neth. J. Zool. 23, 131-167 (1973).

Wada, Y.: Beiträge zur vergleichenden Physiologie des Gehörorgane. Pflügers Archiv. 202, 46-69 (1923).

Wallach, H., Newman, E. B., Rosenzweig, M. R.: The precedence effect in sound localization. Amer. J. Psychol. 62, 315-336 (1949).

Weeden, J. S., Falls, J. B.: Differential responses of male ovenbirds to recorded songs of neighboring and more distant individuals. Auk. 76, 343-351 (1959).

Wever, E. G., Bray, C. W.: Hearing in the pigeon as studied by the electrical responses of the inner ear. J. Comp. Psychol. 22, 353-363 (1936).

Wever, E. G., Vernon, J. A.: Auditory responses in the spectacled caiman. J. Cell. Comp. Physiol. 50, 333-339 (1957).

Winter, P.: Vergleichende qualitative und quantitative Untersuchungen an der Hörbahn von Vögeln. Z. Morph. Ökol. Tiere. 52, 365-400 (1963).

Winter, P., Schwartzkopff, J.: Form und Zellzahl der akustischen Nervenzentren in der Medulla Oblongata von Eulen (*Striges*). Experientia. 17, 515-516 (1961).

Woodworth, R. S.: Experimental Psychology. New York: Rinehart & Winston, 1962, pp. 349-361.

Wright, D., Hebrank, J. H., Wilson, B.: Pinna reflections as cues for localization. J. Acoust. Soc. Am. 56, 957-962 (1974).

Yost, W. A., Wightman, F. L., Green, D. M.: Lateralization of filtered clicks. J. Acoust. Soc. Am. 50, 1526-1531 (1971).

Zwislocki, J., Feldman, R. S.: Just noticeable differences in dichotic phase. J. Acoust. Soc. Am. 28, 860-864 (1956).

Chapter 11

Response Properties of Neurons in the Avian Auditory System: Comparisons with Mammalian Homologues and Consideration of the Neural Encoding of Complex Stimuli

MURRAY B. SACHS, NIGEL K. WOOLF and JOAN M. SINNOTT*

1 Introduction

The neural encoding of so-called "biologically relevant" sounds has been one focus for the efforts of auditory neurophysiologists in recent years (e.g., Woorden and Galambos 1972). Studies on amphibians have shown that the peripheral auditory systems of these animals are highly specialized for the processing of species-specific vocalizations (Frishkopf, Capranica, and Goldstein 1968). Cells have been described in the auditory cortex of squirrel monkeys that respond only to a very limited set of the vocalizations produced by these species (Newman and Wollberg 1973); the responses of such cells to these vocalizations are not easily explained in terms of their response to "simple" stimuli such as tones. Similarly, Leppelsack and Vogt (1976) and Scheich, Langner, and Koch (1977) have found cells in the avian field L and nucleus magnocellularis lateralis pars dorsalis whose selective responsiveness to vocalizations are not easily explained in terms of a relationship between those single frequencies that excite the neuron and the spectral content of the vocalizations. Suga (1978) has described a neural organization in the auditory cortex of bats that is highly specialized for the echolocation functions of these animals. Knudsen and Konishi (1978) have also demonstrated an exquisite neural substrate of sound localizations in the MLD of the barn owl, *Tyco alba*. As a result of these various lines of research, it now appears that at least a portion of the auditory system is a hierarchically organized analyzing network in which is found a progressive degree of abstraction of the acoustic signal as one proceeds centrally along the auditory pathway (Bullock 1977, p. 300). At present, however, there is little understanding of how this degree of abstraction comes about, on the basis of response properties at various stages along the pathway. The goal of the present authors' work has been to delineate the hierarchal transformations that occur as a complex stimulus progresses along the auditory pathway in birds.

Birds are an appealing group for the study of the neural processing of acoustic

*Department of Biomedical Engineering, Johns Hopkins University School of Medicine, 506 Traylor Research Building, 720 Rutland Ave., Baltimore, Maryland 21205.

stimuli for several reasons. First, many field studies of birds have revealed that they possess a complex, highly organized vocal communication system; the functional significance and behavioral manifestations of this system have been and continue to be carefully studied (see, for example, Orians and Christman 1968). Second, new techniques have been developed that allow laboratory studies of the auditory capacities of birds under operant conditioning paradigms (Dooling and Saunders 1975, Hienz, Sinnott, and Sachs 1977). The auditory detection and discrimination capabilities of several species are now being worked out in considerable detail (Dooling, Chapter 9); operant techniques are also being applied to investigations of the perception of species-specific vocalization in birds (Sinnott 1978). The anatomical organization and structure of the avian central auditory system is being worked out in some detail (Boord and Rasmussen 1963, Karten 1967, 1968, Leibler 1975, Jhaveri and Morest 1977). Third, although there are structural differences between the avian and mammalian auditory peripheries (Takasaka and Smith 1971), there do not appear to be gross qualitative differences in neural encoding at the level of the auditory nerve (Sachs, Lewis, and Young 1974). This similarity in peripheral encoding is contrasted with that in amphibians where a highly species-specific encoding, closely related to species-specific vocal spectra, has evolved (Capranica 1978). (Because of these differences at the periphery, it may be of considerable interest to compare the central processing strategies of birds with those of amphibians.)

The outline of this chapter will follow attempts to describe the transformations undergone by acoustic stimuli at various levels of the avian auditory pathway. We shall first consider neural encoding in the avian auditory nerve, with emphasis on comparisons with the encoding in mammalian auditory nerve. We shall then review some recent studies of neural encoding in nucleus magnocellularis and nucleus angularis, and compare response properties in these avian structures with properties of cells in their mammalian homologues—the anteroventral cochlear nucleus and the posteroventral and dorsal cochlear nuclei, respectively. Finally, we shall consider some of the results referred to above that indicate that a high level of processing occurs by the time a stimulus reaches MLD or field L. This review will conclude with an illustration of how some of this processing may be directly related to processing at the earliest levels of the avian brainstem.

2 Neural Encoding in the Avian Auditory Nerve

There have been a number of studies of neural encoding at the level of single fibers in the avian auditory nerve (Sachs et al. 1974, Gross and Anderson 1976, Manley and Leppelsack 1977, Woolf and Sachs 1977, Manley 1978). The results of these studies have shown that qualitatively the peripheral encoding in birds is quite similar to that in mammals. There are, however, a number of quantitative differences that may be important in terms of models of psychophysical performance and are certainly intriguing in terms of the underlying cochlear mechanisms.

In the absence of controlled acoustic stimulation, all avian auditory-nerve fibers appear to be spontaneously active. Figure 11-1A shows the rates of spontaneous activity encountered in 183 fibers in the auditory nerve of pigeon. The distribution of

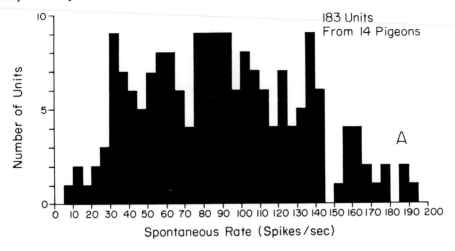

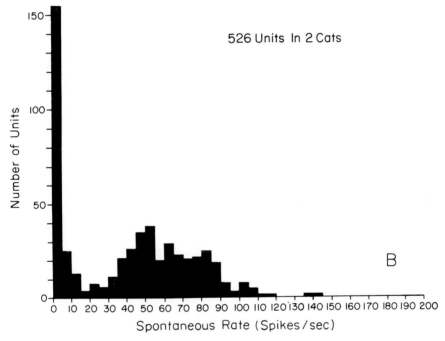

Figure 11-1. Histograms showing numbers of fibers with various rates of spontaneous activity. The height of each bar in the histogram equals the number of fibers with spontaneous rates between the rate at the left and the rate at the right of the bar. (A) Data from 183 fibers in 14 pigeons (replotted from Sachs et al. 1974). (B) Data from 526 fibers in 2 cats (data from Sachs and Young 1978).

spontaneous rates in redwing blackbirds is virtually identical with those in pigeons (Woolf and Sachs 1977). Figure 11-1B shows a histogram of the spontaneous rates from 526 fibers in 2 cats from a recent study by Sachs and Young (1978). The cat results are similar to those published recently by Liberman (1978) and Kim and Molnar (1979). There are several notable differences between results for pigeons and cats. First, there is a large group of fibers in the cat that have very low rates of spontaneous activity. For example, in the data shown in Fig. 11-1B, 13% of the fibers had spontaneous rates of less than 1 spike/sec. In Liberman's study, 16% of the fibers studied in four cats raised in a low-noise chamber had spontaneous rates less than 0.5 spikes/sec. Clearly, from Fig. 11-1A, there is no such low spontaneous rate population in the avian auditory nerve. No pigeon auditory nerve fibers were found to have spontaneous rates of less than 5 spikes/sec; in redwing blackbirds (*Agelaius phoeniceus*) no fibers were found with spontaneous rates below 2 spikes/sec. In the cat, there are few fibers with spontaneous rates between 1 spike/sec and 40 spikes/sec, whereas there is no such "gap" in the distribution for avian species studied to date. There are more high spontaneous rate fibers in birds than there are in cats. The median rate for the pigeon and the blackbird is 90 spikes/sec. The median rate shown in the data for cats in Fig. 11-1B is 40 spikes/sec. Manley (1978) reports higher spontaneous rates in starlings (*Sturnus vulgaris*). Spontaneous rates were in the range 4.9 to 149 spikes/sec.

The absence of a low spontaneous rate population in the avian auditory nerve may be of considerable importance in the understanding of the mechanisms that lead to the spontaneous activity. In the cat, these low spontaneous rate units have a number of distinguishing characteristics that seem to set these units apart as an inherently separate population. Liberman (1978) has shown that these fibers tend to have higher thresholds for responses to tones than other fibers and a wider spread of saturation rates. For broadband and multicomponent stimuli, these low spontaneous fibers in cats are more susceptible to the influences of two-tone suppression (Schalk and Sachs 1978, Sachs and Young 1978). There are a number of possible reasons why we and others might have missed a low spontaneous population in birds, and it is worth considering these reasons briefly. Liberman found many of his low spontaneous fibers through direct electrical stimulation of the auditory nerve. Neither we nor Manley used electrical stimulation. However, we have found almost as large a percentage of low spontaneous fibers in cats with a noise search stimulus as did Liberman with electrical stimulation. It is possible, though unlikely, that a whole population of low spontaneous units was missed in birds with the same noise search stimuli. A second possibility is that the avian fibers are more sensitive to room noise than are the cat fibers, and what is being called spontaneous activity is actually driven activity in response to ambient noise. This hypothesis has been tested by artificially inducing a 40 dB increase in fiber thresholds by filling the external ear canals of a number of birds with silicone impression rubber (Woolf and Sachs 1979). Even with this 40 dB conductive loss (as measured from fiber rate-versus-level functions), the distribution of spontaneous rates is virtually the same as that shown in Fig. 11-1A (Woolf and Sachs 1977). Thus, if the spontaneous activity in avian fibers is generated as a response to some noise source, it is quite unlikely that the source is external to the animal. It could, of course, be an internally generated noise that is somehow coupled into the cochlea. It is important to note here that in some avian auditory-nerve fibers, spontaneous activity can be clearly suppressed by single tones, whereas such single-tone suppression is rarely if ever observed in the auditory-nerve fibers of cats (see also Sachs and Kiang 1968, Gross and Anderson 1976).

Figure 11-2 gives another indication that the higher spontaneous rates in birds are not an artifact of environmental noise. The figure shows that driven rates in avian auditory-nerve fibers are also considerably higher than those in cats. The figure compares saturation rates for pigeon and cats. Median saturation rates for pigeon fibers is about 300 spikes/sec as compared with 190 spikes/sec for cat fibers (both measured with 200 msec tone bursts). Thus, in general, rates of activity in avian auditory-nerve fibers are higher than those in cats. These higher rates of activity may be related to the higher body temperature and higher metabolic rate in the bird.

Another important dimension along which to compare the properties of avian and mammalian auditory-nerve fibers is their frequency selectivity as revealed in their tuning curves. Figure 11-3 shows tuning curves for 6 fibers in the redwing blackbird (on the left) and tuning curves for 6 fibers in the cat, from the study of Kiang and Moxon (1974) (on the right). Both sets of curves show a sharp "tip" centered at fiber characteristic (most sensitive) frequency. The cat fibers have a low-frequency "tail"; the minimum threshold in the tail can be within 40 dB of the threshold at the fiber's characteristic frequency (CF). No such low-frequency tail is seen in the avian auditory-nerve tuning curves, even at levels 80 dB higher than the characteristic frequency threshold (Woolf and Sachs 1979). Manley and Leppelsack (1977) found no tails in the tuning curves of auditory-nerve fibers in starlings.

The sharpness of the tip portion of the avian tuning curves are comparable to those in cats. The sharpness is often described in terms of a measure Q, defined as characteristic frequency (CF) divided by bandwidth of the tuning curve 10 dB above tone threshold at CF. Figure 11-4 shows a plot of Q versus CF for 145 pigeon auditory-nerve fibers and 61 cat fibers, measured in the same experimental set-up. The Qs of the two animals are quite similar. If there is a difference, it is that the Qs for the pigeon fibers are higher than those for the cat. A similar distribution of Qs is found for auditory-nerve fibers in the blackbird (Woolf and Sachs 1977) and for the starling (Manley and Leppelsack 1977).

The differences in spontaneous and saturation rates for auditory-nerve fibers in birds and cats have been discussed. Figure 11-5 shows how the discharge rate grows from spontaneous to saturation as a function of sound level for CF tones. On the right of Fig. 11-5 rate-versus-level functions are shown from five auditory-nerve fibers in the blackbird (Woolf and Sachs 1979); on the left are five functions for the cat from a current study (Schalk and Sachs 1978). The general appearance of the avian and cat functions are similar, and there appears to be a continuum of behaviors. The functions at the top of Fig. 11-5 grow from the spontaneous rate to a saturation rate over a range of stimulus levels of 20 dB to 40 dB. Further increases in stimulus level produce no significant increase in discharge rate. We have previously called such saturation "flat saturations" (Sachs and Abbas 1974). The curves at the bottom of Fig. 11-5 represent the other end of the continuum, what we have called "sloping saturations." Here, discharge rate increases rapidly above threshold for a range of levels of 20 dB to 40 dB. There is a bending over of the function at the high end of this range, but the function does not become flat. Instead, rate continues to increase, often to the upper limit of

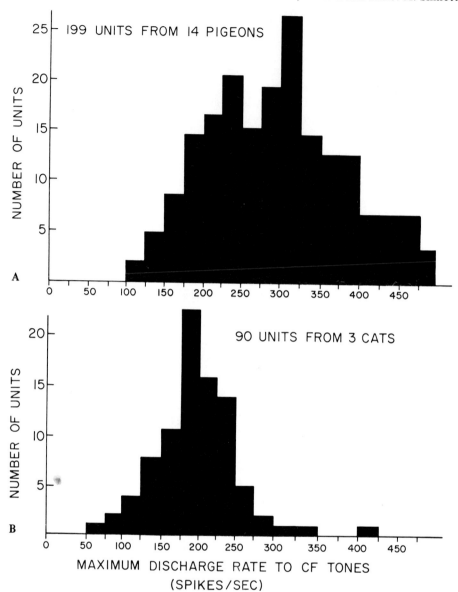

Figure 11-2. Histograms showing numbers of fibers with various saturation rates. Saturation rate is the maximum discharge rate to 200 msec tone bursts at fiber CF. The height of each bar equals the number of fibers with saturation rates within 25 spikes of the rate given on the abscissa at the left of the bar. (A) Data from 199 units in 14 pigeons. (B) Data from 90 units in 3 cats (from Sachs et al. 1978). Reprinted from *Federation Proceedings*, 37, 2329-2335, 1978.

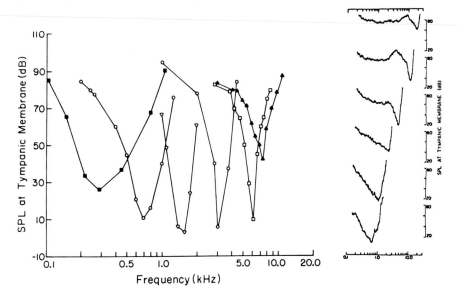

Figure 11-3. Tuning curves for 6 auditory-nerve fibers from several redwing blackbirds (left) and for 6 fibers from cats (right) (data on birds from Woolf and Sachs 1979, data on cats from Kiang and Moxon 1974).

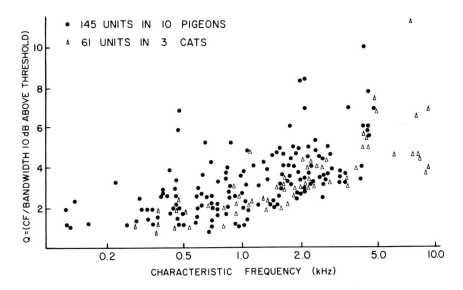

Figure 11-4. Q plotted versus CF for 145 fibers from 10 pigeons (●) and from 61 fibers from 3 cats (△). For cats, data for CFs greater than 10 kHz are not shown. Q is defined as CF divided by the bandwidth of the tuning curve 10 dB above threshold (from Sachs et al. 1974).

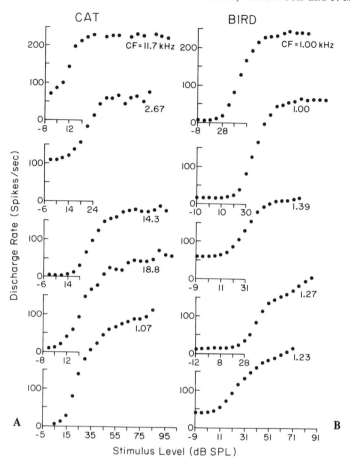

Figure 11-5. Rate-versus-level functions for tones at fiber characteristic frequency. (A) 5 units from one cat. (B) 5 units from several blackbirds (from Woolf and Sachs 1979).

stimulus levels used. For the cat, this continuum of behaviors is related to the absolute threshold of fibers (Sachs and Abbas 1974, Schalk and Sachs 1978); for fibers with approximately the same CF in the same cat, rate level functions change from flat to sloping saturations as fiber thresholds increase. In fact, if threshold is high enough, the "bend" in the rate level function can occur at a low enough stimulus level that the function simply appears to have a lower slope than do functions for low threshold units. We have not yet tested this aspect of the rate-level behavior in auditory-nerve fibers with similar CFs in the same individual bird.

Sachs and Abbas (1974) have suggested that in the cat, the sloping saturation rate-level functions might be related to a nonlinearity similar to that described by Rhode (1971) in his measurement of basilar membrane motion in the squirrel monkey. The appearance of similar sloping saturations in avian auditory-nerve fibers might indicate the existence of a similar nonlinearity in the avian cochlea. Figure 11-6 shows another

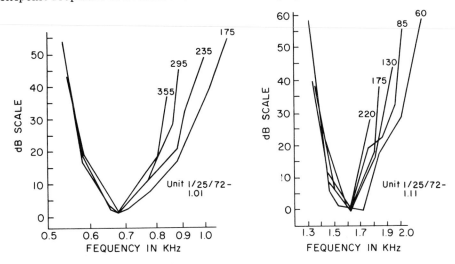

Figure 11-6. Iso-rate contours for 2 fibers. The contours for each fiber have been shifted vertically to coincide at the CF. The numbers associated with each contour indicate the corresponding rates. In all cases the lowest rate is 50% greater than the fiber's rate of spontaneous activity. The dB scales are relative to the sound level corresponding to the tip of each iso-rate contour. In order of increasing rate the zero levels (dB SPL) are for unit 1/25/72−1.01: 9, 14, 19, 24; for unit 1/25/72−1.11: 3, 10, 15, 22 (from Sachs et al. 1974).

aspect of the nonlinear rate behavior of avian fibers that have previously been shown to be qualitatively similar to the nonlinearity described by Rhode. Here are plotted several iso-rate contours for two pigeon auditory-nerve fibers. These are plots of the stimulus levels needed to produce different discharge rates as a function of stimulus frequency. The curves for each fiber have been shifted vertically to coincide at fiber CF. The low frequency branches coincide approximately; there is no systematic deviation. However, there is a systematic deviation in the high frequency branches. The branches corresponding to the higher rates rise more rapidly than do those corresponding to the lower rates. A similar result has been described by Geisler et al. (1974) for the squirrel monkey. As they have pointed out, Rhode has found very similar behavior in iso-amplitude contours for basilar membrane displacement in the squirrel monkey. It is easily shown that the behavior seen in Fig. 11-6 is equivalent to a decrease in slope of rate-versus-level functions as frequency increase above fiber CF. Such a decrease in slope is seen in rate-versus-level functions for both cats (Sachs and Abbas 1974) and squirrel monkeys (Geisler et al. 1974).

Thus far only the aspects of avian auditory-nerve firing patterns that are related to average discharge rates have been discussed. There is, however, considerably more information carried in these patterns than can be described in terms of average rate. It has been known for many years that the fine temporal structure of the auditory-nerve fiber spike trains are closely related to temporal patterns of the acoustic stimulus. For example, in response to tones at frequencies less than about 6.0 kHz, auditory-nerve fibers are "phase-locked" to the tones. That is, although they do not fire during every cycle of the tone, when they do fire, the spikes occur with a distribution about some preferred phase of the tone. Indeed, the probability of a spike occurring at any time during a cycle of the tone is directly related to the stimulus amplitude at that time. The probability of firing as a function of time through a stimulus cycle is usually displayed in terms of a period histogram, which plots the number of spikes occurring at a particular tone phase during a long presentation of the tone. Figure 11-7 shows several period histograms for responses to CF tones from a single auditory-nerve fiber in the redwing blackbird. As in the cat, these histograms appear to be rectified versions of the stimulating sinusoid. At the highest levels shown, the peak of the period histogram splits into two peaks. Such "peak-splitting" is also seen in cat auditory-nerve fibers (see Johnson 1974).

There are a number of reasons why it is of interest to compare the phase locking of avian auditory-nerve fibers, as displayed in the period histogram of Fig. 11-7, with that in mammals. There have been a number of suggestions that the avian cochlea is capable of finer temporal resolution than is the mammalian (for example, see Konishi 1970). However, it has now been shown by several groups that frequency discrimination in birds is considerably poorer than that in cats, monkeys and human beings (Dooling and Saunders 1975, Sinnott, Sachs, and Hienz 1976; Dooling, Chapter 9). Frequency

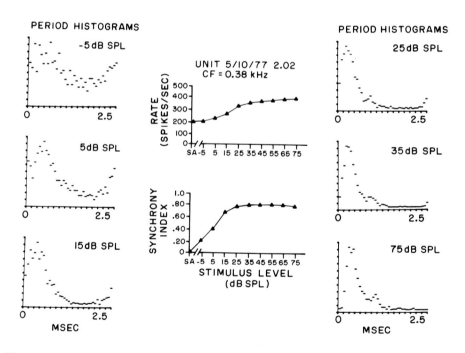

Figure 11-7. Period histograms for a single blackbird auditory-nerve fiber responses to characteristic frequency tones at six sound levels. The abscissa of each histogram corresponds to one cycle of the CF tone (0.38 kHz). Each histogram is scaled so that its maximum value is full scale. The graphs at center show average discharge rate and synchrony index plotted as functions of stimulus levels (data from Woolf and Sachs 1979).

discrimination can be mediated by two mechanisms. "Place" mechanisms depend on the frequency selectivity of auditory-nerve fibers; the frequency of a stimulus is determined by which auditory-nerve fibers respond to it. "Periodicity" mechanisms rely on auditory-nerve phase locking; frequency is determined by the average period of phase locking over some population of the auditory-nerve fibers. The frequency selectivity in the avian auditory nerve is certainly as good as that in mammals (see Fig. 11-4). It seems unlikely, then, that the poorer frequency discrimination by birds can be related to poorer frequency resolution in the avian cochlea. Let us, then, compare the second type of frequency related information, phase-locking, between birds and cats. In order to make this comparison a measure of phase-locking is needed. Johnson (1974) has defined a synchronization index for period histograms like those in Fig. 11-7 as follows: the Fourier series for the period histogram is computed; the synchrony index is defined to be the amplitude of the fundamental component in this Fourier series, divided by the average discharge rate of the fiber to the tone (the DC term in the Fourier series). The synchrony index is thus the normalized amplitude of the best sinusoidal approximation to the period histogram. The two graphs in Fig. 11-7 compare the growth of discharge rate as a function of sound level for this blackbird fiber to the growth of the synchrony index. As has been pointed out before (Rose, Hind, Anderson, and Brugge 1971), the synchrony index (and, hence, phase-locking) becomes significant at stimulus levels well below those that cause an increase in average discharge rate. The synchrony index reaches a maximum over a range of levels of about 30 dB and, in this case, remains at about that maximum value as sound level is increased further. Figure 11-8 shows the synchrony index plotted versus sound level for four other redwing blackbird auditory-nerve fibers. Johnson has compared synchronization across frequencies in the cat by plotting the maximum value attained by the synchrony index (level-maximum) as a function of frequency for a large number of fibers. Figure 11-9 compares his results with a similar plot for the blackbird (Woolf and Sachs 1979). The straight line in both plots represents a simple functional description used by Johnson for his results. The line shown is exactly the same for the bird and the cat; the good fit to the data obtained for both species indicates that in this frequency range, the ability to phase-lock (and hence the temporal resolution) of the bird and the cat are virtually identical. For frequencies less than 1.0 kHz, it appears that the maximum synchrony for birds may be somewhat higher than that for cats, although the data for birds are somewhat sparse in this region. It is, therefore, unlikely that the differences in frequency discrimination between birds and cats is related to poorer phase-locking in the avian auditory periphery. Thus, it is unlikely that this degraded avian frequency discrimination is related to any degradation in the processing of information in the avian periphery.

To briefly summarize this comparison of the avian and mammalian auditory peripheries, although the structure of the avian cochlea is somewhat different from that in mammals, the discharge patterns in the auditory nerves of these two classes are qualitatively similar. Tuning curves have comparable sharpness; rate-versus-level functions for CF tones are similar; and the ability to phase-lock to frequencies between 1.0 kHz and 6.0 kHz appears to be nearly identical. There are, however, some quantitative differences. Discharge rates, both spontaneous and driven, are higher in

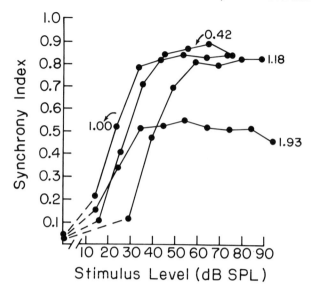

Figure 11-8. Synchrony index plotted versus stimulus level for four auditory-nerve fibers in redwing blackbirds. Tones are at fiber characteristic frequency. Synchrony index is the amplitude of the fundamental component of the Fourier transform of the period histogram divided by the average rate to the tone. Numbers beside each plot give characteristic frequency (data from Woolf and Sachs 1979).

birds than in cats, even if a 40 dB conductive loss is artificially induced by blocking the ear canal in birds. Spontaneous activity seems to be suppressed in some avian auditory-nerve fibers. There do not appear to be low frequency tails on the avian auditory-nerve tuning curves as there are in the cat.

3 Response Properties in the Avian Brainstem Nuclei

Next we discuss response patterns in the first two avian brainstem nuclei: nucleus magnocellularis and nucleus angularis. Nucleus magnocellularis is thought to be homologous with the anterior ventral cochlear nucleus in mammals; nucleus angularis is thought to correspond with the mammalian dorsal and posteroventral cochlear nuclei (Boord and Rasmussen 1963, Boord 1969). We shall compare the properties of cells in nucleus magnocellularis with those in nucleus angularis, and will also compare the response properties of these avian nuclei with the corresponding mammalian nuclei.

Boord and Rasmussen (1963) have demonstrated the spatial arrangement and termination of auditory nerve fibers within nucleus magnocellularis and nucleus angularis with axonal degeneration techniques. These results are summarized in Fig. 11-10. Within the cochlear portion of nucleus magnocellularis, fibers from the basal third of the cochlea terminate most rostral and medial, apical third fibers most caudal and

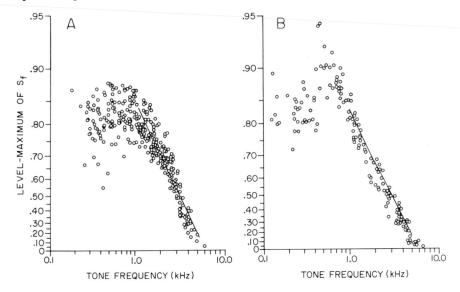

Figure 11-9. Level-maxima attained by synchrony index from cat and blackbird auditory-nerve fibers (data on birds from Woolf and Sachs 1979, data on cats from Johnson 1974). The level maximum is the largest value of synchrony obtained as a function of stimulus level (see, for example, Fig. 11-8). For birds, all tones are at fiber CF, so that each point represents a different fiber. For cats, tones were not necessarily at CF. The lines correspond to the equation $(1 - f/6)$, used by Johnson to represent trend of the data (Johnson 1974). (From Woolf and Sachs 1979).

lateral and middle third fibers occupy an intermediate position. Within nucleus angularis, basal third fibers terminate most caudal and dorsal and middle and apical third fibers progressively most rostral and ventral. A number of investigators have found a corresponding organization of the characteristic frequencies (CF) of cells recorded in these two nuclei (Konishi 1970, Rubel and Parks 1975, Sachs and Sinnott 1978). Figure 11-11 shows the relation of CF to position in nucleus magnocellularis in the chicken (Rubel and Parks 1975). CFs increase from posterior to anterior and from lateral to medial. Figure 11-12 shows the distribution of CFs in part of nucleus angularis of the house sparrow (Konishi 1970). Each column of numbers (frequency in hundreds of Hz) shows the vertical sequence of units encountered in a single penetration at the point occupied by the top number. In nucleus angularis, units are arranged tonotopically in three dimensions. CFs increase systematically in the medial-to-lateral, rostral-to-caudal and ventral-to-dorsal directions. The presence of a distinct vertical tonotopic organization distinguishes nucleus angularis from nucleus magnocellularis. These results are consistent with the anatomical considerations of Boord and Rasmussen (1963). Rubel and Parks (1975) show that the tonotopic organization of nucleus magnocellularis is preserved in its projection to nucleus laminaris (Fig. 11-11).

As one might expect from what is known about their mammalian homologues (see, for example, Goldberg and Brownell 1973), the response properties of cells in nucleus magnocellularis are quite different from those in nucleus angularis. For example, the distribution of rates of spontaneous activity are quite different for these two nuclei. The primary difference illustrated is that the mode of the distribution for angularis occurs at rates less than 10 spikes/sec, whereas no magnocellularis units had rates that low. The mean rate for angularis is 45.5 spikes/sec with a range of 0 spikes/sec to 174 spikes/sec; the mean rate for magnocellularis is 115.7 spikes/sec with a range of 16 to 237.

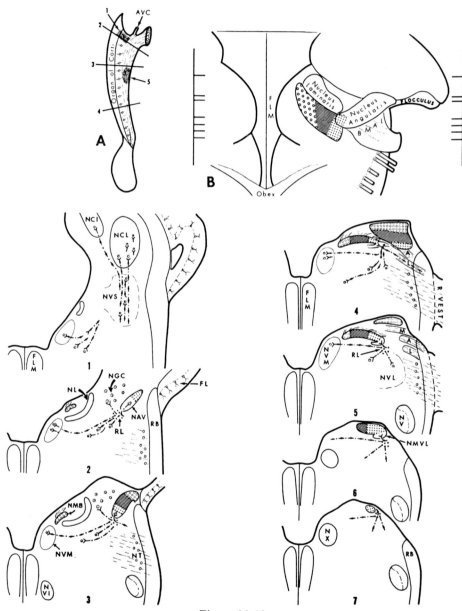

Figure 11-10

Within the central nervous system, it is not possible to characterize the frequency characteristics of cells by simple tuning curves such as those described for the auditory nerve, because responses to single tones can be either excitatory or inhibitory. Cells in the central nervous system are, therefore, characterized by so-called "response maps." Such maps show the areas in the frequency-sound level plane in which pure tones cause excitation or inhibition of spontaneous activity. In all response maps to be presented here, excitatory areas are shaded; inhibitory areas are unshaded regions enclosed by solid lines. The types of response maps found in magnocellularis and angularis will now be considered, and some distinct differences between the two will be highlighted.

In a recent study (Sachs and Sinnott 1978) of cells in the blackbird cochlear nuclei, all units in a sample of 66 from nucleus magnocellularis had response maps similar to those shown in Fig. 11-13. They are characterized by a roughly V-shaped excitatory region whose tip is at the characteristic frequency of the unit; inhibitory areas can occur on one or both sides of the central excitatory area.

Two types of response maps are found in nucleus angularis. The most common type of response map in angularis is similar to those found in magnocellularis. Of 87 units with spontaneous rates greater than 10 spikes/sec (for which inhibitory areas are expected to be found, if they exist), 76 showed the type of response map illustrated in Fig. 11-14. The primary differences between this type of map and those found in magnocellularis is that the inhibitory areas are generally more extensive in angularis, and the maximum amount of inhibition is considerably less in magnocellularis than in angularis. Typically, discharge rate is suppressed to almost zero by some frequency-level combinations for cells in angularis. For cells in magnocellularis, on the other hand, discharge rate is rarely maintained at less than 40% of the spontaneous rate and the inhibitory areas can be quite weak.

A second type of response map, shown in Fig. 11-15, was found in 11 units in nucleus angularis. The detailed organization of the response map for this type varies considerably from unit to unit. One aspect is consistent for all maps of this type, however. Tones at the unit's characteristic frequency are excitatory at sound levels near

Figure 11-10. (A) Diagram of the cochlear nerve showing the locations of lesions used to determine the projection of primary cochlear and lagenar fibers. Lesions of type 1 disrupt most of the extreme basal cochlear fibers and their ganglion cells; lesions of type 2 transect all lagenar fibers and all cochlear fibers except those arising in the most basal part of the cochlear ganglion; lesions of type 3 transect the nerve two-thirds of the distance from the apical end, therefore disrupting all cochlear fibers distal to that point, as well as all lagenar fibers; lesions of type 4 transect the cochlear nerve one-third of the distance from the apical end interrupting all cochlear fibers distal to that point and approximately one-half the fibers of the lagenar nerve; lesions of type 5 disrupt the majority of lagenar fibers but only a few apical cochlear fibers. (B) Diagram of the dorsal aspect of the pigeon medulla showing the location, size, and shape of the nucleus angularis and nucleus magnocellularis. The distribution of basal (circles), middle (oblique lines), and apical (dots) cochlear fibers are projected onto the dorsal surface of nucleus magnocellularis. The ventrolateral part of the latter nucleus and the ventral part of nucleus angularis are not represented here. The plane of each figure's cross section is indicated by numbers 1-7 (from Boord and Rasmussen 1963).

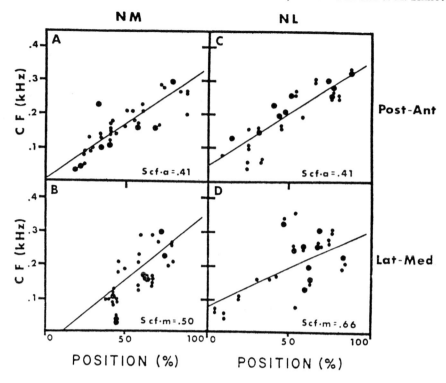

Figure 11-11. Scatter plots, linear regressions, and standard errors of estimate in kHz (Scf), relating characteristic frequency (CF) of units to posterior-anterior and lateral-medial percentile position of electrode tip in n. magnocellularis (NM) and n. laminaris (NL). Large dots show positions at which more than one unit having the same CF was found. (A) Regression of CF on posterior-to-anterior percentile position in NM (48 units). (B) Regression of CF on lateral-to-medial percentile position in NM (48 units). (C) Regression of CF on posterior-to-anterior percentile position to units in NL (45 units). (D) Regression of CF on lateral-to-medial percentile position of units in NL (45 units). Abbreviations: a-anterior position, m-medial position (from Rubel and Parks 1975).

threshold but are always inhibitory at higher levels. Inhibitory regions at higher stimulus levels extend to frequencies both above and below the characteristic frequency. Excitatory areas were also found at higher levels. In some cases these areas were continuous with the excitatory area at characteristic frequency (see Figs. 11-15, 11-16); in other cases, the excitatory areas at higher levels were isolated from the excitatory area near CF (see Fig. 11-20). The resulting interleaved arrangement of excitatory and inhibitory response areas can be quite complex; at a fixed sound level above threshold, the response of a unit can change from inhibitory to excitatory and vice-versa several times as frequency is changed from low to high. (See the plots of rate versus frequency in Fig. 11-15.)

The primary difference between response maps like that in Fig. 11-14 and those in

```
15         18      14        13        15
 9                 12        15         5            ANTERIOR
 7                           11
                              7


32         25      18        21        17        14
15         11      12        17        16         7
            3      13         7        10         5
                    8         6         4         4
                    8         2         3
                                                            —50(=1mm)
37  +      29 +    24        22        20        20
35   8     18  7  15        20         8 #      14
37  27 #   17  8   8        11        14         6
 4  34     14               9         10         4
 2  37     17               7         10         4
 4         14               7          6
 6         10

38         35      30        24        25 +      14
13         18      23        26        23  6     12
10          7      12        23        20        12
36#        10       7         9        18         9
35         12       8                  15          —60
38                                     13
                                        9

        |               |                 |
LATERAL  100(=2mm)      90                80
```

Figure 11-12. Tonotopic organization in part of nucleus angularis of the house sparrow. Numbers indicate characteristic frequencies in hundreds of Hz. Each column of numbers indicates the vertical sequence in which the units were encountered in one unrepeated penetration at the coordinate point occupied by the topmost number. In nucleus angularis (NA) CFs increase caudal, lateral, and dorsal. The ordinate is the distance from the ventral lip of the aqueduct of Sylvius; the abscissa is the distance from the center of the medulla. Unit of scale 1/50 mm. + denotes the continuation of the column from left to right for lack of space. Irregularities in the tonotopic organization are indicated by "#". (This is a portion of a larger plot redrawn from Konishi 1970.)

Figs. 11-15 and 11-16 is illustrated in Fig. 11-17. Here we plot discharge rate versus stimulus level for best-frequency tones for 8 units from nucleus angularis. The stimuli in each case were 200 ms tone bursts at the best frequency of the unit. Discharge rate is averaged over the total duration of the 200 ms tone bursts. The response maps for units on the right were similar to those in Fig. 11-14. For these units, rate is a monotonic function of sound level for best frequency tones. The response maps for units on the left of Fig. 11-17 are similar to those in Figs. 11-15 and 11-16. For units with this type of response map, rate-level functions for best frequency tones are strongly nonmonotonic, with discharge rate decreasing to zero at high stimulus levels.

Similar response maps to those shown here have been studied in some detail in the cat cochlear nuclei (Evans and Nelson 1973, Young and Brownell 1976). Evans and Nelson described five types of response maps in the dorsal cochlear nucleus in unanesthetized and chloralose anesthetized cats. Their Type I and Type II units showed only excitatory responses to tones. The differences between Type I and

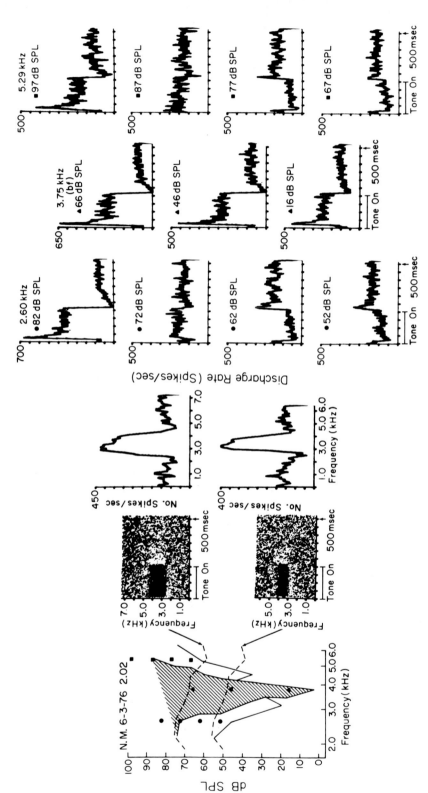

Figure 11-13. Response map and PST histograms from a unit in nucleus magnocellularis of the redwing blackbird. At center are dot displays showing responses to tones at two different stimulus levels. Each line in the display gives responses to the frequency given by the ordinate. For each dot display, the voltage into the earphone is constant, but the resulting acoustic calibration is not flat and is indicated by a dashed line in the response map. Rate-versus-frequency plots corresponding to the dot displays are shown at right. The positions of stimuli for the PST histograms are indicated by the corresponding symbols in the response map (from Sachs and Sinnott 1978).

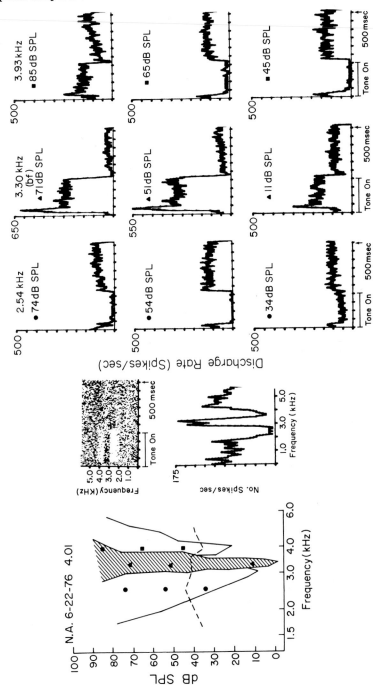

Figure 11-14. Response map, dot display, rate-versus-frequency plot, and PST histograms for a Type III unit from nucleus angularis. The positions of various stimuli are indicated by the corresponding symbols on the response map (from Sachs and Sinnott 1978).

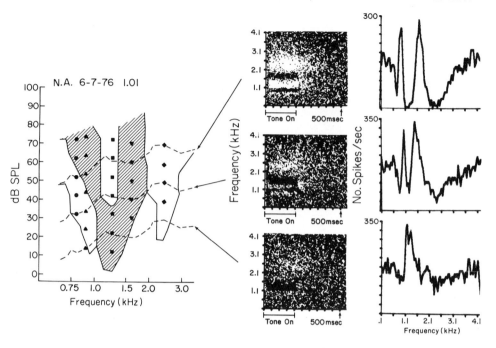

Figure 11-15. Response map, dot displays, and rate-versus-frequency plots for a Type IV unit from nucleus angularis (from Sachs and Sinnott 1978).

Type II depended on whether or not spontaneous activity was suppressed following tonal stimulation. The only units in the avian cochlear nuclei that have no inhibitory areas in their response maps, had little or no spontaneous activity, so that this distinction could not be made. The Type III units of Evans and Nelson had response maps similar to those shown in Figs. 11-13 and 11-14 for birds. Young and Brownell (1976) did not distinguish between Type I, Type II, and Type III units in decerebrate cats because 43% of their units of these types had no spontaneous activity. They combined all these units into one type that they designated Type II/III. Both of these cat studies described response maps similar to those shown in Figs. 11-15 and 11-16, which they designated Type IV maps.

It is difficult to compare the proportions of Type III and Type IV units found in bird cochlear nuclei with the proportions found in cats. As was pointed out, nucleus angularis is thought to be homologous with both the mammalian dorsal cochlear nucleus and posteroventral cochlear nucleus (Boord and Rasmussen 1963, Boord 1969). The medial subdivision of angularis is probably homologous with PVCN, while the lateral division is homologous with DCN. Unit recordings used by the authors of this chapter came from tracks that were almost equally divided between the lateral and medial divisions of nucleus angularis. All eleven units in our sample that would be classified as Type IV (in the Evans and Nelson or Young and Brownell schemes) came from the lateral third of nucleus angularis, which would place them in the homologue of the DCN. Type IV units have only been found in the DCN of cats.

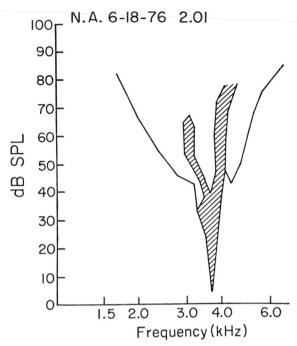

Figure 11-16. Response map of a Type IV unit from nucleus angularis (from Sachs and Sinnott 1978).

All cells in magnocellularis appear to have Type II/III response maps (in the Young and Brownell scheme), although the inhibitory side bands can be quite weak in these cells. Brownell (1975) has found that only one of eight axons in the cat trapezoid body that originated from spherical cells in the anteroventral cochlear nucleus (AVCN) had inhibitory side bands. Goldberg and Brownell (1973) found that only 3 of 34 cells recorded in the spherical cell region of cat AVCN had such side bands. Cells in this spherical cell region are innervated predominately by a few auditory-nerve fibers ending in a calycoid fashion (Osen 1969). It might be expected that these cells would thus closely reflect their primary input and thus would not show inhibitory side bands. Boord and Rasmussen (1963) describe innervation patterns similar to those of these spherical cells in the medial and lateral parts of nucleus magnocellularis (see also Jhaveri and Morest 1977). On the basis of electrophysiological results from cats and the neuroanatomical similarity between the cat spherical cell region and nucleus mag-nocellularis, even the relatively weak inhibitory side bands seen in magnocellularis represent a clear difference between the two species. As was mentioned above, there is some evidence of inhibition of spontaneous activity for some avian auditory-nerve fibers (Woolf and Sachs 1977). Inhibition of spontaneous activity in cat auditory-nerve fibers is rarely if ever seen (Sachs and Kiang 1968). Thus the source of some inhibitory side bands in magnocellularis could be the result of inhibition in the auditory nerve, which is apparently not seen in cats.

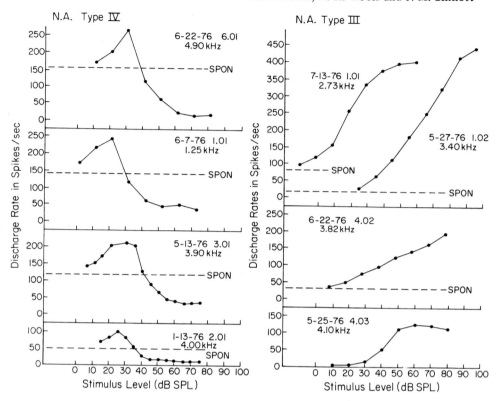

Figure 11-17. Discharge rate versus sound level functions for 8 nucleus angularis units responding to best frequency tones. Units on the left are Type IV; those on the right are Type III. Dashed lines show rates of spontaneous activity.

Response maps, then, in nucleus angularis can be considerably more complex than maps in nucleus magnocellularis. This increase in complexity corresponds well to the increase in complexity between anteroventral and dorsal cochlear nuclei in cats, the mammalian homologues. In terms of temporal patterns of response, it has been shown elsewhere (Sachs and Sinnott 1978) that angularis also shows more complexity than does magnocellularis. Three types of post-stimulus time histograms to best frequency tones are found in nucleus angularis—primary-like, on, and pauser. Similar histograms are seen in the posteroventral and dorsal cochlear nuclei in cats (Godfrey, Kiang, and Norris 1975a, 1975b). Only primary-like histograms are found for best frequency tones in nucleus magnocellularis.

4 Response Properties in Higher Auditory Centers: Responses to Species-Specific Vocalizations

As was pointed out in the introduction to this chapter, there has been much recent interest in the neural encoding of "biologically relevant" sounds. In this light, recent studies of the avian central nervous system have emphasized the responses of cells to species-specific vocalizations. Two of these studies shall be reviewed here, and an attempt will be made to relate the results to the response properties of cells at lower levels.

Scheich, Langner, and Koch (1977) have recorded single units in the auditory midbrain nucleus of guinea fowl (*Numida meleagris*). They employed an awake, chronic preparation. Of the neurons in their material, 60% showed responses to complex stimuli "not simply predictable from pure tone responses" (p. 245). They referred to these as "complex" neurons. Posterior and dorsal areas of MLD appeared to contain a high proportion of such neurons.

One type of complex neuron was characterized by "multiple bands of sensitivity as revealed by the pure tone response." An example is shown in Fig. 11-18. The responses to a number of calls are shown. Each is laid out as follows:

The top part of each response diagram consists of a PST (post-stimulus time) histogram and the name of the call—Kecker, Iambus, Chicken.

The middle part shows the sonagram of the call. A plot of rate versus frequency for tones at 65 dB SPL (the "iso-intensity response," IR) is shown to the right of the sonagram. The IR uses the same frequency scale as the sonagram.

The lower part of each diagram shows the amplitude envelope of the call as a rectified signal.

In the figure, the IR shows three narrow excitatory bands alternating with inhibitory bands. This neuron responds strongly to the Kecker calls whose main energy corresponds to the peaks in the IR and which have little or no energy in the broad low-frequency inhibitory region. The neuron does not respond as strongly to the call Iambus 5. The sonagram for this call is not shown; the main energy of the call was located below 2.0 kHz, in the low frequency inhibitory region. This neuron does not respond well to a Kecker synthesized from broad band noise (S4) with the same temporal structure as the natural Kecker. Scheich et al. (1977) speculates that the decreased response to S4 was due to the simultaneous activation of inhibitory mechanisms below 2.0 kHz and above 3.0 kHz. Figure 11-18 shows the results of other manipulations of the call spectrum. This is one neuron that Scheich et al. called complex even though the responses to calls could be roughly predicted from the shape of the IR.

Other neurons showed wide inhibitory bands in the IR. Figure 11-19 shows an example. Such units had in common a strong excitatory response to wide band calls and other wide-band stimuli that fell into their inhibitory bands. In the unit illustrated in Fig. 11-19, the inhibitory band extended from 1.0 kHz to 6.0 kHz. Upon presentation of calls with broad spectra, excitatory responses were seen (Kecker 4, Iambus 5 and 6). If, on the other hand, a call had primarily a narrow frequency band between 1.0 kHz and 6.0 kHz, the inhibition predominated (Chicken 1). However, a steep FM entirely within the inhibitory range produced an excitation (Chicken 1). Iambus 7, which had only two harmonic bands within the inhibitory region, elicited a weak excitatory response.

Forty percent of the neurons in the sample were classified by Scheich et al. (1977) as "simple." These neurons were typically found in the ventral area of MLD. They were commonly excited by a single frequency band. This excitatory band could be flanked by inhibitory bands. In such units, the qualitative response to calls was pre-

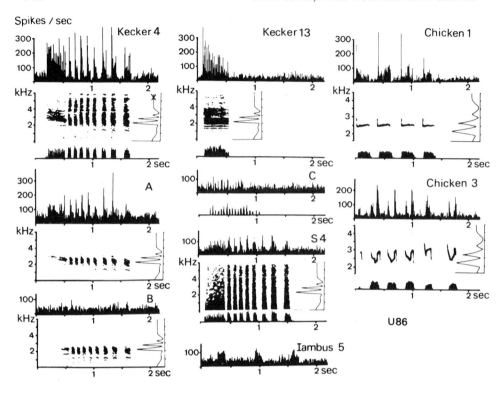

Figure 11-18. Call preferences of a multiple peak neuron that can be correlated to spectral properties. The illustration of each call response consists of a post-stimulus time histogram on the top trace, a sonagram on the middle trace below and a bottom trace that contains the rectified envelope of the analogue signal without ordinate scale. Calls are identified by the name in the right hand corner of the PSTH. To the right of the sonagram the tonal response of the neuron is shown in terms of spikes/sec (Iso-intensity Response=IR). The IR is rotated by 90° so that the abscissa corresponds to the frequency scale of the sonagram. Increase of spike rate is to the left. The ordinate starts at 0 spikes/sec and the scale marks indicate 50 spikes/sec. The spontaneous activity of the unit marked by a cross. The response to Kecker 4 is extremely vigorous. Low-pass filtering of Kecker 4 down to 3 kHz corner frequency in A weakens and down to 2.5 kHz in B abolishes the response. A strong response is obtained with Kecker 13. The response to Kecker 13 is blocked in C by slowing the tape speed to one half. (The sonagram is not shown.) In S4, an artificial Kecker is synthesized from white noise that diminishes the excitatory effect. At the bottom of this column the weak response to Iambus 5 is illustrated without sonagram and sound trace. To the right the steep FM parts of Chicken calls that overlap only with a small part of the IR elicit relatively strong responses. The frequency scale is expanded in these cases (from Scheich et al. 1977).

dictable in most details from knowledge of the call spectra and the IR. If calls with several frequency bands overlapped the excitatory and inhibitory bands, "an interaction between excitation and inhibition was usually seen in the call response" (Scheich et al. 1977, p. 251).

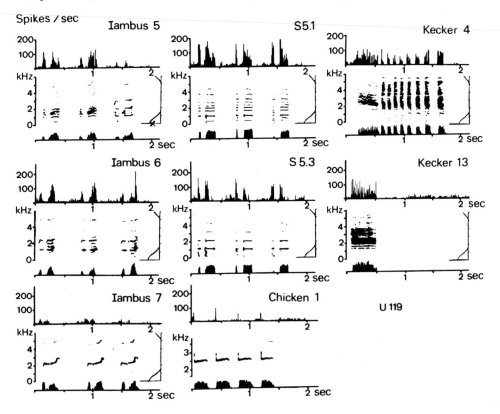

Figure 11-19. Responses of a neuron that is broadly inhibited by pure tones but excited by various wide band calls. The sonagram of Iambus 5 with its many spectral lines fits the inhibitory range of the tone response, yet the call elicits a strong excitatory response. Only to the third call in the sequence does the response fail, which appears to be correlated to the missing sidebands. Iambus 6, like Iambus 5, produces an excitatory response. Iambus 7, a narrow band call within the inhibitory range, yields mainly inhibition. Among the two synthetic Iambi, S5.1 and S5.3, S5.1 is more excitatory than Iambus 5. S5.3 contains some low-frequency noise at the beginning of each element, and this may be correlated with the onset excitation. The response adapts rapidly when only widely spaced bands are present in S5.3. The narrow band Chicken 1 call has an inhibitory effect except for the initial FM part. Kecker 4, which also broadly covers the inhibitory range of the unit, produces a sizeable response. More excitation is produced, however, by Kecker 13, which shows more of a band structure and extends to lower frequencies (from Scheich et al. 1977).

Let us compare these results for MLD with response properties in nucleus angularis. The so-called "simple" neurons of Scheich et al. appear to have response maps similar to those of the Type II/III cells recorded in angularis. The type of "complex" cell shown in Fig. 11-18 has a rate-versus-frequency plot (IR) quite similar to those of the angularis cells shown in Figs. 11-15 and 11-16, if measured at levels above about 45 dB SPL. Figure 11-20 shows another response map from nucleus angularis. This map has a broad inhibitory band at levels above about 50 dB SPL. The rate-versus-frequency plots for this unit at levels above 50 dB SPL are quite similar to that from the MLD neuron shown in Fig. 11-19.

Thus, response maps similar to those of both the simple and complex neurons found in MLD are also found at the lower level in nucleus angularis. A distinctive feature of the MLD neuron in Fig. 11-19 is that broadband stimuli with energy only in the inhibitory region cause strong excitatory responses. Responses to broadband stimuli have not been measured in the avian cochlear nuclei. However, Young and Brownell (1976) have explored in some detail the responses to broadband noise in the dorsal cochlear nucleus. Their analysis may lend insight into both the neural organization responsible for the Type IV response map organization (See Figs. 11-15, 11-16, 11-18, 11-19, and 11-20) as well as for the mechanisms underlying the broadband responses shown in Fig. 11-19. Figure 11-21 shows an example of the relevant result from Young and Brownell (1976). As is illustrated in this figure, Type IV units in their sample responded to broadband stimuli with an increase in discharge rate. For many units, like the one in Fig. 11-21, the response to broadband noise was considerably stronger than the response to any tone. The response map shown in Fig. 11-21 is typical of those seen in the cat DCN. This unit's responses to tones were quite weak. The lower plot in this figure shows the unit's discharge rate as a function of spectral level for four noise bands. The bands were centered slightly above best frequency at the frequency shown

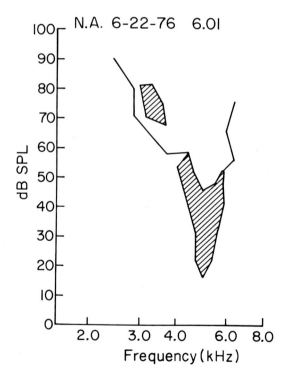

Figure 11-20. Response map from Type IV unit from nucleus angularis (from Sachs and Sinnott 1978).

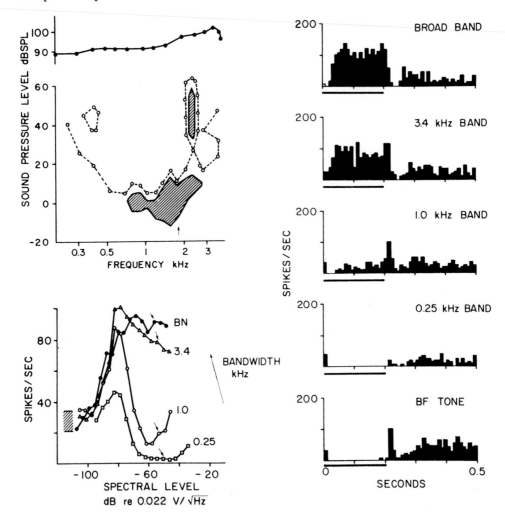

Figure 11-21. Responses to noise of unit 74-11-2. The unit's response map and acoustic calibration (sound level at the mouth of the coupler at 0 dB attenuation) are shown at upper left. The graph at lower left shows discharge rate versus spectral level plots for noise bands of four different widths. The band widths are given by the parameters to the right of the plots. BN means broad-band noise. Bands were centered arithmetically at 1.8 kHz, slightly above best frequency (arrow on the response map). The shaded region next to the ordinate shows the range of spontaneous discharge rate during acquisition of these data. The response histograms in the right-hand column show the unit's response to noise bands of various widths at the spectral levels indicated by arrows next to the discharge rate curves. The bottom histogram shows the unit's response to best-frequency tones whose energy was equal to that of the 0.25-kHz noise band used to obtain the histogram immediately above it (from Young and Brownell 1976).

by the arrow in the response map. The narrowest noise band (0.25 kHz) produced weak excitation at low levels and inhibition at higher levels, as did best frequency tones. The rate-versus-level function for this band is similar to those for tones in nucleus angularis shown in Fig. 11-17. The responses of the DCN Type IV (Fig. 11-21) to successively wider bands were successively stronger with successively less inhibition at higher levels; broadband noise (BN) produced a monotonic rate-versus-level function.

Young and Brownell (1976) present a proposal that might explain the excitatory response to noise of the Type IV cells. They suggest that the Type II/III units in the DCN are actually inhibitory interneurons that are responsible for the massive inhibitory inputs acting on the Type IV neurons. They produce quite convincing arguments for this hypothesis on the basis of a number of properties of their Type II/III and Type IV neurons. Perhaps the most convincing evidence comes from more recent experiments (Young 1978, Voigt and Young 1978) that show: (1) Type IV neurons project out of the DCN via the dorsal acoustic stria, whereas only a small percentage (if any) of the Type II/IIIs project out; and (2) responses of simultaneously recorded Type II/III and Type IV neurons are negatively correlated. The excitatory region of a Type II/III neuron fits into the inhibitory area of a simultaneously recorded Type IV. Now, Type II/III cells respond weakly if at all to broadband noise (presumably because of the strong inhibitory sidebands seen in these units). Thus, under the Young and Brownell hypothesis, the inhibitory inputs to Type IV cells are not activated by broadband noise. If the excitatory inputs, say directly from the auditory nerve, are activated by the noise, then the Type IV would give an excitatory response to the noise, rather than the inhibitory responses that might be predicted on the basis of the pure tone response map. The properties of complex MLD cells like those shown in Fig. 11-19 might thus be explained in one of two ways (or perhaps both). They could simply represent an input from nucleus angularis Type IV neurons that respond to noise as do the DCN Type IVs; or the kind of mechanism proposed by Young and Brownell to explain the properties of Type IV cells could be operating at the level of the MLD.

This proposal of Young and Brownell has been presented in order to underscore the potential insights into the complex properties of higher auditory system neurons that can be gained by a careful examination of the transformations that occur at lower levels in the system. There are now numerous reports of cells with very complex response properties in higher auditory centers. Leppelsack (1978), for example, has reported a very careful study of responses from cells in field L in the forebrain of starlings. He presented each cell with a battery of 80 natural sounds. Most cells responded to only a very limited subset of these sounds. For some units there was no apparent relation between the tonal responses of cells and responses to the natural sounds. For example, 30% of the cells did not respond to pure tones at the levels tested. Many did not respond to white noise. They did, however, respond to a few of the 80 natural sounds. Leppelsack suggests that "these neurons seem to be specialized to respond to very specific sounds or to combinations of acoustic features that are only represented by these sounds."

There continue to be differences of opinion about the use of such terms as "specialized" or "acoustic features" (see, for example, Manley and Mueller-Preuss 1978, Bullock 1977, Bullock, Chapter 16). Nonetheless, the existence of cells such as these that represent the output of a chain of quite complex transformations has been established by studies such as Leppelsack's in birds as well as those in other animals (e.g., Newman and Wollberg 1973). The immediate challenge is not to find more and more complex transformations. It is to analyze carefully the signal transformations at lower system levels that might lead to the complex properties at the higher levels. Only in this way can we hope to arrive at general principles of the neuronal organization underlying the perception of such complex stimuli as species-specific vocalizations.

Acknowledgments. This work was supported by a grant from The National Institute of Neurological and Communicative Disorders and Stroke. J. Sinnott and N. Woolf were National Institutes of Health Postdoctoral Fellows and Mellon Postdoctoral Fellows in Biomedical Engineering. M. Sachs was a National Institute of Health Research Career Development Awardee.

References

Boord, R. L.: The anatomy of the avian auditory system. Ann. N. Y. Acad. Soc. 167, 186-198 (1969).

Boord, R. L., Rasmussen, G. L.: Projection of the cochlear and lagenar nerves on the cochlear nuclei of the pigeon. J. Comp. Neurol. 120, 463-475 (1963).

Brownell, W. E.: Organization of the cat trapezoid body and the discharge characteristics of its fibers. Brain Res. 94, 413-433 (1975).

Bullock, T. H. (ed.). Recognition of Complex Acoustic Signals. Berlin: Dahlem Konferenzen, 1977.

Capranica, R. R.: Auditory processing in anurans. Fed. Proc. 37, 2324-2328 (1978).

Dooling, R. J., Saunders, J. C.: Hearing in the parakeet (*Melopsittacus undulatus*): absolute thresholds, critical ratios, frequency difference limens and vocalizations. J. Comp. Physiol. Psychol. 88, 1-20 (1975).

Evans, E. F., Nelson, P. G.: The responses of single neurones in the cochlear nucleus of the cat as a function of their location and the anesthetic stata. Exp. Brain. Res. 17, 402-427 (1973).

Frishkopf, L. S., Capranica, R. R., Goldstein, M. H., Jr.: Neural coding in the bullfrog's auditory system—a teleological approach. Proc. IEEE 56, 969-980 (1968).

Geisler, C. D., Rhode, W. S., Kennedy, D. T.: Responses to tonal stimuli of single auditory-nerve fibers and their relationship to basilar membrane motion in the squirrel monkey. J. Neurophysiol. 37, 1156-1172 (1974).

Godfrey, D. A., Kiang, N. Y. S., Norris, B. E.: Single unit activity in the posteroventral cochlear nucleus of the cat. J. Comp. Neurol. 162, 247-268 (1975a).

Godfrey, D. A., Kiang, N. Y. S., Norris, B. E.: Single unit activity in the dorsal cochlear nucleus of the cat. J. Comp. Neurol. 162, 269-284 (1975b).

Goldberg, J. M., Brownell, W. E.: Discharge characteristics of neurons in anteroventral and dorsal cochlear nuclei of cat. Brain Res. 64, 35-54 (1973).

Gross, N. B., Anderson, D. J.: Single unit responses recorded from the first order neuron of the pigeon auditory system. Brain Res. 101, 209-222 (1976).

Hienz, R. D., Sinnott, J. M., Sachs, M. B.: Auditory sensitivity of the redwing blackbird (*Agelaius phoeniceus*) and brown-headed cowbird (*Molothrus ater*). J. Comp. Physiol. Psychol. 91, 1365-1376 (1977).

Jhaveri, S. R., Morest, D. K.: Morphological transformations of cells and axons in n. magnocellularis of developing chicken embryos. Soc. Neurosci. Abstr. 3, 109 (1977).

Johnson, D. H.: The response of single auditory-nerve fibers in the cat to single tones: synchrony and average discharge rate. (Ph.D. Thesis) Cambridge: Massachusetts Inst. Technology, 1974.

Karten, H. J.: Organization of the ascending auditory pathway in the pigeon (*Columba livia*). Diencephalis projections of the inferior colliculus (nucleus mesencephalicus lateralis pars dorsalis). Brain Res. 6, 409-427 (1967).

Karten, H. J.: The ascending pathway in the pigeon (*Columba livia*). II. Telencephalic projections of the nucleus oviodalis thalami. Brain Res. 11, 134-153 (1968).

Kiang, N. Y. S., Moxon, E. C.: Tails of tuning curves of auditory-nerve fibers. J. Acoust. Soc. Am. 55, 620-630 (1974).

Kim, D. O., Molnar, C. E.: A population study of cochlear nerve fibers: comparison of the spatial distributions of average rate and phase-locking measures of responses to single tones. J. Neurophysiol. 42, 16-30 (1979).

Konishi, M.: Comparative neurophysiological studies of hearing and vocalizations in songbirds. S. Vergl. Physiol. 66, 257-272 (1970).

Knudsen, E. I., Konishi, M.: Space and frequency are represented separately in auditory midbrain of the owl. J. Neurophysiol. 41, 870-884 (1978).

Leibler, L. M.: Ascending binaural and monaural pathways to mesencephalic and diencephalic nuclei in the pigeon (*Columba livia*). (Ph.D. Thesis) Cambridge: Massachusetts Institute of Technology, 1975.

Leppelsack, H. J.: Unit responses to species specific sounds in the auditory forebrain centers of birds. Fed. Proc. 37, 2336-2341 (1978).

Leppelsack, H. J., Vogt, M.: Responses of auditory neurons in the forebrain of a songbird to stimulation with species-specific sound. J. Comp. Physiol. 107, 263-274 (1976).

Liberman, M. C.: Auditory-nerve response from cats raised in a low-noise chamber. J. Acoust. Soc. Am. 63, 442-455 (1978).

Manley, G. A.: Response characteristics of auditory neurons in the cochlear ganglion of the starling. Proc. XVII Int. Ornith. Congress, Berlin, 1978.

Manley, G. A., Leppelsack, H. J.: Preliminary data on activity patterns of cochlear ganglion neurons in the starling. Collogues Inst. Nat. Sante Rech. Med. 68, 117-136 (1977).

Manley, J. A., Mueller-Preuss, P.: Response variability in the mammalian auditory cortex: an objection to feature detection? Fed. Proc. 37, 2355-2359 (1978).

Newman, J. D., Wollberg, Z.: Multiple coding of species-specific vocalizations in the auditory cortex of squirrel monkeys. Brain Res. 54, 287-304 (1973).

Orians, G. H., Christman, G. M.: A comparative study of the behavior of red-winged, tricolored, and yellow-headed blackbirds. Univ. Calif., Berkeley. Publ. Zool. 84, 1-85 (1968).

Osen, K. K.: Cytoarchitecture of the cochlear nuclei in the cat. J. Comp. Neurol. 136, 453-483 (1969).

Rhode, W. S.: Observations of the vibration of the basilar membrane in squirrel monkeys using the Mössbauer technique. J. Acoust. Soc. Am. 49, 1218-1231 (1971).

Rose, J. E., Hind, J. E., Anderson, D. J., Brugge, J. F.: Some effects of intensity on responses of auditory nerve fibers in squirrel monkey. J. Neurophysiol. 34, 685-699 (1971).

Rubel, E. W., Parks, T. N.: Organization and development of the brainstem auditory nuclei of the chicken: tonotopic organization of n. magnocellularis and n. laminaris. J. Comp. Neurol. 164, 411-434 (1975).

Sachs, M. B., Abbas, P. J.: Rate versus level functions for auditory-nerve fibers in cats: tone-burst stimuli. J. Acoust. Soc. Am. 56, 1835-1847 (1974).

Sachs, M. B., Kiang, N. Y. S.: Two-tone inhibition in auditory-nerve fibers. J. Acoust. Soc. Am. 43, 1120-1128 (1968).

Sachs, M. B., Lewis, R. H., Young, E. D.: Discharge patterns of single fibers in the pigeon auditory nerve. Brain Res. 70, 431-447 (1974).

Sachs, M. B., Sinnott, J. M.: Responses to tones of single cells in nucleus magnocellularis and nucleus angularis of the redwing blackbird (*Agelaius phoeniceus*). J. Comp. Physiol. 126, 347-361 (1978).

Sachs, M. B., Sinnott, J. M., Hienz, R. D.: Behavioral and physiological studies of hearing in birds. Fed. Proc. 37, 2329-2335 (1978).

Sachs, M. B., Young, E. D.: Patterns of auditory-nerve fiber responses to steady-state vowels. J. Acoust. Soc. Am. 63, S76 (1978).

Schalk, T. B., Sachs, M. B.: Rate versus level functions for auditory-nerve fibers in cats: Bandlimited noise bursts. J. Acoust. Soc. Am. 64, S135 (1978).

Scheich, H., Langner, G., Koch, R.: Coding of narrow-band and wide-band vocalizations in the auditory midbrain nucleus (MLD) of the guinea fowl (*Numida meleagris*). J. Comp. Physiol. 117, 245-265 (1977).

Sinnott, J. M.: Species-specific coding strategies in bird song. J. Acoust. Soc. Am. 64, S86 (1978).

Sinnott, J. M., Sachs, M. B., Hienz, R. D.: Differential sensitivity to frequency and intensity in songbirds. J. Acoust. Soc. Am. 60, S87 (1976).

Suga, N.: Specialization of the auditory system for reception and processing of species-specific sounds. Fed. Proc. 37, 2342-2354 (1978).

Takasaka, T., Smith, C. A.: The structure and innervation of the pigeon's basilar papilla. J. Ultrastruct. Res. 35, 20-65 (1971).

Voigt, H. F., Young, E. D.: Interactions of two types of neurons in the DCN. J. Acoust. Soc. Am. 64, S137 (1978).

Woolf, N. K., Sachs, M. B.: Phase-locking to tones in avian auditory-nerve fibers. J. Acoust. Soc. Am. 62, S46 (1977).

Woolf, N. K., Sachs, M. B.: Response properties of auditory-nerve fibers in the redwing blackbird (*Agelaius phoeniceus*). To be submitted (1980).

Worden, F. G., Galambos, R.: Auditory Processing of Biologically Significant Sounds: A Report Based on an NRP Work Session, Neurosciences Research Program Bulletin 10 (1972).

Young, E. D.: Identification of the response properties of efferent axons of the dorsal cochlear nucleus. J. Acoust. Soc. Am. 63, S76 (1978).

Young, E. D., Brownell, W. E.: Responses to tones and noise of single cells in dorsal cochlear nucleus of unanesthetized cats. J. Neurophysiol. 39, 282-299 (1976).

PART FIVE

Mammals

Volumes can be (and have been) written regarding mammalian audition. Rather than attempt to sample all areas, we have chosen a limited number of topics to emphasize selected comparative issues. Gourevitch (Chapter 12) considers the psychophysics of sound localization among the mammals, and Goldstein and Knight (Chapter 13) discuss recent data and theory on the organization of mammalian auditory cortex. Yost (Chapter 14) presents recent views of selected human psychoacoustic phenomena that may have considerable bearing on the kinds of questions that we will soon be asking of other vertebrates. Finally, Stebbins (Chapter 15) considers the historical and evolutionary development of the mammalian ear and capacities of auditory processing.

Chapter 12

Directional Hearing in Terrestrial Mammals

GEORGE GOUREVITCH*

1 Introduction

Whether a sound is made by prey or predator, or is that of a conspecific courting or signaling danger, there is little doubt that being able to locate its source in space is of great utility to an animal. One might expect, therefore, that a wide-ranging research interest in animal auditory localization would exist and would have spawned a vast literature. The contrary is true if the concern is with psychophysical determinations of sound localization and sound lateralization in terrestrial mammals, as is the case with this chapter. (Judgements about the position in space of a sound source are referred to as localization. Lateralization indicates "localization" within the head of a fused sound image that occurs when separate acoustic signals are delivered to each ear through headphones.)

Numerous studies have examined physiological and anatomical aspects of directional hearing, and some of them even used behavioral procedures in conjunction with ablations of the auditory pathways (reviewed by Erulkar 1972, Neff, Diamond, and Casseday 1975); however, such studies are not included here, for their research strategies were not primarily concerned with determining the capacities of intact organisms to locate sound, and, generally, they have made few basic measurements of this auditory function.

The one mammal in whom localization and lateralization have been examined extensively with psychophysical procedures is man. This research has identified some of the acoustic cues and auditory mechanisms involved in directional hearing. To the extent that a morphological and physiological resemblance exists between man and other mammals, the work on man may give a good indication of what occurs during localization in nonhuman mammals.

*Department of Psychology, Hunter College of the City University of New York, 695 Park Ave., New York, New York 10021.

2 Interaural Disparities

For precise spatial localization of a sound, knowledge of the azimuth, elevation, and distance of the sound source is necessary. However, in most investigations of localization in terrestrial mammals other than man, the principal concern has been with azimuthal accuracy, i.e., the discriminability between sound sources in the horizontal plane.

2.1 The Binaural Arrangement

A general analysis of the placement of two ears across a head reveals that interaural, temporal, and amplitude differences emerge when the origin of an acoustic signal is to one side of the listener. The temporal disparity is due to the longer path across the head that the sound must traverse to reach the distal ear. Since the length of this path changes with the azimuth[1] of the source, information about the arrival times of the sound at the two ears is available to the listener and can be used to determine the direction of the sound. (Ambiguity between sounds originating at symmetrical sites in the front and rear quadrants of the horizontal plane will persist (Blauert 1969/70, Mills 1972), but the pinna serves to diminish this source of error.) Differences in amplitude of the signals at the two ears results from the interference with sound propagation that is primarily produced by the head and, to a lesser extent, by the pinnae.

2.2 Cues for Localization

Binaural disparities give rise to particular acoustic cues whose existence and magnitude depend on factors such as the spatial location of the sound source, the type of signal, and the size of the head and pinnae. Among these binaural cues are:

1. *On-going time (phase) differences.* When the signal is a tone, a phase difference will develop at the ears that corresponds to the time it takes for identical portions of the waves to travel the additional distance to the far ear. The usefulness

---

[1]With the assumption that the head is a sphere and that the ears are diametrically opposed within the horizontal plane, the dependence of the time difference on the azimuth is given by

$$\Delta t = \Delta d/c = r/c \, (\theta + \sin \theta)$$

where Δt is the time difference, Δd is the difference in path length, c is the speed of sound, r is the radius of the head, θ is the angle of incidence of the sound with the midline ($0°$ azimuth). This geometric model presented by Woodworth (1938), and extensively applied in human and animal studies, predicts interaural arrival time differences for acoustic signals irrespective of their frequencies. Recently, a frequency dependent model that describes interaural time differences at low and intermeidate frequencies more accurately than Woodworth's was proposed by Kuhn (1977). This model has been used in studies of animal directional hearing to establish the correspondence between interaural time difference thresholds of monkeys and their minimum audible angle thresholds (see Fig. 12-7; Houben and Gourevitch 1979, Brown 1978a).

of this cue vanishes for frequencies whose wavelengths are shorter than the distance between the ears since the same interaural phase difference will occur for numerous sound sources on either side of the listener.

2. *Intensity differences.* When the wavelengths of tones are of comparable or shorter dimension than the head or pinnae, these structures impede sound propagation to the distal ear and create a difference in intensity at the two ears.

3. *Transient arrival time difference.* This cue is universal in that it occurs when the initial wave front of any acoustic signal (tones or complex waves) travels from one ear to the other. Compared to long-lasting disparities of a continuing signal, its effectiveness is minor (Tobias and Schubert 1959); however, when the signal is brief, as with a click, the role of this cue is probably much more important.

4. *Time differences for high frequency complex signals.* Man is unable to discriminate on-going time differences between binaural tones above approximately 1 200 Hz (Zwislocki and Feldman 1956). However, time information is still available to man at high frequencies in the interaural differences of the relatively slow fluctuations of some complex signal envelopes (Henning 1974, McFadden and Pasanen 1976). This cue is probably also important to other mammals; however, no measurements are yet available.

5. *Interaural spectral difference.* As the angle of incidence of a complex signal changes, the spectrum of the signal at one ear is altered by the shadowing of the pinna. Because of the presence of the head and of the distal pinna, other spectral changes occur at the far ear. It is the difference in the spectra at the two ears that can serve to locate sound (Blauert 1969/70, Searle, Braida, Cuddy, and Davis 1975).

In most natural settings, localization is not dependent on any one of the cues that have been discussed but rather on the occurrence of a number of them together, e.g., transient time difference and on-going time difference. Typically, the relative contribution of each to localization will not be the same in all instances and will depend on the type of signal, the physical characteristic of the head, and the sensitivity of the organism to each cue.

Two of the cues listed above, on-going time difference and intensity difference, have been incorporated into the duplex theory that well describes how man localizes pure tones (Stevens and Newman 1936). This theory also appears to be applicable to some other mammalian species. According to the duplex theory, on-going time differences serve as cues for directional hearing in the lower portion of the audible spectrum, while intensity differences serve as cues in the higher portion of the spectrum. The boundary between these cues consists of a segment of intermediate frequencies for which neither cue is very effective and, therefore, localization is poorer at these frequencies than elsewhere (Green and Henning 1969).

3 Head and Pinna Effects on Localization

Of numerous factors that determine the magnitude of interaural amplitude and temporal differences, those that are especially important for a comparative examination of localization include head size, pinna mobility, and, in general, head movement. The effects on localization of such dynamic factors as pinna mobility and head movement, however, have not been measured systematically in terrestrial mammals.

Interaural distances, on the other hand, have been reported although for only a small number of animals. Even in the small sample of available measurements (Table 12-1), an almost six-fold range of distances exists. Furthermore, if the sample were to include determinations on large animals such as gorillas and elephants, the range would extend considerably. It should also be noted that the dispersion of interaural distances is large within an order. For the primates listed in Table 12-1, the differences in interaural distance are more than three fold and would be even greater if measurements were available for tree shrews, marmosets, and gorillas. Thus, the differences in the magnitude of cues due to head size may be quite striking within an order.

Because of the diversity in the morphology of animal heads, instances exist where the relative contribution to localization of interaural temporal and amplitude cues may be lopsided. Masterton, Heffner, and Ravizza (1969) argued that in small animals the time cue is not effective but that accurate sound localization would be achieved with nontemporal cues. Furthermore, they proposed that high frequency hearing, common to small mammals, reflects selective pressure for localization. Interaural amplitude differences would be the predominant, or for some animals, the only directional cue and would emerge at those frequencies that are short relative to the size of the animal's head.

Pertinent to this argument are direct measurements of interaural intensity differences that develop across the heads of animals. In man, the dependence of intensity differences on frequency and azimuth is complex and, thus, binaural intensity differences cannot serve as unambiguous indicators of tone sources (Feddersen, Sandel, Teas, and Jeffress 1957).

A similar finding was observed in other animals by Harrison and Downey (1970). Their measurements for squirrel monkeys, *Saimiri sciureus,* appear in Fig. 12-1, and for the albino rat in Fig. 12-2. In these instances, intensity differences are related in an

Table 12-1. Mammalian Interaural Distances

Mammal	Distances in cm
Primate	
Man	17.5 [a]
Monkey (*M. nemestrina*)	9-10
Squirrel monkey (*Saimiri sciureus*)	~5
Carnivore	
Red fox	3 [b]
Cat	10 [a]
Rodents	
Chinchilla	~7 [a]
Guinea pig	~4 [a]
Rat	3

[a] Shaw (1974)
[b] Isley and Gysel (1975)

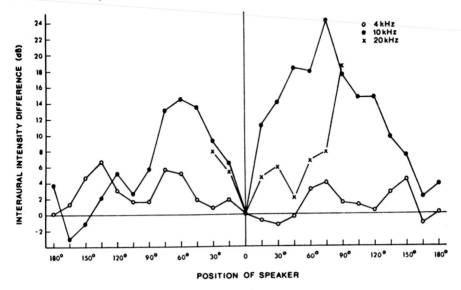

Figure 12-1. Interaural intensity differences in a squirrel monkey (*Samiri sciureus*) for three frequencies originating at different azimuths (from Harrison and Downey 1970).

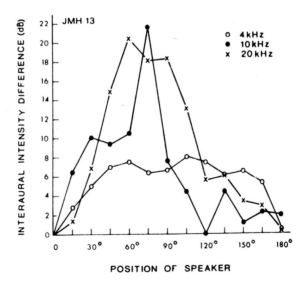

Figure 12-2. Interaural intensity differences in a rat for three frequencies originating at different azimuths (from Harrison and Downey 1970).

even more complicated way to frequency and azimuth than in man. In both species, low frequency tones can produce greater interaural intensity differences than higher ones. Furthermore, angular segments exist over which little change in intensity difference occurs. These segments are most obvious for 4 kHz tones but are also visible at other frequencies. Some examples are the segment between 30° and 60° at 10 kHz in the rat or the segment 15° to 75° at 20 kHz in the monkey. Finally, within 15° or so from the midline, interaural intensity differences for most of the test frequencies are less than 6 dB. As will be seen, such differences may be very difficult for the animals to detect.

4 Localization in Terrestrial Animals

4.1 Behavioral Methods

Typically, two procedures have been used in investigating localization in mammals. In one, the approach-to-target method, the animal is placed at a starting site to listen for a signal and locate its origin. The animal is then released and proceeds to the source where positive reinforcement is delivered. The other procedure, the identification method, has two forms. In the first, the animal is situated in one location and indicates when the position of the sound source has changed by manipulation of a lever. He is rewarded for correct identification. In the second, which incorporates the conditioned suppression technique, the animal licks a spout, for example, while a signal originates from a "safe" location; he stops or suppresses licking when he detects a shift in the signal source, since an inevitable shock follows the shift.

The approach-to-target procedure does not prevent head movements. The animal is free to scan the sound field and thus maximize interaural differences. On the other hand, the identification procedures limit head movement by having the animal lick the spout (Heffner 1973), or in some other experimental arrangements, eliminate head movement completely by anchoring the skull (Brown 1976). Many of the localization studies to be mentioned used the approach-to-target procedure. The results from these studies that allow head movements probably reflect the best performance the animal is capable of producing.

4.2 Rodents

The mammalian order Rodentia includes the smallest terrestrial mammals in which directional hearing has been tested. Heffner (1978b) used an approach procedure to determine localization thresholds in the kangaroo rat (*Dipodomys merriani*) for single clicks and for trains of 2 clicks/sec. The thresholds for the smallest azimuthal separations between the click sources that these animals could detect (minimum audible angle, MAA) were 22° and 24°, respectively. In a screening experiment in which two speakers were separated by 60° and placed 5 feet in front of the listening position, kangaroo rats differentiated effectively (85% or better correct) the two sources of tone for frequencies of 250 Hz to 32 kHz, except at 4 kHz (Heffner 1978b).

Another small rodent, the albino rat, was investigated by means of the identification method combined with the conditioned suppression technique. Kelly and Glazier (1978) determined that for 1 click/sec, the mean localization threshold for five rats was approximately 28°. Elsewhere, Masterton, Thompson, Bechtold, and RoBards (1975) conducted a screening test for the locatability of pure tones, 250 Hz to 32 kHz, originating from one of two speakers 60° apart. Rats readily achieved the discrimination except at 8 kHz.

Thus, the findings in these rodents are similar. The screening tests indicated that both species could differentiate fixed, widely separate sources of low and high frequency tones. Presumably, intensive cues served to localize signals belonging to the upper spectral region (16 kHz and 32 kHz), and temporal cues did the same for the lower portion of the spectrum. The inability of the kangaroo rat and the white rat to localize 4 kHz and 8 kHz, respectively, suggests that these frequencies are within the spectral segment where neither cue is effective. Since no minimum audible angle function has been determined for the wide range of tones used in the screening experiment, it is difficult to estimate the effectiveness of temporal and intensive cues in localization by these species. However, the large localization thresholds obtained with clicks suggest that minimum audible angles for tones would be wide in these species.

4.3 Carnivores

Carnivores are well-known predators that would be expected to localize sound particularly well. Species in the canid and feline families have been examined. The red fox (*Vulpes vulpes*), for example, was tested with the approach-to-target method using two fixed speakers 35.5° apart, placed in front of the animal (Isley and Gysel 1975). At low and high frequencies, the discrimination accuracy decreased significantly from better than 90% correct localization achieved at intermediate frequencies. Only about 75% correct localization occurred at 600 Hz and 65% at 300 Hz; at 18 kHz and 34 kHz, no better discrimination than 70% correct was attained.

Another carnivore of particular interest is the domestic dog. Although this animal is widely believed to localize sound well, very little research has been done, and no minimum audible angle function has been reported. According to Heffner (1978b) mongrel dogs can discriminate click sources about 4° apart. Busnel (1978) conducted a screening test in an open field with five speakers uniformly spaced 20° apart and situated 300 m from the dog. He found that dogs discerned easily which speaker was the source of the signal (e.g., footsteps, breaking twigs). In spite of these studies, no extensive investigation of directional hearing in intact dogs have been conducted that would examine the effects of different acoustic signals, of head size, or of pinna configuration. At this time little can be concluded about localization prowess of canids.

The favorite carnivore for auditory research is the domestic cat; so it is not surprising that more extensive measurements have been made on this animal. Recently, Casseday and Neff (1973) used an approach-to-target procedure to determine, on the cat, a minimum audible angle function for frequencies from 250 Hz to 8 kHz. This experiment was not conducted in an anechoic environment. The results, depicted in Fig. 12-3, show that in comparison to human subjects tested in the same apparatus,

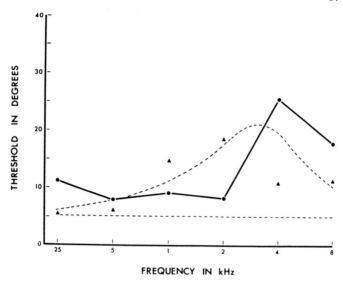

Figure 12-3. Mean localization thresholds for tones in cats, ●, and man, ▲. The dashed horizontal line represents the average localization threshold in cats for wideband noise. The dashed curve is the localization for man obtained by Stevens and Newman (1936) (from Casseday and Neff 1973).

the cat's localization performance was poorer than man's at the highest and lowest frequencies, but was better at 1 kHz and 2 kHz. To the extent that high frequencies were used, cats were not effective localizers. Casseday and Neff noted that cats had more difficulty learning to locate higher frequencies than lower frequencies. It is difficult to know how minimum audible angle thresholds would behave at frequencies higher than 8 kHz since neither binaural sensitivity to intensive differences nor the magnitude of these differences for different azimuths is known. Nevertheless, the similarity in shape of the MAA functions in cat and man (Fig. 12-3) suggests that the duplex theory may be applicable to cats with the distinction that the boundary between time and intensity cues is shifted to higher frequencies than in man.

Casseday and Neff (1973) also reported that in cats, angular thresholds for wideband noise signals were a little over 5° (Fig. 12-3). In a similar experiment, Strominger (1969) showed that the angular threshold for trains of clicks from a 6 volt AC buzzer was 8.4°. As might be expected, some improvement in localization occurred with complex signals. Thus, the cat is a better localizer of sound than rodents; yet, in comparison to man, the cat's overall accuracy is poorer except at two frequencies (Fig. 12-3).

4.4 Marsuipals

In the only study on these neurologically primitive mammals, Ravizza and Masterton (1972) reported that the localization threshold for 1 noise burst/sec in two opossums (*Didelphis virginiana*) was about 5°.

4.5 Primates

Recently, interest has increased in the directional hearing of primates other than man. The research has been conducted primarily on Old World monkeys (*Macaca mulatta* and *Macaca nemestrina*).

In a screening experiment with two speakers fixed 60° apart, Heffner and Masterton (1978) showed that monkeys could correctly locate (90% level) the source of tones ranging from 125 Hz to 27 kHz.

An extensive examination of the minimum audible angle function for Old World monkeys was conducted by Brown, Beecher, Moody, and Stebbins (1978a) in an anechoic room, using an identification method. Their experimental procedure resembled quite closely the one used by Mills (1958) in his meticulous study on humans. The means of the MAA thresholds for the monkeys (*M. mulatta, M. nemestrina*) appear in Fig. 12-4. The smallest detectable angle was about 4° at 1000 Hz. At lower frequencies locatability decreased rapidly, while at higher frequencies it also declined, but not monotonically. (The range of thresholds among animals, not shown in the figure, was quite large at some of the frequencies above 2000 Hz, e.g., at 4000 Hz, it was about 20°).

The MAA function of the monkey is approximately V shaped between 250 Hz and 16000 Hz. The contour of this function does not resemble the one for the only other primate that has been investigated, man, even over the central portions of their respective audible spectra; and clearly, the performance of Macaques is less accurate. Thus, the band of frequencies that are particularly well localized is much narrower than in man. In addition, the shape of the Macaque function does not resemble closely the one for the domestic cat, whose head is comparable in size. Over the middle frequencies (400 Hz to 2000 Hz) the localization accuracy is similar in both species; yet, at the lowest and highest frequencies tested on both animals, the thresholds differ by as much as 5° to 20° (compare Figs. 12-3, 12-4).

Although the shape of the monkeys' MAA function is unusual and does not possess a frequency interval of poor localization as seen in man and cat, it appears that this function is best summarized by the duplex theory (see Section 5.2).

Localization of complex signals has also been examined in Old World monkeys. By means of the identification method, Heffner (1973) found that the MAA for trains of 3 click/sec was about 7°. In a parallel experiment to the one conducted by Brown et al. (1978a), with pure tones, Brown (1976) showed that the MAA for wide noise bands centered at high frequencies were lower than for the center frequencies alone. With a 4 kHz noise band centered at 8 kHz, 11.2 kHz, and 16 kHz, the thresholds were all about 5° and improved localization by 6° to 13° relative to the MAA's for the tones alone.

The locatability of more natural sounds such as Macaque vocalizations were also tested by Brown et al. (1978b). They found that the directionality of wide band vocalizations referred to as "harsh calls" could be discriminated when the calls originated approximately 5° apart.

These findings on monkeys agree with observations made of the cat (Casseday and Neff 1973), that wide band signals yield the greatest localization accuracy for the ani-

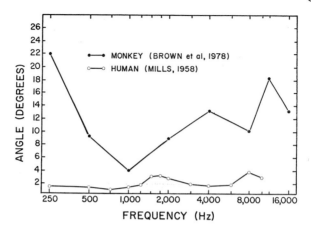

Figure 12-4. Mean localization thresholds for tones in Old World monkeys (*M. mulatta* and *M. nemestrina*) and man (adapted from Brown et al. 1978 and Mills 1958).

mal. It should be noted, however, that the accuracy is approximately the same as the best performance with a tone, e.g., at 1000 Hz in the Macaque, and that with neither tones nor noise does the monkey localize as precisely as man.

5 Lateralization in Terrestrial Mammals

A convenient way to individually examine the effects of some of the cues that contribute to sound localization is by means of lateralization experiments. In animal studies, as with humans, identical (diotic) or different (dichotic) signal pairs are delivered to the ears through earphones. (Human subjects typically report a sound image in the center of the head under diotic conditions and a sound image lateralized toward one ear under dichotic conditions.)

In most procedures, used on mammals, and especially with monkeys, two levers are available to the animal, and reward is given for pressing the lever that is on the same side to which the more intense or the leading member of the binaural signal pair is delivered. During testing, binaural disparities of different magnitudes are presented randomly, and the accuracy of responses is recorded. The binaural difference thresholds can then be estimated from these measurements.

5.1 Interaural Intensity Difference Thresholds

The usefulness of intensity disparities for localization depends on the size of the signal wavelength relative to head size and also on the sensitivity of the binaural system to these differences. Houben and Gourevitch (1976a, 1979) determined interaural intensity difference thresholds in monkeys (*M. nemestrina*) at frequencies ranging from 125 Hz to 8000 Hz. As depicted in Fig. 12-5, the sensitivity of monkeys below 1000 Hz

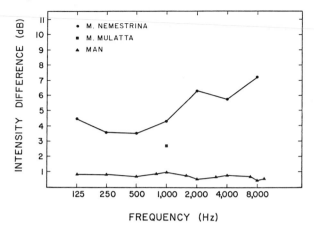

Figure 12-5. Mean interaural intensity difference thresholds in three monkeys (*M. nemestrina*) and man (from Houben and Gourevitch 1979).

is approximately constant at 3.5 dB to 4.5 dB and is about 2.5 dB to 3.5 dB poorer than man's. Above 1000 Hz, the intensity difference function increases and reaches its maximum value of 7 dB at 8000 Hz. Since man's binaural intensive thresholds remain relatively constant above 1000 Hz, the divergence between the two functions is due to the monkey's decreasing sensitivity. In another study Wegener (1974) found that at 1000 Hz, the mean intensity difference threshold for nine monkeys (*M. mulatta*) was 2.7 dB, which agrees well with the measurements by Houben and Gourevitch.

Other lateralization studies in monkeys used clicks. Don and Starr (1972) found that squirrel monkeys (*Samiri sciureus*) were able to detect at an 85% correct level, interaural intensity differences of 6 dB to 10 dB in trains of 32 clicks/sec. (The binaural thresholds ranged between 3 dB and 6 dB for a 75% correct level; however, these values may not be accurate since the animals would not respond consistently at smaller intensity differences.) The only other threshold measurement was made by Heffner (1973) on one monkey (*M. mulatta*) in which the difference threshold for trains of 3 clicks/sec was 2.55 dB.

At best, the click studies indicate that the difference thresholds of monkeys are somewhat higher than the approximately 1.0 dB threshold reported for man (Hall 1964). However, the methodological differences among these studies are too great to establish any firm conclusions.

Lateralization of tones in cats was examined by Wakeford and Robinson (1974a) who reported interaural intensity thresholds for 500 Hz, 1000 Hz, and 3000 Hz tones that were between 1.5 dB and 0.5 dB. In contrast to other investigators, Wakeford and Robinson used shock instead of food in their conditioning procedure; they also required that the animal detect in on-going sequences of dichotic tone bursts an occasional introduction of a three tone burst in which the more intense signal was reversed to the opposite ear. By using shock, and by only requiring detection of a change in stimulus, this study yielded unusually low threshold estimates which, most likely, reflect the methodology rather than exceptional sensitivity in the cat.

5.2 Interaural Time Difference Thresholds

In small headed animals, the maximum interaural time difference can be so short that its effectiveness in localization is questioned. It is of importance, therefore, to examine the sensitivity of animals to this cue.

Detectability of transient time differences by squirrel monkeys (*Samiri sciureus*) was investigated by Don and Starr (1972). They reported that interaurally delayed click pairs presented at a rate of 32 clicks/sec were lateralized with 85% accuracy when the time disparity was between 60 μsec and 180 μsec. (The psychometric functions did not yield dependable time difference estimates for the 75% correct level.) The only data for Macaques show the difference threshold on one rhesus monkey to be 27 μsec for trains of 3 clicks/sec (Heffner 1973).

An indirect measure of the rat's sensitivity to interaural time differences was made by Kelly (1974) in a study on the precedence effect. He estimated, on the basis of "three rats with the most reliable performance" (p. 1283) that the minimum detectable interaural time disparities for clicks was approximately 46 μsec.

With the exception of Heffner's monkey, the above mentioned animals are less sensitive than man whose single click threshold is about 30 μsec (Hall 1964) and whose multiple click threshold (30/sec) is around 11 μsec (Klumpp and Eady 1956).

The cue for localization of low frequency sinusoids is on-going time difference. Sensitivity to this time difference has been examined in monkeys and in cats. The same procedures were used for these measurements as for intensity differences.

Wegener (1974) found that the threshold for on-going time differences in rhesus monkeys was 54 μsec at 1000 Hz, the only tone that was examined. Houben and Gourevitch (1976b, 1979) reported time difference thresholds for tones ranging from 250 Hz to 2000 Hz (Fig. 12-6). At low frequencies the monkeys are quite insensitive to time disparities. The thresholds decrease from a high of 120 μsec at 250 Hz to less

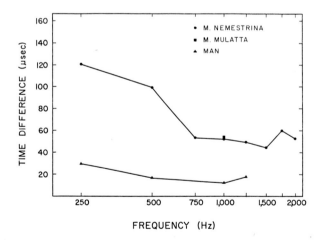

Figure 12-6. Mean interaural time difference thresholds in three monkeys (*M. neme-strina*) and man (from Houben and Gourevitch 1979).

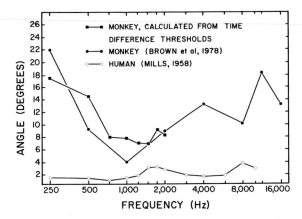

Figure 12-7. Minimum audible angles in monkeys and man (from Houben and Goure-vitch 1979).

than 60 μsec at 750 Hz and remain relatively constant at about 55 μsec beyond 750 Hz. These thresholds are at least 35 μsec greater than those of man (Zwislocki and Feldman 1956) measured at common frequencies (Fig. 12-6).

The correspondence between binaural sensitivity to on-going time differences and localization acuity can be seen in Fig. 12-7. Houben and Gourevitch (1979) calculated the angular separations that would give rise to the observed time difference thresholds and found that this curve is in good agreement with the minimum audible angle function determined directly on monkeys (Brown et al. 1978a). Thus, the mechanism proposed by the duplex theory for directional hearing at the low end of the spectrum, although less sensitive in monkeys, appears to bear some similarity with man's.

Wakeford and Robinson (1974a) examined the sensitivity of cats to on-going time differences at 500 Hz, 1000 Hz, and 2000 Hz and applied the same method and definitions they had used with intensity differences. The time difference threshold at the lowest two frequencies was about 25 μsec and increased to more than 80 μsec at 2000 Hz. The cats were only 10 μsec less sensitive than man at corresponding frequencies. The same methodological problems existed in this case as with measurements of intensity difference thresholds (see Section 5.1), and consequently, time difference thresholds for cats may, in fact, be larger and more in line with those of monkeys.

It is important to note that on-going time difference discriminations occur up to 2000 Hz in both Old World monkeys and in cats. Further indication that this is a characteristic of the binaural system of these animals comes from studies on masking level differences (MLD).

5.3 Masking Level Differences

The MLD phenomenon refers to an improved detectability of signal in noise that appears under certain circumstances of binaural disparity. For example, when identical noises are delivered to two ears and a sinusoid to one ear, recognition of the signal is easier than when both the signal and noise are presented to the two ears. Binaural processes that give rise to MLDs and to localization of time differences are thought to be closely related (Green and Yost 1975; Yost, Chapter 14), and both appear to depend on analysis of low frequency interaural temporal information. In man, MLDs decrease between 500 Hz and 1 500 Hz or so, and become very small or disappear above 1 500 Hz, depending on the conditions of binaural stimulation (Durlach and Colburn 1978). In contrast, MLDs of cats are relatively constant between 500 Hz and 1 500 Hz (Wakeford and Robinson 1974b), and they are greater than man's at 1 500 Hz. These results suggest that cats continue to process MLDs at higher frequencies than man.

Similar results were also found in Macaques. Flammino and Clopton (1975) determined MLDs in monkeys from 500 Hz to 8000 Hz. Between 500 Hz and 4000 Hz MLDs declined while they remained constant above 4000 Hz. In comparison with a human subject tested in the same experimental arrangement, MLDs for monkeys were greater at 1000 Hz and 2000 Hz.

These measurements, as well as those of time difference thresholds, clearly indicate that certain mammals process time disparities at higher frequencies than man. If a small headed mammal were only capable of time difference discriminations up to the same, relatively low frequency limit as man, the intermediate frequency interval noted for ineffective time and intensity differences would be wider in the small mammal; this would occur because, relative to man, the frequency above which intensity differences are effective is, of physical necessity, displaced upward in the smaller organism. On the other hand, if the frequency limit for the time cue is shifted to higher frequencies by about the same increment as the shift of the intensity difference cue, the intermediate frequency range is not enlarged. Thus, extension of sensitivity for the time difference cue toward higher frequencies in small mammals may reflect an adaptation that keeps the interval of poor localization (intermediate frequencies) narrow, and comparable to similar intervals in large animals such as man.

Whether extension of sensitivity to the time difference cue toward spectral regions where intensity differences gain effectiveness is a general characteristic of smaller headed mammals must await results gathered on a greater range of species.

6 Conclusion

To the extent that localization has been sampled in different terrestrial mammalian species, none has shown the same directional hearing accuracy as man. The few available MAA functions are generally well above his, even when determined under the most favorable acoustic and behavioral conditions. This difference between man and other mammals does not appear to depend only on the smaller magnitude of binaural disparities due to the narrower head of the animal. (In species with mobile pinnae the disparities are, presumably, not as small as in those with fixed pinnae.) The difference appears to depend also on the animal's lower sensitivity to these cues.

It should also be noted that the pattern of binaural sensitivity observed in animals is not always advantageous for precise localization. For example, the effect of diminishing sensitivity for intensity differences seen in Old World monkeys between 1000 Hz and 8000 Hz is in the opposite direction to the one that would compensate for smaller intensive disparities that exist across a narrow head relative to a wide head. It remains to be seen whether in larger animals than man, such as elephants, discriminability of interaural differences is also relatively poor. Presumably, they could afford it without great loss in precision of localization.

Finally, one might speculate that the differences in directional hearing seen among mammals including man reflect the relative contribution of the auditory system to that of other sensory modalities in identifying accurately the position of a stimulus in space. Hafter and DeMaio (1975) suggested that the principal role of localization in humans is to establish an approximate direction of the target, which becomes precisely located by vision. In other animals comparable arrangements may exist which, for example, could include additional modalities such as olfaction so that fewer demands would be placed on the auditory system during attempts to find the position of the target. In this case, various levels of auditory localization accuracy would probably exist among the numerous mammalian species, and there would be little selective pressure for exceptionally precise directional hearing. A better understanding of comparative sound localization will require examination of numerous species' capacities to localize different acoustic signals, their sensitivity to particular acoustic cues acting alone and cooperatively, and the behavioral contexts in which the individual species spatially locate targets.

References

Blauert, J.: Sound localization in the median plane. Acustica. 22, 205-213 (1969/70).

Brown, C. H.: Auditory localization in primates: The role of stimulus bandwidth. Doctoral dissertation, The University of Michigan (1976).

Brown, C. H., Beecher, M. D., Moody, D. B., Stebbins, W. C.: Localization of pure tones by Old World monkeys. J. Acoust. Soc. Am. 63, 1484-1492 (1978a).

Brown, C. H., Beecher, M. D., Moody, D. B., Stebbins, W. C.: Localization of primate calls by Old World monkeys. Science. 201, 753-754 (1978b).

Busnel, R. G.: Personal communication (1978).

Casseday, J. H., Neff, W. D.: Localization of pure tones. J. Acoust. Soc. Am. 54, 365-372 (1973).

Don, M., Starr, A.: Lateralization performance of Squirrel monkey (*Samiri sciureus*) to binaural click signals. J. Neurophysiol. 35, 493-500 (1972).

Durlach, N. I., Colburn, H. S.: Binaural phenomena. In: Handbook of Perception, vol. IV, Hearing. Carterette, E. C., Friedman, M. P. (eds.). New York: Academic Press, 1978, pp. 365-466.

Erulkar, S. D.: Comparative aspects of spatial localization of sound. Physiol. Rev. 52, 237-360 (1972).

Fedderson, W. E., Sandel, T. T., Teas, D. C., Jeffress, L. A.: Localization of high frequency tones. J. Acoust. Soc. Am. 29, 988-991 (1957).

Flammino, F., Clopton, B. M.: Binaural interaction in the Macaque monkey. J. Acoust. Soc. Am. 57, S41 (1975).

Green, D. M., Henning, G. B.: Audition. Ann. Rev. Psychol. 20, 105-128 (1969).

Green, D. M., Yost, W. A.: Binaural analysis. In: Handbook of Sensory Physiology, vol. V/2. Keidel, W. D., Neff, W. D. (eds.). New York: Springer-Verlag, 1975, pp. 461-480.

Hall, II, J. L.: Minimum detectable change in the interaural time or intensity difference for brief impulsive stimuli. J. Acoust. Soc. Am. 36, 2411-2413 (1964).

Hafter, E. R., DeMaio, J.: Difference thresholds for interaural delay. J. Acoust. Soc. Am. 57, 181-187 (1975).

Harrison, J. M., Downey, P.: Intensity changes at the ear as a function of the azimuth of a tone source: A comparative study. J. Acoust. Soc. Am. 47, 1509-1518 (1970).

Heffner, H. E.: The effect of auditory cortex ablation on sound localization in the monkey (Macaca mulatta). Doctoral dissertation, Florida State University (1973).

Heffner, H. E.: Personal communication (1978).

Heffner, H., Masterton, B.: Contribution of auditory cortex to hearing in the monkey (Macaca mulatta). In: Recent Advances in Primatology, vol. I/ Behaviour. Chivers, D. J., Herbert, J. (eds.). London: Academic Press, 1978, pp. 735-754.

Henning, G. B.: Detectability of interaural delay in high-frequency complex waveforms. J. Acoust. Soc. Am. 55, 84-90 (1974).

Houben, D., Gourevitch, G.: Sensitivity to interaural intensity differences in monkeys. Bull. Psychon. Soc. 8, 270 (1976a).

Houben, D., Gourevitch, G.: Interaural time difference threshold in pig-tailed monkeys. J. Acoust. Soc. Am. 60, 89 (1976b).

Houben, D., Gourevitch, G.: Auditory lateralization in monkeys: An examination of two cues serving directional hearing. J. Acoust. Soc. Am. 66, 1057-1063 (1979).

Isley, T. E., Gysel, L. W.: Sound localization in the Red Fox. J. Mammalogy. 56, 397-404 (1975).

Kelly, J. B.: Localization of paired sound sources in the rat: small time differences. J. Acoust. Soc. Am. 55, 1277-1284 (1974).

Kelly, J. B., Glazier, S. J.: Auditory cortex lesions and discriminations of spatial location by the rat. Brain Res. 145, 315-321 (1978).

Klumpp, R. G., Eady, H. R.: Some measurements of interaural time difference thresholds. J. Acoust. Soc. Am. 28, 859-860 (1956).

Kuhn, G. F.: Model for the interaural time differences in the azimuthal plane. J. Acoust. Soc. Am. 62, 157-167 (1977).

Masterton, B., Heffner, H. E., Ravizza, R.: The evolution of human hearing. J. Acoust. Soc. Am. 45, 966-985 (1969).

Masterton, B., Thompson, G. C., Bechtold, J. K., RoBards, M. J.: Neuroanatomical basis of binaural phase-difference analysis for sound localization: A comparative study. J. Comp. Physiol. Psychol. 89, 379-386 (1975).

McFadden, D., Pasanen, E. G.: Lateralization at high frequencies based on interaural time differences. J. Acoust. Soc. Am. 59, 634-639 (1976).

Mills, A. W.: On the minimum audible angle. J. Acoust. Soc. Am. 30, 237-246 (1958).

Mills, A. W.: Auditory localization. In: Foundations of Modern Auditory Theory, vol. 2. Tobias, J. V. (ed.). New York: Academic Press, 1972, pp. 303-348.

Neff, W. D., Diamond, I. T., Casseday, J. H.: Behavioral studies of auditory discrimination: central nervous system. In: Handbook of Sensory Physiology, vol. V/2. Keidel, W. D., Neff, W. D. (eds.). New York: Springer-Verlag, 1975, pp. 307-400.

Ravizza, R. J., Masterton, B.: Contribution of neocortex to sound localization in Opossum (Didelphis virginiana). J. Neurophysiol. 35, 344-356 (1972).

Searle, C. L., Braida, L. D., Cuddy, D. R., Davis, M. F.: Binaural pinna disparity: another auditory localization cue. J. Acoust. Soc. Am. 57, 448-455 (1975).

Shaw, E. A. G.: The external ear. In: Handbook of Sensory Physiology, Auditory System, V/1. Keidel, W. D., Neff, W. D. (eds.). New York: Springer-Verlag, 1974, pp. 455-490 (1974).

Stevens, S. S., Newman, E. B.: The localization of actual sources of sound. Amer. J. Psychol. 48, 297-306 (1936).

Strominger, N. L.: Localization of sound in space after unilateral and bilateral ablation of auditory cortex. Exp. Neurol. 25, 521-533 (1969).

Tobias, J. V., Schubert, E. D.: Effective onset duration of auditory stimuli. J. Acoust. Soc. Am. 31, 1595-1605 (1959).

Wakeford, O. S., Robinson, D. E.: Lateralization of tonal stimuli by the cat. J. Acoust. Soc. Am. 55, 649-652 (1974a).

Wakeford, O. S., Robinson, D. E.: Detection of binaurally masked tones by the cat. J. Acoust. Soc. Am. 56, 952-956 (1974b).

Wegener, J. G.: Interaural intensity and phase angle discrimination by Rhesus monkey. J. Sp. Hear. Res. 17, 638-655 (1974).

Woodworth, R. S.: Experimental Psychology. New York: Holt (1938).

Zwislocki, J., Feldman, R. S.: Just noticeable differences in dichotic phase. J. Acoust. Soc. Am. 28, 860-864 (1956).

Chapter 13

Comparative Organization of Mammalian Auditory Cortex

Moïse H. Goldstein, Jr.* and Paul L. Knight**

1 Introduction

A range of species will be discussed here. Domestic cats, however, will receive the most attention since most of the research on auditory cortex has used them as experimental animals.

The auditory cortex shall be construed to be all of the neocortex involved in audition. Defining the scope of the paper so broadly allows the consideration of not only AI, the so-called "primary auditory cortex," but also several cortical fields that lie outside of AI and yet are associated with auditory functions.

The tonotopic organization of the auditory cortex will receive the largest share of attention, primarily because it has been studied carefully in a range of vertebrate species. Binaural representation and the neural encoding of stimulus location in space will not be treated, principally because of the paucity of comparative study of binaural coding in auditory cortex.

We will consider, although briefly, the phylogenetic development of the auditory cortex. In all but the most primitive mammals a specialized AI region can be distinguished (Diamond and Hall 1969). In ascending the phyletic scale, AI and its surrounding secondary belt tend to remain constant while the association areas increase in size.[1] Finally, a few comments will be made about the anatomical and functional assymetries in auditory regions of cerebral cortex.

*Department of Electrical Engineering and Department of Biomedical Engineering, Johns Hopkins University, 720 Rutland Ave., Baltimore, Maryland 21205.
**Health Care Technology Center, University of Missouri, Columbia, Missouri 65201.
[1]The labels "association," "integration," "primary," and "secondary" are used in a purely descriptive sense although they seem to imply function.

2 Functional Localization of Auditory Cortex

2.1 Early Behavioral, Anatomical, and Physiological Observations

The first observations of the location of the cortical projections of the mammalian auditory system were made by David Ferrier (1876). He showed that electrical stimulation of restricted regions of the cortex of various species was followed by behavioral responses indicative of auditory activity. In the cat, stimulation of the middle and posterior ectosylvian gyri (regions marked 14 in Fig. 13-1) was followed by retraction of the ears and turning of the head and eyes to the side of the body contralateral to the stimulated cortex. Ferrier's pioneering work first demonstrated the general location of auditory cortex in the cat and other species. All subsequent studies of auditory function in the cortex have been in general agreement with his observations.

Confirmation of Ferrier's observations of a possibly homologous auditory region in the ectosylvian cortex of the dog came from Munk (1881). He showed that lesions involving part of the dog's ectosylvian cortex caused temporary behavioral deficits in understanding verbal commands.

Just before the turn of the century, Larionow (1899) stimulated the cochlea of the dog using sounds produced by tuning forks and crudely recorded (with a galvanometer) electrical response from the cortical region identified by Ferrier. He suggested that an S-shaped belt of cortex on the middle and posterior ectosylvian gyri was the faithful representation of the unfurled cochlea (Fig. 13-2).

One of the early neuroanatomical observations by Vogt (1900) pointed out a discrete region of early myelinization in the middle ectosylvian gyrus that was interpreted as the location of the primary auditory cortex in the cat. This area was anatomically

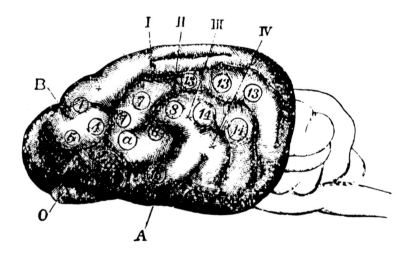

Figure 13-1. Location of the auditory cortex in the cat by Ferrier (1876). Behavioral responses indicative of auditory responses were observed for electrical stimulation of points labeled "14."

similar to the area described as the primary auditory cortex in man (Vogt and Vogt 1919, 1926). Subsequently, Campbell (1905), Winkler and Potter (1914), and Kornmuller (1937) described temporal auditory regions in the cat and other species on the basis of cytoarchitecture. Each recognized a region (koniocortex) in which there was a conspicuous radial orientation of cells, particularly in laminae 2 and 3 (the "rainshower formation" of von Economo and Koskinas 1925).

Campbell also introduced the concept of hierarchical processing in the auditory cortex. He described two auditory cortical regions, a central "primary" acoustic region (ectosylvian A in the cat, "audito-sensory" cortex) receiving projections from the medial geniculate body and a secondary auditory region (ectosylvian B in the cat) surrounding the primary region and subserving "higher" auditory function ("auditopsychic" cortex), as shown in Fig. 13-3. On the basis of cytoarchitecture, he also described a similar organization of the auditory cortex in primates, including man (Fig. 13-4). The central region of koniocortex in man bears a striking resemblance to the central region in other primate auditory cortices (Campbell 1905, von Economo 1929, Beck 1929, Sanides 1972, Pandya and Sanides 1973), which has been shown to be primate AI (Merzenich and Brugge 1973, Imig, Ruggero, Kitzes, Javel, and Brugge 1977).

Technical improvements in electrical recording allowed Kornmuller (1937) to produce a map of cat auditory cortex based on the cortical responses to sound stimulation. He observed surface evoked potentials restricted to the middle ectosylvian gyrus. In their later study of surface evoked potentials, Bremer and Dow (1938) mapped strongly driven auditory responses in the area described by Kornmuller, but

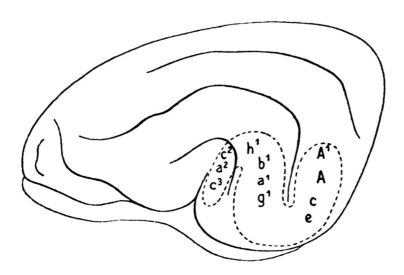

Figure 13-2. Representation of the unfurled cochlea within the auditory cortex of the dog described by Larionow (1899) from galvanometric measurements of cortical responses to tonal stimuli. Letters indicate musical scales of tuning forks used for stimulation.

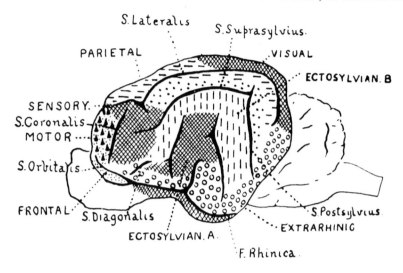

Figure 13-3. Auditory cortex in the cat described by Campbell (1905) on the basis of cytoarchitecture. Region ectosylvian A was considered to be audito-sensory, while the region ectosylvian B was considered to be audito-psychic.

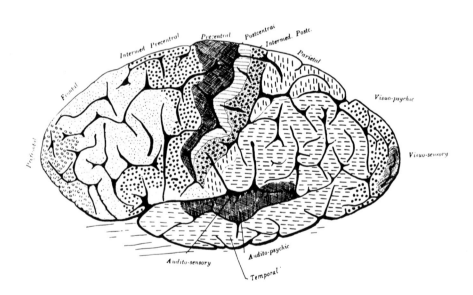

Figure 13-4. Parcellation of the human auditory cortex by Campbell (1905) into a central audito-sensory field and a more peripheral audito-psychic area.

also found weaker responses in surrounding areas. They argued that the auditory cortex may be physiologically divided as indicated by the cytoarchitectonic parcellation.

There have been many differences between the parcellation of auditory cortex by different authors. One reason for these differences was pointed out by Ades (1941) who cautioned that the variability of sulcal patterns in the auditory region of the cat could cause considerable difficulty in generalizing maps of auditory cortex from one animal to another.

Woolsey and Walzl (1942) were the first investigators to demonstrate a topographic representation of restricted sectors of the cochlear partition (basilar membrane) within the auditory cortex of the cat. While electrically stimulating short sectors of eighth nerve fibers within the osseous spiral lamina in the dissected cochlea, they recorded the stimulus-locked evoked potential from a large number of points on the cortical surface of the anterior, middle, and posterior ectosylvian gyri as shown in Fig. 13-5. They interpreted their results as showing double representations of the cochlea within each hemisphere. The dorsal area, AI (Woolsey and Fairman 1946), had a topographic map of the cochlea with the cochlear base represented rostrally near the dorsal tip of the anterior ectosylvian sulcus and the cochlear apex represented caudal to the dorsal tip of the posterior ectosylvian sulcus. Woolsey and Walzl described a second region, AII, immediately ventral to AI, as having a reversed cochlear map with the cochlear base represented caudally on the posterior ectosylvian gyrus and the cochlear apex represented rostrally on the ventral portion of the anterior ectosylvian gyrus.

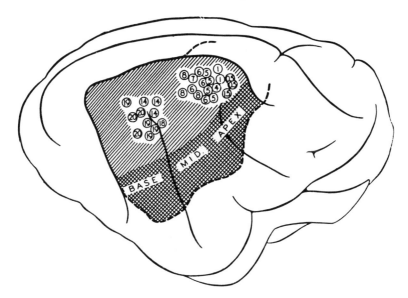

Figure 13-5. Two complete representations of the cochlea in the cat auditory cortex as determined by Woolsey and Walzl (1942). In the dorsal representation, AI, numbers indicate stimulation positions along the cochlea as distance in millimeters from the cochlear base. Regions of cochlear representation are shown in the ventral AII. Note reversed order of the two representations crossing the cortical surface.

Ades (1943) showed that the acoustically responsive area on the posterior ectosylvian gyrus was separate from the rest of Woolsey and Walzl's (1942) AII. He identified AI physiologically by strong responses to click stimuli. When strychnine-treated
patches were placed on AI, causing strong synchronous firing of neural units there
(see electrically as "strychnine spikes"). Electrical responses were observed in the posterior ectosylvian gyrus, demonstrating projections from AI to the area that is now
called Ep.

Rose (1949) was able to divide the auditory cortex of the cat into four regions on
the basis of cytoarchitecture: the dorsal AI (koniocortex); a ventral AII, corresponding to the rostral two-thirds of the AUU described by Woolsey and Walzl (1942) and
Ades (1943); the posterior ectosylvian (Ep); and the suprasylvian fringe (SF), a field
partially exposed on the ventral anterior ectosylvian gyrus but mostly buried on the
ventral bank of the suprasylvian sulcus. In a companion retrograde degeneration study,
Rose and Woolsey (1949) showed that AI received an orderly and essential (i.e.,
degenerating with lesions restricted to AI) projection from the principal division
(Rioch 1929) of the medial geniculate body (MGB). Auditory areas outside of AI were
said to receive sustaining projections from the MGB (i.e., projections that do not
degenerate with lesions restricted to one of these other fields, but that degenerate with
lesions in both AI and surrounding fields).

The notion of a single primary acoustic area hierarchically connected to adjacent
secondary fields in the cat received support from Bremer (1953), Mickle and Ades
(1953), and Downman, Woolsey, and Lende 1960). They showed that Woolsey's AII
and Ep had different response characteristics than AI (e.g., higher thresholds, longer
response latencies) and were dependent on an intact AI for activation.

2.2. Topographic Representations of the Cochlea
in the Auditory Cortex

In recent years attention has turned to study of the mapping of the cochlea on the cortical surface, called cochleotopic mapping. It is also possible to map tone frequency on
the cortical surface, called tonotopic mapping. Since tones at low intensity excite a restricted segment of the cochlea, the two maps are closely related (Greenwood 1961).

2.2.1 Mapping Using Gross Responses

Recording surface evoked potentials from small strychnine-treated patches of the auditory cortex of the dog, Tunturi (1945, 1950a, 1950b, 1952) showed that AI of the
dog has a complete and orderly representation of the cochlea with the apical cochlea
represented caudally and the basal cochlea represented rostrally. His maps showed that
a sector of the cochlear partition was represented by a dorsoventral strip of cortex.
Furthermore, he defined another partial cochlear representation, AIII, located ventrally on the anterior ectosylvian gyrus. Hind (1953), using the same strychnine-patch
technique, determined tuning curves for restricted loci in cat auditory cortex. In his

most extensive map of a single animal, Hind showed a clear caudal-to-rostral progression of higher frequencies within AI and a reversal in the sequence of represented frequencies on the anterior ectosylvian gyrus. However, in his composite map, Hind combined data from several animals to give an overall summary that obscured the reversal.

Woolsey (1960, 1961) evaluated these and other studies in an effort to synthesize a complete map of auditory cortex in the cat (Fig. 13-6). This map shows a central auditory region comprising four central fields plus five more peripheral regions. Within each of the central fields—AI, AII, SF and Ep—Woolsey described a complete and orderly representation of the cochlea. In Fig. 13-6, "A" denotes areas of representation of the cochlear apex, and "B" denotes areas of representation of the cochlear base. Long latency responses (under chloralose anesthesia) to click stimulation were recorded in all association areas except AIII, which was inferred to be auditory from Tunturi's (1950a) work in dogs, but was never clearly demonstrated in the cat. Woolsey's belief in complete representations of the cochlea within each of the central fields led him to draw the low frequency region on the ventral bank of the anterior ectosylvian sulcus as being part of a larger field, the "suprasylvian fringe" (SF), that was continuous with a high frequency area dorsocaudal to AI.

This map has been considered to be the "standard" map of auditory cortex in cats and has been used as a reference to determine the locations of electrode penetrations and ablations within the multiple representations of the cochlea in the cat's auditory cortex. Recent mapping studies (Merzenich, Knight, and Roth 1973, Merzenich, Kaas, and Roth 1976, Knight 1977) indicate that there is some variability in the location of auditory cortical fields among individuals of a given species. However, maps that are typical can be drawn, and we shall see that recent work has led to some revision of Woolsey's map.

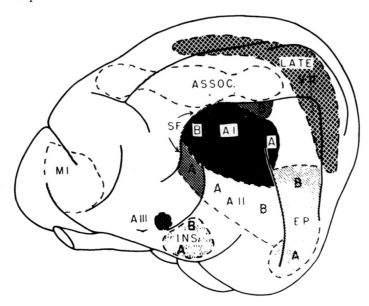

Figure 13-6. Woolsey's (1960) map of auditory regions in the cortex of the cat.

2.2.2 Mapping Using the Single Unit Technique

The descriptions of cochlear representation in the cerebral cortex based on the recording of surface evoked potentials are limited by the resolution of the stimulation and recording techniques used in these earlier studies. Microelectrode mapping techniques
have a finer spatial resolution than evoked potential recording methods and have been
used to determine the representation of the sensory epithelium in visual (e.g., Kaas,
Hall, and Diamond 1970) and somatosensory (e.g., Paul, Merzenich, and Goodman
1972, Welker, Johnson, and Pubols 1964) cortices before being applied to the mapping
of the auditory cortex.

    There are three reasons for the potentially greater resolution of microelectrode
mapping in studies of the auditory cortex.

1. Response properties of single cortical neurons (e.g., latency, tuning curves, response consistency) that may be used to differentiate units in different cortical
   fields are evident in single unit recording but may be obscured in the surfaced
   recorded evoked potentials.
2. The "slow-wave" evoked potentials recorded by gross electrodes from the cortical surface do not, by their nature, have the spatial resolution of single-unit responses.
3. Gliosis and electrolytic lesions produced by microelectrodes can be located in
   tissue sections to give accurate cyto- and myeloarchitectonic correlations with
   the representational boundaries determined physiologically.

    The use of microelectrode mapping for determining the boundaries of cortical
fields and representation of functions within them is potentially far more accurate
than determinations made on the bases of cyto- and/or myeloarchitectonics alone,
techniques that depend on the skillful observation of changes in the patterns of
organization of cells and/or fibers. These changes are generally very subtle, and their
appearance can be distorted by curvature of the cortical surface. Analysis of architectonics alone is, at present, not as quantitative as analysis of microelectrode recordings.
Ideally, cortical fields should be determined from combined physiological/architectonic
studies.

2.2.2.1 *Initial Microelectrode Mapping of Cat AI.* The first applications of microelectrode mapping in auditory cortex of the cat (Hind, Rose, Davies, Woolsey, Benjamin,
Welker, and Thompson 1960, Evans and Whitfield 1964, Evans, Ross, and Whitfield
1965, Goldstein, Abeles, Daly, and McIntosh 1970) did not show the highly-ordered
cochlear representation suggested by the earlier evoked potential studies. While each of
these studies showed a tendency for units of higher best frequency to be located more
rostrally in AI, they did not indicate the precise and orderly mapping observed in more
peripheral loci in the auditory pathway. On the basis of these and other experiments,
spatial cochlear representation has been described as being progressively degraded in
ascending auditory pathways (Clopton, Winfield, and Flammino 1973, Evans 1968,
1974, Evans and Whitfield 1964, Whitfield 1971).

In each of these studies, partial maps with few penetrations were made in each animal. Assuming a consistent relationship between sulcal patterns and the position of AI on the cortical surface, each of these investigators combined data points from several animals to give composite maps of AI. However, it is clear that the sulcal patterns of the cortex are highly variable among individual animals (Ades 1941, 1959, Kawamura 1971, Rose 1949); furthermore the locus of cortical fields in relation to the sulci is variable (Merzenich et al. 1973, 1974, 1975). Thus, considerable error could result from trying to combine data from several animals using sulci as references to obtain composite maps. Extensive maps of AI in individual cats were needed to confirm the complete and orderly representation of the cochlea in that field.

2.2.2.2 *Confirmation of Cochleotopic Organization of Cat AI.* Merzenich et al. (1975) reexamined the representation of the cochlea within AI of individual cats anesthetized with Ketamine. For each animal in this study, extensive, fine-grained maps of single unit and unit cluster best frequencies were made in AI and, to a limited extent, in adjoining fields. Several important conclusions were drawn from this study concerning the pattern of the representation of the cochlea in AI of the cat.

Single units in AI were found to be sharply tuned near threshold allowing unequivocal assignment of a value for best frequency. When recording multiple neurons at a given cortical point, units driven to discharge to the tonal stimuli had close best frequencies. AI was found to be radially[2] organized with units of similar best frequencies being found through the middle and deep cortical layers in penetrations normal to the cortical surface. Similar results had been noted previously by other workers (Abeles and Goldstein 1970, Gerstein and Kiang 1964, Hind et al. 1960, Parker 1965, Oonishi and Katsuki 1965).

In extensive maps of AI, such as that shown in Fig. 13-7, it was demonstrated that there was a complete and orderly representation of the cochlea within AI. The lowest best frequencies recorded were found caudally in AI, and there was an orderly progression to higher best frequencies rostrally. Furthermore, penetrations of similar best frequencies were arranged along a line of dorso-ventral orientation. Long penetration directed down the banks of the posterior ectosylvian sulcus also exhibited an orderly progression of best frequencies (see tracks A, B, C, and D in Fig. 13-7).

Further analysis of the spatial distribution of best frequency in AI revealed that:

1. A given frequency band is represented by a nearly straight belt of cortex crossing AI with a predominantly dorso-ventral orientation.
2. Such a sector must be of nearly constant width across AI.
3. The proportionality of representation of different frequencies must be maintained across AI.
4. There is a proportionately larger representation of higher frequency octaves along the rostro-caudal dimension of AI.

Furthermore, the spatial distribution of frequency representation is similar for AI of

---

[2] A radial electrode track is perpendicular to the cortical surface.

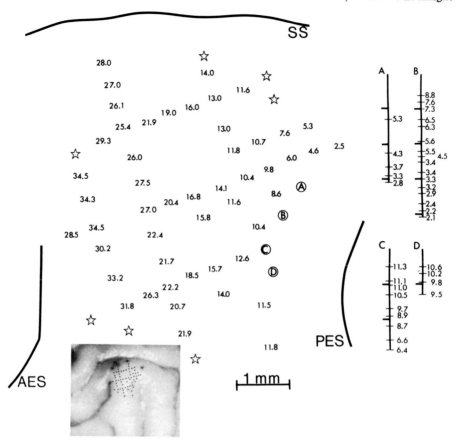

Figure 13-7. Map of best frequencies (kHz) within AI of the cat. Penetration sites are also shown on the inset photograph. Penetrations down the bank of the posterior ecto-sylvian sulcus (PES) are shown on the right. Broad bars indicate cortical depth in mm. Narrow bars indicate recording positions. Stars indicate penetrations outside AI (from Merzenich, Knight, and Roth 1975).

most cats; however, the relationship of this representation to the sulcal pattern varies from cat to cat.

Features of the cochleotopic organization of AI can be determined by conversion of best frequencies to represented cochlear positions using the function of Greenwood (1961, 1974). The total areas of representation of best frequency bands and cochlear sectors in AI in the cat are shown in Fig. 13-8. From this figure, it can be appreciated that there is a proportionately larger total area devoted to the representation of pro-gressively higher octaves, or equivalently, more basal equal-length sectors.

In comparing the orderly mapping seen by Merzenich et al. (1975) and the less orderly tonotopic representation reported in earlier studies (Evans et al. 1965, Gold-stein et al. 1970), a number of differences in methodology should be noted, as discussed

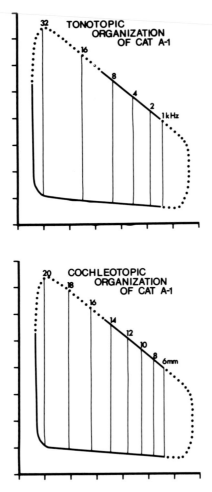

Figure 13-8. Tonotopic (upper graph) and cochleotopic (lower graph) organizations of AI in the cat. Upper graph shows the approximate area of representation of frequency bands (kHz) in AI. Lower graph shows the approximate area of representation of cochlear sectors (in mm from cochlear apex) in AI. Axes represent mm across the cortical surface (from Merzenich, Knight, and Roth 1975).

in Merzenich et al. (1975) and Goldstein and Abeles (1975). These include use of recording from clusters of units versus exclusive use of single-unit recordings, and the anesthetic state of the animals. When unanesthetized preparations were used, units were found in layers II-VI; but in the anesthetized preparations, most recordings were from layer IV.

2.2.2.3 *Anterior Auditory Field (AAF) in Cat.* Strongly-driven responses to tonal stimuli can be recorded in some of the "belt" fields adjacent to AI. One of these fields is located immediately rostral to AI. To determine the pattern of representation of the cochlea within this rostral field, fine-grained microelectrode maps of units' best frequencies were made within it (Knight 1977). In order to avoid implying homology with the rostral field in primates, this field is renamed by Knight (1977) the "anterior auditory field" (AAF). Results from one experiment are shown in Fig. 13-9. The AAF is anterior to the dashed line.

Important features of the AAF's organization are the following:

1. Most units in AAF have sharp tuning curves and unequivocal best frequencies at thresholds of stimulation similar to that in AI of the same animal.
2. In the ketamine-anesthetized cat, units of similar best frequency are found to be radially-aligned in the middle cortical layers. As with AI, units radially aligned tend to have close best frequencies.
3. There is a highly-ordered representation of the cochlea within AAF, where a restricted portion of the cochlear partition is represented by a nearly straight belt of cortex crossing the entire field.

The cochleotopic organizations of AAF and AI are shown schematically in Fig. 13-10. Comparison of the properties of AAF and AI show that these two fields are remarkably similar in many important features including unit response properties, short latency to earliest unit discharge, radial organization in depth, spatial representation of

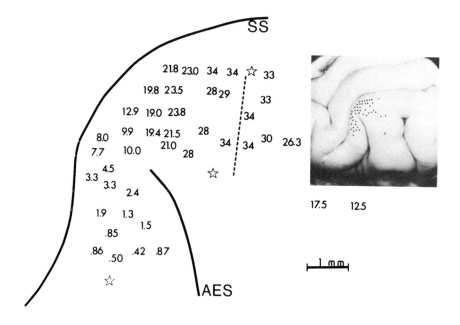

Figure 13-9. Map of best frequencies (kHz) of 44 penetrations into AAF and AI of the left hemisphere of the cat 75-67 (anterior is to the left). Stars indicate penetrations in which no cells responded to tones. Penetration sites are shown on the inset photograph. Abbreviations: SS-suprasylvian sulcus; AES-anterior ectosylvian sulcus. Dashed line indicates boundary between AAF (left) and AI (right) (from Knight 1977).

frequency, and a proportionately greater representation of the higher frequency octaves.

Neuroanatomical experiments undertaken to ascertain the origin of the thalamocortical and corticothalamic projections to and from AI (Colwell and Merzenich 1975) and AAF (Andersen, Patterson, Knight, Crandall, and Merzenich 1977) used the technique of injection of tracers actively transported by neurons (horseradish peroxidase and tritiated amino acids) into cortical sites with physiologically defined location. These experiments revealed that AAF and AI share some inputs from the thalamus and project reciprocally back to the thalmus.

These similarities, particularly the similar short latency to discharge and the presence of some common thalamic inputs, suggest that AAF is not a "secondary" cortical field. On the contrary, it suggests that AAF and AI are virtually mirror images of one another and may be coparticipants in the earliest and fundamental processing of acoustic information at the cortical level in the cat. Other fields surrounding AI in the cat may also have similar properties (Reale and Imig 1977). These observations suggest that the concept of a strictly hierarchical organization of the auditory cortex, i.e., a single "primary" field, AI, that processes information before the adjacent fields and passes process information to them, may require some modification.

2.2.2.4 *Mapping in Species Other than Cat.* The technique of fine-grained microelectrode mapping has been used to determine the cochlear representations on the superior temporal plane of two species of macaques, *Macaca mulatta* and *Macaca arctoides* (Merzenich and Brugge 1973). By mapping the best frequencies of neurons found in a large number of closely spaced penetrations into auditory cortex, they were able to

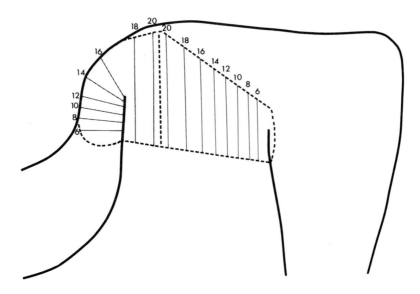

Figure 13-10. A view of AAF and AI. Solid lines indicate the approximate length and orientation of the lines of representation of cochlear position indicated by numbers representing mm from the cochlear apex (from Knight 1977).

demonstrate the presence of at least five distinct auditory representations on the superior temporal plane. The single unit technique allowed more detailed mapping than earlier evoked potential studies (Kennedy 1955, Licklider and Kryter 1942, Walzl 1947, Woolsey 1943, Woolsey and Walzl 1944, Bailey, von Bonin, Garol, and McCulloch 1943, Woolsey 1971, Pribram, Rosner, and Rosenblith 1954).

Merzenich and Brugge (1973) found that the auditory cortex of these monkeys includes the following fields: AI, a field of sharply tuned units having a complete and orderly representation of the cochlea (see Fig. 13-11); the caudomedial field (CM) with broadly-tuned units; the rostro-lateral field (RL) with an orderly but only partial representation of the apical cochlea; the lateral field (L), apparently containing a complete cochlear representation; and one or two adjacent fields that were not completely mapped.

Similar microelectrode mapping studies have been carried out on the auditory cortex of the owl monkey, *Actus trivirgatus* (Imig et al. 1977). Imig et al. demonstrated complete and orderly representation of sharply tuned neurons within AI and the rostral field (R) as well as auditory responses in the surrounding fields, the caudomedial (CM), the anterolateral (AL), and the posterolateral (PL).

In the grey squirrel, *Sciurus carolinensis,* Merzenich, Kaas, and Roth (1976) found that there was a central AI field of sharply-tuned units with a complete and orderly representation of the cochlea. Surrounding AI, they found an active belt of cortex in which neurons were sharply tuned as well as a more peripheral belt of less responsive

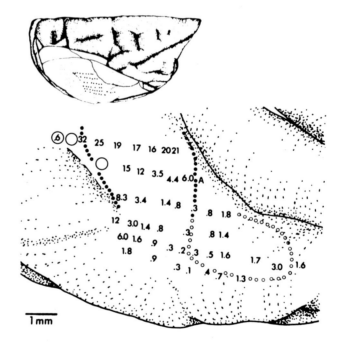

Figure 13-11. Map of best frequencies recorded in three cortical auditory fields (AI, RL, and L) in a single *Macaca arctoides* by Merzenich and Brugge (1973). Closed circles denote boundary of AI; open circles denote boundary of the rostrolateral field (RL).

neurons. Similarly, a strictly tonotopic (cochleotopic) organization has been shown within AI of the tree shrew (*Tupaia glis*) using microelectrode mapping techniques (Oliver, Merzenich, Roth, Hall, and Kaas 1976).

Two highly responsive auditory cortical fields have been mapped in the guinea pig (*Cavia cobaya*) (Hellweg, Koch, and Vollrath 1977). These two fields, "area I" and "area II" have reversed frequency progressions crossing the cortical surface in the rostrocaudal direction, and the boundary between them appears to be the high frequency representation in both fields. This organization is similar to that of AI and the anterior field in the cat (Knight 1977).

A notable exception to this pattern of cochlear representation in AI is found in the echo-locating mustache bat (*Pteronotus parnellii rubiginosus*). While most of AI appears to be similar to that in other species, there are two disproportionately large high frequency representations. One specialized region is thought to subserve processing of constant-frequency echo location information (61000 Hz to 63000 Hz) and the other for processing frequency-modulated sounds (50000 Hz to 60000 Hz) (Suga 1977, Suga and Jen 1976, Manabe, Suga, and Ostwald 1978).

2.2.2.5 Comparison of Cochlear Representation in AI for Three Species. Merzenich et al. (1976) compared the representations of the cochlea in AI of the cat (*Felis catus*), monkey (*Macaca mulatta*), and grey squirrel (*Sciurus carolinensis*) (as shown in Fig. 13-12). In Fig. 13-12D the cat and grey squirrel dimensions were scaled upward, as

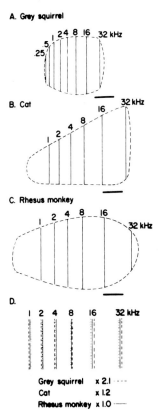

Figure 13-12. Tonotopic organization of primary auditory cortex in the squirrel (A), cat (B), and rhesus monkey (C). Isofrequency contours for frequencies one octave apart are represented by thin vertical lines in each of the three species. The boundaries and shapes of the fields are highly schematized. The bar below each drawing is one millimeter in length. The orientation of the field of the cat has been reversed in the drawings, so it can be compared with the field organization in the monkey and squirrel. In (D), the octave frequency-band strips are scaled in size so that the proportional representation of different frequencies in the three species can be directly compared (from Merzenich et al. 1976).

indicated by the figures below the graph. From this figure, it is apparent that with appropriate scaling, the spacings of the iso-frequency contours for these species match closely.

3 AI and Beyond: Phylogenetic and Functional Considerations

The primary auditory cortex has received by far the most attention in single-unit studies. To bring some perspective to this chapter, something will be said in this section about the belt or secondary area and more will be said about the regions called the auditory association area or auditory integration cortex.

3.1 Stability of AI

The most primitive mammals, such as hedgehogs or opossums, seem to lack a primary auditory cortex. Instead, at the site of the projection of the medial geniculate nucleus, one finds a poorly laminated cortex with reduced granularization rather than koniocortex (Sanides 1967). However, studies on rodents and other animals demonstrate the phylogenetically early appearance of primary sensory regions characterized by koniocortex. Diamond and Hall (1969) propose that the belt areas "may have come first in evolution and may have given rise to both the primary cortex and the association cortex found in advanced animals."

The great stability of the primary cortex across quite diverse mammalian species is noteworthy. In the foregoing, the studies of Merzenich et al. (1976) were cited showing clearly the similarity of the cortices of grey squirrel, cat, and monkey (Fig. 13-12).

It seems that the early appearance of AI is accompanied by the appearance of the secondary areas, arranged in a belt-like fashion around AI. The belt region, with the exception of the anterior auditory field, AAF, in cats (Knight 1977) has not been studied in great detail. There is not enough comparative information to say to what extent the belt areas remain unchanged across species.

3.2 Phylogenetic Development of Auditory Neocortex

The secondary belt area seems to be phylogenetically more ancient than the more highly differentiated primary area, AI (Diamond and Hall 1969, Sanides 1970). Sanides refers to growth rings of the neocortex. In this scheme the first growth ring, periallocortex, is a common derivation of both older cortices, achicortex, and paleocortex. The second growth ring along with the first is called association or integrative cortex. The third ring is the belt area and the fourth is the primary cortex (Sanides 1970).

Since each successive growth ring is derived from the previous one, it is not surprising to find the auditory cortex organized with the primary area, AI, surrounded by the belt, and with the integration cortex constituting the next neocortical rings (Sanides 1975). The newest cortex, AI, is characterized by extreme granularization in layer IV, anatomic columns of cells, and heavy myelination. Functionally, the newest cortex is the most selectively auditory, and the integration cortex, although capable of great selectivity, seems to be polysensory.

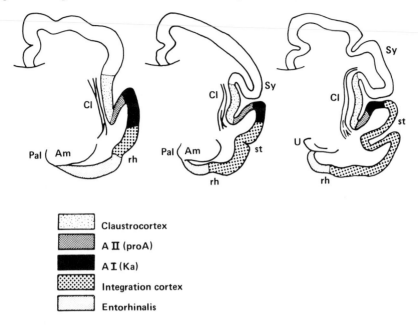

Claustrocortex

A II (proA)

A I (Ka)

Integration cortex

Entorhinalis

Figure 13-13. Stages of the temporal lobe and temporal operculum in primates as seen in coronal diagrams from prosimian (*Loris gracilis*) (left), through small New World monkey (*Saimiri sciureus*) (middle), to Old World monkey (*Macaca mulatta*) (right). Note the partly exposed claustrocortex in the slow loris, and the shift of AI into the developing supratemporal plane in the rhesus monkey, correlated with the expanding integration cortex. The medial belt area of the auditory region AII (pro A) persists in parinsular location; the lateral belt area (par A) is not delineated in the figure. Abbreviations: Am-amygdaloid nucleus; Cl-claustrum; Pal-palecorticoid; rh-rhinal sulcus; st-superior temporal sulcus; Sy-Sylvian fissure; U-uncus. The diagrams are not to scale (from Sanides 1975).

3.2.1 Growth of Association Areas (Integrative Cortex)

Earlier was stressed the stability of the primary auditory cortex (AI) and its belt in all but the most primitive mammalian species. Here two points will be made: first, progress of mammalian species along the phylogenetic scale is characterized by an increase in neocortex; and second, the part of the neocortex that expands the most by far is the association or integrative cortex. The terms association area and integrative cortex are used interchangeably here and are defined solely by cytoarchitectonic criteria. Functional characterization will have to await the appropriate set of physiological experiments.

It is interesting that the most recently developed mammalian species are characterized by expansion of the most primitive part of their neocortex: a region bounded by claustrocortex and entrorhinal cortex. The result of this expansion is the formation of the temporal lobe that occurs only in primates. Figure 13-13 illustrates further expansion within primates as seen in coronal diagrams of a prosimian (left), small New World monkey (middle), and Old World monkey (right) (Sanides 1975).

Stephan and Andy (1969) present an interesting quantitative comparative neuro-anatomy of primates. They point out that all primates and most mammals have their phylogenetic origin in insectivore-like ancestors. The recent representatives of these primitive forms—including tenrecs, hedgehogs, and shrews—are called basal insectivores. A normalized "progression index" was obtained for different brain structures. The index gives the ratio of the volume of some part of the brain in a given animal to the volume of the same part in basal insectivores, taking into account the differences in body size.

The progression indices for the neocortex of primates of interest here are: man = 156; chimpanzee = 58; rhesus monkey = 46; tree shrew = 7.7 (Stephan and Andy 1969). It is evident that this great increase in neocortex represents expansion of the in-tegration cortex since primary and belt sensory cortices remain constant. It is note-worthy that the index in man is over twice that of the nearest nonhuman primate.

The relative growth of the neocortex far exceeds the growth of other brain struc-tures. The next highest progression indices are for striatum, with the index for man equal to about sixteen and one-half. Nonhuman primates have indices for striatum up to almost fourteen.

3.2.2 Lateralization of Brain Function

There is no doubt that there is asymmetry of functional organization of the two hemi-spheres of the human brain. Evidence may be found in anatomy (Geschwind and Levitsky 1968, Wada, 1977, Rasmussen and Milner 1977), psychophysics (Kimura 1967), and studies of patients with commissurotomies (Gazzaniga and Sperry 1967). On the other hand, behavioral testing of split-brain monkeys (Hamilton 1977) and anatomical studies (Yeni-Komshian and Benson 1976) give no indication of asym-metry in that species.

In the neurological clinic, the elegant work of Penfield and Roberts (1959), in which the cortical surface was stimulated directly in patients prior to brain surgery, gave direct evidence of left hemisphere localization of language. The patients were con-scious and awake allowing the researchers to learn a great deal from their verbal reports.

Another technique used with patients who are to undergo brain surgery has yielded even more direct information about cerebral dominance. This is the intracarotid injec-tion of sodium Amytal in such a way that only one hemisphere or the other would be anesthetized (Wada and Rasmussen 1960). A recent study (Rasmussen and Milner 1977) of patients without clinical evidence of early damage to the left cerebral hemi-sphere yielded the following results: in 140 right-handed patients, 134 (96%) had speech lateralized left, and 6 (4%) had speech lateralized right; in 122 left-handed patients, 86 (70%) had speech lateralized left, 18 (15%) had speech lateralized right, and 18 (15%) had bilateralized speech.

Examination of 100 adult brains shows 65 to have larger planum temporale on the left, 11 to have larger planum temporale on the right, with equality in 24 (Geschwind and Levitsky 1968). The planum is adjacent to the transverse gyrus of Heschl (primary auditory cortex) and, thus, should be considered as part of the auditory belt region.

This study was followed by one (Yeni-Komshian and Benson 1976) in which 25 specimens each of human, chimpanzee, and rhesus monkey brains were examined. The measure was a comparison of the length of the Sylvian fissure on the left and the right sides. Both the human and chimpanzee specimens had statistically significant longer Sylvian fissures in the left hemispheres. In the monkey specimens, there was no statistically significant asymmetry. The study is of interest in that there are reports of some degree of language learning in chimpanzee (Gardner and Gardner 1969, Premack 1971, Rumbaugh and Gill 1976) but no similar successes with monkeys. It seems clear that the asymmetry in human brains is inborn. The planum asymmetries are present and visible in the twentieth week of gestational age and can be measured objectively at the twenty-ninth week (Wada 1977).

Although the general correspondence of functional and anatomical asymmetry is reassuring, the fit is not complete (Witelson 1977). From neurological studies, one would estimate a 90% to 95% left hemisphere representation of speech; yet, in the anatomical studies, a smaller portion of specimens exhibit larger left planum than right. Also, attempts to teach chimpanzees a vocalization-based language have not been successful. Sign language has been used. However, the asymmetry demonstrated anatomically was for the belt region of the auditory cortex.

Compared to the available knowledge of primary auditory cortex, knowledge of the association or integration cortex is rudimentary. It is also evident that integration cortex has many subregions with different functions. Although it is difficult to study functionally, the integration cortex presents one of the most exciting challenges in brain research. How anatomical and neurological approaches have increased understanding of the integration cortex has been indicated briefly. Two other methodologies are used in studies of this part of the mammalian brain. One is ablation coupled with behavioral techniques, the other is single-unit recording from a behaving animal. One may hope that over the next decade or two the association or integration cortex will emerge from the realm of brain regions which are poorly understood.

Acknowledgments. We would like to acknowledge support from the National Science Foundation and from the Mellon Foundation. Our colleagues in the Neural Encoding Laboratory made many helpful comments and suggestions. We are especially indebted to Dennis Benson.

References

Abeles, M., Goldstein, M. H., Jr.: Functional architecture in cat auditory cortex: columnar organization and organization according to depth. J. Neurophysiol. 33, 172-187 (1970).

Ades, H. W.: Connections of the medial geniculate body in the cat. Arch. Neurol. Psychiat. 45, 138-144 (1941).

Ades, H. W.: A secondary acoustic area in the cerebral cortex of the cat. J. Neurophysiol. 6, 59-63 (1943).

Ades, H. W.: Central auditory mechanisms. In: Handbook of Physiology, Section 1, Vol. 1, Neurophysiology. Washington: Amer. Physiolog. Soc., pp. 585-716, 1959.

Andersen, R. A., Patterson, H., Knight, P., Crandall, B., Merzenich, M. M.: Thalamocortical, corticothalamic and corticotectal projections to and from physiologically identified loci within the auditory cortical fields: AAF, AII and AI. Soc. for Neurosci. Abstracts. 3, No. 2 (1977).

Bailey, P., von Bonin, G., Garol, H. W., McCulloch, W. S.: Functional organization of temporal lobe of monkey (*Macaca mulatta*) and chimpanzee (*Pan satyrus*). J. Neurophysiol. 6, 121-128 (1943).

Beck, E.: Die myeloarchitektonische Bau des in der Sylvischen Furche gelegenen Teiles des Schlafenlappens beim Schimpansen (Troglodytes niger). J. Psychol. Neurol. (Leipzig). 38, 309-420 (1929).

Bremer, F.: Some Problems in Neurophysiology. London: Athlone Press, 1953.

Bremer, F., Dow, R. S.: The cerebral acoustic area of the cat. A combined oscillographic and cytoarchitectonic study. J. Neurophysiol. 2, 308-318 (1939).

Campbell, A. W.: Histological Studies on the Localization of Cerebral Function. Cambridge: Cambridge University Press, 1905, 360 pp.

Clopton, B. M., Winfield, J. A., Flammino, F. J.: Tonotopic organization over levels of the auditory system. J. Acoust. Soc. Amer. 54, 284 (1973).

Colwell, S. A., Merzenich, M. M.: Organization of the thalamocortical and corticothalamic projections to and from physiologically defined loci within primary auditory cortex of the cat. Anat. Rec. 181, 336 (1975).

Diamond, I. T., Hall, W. C.: Evolution of neocortex. Science. 164, 251-262 (1969).

Downman, C. B. B., Woolsey, C. N., Lende, R. A.: Auditory areas I, II, and EP: cochlear representation, afferent paths and interactions. Bull. Johns Hopkins Hosp. 106, 127-142 (1960).

Evans, E. F.: Cortical representation. In: Ciba Foundation Symposium on Hearing Mechanisms in Vertebrates. de Reuck, A. V. S., Knight, J. (eds.). London: Churchill, 1968.

Evans, E. F.: Neural processes for the detection of acoustic patterns and for sound localization. In: The Neurosciences. Schmidt, F. O., Worden, F. G. (eds.). Cambridge, Mass.: The MIT Press, 1974.

Evans, E. F., Ross, H. F., Whitfield, I. C.: The spatial distribution of unit characteristic frequency in the primary auditory cortex of the cat. J. Physiol. (London). 179, 238-247 (1965).

Evans, E. F., Whitfield, I. C.: Classification of unit responses in the auditory cortex of the unanesthetized and unrestrained cat. J. Physiol. (London). 171, 476-495 (1964).

Ferrier, D.: The Functions of the Brain. London: Smith, Elder and Co., 1876, pp. 138-171.

Gardner, R. A., Gardner, B. T.: Teaching sign language to a chimpanzee. Science. 165, 664-672 (1969).

Gazzaniga, M. S., Sperry, R. W.: Language after section of the cerebral commissures. Brain Res. 90, 131-148 (1967).

Gerstein, G. L., Kiang, N. Y. S.: Responses of single units in the auditory cortex. Exp. Neurol. 10, 1-18 (1964).

Geschwind, N., Levitsky, W.: Human brain: left-right asymmetries in temporal speech region. Science. 161, 186-187 (1968).

Goldstein, M. H., Jr., Abeles, M.: Note on tonotopic organization of primary auditory cortex in the cat. Brain Res. 100, 188-191 (1975).

Goldstein, M. H., Jr., Abeles, M., Daly, R. L., McIntosh, J.: Functional architecture in cat primary auditory cortex: tonotopic organization. J. Neurophysiol. 33, 188-197 (1970).

Greenwood, D. D.: Critical bandwidth and the frequency coordinates of the basilar membrane. J. Acoust. Soc. Amer. 33, 1344-1356 (1961).

Greenwood, D. D.: Critical bandwidth in man and in some other species in relation to the travelling wave envelope. In: Sensation and Measurement. Moskowitz, H. R., Scharf, B., Stevens, J. C. (eds.). Dordrecht: D. Reidel, 1974, pp. 231-239.

Hamilton, C. R.: An assessment of hemispheric specialization in monkeys. Ann. N. Y. Acad. Sci. 299, 222-232 (1977).

Hellweg, F. C., Koch, R., Vollrath, M.: Representation of the cochlea in the neocortex of guinea pig. Exp. Brain Res. 29, 467-474 (1977).

Hind, J. E.: An electrophysiological determination of tonotopic organization in auditory cortex of the cat. J. Neurophysiol. 16, 475-489 (1953).

Hind, J. E., Rose, J. E., Davies, P. W., Woolsey, C. N., Benjamin, R. M., Welker, W., Thompson, R. F.: Unit activity in the auditory cortex. In: Neural Mechanisms of the Auditory and Vestibular Systems. Rasmussen, C., Windle, W. (eds.). Springfield, Ill.: Charles C. Thomas, 1960, pp. 201-210.

Imig, T. J., Ruggero, M. A., Kitzes, L. M., Javel, E., Brugge, J. F.: Organization of auditory cortex in the owl monkey (*Aotus trivirgatus*). J. Comp. Neurol. 171, 111-128 (1977).

Kaas, J. H., Hall, W. C., Diamond, I. T.: Cortical visual areas I and II in the hedgehog: relation between evoked potential maps and cytoarchitectonic subdivisions. J. Neurophysiol. 33, 595-615 (1970).

Kawamura, K.: Variations of the cerebral sulci in the cat. Acta Anat. 80, 204-221 (1971).

Kennedy, T. T. K.: An electrophysiological study of the auditory projection on areas of the cortex in monkey (*Macaca mulatta*). (Thesis) Chicago: University of Chicago (1955).

Kimura, D.: Functional asymmetry of the brain in dichotic listening. Cortex. 3, 163-178 (1967).

Knight, P. L.: Representation of the cochlea within the anterior auditory field (AAF) of the cat. Brain Res. 130, 447-467 (1977).

Kornmuller, A. E.: Die bioelektrischen Erscheinungen der Hirnrindfelder. Leipzig: G. Thieme, 1937, 118 p.

Larionow, W.: Ueber die musicalischen Centren des Gehirns. Arch. ges. Physiol. 76, 608-625 (1899).

Licklider, J. C. R., Kryter, K. D.: Frequency localization in the auditory cortex of the monkey. Fed. Proc. 1, 51 (1942).

Manabe, T., Suga, N., Ostwald, J.: Aural representation in the Doppler-shifted-CF processing area of the auditory cortex of the mustache bat. Science. 200, 339-342 (1978).

Merzenich, M. M., Brugge, J. F.: Representation of the cochlear partition on the superior temporal plane of the macaque monkey. Brain Res. 50, 275-296 (1973).

Merzenich, M. M., Knight, P. L., Roth, G. L.: Cochleotopic organization of primary auditory cortex in the cat. Brain Res. 63, 343-346 (1973).

Merzenich, M. M., Knight, P. L., Roth, G. L.: Orderly representation of the cochlea within primary auditory cortex in the cat. J. Acoust. Soc. Amer. 55, 86 (1974).

Merzenich, M. M., Knight, P. L., Roth, G. L.: Representation of cochlea within primary auditory cortex in the cat. J. Neurophysiol. 38, 231-249 (1975).

Merzenich, M. M., Kaas, J. H., Roth, G. L.: Auditory cortex in the grey squirrel: tonotopic organization and architectonic fields. J. Comp. Neurol. 166, 387-402 (1976).

Mickle, W. A., Ades, H. W.: Spread of evoked cortical potentials. J. Neurophysiol.

16, 608-633 (1953).

Munk, H.: On the functions of the cerebral cortex. Translated by G. von Bonin. In: The Cerebral Cortex. von Bonin, G. (ed.). Springfield, Ill.: Charles C Thomas, 1881, pp. 107-108.

Oliver, D. L., Merzenich, M. M., Roth, G. L., Hall, W. C., Kaas, J. H.: Tonotopic organization and connections of primary auditory cortex in the tree shrew, *Tupaia glis*. Anat. Rec. 184, 491 (1976).

Oonishi, S., Katsuki, Y.: Functional organization and integrative mechanisms of the auditory cortex of the cat. Jap. J. Physiol. 15, 342-365 (1965).

Pandya, D. N., Sanides, F.: Architectonic parcellation of the temporal operculum in rhesus monkey and its projection pattern. Z. Anat. Entwickl. Gesch. 139, 127-161 (1973).

Parker, D. E.: Vertical organization of the auditory cortex of the cat. J. Audit. Res. 2, 99-124 (1965).

Paul, R. L., Merzenich, M. M., Goodman, H.: Representation of slowly and rapidly adapting cutaneous mechanoreceptors of the hand in Broadmann's areas 3 and 1 of *Macaca mulatta*. Brain Res. 36, 229-249 (1972).

Penfield, W., Roberts, L.: Speech and Brain-Mechanisms. Princteon: Princeton University Press, 1959.

Premack, D.: Language in chimpanzee? Science. 172, 808-822 (1971).

Pribram, K. H., Rosner, B. S., Rosenblith, W. A.: Electrical responses to acoustic clicks in monkey: extent of neocortex activated. J. Neurophysiol. 17, 336-344 (1954).

Rasmussen, T., Milner, B.: The role of early left-brain injury in determining lateralization of cerebral speech functions. Diamond, S. J., Bilzard, D. A. (eds.), Ann. N. Y. Acad. Sci. 299, 1977, pp. 355-369.

Reale, R. A., Imig, T. J.: An orderly frequency representation in the posterior ectosylvian sulcus of the cat. Soc. for Neurosci. Abstracts. 3, No. 2 (1977).

Rioch, D.: Studies on the diencephalon of carnivora. I. The nuclear configuration of the thalamus, epithalamus, and hypothalamus of the dog and cat. J. Comp. Neurol. 49, 1-120 (1929).

Rose, J. E.: The cellular structure of the auditory region of the cat. J. Comp. Neurol. 91, 409-440 (1949).

Rose, J. E., Woolsey, C. N.: The relations of the thalamic connections, cellular structure and evocable electrical activity in the auditory region of the cat. J. Comp. Neurol. 91, 441-466 (1949).

Rumbaugh, D. M., Gill, T. V.: The mastery of language-type skills by the chimpanzee (*Pan*). In: Origins of Evolution of Language and Speech. Ann. N. Y. Acad. Sci. 280, 562-578 (1976).

Sanides, F.: Comparative architectonics of the neocortex of mammals and their evolutionary interpretation. Ann. N. Y. Acad. Sciences. 167, 404-423 (1967).

Sanides, F.: Functional architecture of motor and sensory cortices in primates in the light of a new concept of neocortex evolution. In: The Primate Brain. Noback, N., Montagna, W. (eds.). New York: Appleton-Century Croft (1970).

Sanides, F.: Representation in the cerebral cortex and its areal lamination pattern. In: The Structure and Function of the Nervous Tissue. Bourne, G. H. (ed.). Vol. 5. New York: Academic Press, 1972, pp. 329-453.

Sanides, F.: Comparative neurology of the temporal lobe in primates including man with reference to speech. Brain and Language. 2, 396-419 (1975).

Stephan, H., Andy, O. J.: Quantitative comparative neuroanatomy of primates: an attempt at a phylogenetic interpretation. Ann. N. Y. Acad. Sci. 167, 370-387 (1969).

Suga, N.: Amplitude spectrum representation in the Doppler-shifted-CF processing area of the auditory cortex of the mustache bat. Science. 196, 67-67 (1977).

Suga, N., Jen, P. H. S.: Disproportionate tonotopic representation for processing CF-CM sonar signals in the mustache bat auditory cortex. Science. 194, 542-544 (1976).

Tunturi, A. R.: Further afferent connections of the acoustic cortex of the dog. Amer. J. Physiol. 144, 389-394 (1945).

Tunturi, A. R.: Physiological determination of the boundary of the acoustic area in the cerebral cortex of the dog. Amer. J. Physiol. 160, 395-401 (1950a).

Tunturi, A. R.: Physiological determination of the arrangement of the afferent connections to the middle ectosylvian area in the dog. Amer. J. Physiol. 162, 489-502 (1950b).

Tunturi, A. R.: A difference in the representation of auditory signals for the left and right ears in the isofrequency contours of the right middle ectosylvian cortex of the dog. Amer. J. Physiol. 168, 712-727 (1952).

Vogt, C.: Étude sur la myelinisation des hemispheres cerebraux. Paris: Steinheil, 1900.

Vogt, C., Vogt, O.: Allgemeinere Ergebnisse unserer Hirnforschung. J. Psychol. Neurol. 25, 279-461 (1919).

Vogt, C., Vogt, O.: Die vergleichend-architektonische und die vergleichend-reizphysiologische Felderung der Grosshirnrinde unter besonderer Berücksichtigung der menschlichen. Die Naturwissenschaften. 14, 1190-1194 (1926).

von Economo, C.: The Cytoarchitectonics of the Human Cerebral Cortex. London: Oxford Medical Publications, 1929.

von Economo, C., Koskinas, G. M.: Die Cytoarchitektonic der Hirnrinde der erwachsenen Menschen. Berlin: J. Springer, 1925, pp. 810.

Wada, J. A.: Pre-language and fundamental asymmetry of the infant brain. Ann. N. Y. Acad. Sci. 299, 370-379 (1977).

Wada, J. A., Rasmussen, T.: Intracarotid injection of sodium Amytal for the lateralization of cerebral speech dominance. J. Neurosurg. 17, 266-282 (1960).

Walzl, E. M.: Representation of the cochlea in the cerebral cortex. Laryngoscope. 57, 778-787 (1947).

Welker, W. I., Johnson, J. I., Jr., Pubols, B. H., Jr.: Some morphological and physiological characteristics of the somatic sensory system in raccoons. Amer. Zoologist. 4, 75-94 (1964).

Whitfield, I. C.: Auditory cortex: tonal, temporal or topical? In: Physiology of the Auditory System. Sachs, M. B. (ed.). Baltimore: National Educational Consultants, 1971, pp. 289-298.

Winkler, C., Potter, A.: An Anatomical Guide to Experimental Research on the Cat's Brain. Amsterdam: Versluys, 1914.

Witelson, S. F.: Early hemisphere specialization and interhemisphere plasticity; an empirical and theoretical review. In: Language Development and Neurological Theory. Segalowitz, S., Gruber, F. (eds.). New York: Academic Press, 1977.

Woolsey, C. N.: "Second" somatic receiving areas in the cerebral cortex of cat, dog and monkey. Fed. Proc. 2, 55-56 (1943).

Woolsey, C. N.: Organization of cortical auditory system: a review and a synthesis. In: Neural Mechanisms of the Auditory and Vestibular System. Rasmussen, G.,

Windle, W. (eds.). Springfield, Ill.: Charles C Thomas, 1960, pp. 165-180.

Woolsey, C. N.: Organization of cortical auditory system. In: Sensory Communication. Rosenblith, W. A. (ed.). Cambridge, Mass.: The MIT Press, 1961, pp. 235-257.

Woolsey, C. N.: Tonotopic organization of the auditory cortex. In: Physiology of the Auditory System. Sachs, M. B. (ed.). Baltimore: National Educational Consultants, 1971.

Woolsey, C. N., Fairman, D.: Contralateral, ipsilateral and bilateral representation of cutaneous receptors in somatic areas I and II of the cerebral cortex of pig, sheep and other mammals. Surgery. 19, 684-702 (1946).

Woolsey, C. N., Walzl, E. M.: Topical projection of nerve fibers from local regions of the cochlea to the cerebral cortex of the cat. Bull. Johns Hopkins Hosp. 71, 315-344 (1942).

Woolsey, C. N., Walzl, E. M.: Topical projection of the cochlea to the cerebral cortex of the monkey. Amer. J. Med. Sci. 207, 685-686 (1944).

Yeni-Komshian, G. H., Benson, D. A.: Anatomical study of cerebral asymmetry in the temporal lobe of humans, chimpanzees and Rhesus monkeys. Science. 192, 387-389 (1976).

Chapter 14

Man as Mammal: Psychoacoustics

WILLIAM A. YOST*

1 Introduction

Phenomena such as critical bands, temporal integration, and lateralization have been studied extensively in man. The procedures used to study these phenomena and the data that have been obtained form a significant part of the foundation of mammalian psychoacoustical theory. It is not surprising, then, that these phenomena have been investigated in other animals in order to compare modes of auditory processing across species and to add to our knowledge of auditory functioning.

These capacities are quite often studied after more basic behavioral data have been obtained from an animal. That is, once an audiogram and some indication of the animal's ability to discriminate have been determined, many investigators have gone on to consider how acoustic information impinging on the animal might be organized and processed. Spectral processing usually involves the notion of frequency selectivity. Processing in the time domain is of equal importance, and the animal's ability to locate sounds in space is an important behavioral aspect of the organization of acoustic information.

Critical bands form a major point of departure for the classical theories of frequency selectivity. Temporal integration provides a pivotal concept for our classical theories of frequency selectivity. Temporal integration provides a pivotal concept for our classical theories of temporal encoding. Studies of lateralization have added significantly to the classical "duplex theory" of localization. However, in recent years new research findings have suggested some major renovation of the classical ideas on frequency selectivity, temporal encoding, and localization. These new findings relate not only to theories of hearing in man, but also to theories of hearing in other mammals and in vertebrates generally.

Along with the new data and interpretations have come new stimuli and psychophysical procedures that by themselves should prove interesting to investigators of

*Parmly Hearing Institute, Loyola University of Chicago, 6525 North Sheridan Road, Chicago, Illinois 60626.

comparative auditory processing. In addition, the new research on frequency selectivity, temporal encoding, and lateralization have brought the psychoacousticians and the auditory physiologist many steps closer together. Therefore, investigators and students of comparative auditory processing might find relevant a discussion of some of these recent psychoacoustical developments in frequency selectivity, temporal encoding, and lateralization. In addition to describing the new theoretical implications, some attention will be given to the procedures and stimuli used in these studies.

2 Frequency Selectivity

For years auditory scientists have considered the possibility that neural sharpening or some type of neural nonlinearity plays a significant role in determining man's, and most mammal's, frequency selectivity ability. Differences in the shapes of tuning curves at different levels of the nervous system and between the mechanical activity of the organ of corti and neural-tuning could be explained if there were such neural sharpening. Also, critical bandwidths as measured in psychophysical experiments appear much broader than the widths of the functions obtained from most single-unit tuning curves. Two-tone inhibition as investigated initially by Sachs and Kiang (1968) seemed to indicate a type of inhibitory mechanism that might provide a basis for a neural sharpening mechanism. However, research suggesting that this type of inhibition takes place at the mechanical level (see Legouix, Remand, and Greenbaum 1973, Rhode 1977) complicated the neural site hypothesis for this type of nonlinearity.

Analogies to lateral inhibition in the visual system (Cornsweet 1970, von Békésy 1967) were tempting as a mechanism for neural sharpening, although the supporting anatomical network appears missing in the mammalian cochlea. In 1969 Carterette, Friedman, and Lowell studied the masking effect of a narrowband, computer-generated noise with very steep slopes. They had hoped that this noise would produce a steep gradient of activity in the nervous system much like a sharp change in light intensity (an edge) produces a steep gradient of activity in the visual nervous system. In vision, such a gradient may enhance the perception of edges and lead to the phenomena of Mach bands. Carterette et al. felt their data indicated such an enhancement of spectral edges for auditory masking of sinusoids by the steeply filtered noise. However, research by Houtgast (1972) and Rainbolt and Small (1972) suggested that the Carterette et al. results were due more to the characteristics of the noise and psychophysical method than to auditory neural sharpening.

In his approach to the question of lateral inhibition in the auditory system, Houtgast (1972) made the important argument that the effect of inhibition cannot be measured in simultaneous masking conditions. He reasoned that if there is inhibition, then the signal would inhibit the masker as well as the masker inhibiting the signal, and since masking is measured as a signal-to-masker ratio, the effect of the inhibition would be cancelled. Houtgast went on to suggest that temporally separating the signal and masker might allow one to measure inhibition since the mutual inhibition of signal on masker and masker on signal might not occur under these conditions. Since this observation, many investigators have shown, for a variety of stimuli and in a variety of temporal paradigms, results that suggest there is a type of suppression or inhibition that could be a basis for neural sharpening.

A basic experiment is the two-tone forward masking, suppression condition. The paper by Shannon (1975) provides an excellent example of the procedure. A 1000-Hz, 500-m tone is used as a masker. The signal, also a 1000-Hz tone, follows the masker by some time (let us say, 5 m). The forward masked threshold of the 1000-Hz signal is measured as a baseline condition. A second sinusoid is then added to the 1000-Hz masker. The second sinusoid is the suppression sinusoid. That is, for some suppression-tone frequency, the suppression sinusoid might inhibit the masking effect of the 1000-Hz signal should be easier to detect (its threshold is lower), than when the masker contained only the 1000-Hz tone. Thus, the frequency of the suppression tone is varied, and the threshold of the signal is measured. Data such as those shown in Fig. 14-1 are the results of such an experiment. As the results show, when the 1000-Hz masker is combined with a suppression tone having a frequency of between 1 050 and 1 300 Hz, the signal has a lower thresnold than when the masker is only the 1000-Hz tone. This suggests that frequencies in the region of 1 050 Hz to 1 300 Hz inhibit the masking effectiveness of the 1000-Hz masker.

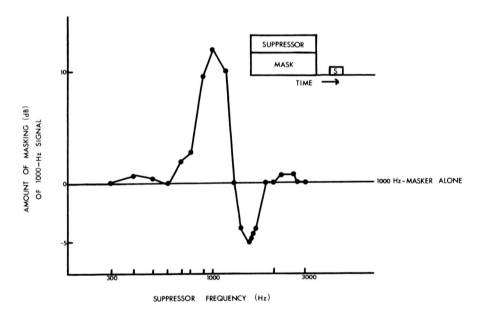

Figure 14-1. The amount of masking is shown for a 1000-Hz signal (S) that is masked by a 1000-Hz masker along with a suppressor tone that is added to the masker at different frequencies. The suppressor tone is also 20 dB greater in intensity than the masker. The solid line at 0 dB represents the amount of masking of the 1000-Hz signal by the 1000-Hz masker. The suppressor frequency region between 1 050 Hz and 1 300 Hz shows the region where the suppressor tone inhibits the masker tone. The inset displays the temporal masking procedure used by Shannon (1975) to obtain these data.

If the assumptions of Houtgast concerning temporal and simultaneous masking are correct, then no evidence of inhibition should exist for a simultaneous masking condition. Fig. 14-2 compares the data of Fig. 14-1 with those of a condition where the stimuli are the same as previously explained for the forward masking procedure, except the signal and maskers are both on at the same time. As can be seen, there is no evidence for suppression in the simultaneous masking condition. The research to date suggests that both the suppression and the excitation areas of these weighting functions are one-sixth of an octave wide. Thus, the entire weighting function is approximately one-third octave wide, which is approximately the width of the traditional critical band when measured in simultaneous masking experiments (see Scharf 1970).

Results similar to those of Fig. 14-2 have been obtained by different investigators (Houtgast 1972, 1973, 1977, Abbas and Sachs 1976, Terry and Moore 1977, Weber and Green 1978) and appear to strongly suggest that the release from masking obtained when the suppression tone is present is due to some type of inhibitory process. Houtgast (1973) has used another type of procedure to temporally separate the masker and signal. In his procedure called "pulsation threshold," there is a repeating train of pulsed maskers. Temporally interleaved between the maskers is a repeating train of pulsed signals. As the intensity of the signal is decreased, the observer reports that the signal begins to appear to be continuous instead of alternating. The intensity at which

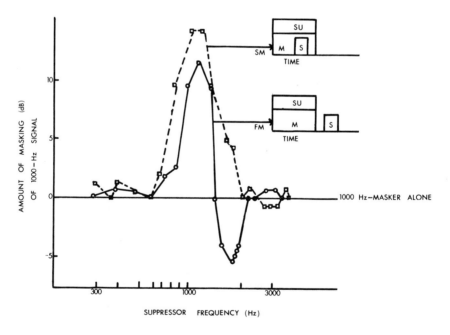

Figure 14-2. The same display as shown in Fig. 14-1. The solid curve is from Fig. 14-1 from the forward masking experiment (FM). The dashed curve represents data from the simultaneous masking procedure (SM), where no suppression is obtained. Abbreviations: SU-suppressor; S-signal. The insets show the temporal alignment of the stimuli used by Shannon (1975) to obtain these data.

this change from "alternating" to "continuous" takes place is called the pulsation-threshold for the signal and is used to study inhibition with conditions similar to those used for Fig. 14-2. Thus, the suppression effect appears to exist for a variety of procedures in which the signal and masker are not presented simultaneously.

If in the forward masking or pulsation threshold experiment there is not mutual inhibition between the masker and the signal, then the masking of one frequency by another should appear without the mutual inhibition. This being the case, the tuning-curves or critical bands obtained in a forward masking experiment should be narrower than those obtained in simultaneous masking, since only the excitation area is measured in the forward masking condition. The data shown in Fig. 14-3 are a comparison obtained by Moore (1978) between simultaneous and forward masking psychophysical tuning curves for otherwise identical stimulus conditions. As can be seen, the forward masking curves indicate much narrower bandwidths or sharper tuning than the simultaneous masking condition. This is further evidence for the notion of suppression or inhibition in the mechanisms responsible for frequency selectivity.

The data of Fig 14-3 were obtained using the psychophysical tuning-curve technique. It is assumed that the procedure is analogous to measuring the tuning curve of a single auditory nerve fiber (see Wightman, McGee, and Kramer 1977). In the psychophysical procedure, the signal is very low in intensity and is fixed in frequency (analogous to finding a fiber with a characteristic frequency (CF) and using a threshold measure of neural activity), while the masker is varied in frequency and its level is adjusted to obtain thresholds (analogous to varying the input frequency and adjusting its

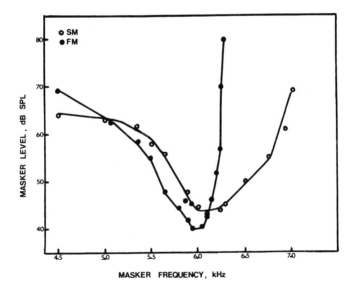

Figure 14-3. A comparison of the masking of a 6000-Hz signal by maskers of different frequencies obtained in forward masking (FM) and simultaneous masking (SM) conditions. These data, using the psychophysical tuning curve technique, show that narrower tuning is obtained in forward masking than in simultaneous masking (from Moore 1978).

level to obtain a neural tuning curve). The results, such as those shown in Fig. 14-3, indicate that the psychophysical forward-masking tuning curves closely resemble neural tuning curves (Moore 1978, Houtgast 1973, Vogten 1978).

Let us assume that the nervous system processes the input stimulus in such a way that both excitation and inhibition operate. That is, the input is not just filtered by a critical band filter, but the contribution of certain high frequencies (in some cases also low frequencies) inhibit the neural activity. This assumption leads to some interesting predictions for complex stimuli. Houtgast (1974, 1977) used a complex stimulus known either as combed-filtered noise, cosine noise, or ripple noise. It is generated by the network shown in Fig. 14-4 and produces a spectrum whose power varies cosinusoidally with frequency (also shown in Fig. 14-4). Houtgast used this noise as a masker for a 1000-Hz tone and varied the number of peaks in the spectrum between D. C. and 1000 Hz by varying the delay, T. Figure 14-5 shows the predicted masked thresholds as a function of changing T, if the noise were weighted by a typical critical band filter (top panel) and by a weighting function such as that shown in Fig. 14-1 (bottom panel of Fig. 14-5). The predictions are quite different for small values of T. The results from Houtgast's (1977) masking experiment are shown in Fig. 14-6. The top panel shows simultaneous masking results and the bottom panel shows forward masking results. As can be seen, these masking data yield results consistent with the assumption that only the forward masking experiment will unveil the effects of inhibition. The results also help support the notion of inhibition as discussed here.

Ripple noise has other interesting properties. One, is that it produces a pitch called repetition pitch which is similar, if not identical, to residue pitch (pitch of the missing fundamental) (for a review see Yost, Hill, and Perez-Falcon 1978). The stimulus is of further interest since a broadband source that is delayed by an echo or reflection and combined with the direct source produces the cosine spectrum of ripple noise. Not only has this provided insights into localization in man (Blauert 1971, Bilsen and Ritsma 1969/70) but it might also be important for echolocation in other animals (Johnson and Titlebaum 1974). Ripple noise is, therefore, an excellent example of the type of complex stimuli psychoacousticians have been using in recent years.

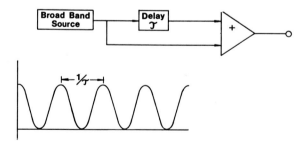

Figure 14-4. A schematic diagram of the network used to generate ripple noise and the resulting power spectrum. A broad band source is delayed by an amount T and added back to the undelayed noise. The resulting power spectrum varies as a cosine function with the spacing between the peaks in the spectrum equal to 1/T.

The recent work described suggests that classical theory of frequency selectivity based on the critical band, might have to be restructured. That is, weight functions with areas of suppression might modify the concept of a simple band-pass filter as the "critical band filter." However, there are some aspects of the experiments discussed and some additional studies that indicate a need to be careful in moving too quickly to accept the concept of suppression. Moore (1978) and Terry and Moore (1977) have indicated that other aspects of the stimulus condition of forward masking, in addition to detecting the presence or absence of the signal, might yield the obtained masking functions. That is, there is a pitch change from masker frequency to signal frequency and a duration change in the masker plus signal condition that might provide cues for discriminating a difference between conditions of a masker plus signal and a masker only. Widin and Viemeister (1978) have pointed out that in simultaneous masking, a

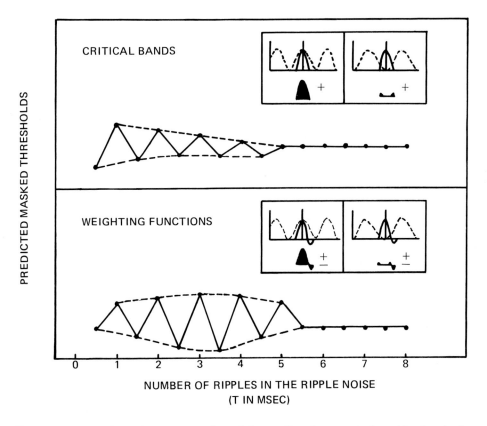

Figure 14-5. A schematic representation of the predicted amount of masking by ripple noise that would be obtained if a typical critical band filter (top panel) or a weighting function with high frequency suppression (bottom panel) were used. The insets show that the ripple noise spectra (dotted lines) are weighted by the critical bands or weighting functions (solid lines) producing an output (solid areas). The magnitude of the outputs ("+" areas, minus "−" areas) form the predicted masked thresholds for a sinusoid masked by ripple noise.

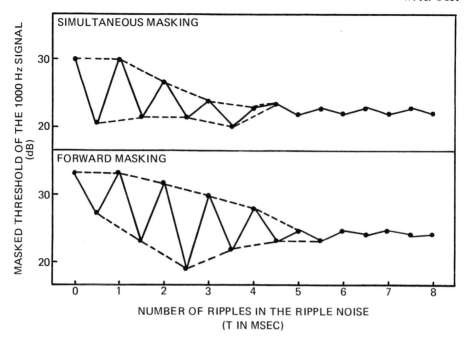

Figure 14-6. The actual masked thresholds of a 1000-Hz signal using ripple noise maskers produced with various values of T. These data show that masking in simultaneous masking is similar to the predicted critical band filtering shown in Fig. 14-5, and masking in forward masking is similar to that predicted by a weighting function as was shown in Fig. 14-5 (from Houtgast 1977).

one-decibel change in the masker causes a one-decibel change in signal threshold. This is not the case in forward masking; a decibel change in the masker level may only cause the signal to change by half a decibel. Widin and Viemeister argue that this type of nonlinearity or compression might underlie the differences between simultaneous and forward masking. Tyler and Small (1977), Weber and Green (1978), and Yost and Bergert (1979) have shown that results obtained in a backward masking paradigm (signal occurring before the masker) are very different than those obtained in forward masking. Among other things, tuning is much broader in backward masking than in forward or simultaneous masking. If simple separation between masker and signal is all that is necessary to measure suppression, the results are unexpected. Thus, the question of suppression affecting frequency selectivity is currently receiving much attention. The new data and procedures are greatly increasing the knowledge of auditory processing and are posing some interesting hypotheses in light of classical notions of the critical band and frequency selectivity.

There is great variety in the auditory anatomy of the mechanical transduction and hydro-mechanical systems across species of animals. The conjecture is that these differences may lead to differences in the type and degree of spectral analysis performed by each species. The forward masking procedures described could provide interesting tests for the various notions of frequency analysis when applied to some of these animals. That is, if some animals appear to encode information largely temporally, rather than spectrally, is there still evidence for suppression? If there is, then this raises interesting questions for the type of mechanisms thought responsible for suppression. If certain animals have poorly developed auditory peripheries in terms of auditory frequency analysis, might the influence of suppression help in developing analysis capability?

3 Temporal Integration

Since the early 1940s with experiments such as that by Munson (1947), the concept of a simple temporal integrator with some fixed period of integration has dominated much of the temporal processing literature. The early measures, usually based on detection of signals of different duration, indicated that the integration time of the system was between 100 msec and 300 msec. Green (1971, 1973) has suggested that this estimate is a long integration time. That is, for detection of a 100 μs click in the middle of a pulsed noise a 100 msec integration time would lead to poorer performance than is actually obtained. Thus, Green (1971, 1973) proposed that for many auditory tasks there is a minimum integration time. Thus, for some tasks a temporal weighting function with a long integration time might be appropriate (e.g., detection of tones of different durations), whereas a short integration time would be required for other tasks (e.g., discrimination between clicks with different temporal gaps).

In attempting to describe temporal acuity, Green (1971) argued that measures of the minimum integration time are important. Most stimulus conditions one might use to study the minimum integration time, probably would produce a confounding between temporal and spectral changes. That is, the temporal changes (e.g., shortening the duration of a stimulus) would generate spectral changes (e.g., energy splatter). Thus, the trick to estimating the minimum integration time is to find stimuli whose spectra (long-term power spectra) remain unchanged as some temporal property is varied.

Patterson and Green (1970) used a stimulus called a Huffman sequence, which appears to meet this criterion. A Huffman sequence is a stimulus generated by exciting an all-pass filter with one click and terminating the filter's response with a second click. An all pass-filter generates a flat amplitude spectra, but a varying phase spectrum. The result is that the Huffman sequence produces a flat amplitude spectrum with 360° jumps in the phase spectrum at spectral locations that are determined by the tuning of the filter. A sample spectrum is shown in Fig. 14-7. Patterson and Green showed that subjects could discriminate between pairs of Huffman sequences when each stimulus was only 1 msec to 3 msec long and differed only in the spectral location of the phase shift. The change in the phase shift in some spectral location means that at some time during the duration of the stimulus, there is a change in the spectrum where the energy is located. That is, there is a short-term energy spectrum (energy spectrum computed during the duration of the signal) difference between the two brief Huffman sequences. Or, in other words, during the 1 msec to 3 msec signal the energy in one stimulus might shift from a high to a low frequency in 1.5 msec and at 2.5 msec for the other stimulus. Since subjects can make this discrimination with stimuli of 1 msec

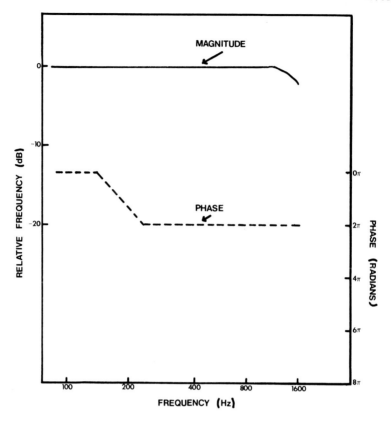

Figure 14-7. The magnitude and phase spectra of a Huffman sequence. The amplitude response is flat out to 1000 Hz, while there is a 2π radian phase change in the region of 150 Hz to 200 Hz.

to 3 msec duration, this implies that minimum integration time is 1 msec to 3 msec.

Weir and Green(1975) suggested that if subjects could discriminate between Huffman sequences with 1 msec to 3 msec durations, then a simple change in the intensity of a sinusoid might be discriminable when the duration of the sinusoid was only 1 msec to 3 msec. In this experiment, the subjects were presented with two sinusoids each with duration T. In one signal, the intensity increased by 10 dB at T/2 and in the other stimulus the intensity decreased by 10 dB at T/2. The stimuli were constructed such that they each had the same long-term power spectrum (thus, they only differed in their short-term energy spectra). The subjects were asked to discriminate between these pairs as T was varied. The results are shown in Fig. 14-8 as the percent correct discrimination versus total duration, T. As can be seen, threshold performance can be obtained between 1 msec to 3 msec. The psychometric function is nonmonotonic, however, and Weir and Green suggested that the first branch of the function might represent aspects of mechanism closer to the minimum temporal integration times and the second branch mechanisms closer to longer temporal integration times. Thus, the

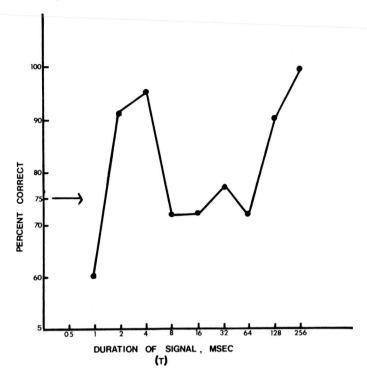

Figure 14-8. Percent correct versus the duration (T) of a tone whose intensity changes by 10 dB at T/2. These data show that threshold performance (percent correct equal 75) is reached at both 1 msec and 128 msec, which might indicate different integration times (from Weir and Green 1975).

results indicated that the minimum integration time is approximately 1 msec to 3 msec.

Viemeister (1977) and Rodenburg (1977) have also studied temporal acuity. Not only did Viemeister hope to measure the minimal integration time, but he hoped to use a linear system's approach in generating a temporal transfer function. Such a temporal transfer function would then enable one to make predictions about other temporal phenomena. His motivation for the experimental and theoretical approach came from vision (Cornsweet 1970) where temporal modulation transfer functions have proved fruitful in attempting to describe visual temporal phenomena.

Viemeister's basic stimulus is a wide-band noise that is amplitude modulated (AM) sinusoidally. The main parameter of his experiment is the modulation frequency. Since the noise is broad band, the AM modulation does not alter the long term power spectra. In his initial experiment, Viemeister asked subjects to discriminate between the AM noise and an unmodulated noise. The modulation depth (peak to valley distance in the AM envelope of the temporally changing noise) of the AM noise was decreased until the observers were at threshold performance. The average results from three subjects are shown in Fig. 14-9. The data are plotted as 20 log m (modulation depth)

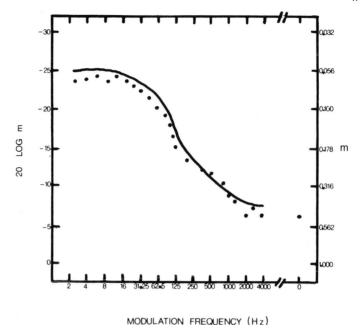

MODULATION FREQUENCY (Hz)

Figure 14-9. The modulation depth (M) of AM modulated noise required to discriminate AM noise from temporally flat noise, is shown as a function of the sinusoidal
modulation frequency. The curve is derived by assuming an integrating system with a
time constant of 3 msec (from Viemeister 1977).

versus modulation frequency. The results show that as the modulation frequency increased, the observers required greater modulation depth to discriminate between the
AM noise and the temporally flat noise. Viemeister argued that if one assumes a
system in which the stimulus is first bandpassed filtered and then integrated, then a
3 msec integration time to this integrator would yield the curve shown in Fig. 14-9.
Thus, a time constant of 3 msec for the integrator seemed capable of accounting for
these results. In addition, the results represent a temporal transfer function of the
auditory system.

   Viemeister then argued that a complete description of the transfer function would
require phase information. He also suggested that if the function in Fig. 14-9 was at
least the amplitude spectrum of the temporal transfer function, then he should be able
to use it to predict other temporal results. In order to study both ideas, Viemeister
used the AM modulated noise as a masker for a short duration click (100 μsec) signal.
The click stimulus were presented at different times relative to the onset of the AM
envelope. The masked threshold of the click was then obtained as a function of the
temporal location (phase) of the click within the AM envelope for different modulation
frequencies. Results such as those shown in Fig. 14-10 (top) were obtained. As can be
seen, the masked thresholds follow the AM envelope modulations. The functions fit to
the data are the best fitting sinusoidal functions. The amplitude and phase of these
best fitting functions were used to describe the amplitude and phase characteristics of

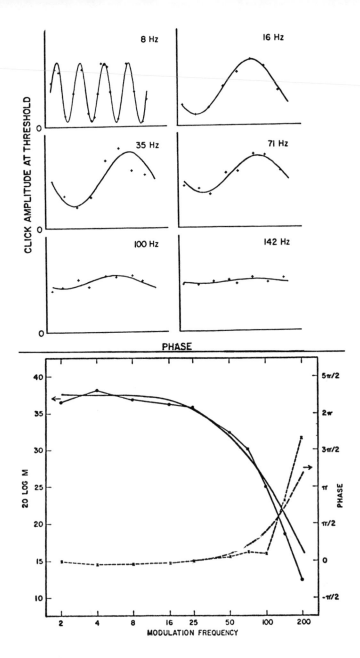

Figure 14-10. In the top figure, the masked thresholds for clicks masked by the AM modulated noise are shown. The data points are the masked thresholds of clicks presented at different times relative to the onset of the envelope of the AM modulated noise. The curves are the best fitting sinusoids to the data (from Viemeister 1977). In the bottom figure, the estimates and predictions of the temporal modulation transfer functions are shown. The data points (connected by straight lines) represent the amplitude and starting phases of the best fitting sinusoids obtained from the data shown above. The curves represent the amplitude and phase characteristics of the temporal modulation transfer function with a time constant of 3 msec, which was used to fit the data of Fig. 14-9.

the theoretical temporal modulation transfer function. These data are also shown in Fig. 14-10 (bottom). The lines without data points represent the phase and amplitudes of the assumed integration system used for Fig. 14-9. The data at the bottom of Fig. 14-10 thus provides a description of the temporal transfer function for the human auditory system and, as such, could be used to predict and account for other temporal phenomena. Viemeister is presently extending this idea to other temporal stimuli and procedures.

The concept of an integration time has been used by many investigators (Robinson and Pollack 1971, 1973, Penner, Robinson, and Green 1972) to attempt to account for the time course of masking in backward, forward, and other types of temporal masking procedures. The temporal integration concept is particularly challenged when a signal is temporally flanked by both a forward and a backward masker. In these combined forward-backward masking studies, there is often considerably more masking when both maskers are present than when either of the single (forward or backward) maskers are present. An example of this is shown in Fig. 14-11. In this experiment (Yost and Bergert 1979) the masker (or maskers) was a 1000-Hz, 500-msec tone and the signal a 1000-Hz, 20-msec tone. As can be seen when both maskers are present (FBM), there is considerably more masking (up to 15 dB) than when either masker is present alone. Simple integrators with fixed time constants are not able to predict these combined forward-backward masking data. In a series of studies, Penner (for a review see Penner 1978) has argued that not only is the integration time an important variable to account for these and other temporal data but the type of integrator that is assumed is critical. Thus, the concept of simple integration with a fixed time constant of between 100 msec to 200 msec to explain temporal phenomena appears no longer to be valid.

The work of Green, Viemeister, Penner, and others has shown that there is a lower limit of temporal acuity in the neighborhood of 1 msec to 2 msec. Thus, the classical notion of temporal integration should perhaps be modified, as Green suggested, to include both estimates of minimum and long integration times. Moreover, the possibility must be considered, as Penner suggests, that different types of integrators may provide the best description of temporal summation data from different experimental paradigms.

The concepts, data, and procedures used to measure integration times and temporal functioning should prove valuable in comparative auditory function. For instance, comparisons of frequency discrimination and AM modulated noise discrimination might help in deciding to what extent an animal uses spectral and temporal information. An estimate of a temporal transfer function might help integrate a variety of temporal data from a species. Since phase locking of neural units to temporal events is a common electrophysiological technique, similar stimuli could be used to plot both this phase locking activity and the behavioral temporal acuity of an animal (see Fay and Popper, Chapter 1). In general, there have been few studies of temporal processing in nonhuman species (see also Dooling, Chapter 9), and it is hoped that this discussion will help alert workers in the field to the various approaches that may be taken in studying this important aspect of auditory system function.

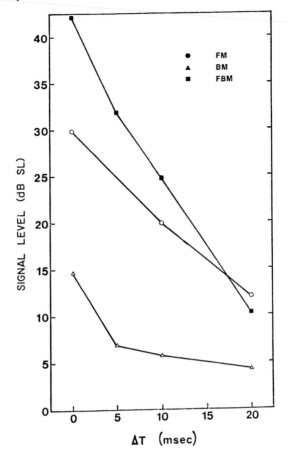

Figure 14-11. The masked thresholds of a 1000-Hz tone masked by a 1000-Hz masker in three temporal procedures: forward masking (FM), backward masking (BM), and combined forward-backward masking (FBM). The abscissa is the time between the signal and the masker. In the FBM task the signal was placed symmetrically between the two maskers. As can be seen in this study by Yost and Bergert (1979), there is more masking in the FBM task than in either the FM or BM tasks.

4 Localization

The early "duplex theory" of localization has remained intact with only minor modifications since Stevens and Newman (1936) first proposed it in the late 1930s. Interaural intensity differences appeared to be a cue for localization only at high frequencies due to the sound shadow produced by the head (see Knudsen, Chapter 10, for the birds' peculiar solution to this problem). Interaural time (or phase) differences were a cue only at low frequencies due either to the "effective time width of the head," or to the limited ability of the nervous system to follow temporal changes in a stimulus, or both. Although there was some evidence to the contrary (studies of the precedence effect, Gardner 1968, or studies of filtered noise, Klumpp and Eady 1956), the basic concept of the ongoing interaural time difference providing a localization cue only at low frequencies was unchallenged.

Since the early 1970s a variety of studies have demonstrated that man can lateralize certain stimuli with only high-frequency spectral content when only an interaural temporal difference is presented. In 1971 Yost, Wightman, and Green, in their study of the lateralization of filtered clicks, showed that humans could lateralize clicks containing only high-frequency information with values of interaural time differences 3 to 5 times larger than that required to lateralize low-frequency clicks. They showed that a low-frequency repetition of the high spectral-frequency clicks could result in lateralization as good as that obtained with low spectral-frequency clicks. Henning (1974a, 1974b) in two excellent papers showed that high-frequency waveforms with low-frequency repetitions could be lateralized with values of interaural time equal to that required to lateralize low spectral-frequency waveforms. Yost (1976) showed that the same type of result could also be obtained for high-frequency filtered clicks repeated at low-frequency rates.

Henning (1974a, 1974b) used amplitude modulated (AM) high-frequency sinusoids and narrowband filtered, high-frequency noises. Figure 14-12 shows the results from his study for a 300-Hz sinusoid compared to a 3 900 Hz sinusoid amplitude modulated at 300 Hz, a noise filtered with a 300-Hz bandpass filter centered at 3 900 Hz, and a 3 600-Hz sinusoid. As can be seen, the value of interaural time required for lateralization is essentially the same for three of the stimuli, but the unmodulated 3 600-Hz tone could not be lateralized for values of interaural time of up to a few hundred microseconds. Henning (1974a, 1974b) and others (McFadden and Pasanen 1976, Yost, Wightman, and Green 1971) suggested that these results indicated that high-frequency channels can process interaural time differences, if the high-frequency stimuli contain low-frequency repetitions.

An experiment by Yost (1976) demonstrated the use of high-frequency transient waveforms to study interaural time discriminations. The experiment indicated the importance of the number of temporal repetitions in lateralization of, at least, high-frequency transients. In this study, which was a follow-up to that done by Yost, Wightman, and Green (1971), trains of either high-pass or low-pass filtered transients (100 μsec D. C. pulses) were presented in an interaural-time, lateral discrimination procedure. As the data in Fig. 14-13 show, the interaural-time threshold for the high-frequency click decreases as the number of repetitions increases. When the high-frequency click is presented 12 times the interaural-time threshold is the same as for a low-frequency click. The repetition rate of the click does not appear to affect this relationship, implying that the binaural systems needs a certain number of "looks" at the temporal fluctuations rather than a certain length of time. This and other studies on AM tones and narrowband noises (see McFadden and Pasanen 1978) and on varying the modulation depth of the AM tones (Henning 1974a, 1974b) all strongly imply that it is the low-frequency temporal modulation of these high-frequency waveforms which is responsible for the binaural system's ability to use interaural time differences at high-frequencies.

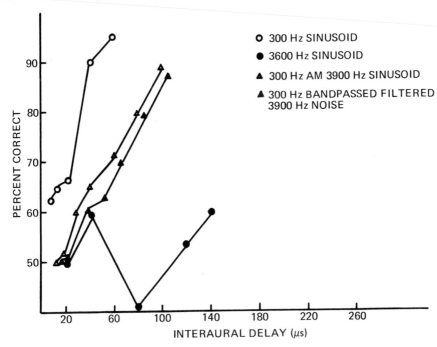

Figure 14-12. Percent correct in a lateral discrimination task is shown as a function of the interaural-time difference, for three high-frequency stimuli (3 600-Hz sinusoid, a 300-Hz amplitude modulated 3 900-Hz tone, and a 300-Hz bandpassed 3900-center frequency noise) and one low-frequency stimulus (a 300-Hz sinusoid). These data show that only the high-frequency waveforms with temporal modulation can be lateralized like the low-frequency tone (from Henning 1974a).

McFadden and Pasanen (1975) pursued these ideas with an interesting prediction and experiment. They argued that if high-frequency channels could follow low-frequency repetitions, then binaural beats should be audible for high-frequency signals. Thus, they presented a 3000-Hz plus a 3 100-Hz tone to the left ear and a 3000-Hz plus a 3 101-Hz tone to the right ear of their subjects. The temporal envelope at the left ear was alternating, therefore, at 100 Hz and the envelope at the right ear was alternating at 101 Hz; the subject reported hearing a 1-Hz binaural beat. Even if the right ear contained different frequencies than the left ear (for instance, 2000 Hz and 2 050 Hz in the right and 3000 Hz and 3 051 Hz in the left) the binaural beats could still be perceived. Although there were some differences between the binaural beats heard for high-frequency waveforms and those heard for low-frequency sinusoids, McFadden and Pasanen argued that the subjects were listening to the envelope fluctuations at high-frequencies. This helped support the original observations of Yost, Wightman, and Green and Henning.

Not only can observers lateralize these high-frequency waveforms, but under certain conditions (see Yost 1975, McFadden and Pasanen 1978), subjects can detect the dichotically presented waveforms at lower (sometimes much lower) signal-to-masker

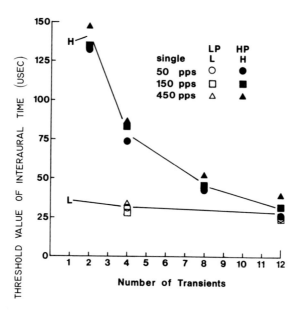

Figure 14-13. The threshold values of interaural time required for interaural discrimination are shown as a function of the number of transients for a high-frequency (HP) and a low-frequency (LP) transient. The transients were either a single transient (H, L) or repeated at 50, 150, or 450 pulses per second (pps). The data from Yost (1976) show that approximately 12 repetitions of a high-frequency transient yields results equivalent to that obtained for a low-frequency transient.

ratios than they can detect the diotically presented waveforms. That is, masking-level differences (see Green and Yost 1975) can also be obtained in some conditions for high-frequency waveforms presented with low-frequency temporal envelopes. Figure 14-14 shows the results from a masking experiment by McFadden and Pasanen (1978) in which the signal was a 4000-Hz sinusoid and the masker was a 50-Hz band of noise centered at 4000 Hz. The interaural time difference of the master wave-form is varied and the signal threshold is plotted. Diotic detection is for the zero interaural delay conditions (NoSo). As can be seen, a large release from diotic masking occurs at integer multiples of half the period of the 4000-Hz signal (250 micro-seconds). The oscillation in the masking function is similar to that obtained by Lang-ford and Jeffress (1964) for low-frequency tones embedded in wide-band noise. These results might not be due to temporal processing at high frequencies, since large inter-aural intensity differences exist at integer multiples of half the period of the signal frequency for these stimulus conditions. But these results do support the other re-search described in that binaural processing at high frequencies is possible under con-ditions that might not have been considered given the original "duplex theory" of localization.

A summary of these results obtained at high frequencies along with those obtained at low frequencies suggest that a stimulus can be localized on the basis of interaural

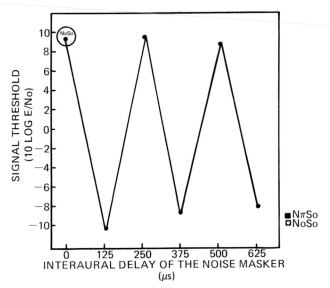

Figure 14-14. The masked thresholds of a 4000-Hz signal masked by a 50-Hz band-passed noise centered at 4000 Hz are shown as a function of the interaural time difference time difference of the noise masker. When there is no interaural time delay the condition is diotic (NoSo). There is a reduction in threshold (an MLD) when the interaural delay is equal to integer multiples of a half period of 4000 Hz (250 µsec). The data are from McFadden and Pasanen (1978).

time if the stimulus contains either low-frequency spectral information or low-frequency temporal repetition or both (see Yost 1977). Binaural temporal processing at high frequencies in animals has been studied by many investigators. The results are often discussed in terms of the "duplex theories" assumption that interaural time is not a cue at high frequency. Since some animals with small heads can localize at higher frequencies than animals with large heads, correlation between waveform period and head size is often used as an explanation for these findings (see Gourevitch, Chapter 12). Given the results cited, it might be of interest to determine if these animals differ in their abilities to process temporal envelope information. For instance, animals with poor temporal acuity might perform poorer at localizing at high frequencies than animals with good temporal acuity. These new ideas might suggest new interpretations of localization at high frequency. In addition, studies of the type of signals animals localize in nature (for instance, vocalizations from their own species) and their temporal structure might lead to interesting new developments in our understanding of localization in animals (see Stebbins, Chapter 15).

5 Conclusion

Recent developments involving frequency selectivity, temporal processing, and localization have been reviewed. It has been suggested that this new research is reshaping our classical views of the critical band, temporal integration, and the "duplex theory" of localization. It is hoped that when scholars of comparative auditory function begin to investigate the way in which various animals organize acoustical information, they will carefully consider these new developments. Hopefully, the research we have described will provide as useful insights into comparative auditory function as they have into the psychoacoustic function of man.

Acknowledgments. This work was partially supported by grants from National Science Foundation and the National Institutes of Health.

References

Abbas, P. J., Sachs, M. B.: Two-tone suppression in auditory nerve fibers: Extension of a stimulus-response relationship. J. Acoustic Soc. Amer. 59, 112-122 (1976).

Bilsen, R. A., Ritsma, R. J.: Repetition pitch and its implication for hearing theory. Acustica. 22, 63-68 (1969/70).

Blauert, J.: Localization and the law of the first wavefront in the median plane. J. Acoustic. Soc. Amer. 50, 446-470 (1971).

Carterette, E. C., Friedman, M. P., Lowell, J. D.: Mach bands in hearing. J. Acoustic. Soc. Amer. 45, 986-998 (1969).

Cornsweet, T. N.: Visual Perception. New York: Academic Press, 1970.

Gardner, M. B.: Historical background of the Hass and/or Precedence Effect. J. Acoustic. Soc. Amer. 43, 1243-1250 (1968).

Green, D. M.: Temporal auditory acuity. Psych. Rev. 78(b), 542-551 (1971).

Green, D. M.: Minimum integration time. In: Basic Mechanisms in Hearing. Mueller (ed.). New York: Academic Press, 1973.

Green, D. M., Yost, W. A.: Binaural analysis. In: Handbook of Sensory Physiology, Vol. V(2). Kridel and Neff (eds.). New York: Springer-Verlag, 1975.

Henning, G. B.: Detectability of interaural delay in high-frequency complex waveforms. J. Acoustic. Soc. Amer. 55, 84-90 (1974a).

Henning, G. B.: Lateralization and the binaural masking level difference. J. Acoustic. Soc. Amer. 55, 1259-1265 (1974b).

Houtgast, T.: Psychophysical evidence for lateral inhibition in hearing. J. Acoustic. Soc. Amer. 51, 1885-1894 (1972).

Houtgast, T.: Psychophysical experiments on "tuning curves" and two-tone inhibition. Acustica. 29, 168-179 (1973).

Houtgast, T.: Masking patterns and lateral inhibition. In: Facts and Models in Hearing. Zwicker and Terhardt (eds.). New York: Springer-Verlag, 1974.

Houtgast, T.: Auditory filter characteristics derived from direct-masking and pulsation threshold data with a ripple noise masker. J. Acoustic. Soc. Amer. 62, 409-415 (1977).

Johnson, R. A., Titlebaum, E. L.: Energy spectrum analysis: A model of cholocation processing. J. Acoustic. Soc. Amer. 56, S39 (1974).

Klumpp, R. G., Eady, H. R.: Some meeasurements of interaural time difference thresholds. J. Acoustic. Soc. Amer. 28, 859-861 (1956).

Langford, T. L., Jeffress, L. A.: Effect of noise crosscorrelation on binaural signal detection. J. Acoustic. Soc. Amer. 36, 1145 (1964).

Legouix, J. P., Remand, M. C., Greenbaum, H. B.: Interface and two-tone inhibition. J. Acoustic. Soc. Amer. 53, 409-412 (1973).

McFadden, D., Pasanen, E. G.: Binaural beats at high frequencies. Science. 190, 394-397 (1975).

McFadden, D., Pasanen, E. G.: Lateralization at high-frequencies based on interaural time differences. J. Acoustic. Soc. Amer. 59, 636-639 (1976).

McFadden, D., Pasanen, E. G.: Binaural detection at high frequencies with time-delayed waveforms. J. Acoustic. Soc. Amer. 63, 1120-1132 (1978).

Moore, B. C. J.: Psychophysical tuning curves measured in simultaneous and forward masking. J. Acoustic Soc. Amer. 63(2), 524-532 (1978).

Munson, W. A.: The growth of auditory sensation. J. Acoustic. Soc. Amer. 19, 584-591 (1947).

Patterson, J. H., Green, D. M.: Discrimination of transient signals having identical energy spectra. J. Acoustic Soc. Amer. 48, 984-905 (1970).

Penner, M. J., Robinson, C. E., Green, D. M.: Critical masking interval. J. Acoustic. Soc. Amer. 52, 1661-1668 (1972).

Penner, M. J.: A power law transformation resulting in a class of short-term integrators that produce time-intensity trends for noise bursts. J. Acoustic. Soc. Amer. 63, 195-202 (1978).

Rainbolt, H., Small, A. M.: Mach band in auditory masking: An attempted replication. J. Acoustic. Soc. Amer. 51, 567-574 (1972).

Rhode, W. A.: Some observations on two-tone interaction measured with the Mossbauer effect. In: Psychophysics and Physiology of Hearing. Evans and Wilson (eds.). London: Academic Press, 1977.

Robinson, C. E., Pollack, I.: Forward and backward masking: Testing a discrete perceptual-moment hypothesis in audition. J. Acoustic. Soc. Amer. 50, 1512-1519 (1971).

Robinson, C. E., Pollack, I.: Interaction between forward and backward masking: A measure of the integrating period of the auditory system. J. Acoustic. Soc. Amer. 53, 1313-1317 (1973).

Rodenburg, M.: Investigation of temporal effects with amplitude modulated signals. In: Psychophysics and Physiology of Hearing. Evans and Wilson (eds.). London: Academic Press, 1977.

Sachs, M. B., Kiang, N. Y-s.: Two-tone inhibition in auditory nerve fibers. J. Acoustic. Soc. Amer. 42, 1120-1128 (1968).

Scharf, B.: Critical bands. In: Foundation of Modern Auditory Theory. Tubias (ed.). New York: Academic Press, 1970, pp. 157-200.

Shannon, R. V.: Two-tone unmasking and suppression in a forward-masking situation. J. Acoustic. Soc. Amer. 59, 1460-1471 (1975).

Stevens, S. S., Newman, E. B.: The localization of actual sources of sound. Am. J. Psych. 48, 297-306 (1936).

Terry, M., Moore, B. C. J.: Suppression effects in forward masking. J. Acoustic. Soc. Amer. 62, 781-784 (1977).

Tyler, R. S., Small, A. M.: Two-tone suppression in backward masking. J. Acoustic. Soc. Amer. 62, 215-217 (L) (1977).

Viemeister, N. F.: Temporal factors in audition: A system analysis approach. In: Psychophysics and Physiology of Hearing. Evans and Wilson (eds.). London: Academic Press, 1977.

Vogten, L. L. M.: Low-level pure tone masking: A comparison of "tuning curves" obtained with simultaneous and forward masking. J. Acoustic. Soc. Amer. 63(5), 1520-1527 (1978).

von Békésy, G.: Sensory Inhibition. Princeton, N. J.: Princeton University Press, 1967.

Weber, D. L., Green, D. M.: Temporal factors in suppression of non-simultaneous masking. J. Acoustic. Soc. Amer. 64, 1408-1414 (1978).

Widen, G. P., Viemeister, N. F.: Intensity effects in forward masking: Implications for psychophysical tuning curves. J. Acoustic. Soc. Amer. 63, S44 (1978).

Weir, C. C, Green, D. M.: Temporal acuity as a function of frequency difference. J. Acoustic, Soc. Amer. 57, 1516-1521 (1975).

Wightman, F. L., McGee, T., Kramer, M.: Factors influencing frequency selectivity in normal and hearing-impaired listeners. In: Psychophysics and Physiology of Hearing. Evans and Wilson (eds.). London: Academic Press, 1977.

Yost, W. A.: Comment on lateralization and the binaural masking level difference (G. B. Henning, J. Acoustic. Soc. Amer. 55, 1259-1262, 1974). J. Acoustic. Soc. Amer. 57, 1214-1215 (1975).

Yost, W. A.: Lateralization of repeated filtered transients. J. Acoustic Soc. Amer. 60, 178-181 (1976).

Yost, W. A.: Lateralization of pulsed sinusoids based on interaural onset, ongoing, and offset temporal differences. J. Acoustic. Soc. Amer. 61, 190-194 (1977).

Yost, W. A., Bergert, B.: Tonal temporal masking in three temporal procedures. J. Acoustic. Soc. Amer. (1980). In press.

Yost, W. A., Hill, R., Perez-Falcon, T.: Pitch and pitch discrimination of broadband signals with rippled power spectra. J. Acoustic. Soc. Amer. 63, 1166-1173 (1978).

Yost, W. A., Wightman, F. L., Green, D. M.: Lateralization of filtered clicks. J. Acoustic. Soc. Amer. 50, 1526-1531 (1971).

Chapter 15

The Evolution of Hearing in the Mammals

WILLIAM C. STEBBINS*

1 Introduction

Of the many successful adaptations that characterize the living mammals and give them an advantage over ancestral forms, perhaps few are as distinctive as their sense of hearing. Although parallel developments have occurred in the living birds and in one or two of the modern reptilian orders, to some extent, in the mammals as a group, the diversity and complexity in auditory form and function, the range of hearing, and the discriminative acuity for the various parameters of acoustic signals surpasses that seen in other animals (Masterton, Heffner, and Ravizza 1969, Stebbins 1975, 1978). Clearly, in the course of mammalian evolution there have been important changes in other sensory systems such as olfaction and vision, but by comparison, the changes that have occurred in hearing have been perhaps the most dramatic. From premammalian ears, which probably functioned as substrate vibration detectors or, at best, receptors of low frequency aerial sound, have evolved such successful adaptations as the high frequency echoranging of the bats and cetacea, the almost uncanny sensitivity of the felids, and the discriminative acuity of the primates for minute changes in the amplitude and spectral composition of the acoustic waveform. It was these developments in the primates that made possible the effective resolution of one of the more complex of all periodic waveforms—human language.

This chapter will reconstruct the evolution of hearing in the mammals, first by considering the morphological evidence from fossil material available for some of the mammal-like reptiles and the early mammals, together with the selective pressures that are thought to have been responsible for the subsequent development of mammalian hearing. The paleontological findings and their implications for the structural evolution, particularly of the mammalian middle ear, have received considerable attention in recent years (Hopson 1966, 1969, Allin 1975, Fleischer 1978, Lombard and Bolt 1979, Henson 1974, Parrington 1979, Tumarkin 1968, Van Bergeijk 1967). Some of the issues discussed by these authors will be considered in this chapter where they are

*Kresge Hearing Research Institute, University of Michigan, Ann Arbor, Michigan 48109.

relevant to the evolution of function (hearing) which, rather than structure, is the primary emphasis of this chapter. There are obvious pitfalls in studying the evolution of function, and these will be discussed and usually heeded. Based on the fossil evidence together with data on hearing obtained from living reptiles and from living mammals, this author will attempt to present a moderately plausible account of the evolution of hearing in the mammals as a group, and further, within a particular group of mammals—the primates.

2 Evolution of the Mammals

The beginning of the great mammalian radiation occurred about 70 million years ago, although the earliest signs of mammalian differentiation from the reptiles appeared more than 100 million years earlier (Romer 1966). In fact, that group from which the mammals were derived split off from the ancestors of modern reptiles very soon after the origin of reptiles (see Hopson 1969, Romer 1959, 1966). Among the earliest reptiles to show mammalian characteristics were the pelycosaurs, representatives of the synapsid reptiles. One of the more specialized among the various taxa was Dimetrodon, the famous "sail back." Although it is likely that the sail subserved a thermoregulatory function (Romer 1959, 1966) or perhaps as a device for intraspecific display behavior (Bakker 1971), the fanciful notion has been put forward that it may have been one of the earliest tetrapod aerial sound detectors (Tumarkin 1968).

Although many of the pelycosaurs were semiaquatic fish eaters or even herbivores, the sphenacodontids, which were ancestral to the therapsids and, subsequently, to the true mammals, were terrestrial predators (Romer 1959, Hopson 1969). The more advanced members of this group were the first reptiles to show clear evidence of adaptive strategies appropriate to a mammalian lifestyle (Hopson 1969). They provided early evidence of new developments in the structures associated with eating, permitting them to dismember their prey before swallowing. The appearance of canines and of a more powerful and efficient jaw articulation (Barghusen 1968) were progressive changes that eventually gave rise, in a series of stages, to the three-bone ossicular chain of the mammalian middle ear in place of the single piston-form columella characteristic of living reptiles and birds. Finally, the therapsid reptiles provided the bridge to the mammals. As efficient, somewhat smaller, and more lightly built carnivores, the therapsids developed larger and more completely differentiated canines than earlier mammals (Hopson 1969). Among the therapsids, the theriodonts, and particularly the more advanced cynodonts showed still further adaptations for eating, including versatile jaw movements and a more precise dental occlusion, that directly set the stage for the mammalian adaptation of the posterior jaw bones for hearing (Hopson 1969). In fact, it is entirely possible that these structures were even in the cynodont reptiles used as aerial sound detectors (Allin 1975). Bones that had been an integral part of the primitive jaw support system in early reptiles gradually, over many millions of years, lost their function in eating as the reptilian jaw became foreshortened with a new form of articulation. These bones became reduced in size and eventually formed the ubiquitous three-bone ossicular chain of the mammals.

3 Selective Pressures on Hearing and Structural Evolution of the Mammalian Ear

In order to understand many of the successful auditory adaptations of the mammals, it is helpful to consider some of the selective pressures that were probably operative in the period of the reptilian-mammalian transition and later during the continued evolution of the mammals. The scope of this chapter does not permit consideration of the varied pressures to which the many different mammalian taxa responded, although some of the unique adaptations of the primates will be considered by way of example.

It is likely that the earliest mammals, or their reptilian progenitors, exploited nocturnal niches (Hopson 1973, Jerison 1973), for these would have been relatively free of many of the large, diurnal, predacious reptiles (Bakker 1971). Several adaptations were necessary before nighttime living could become a reality. Certainly homoiothermy had to be at or near the top of the list, if anything was to be accomplished after the sun went down (Hopson 1973); and homoiothermy, together with superficial insulation, would have allowed better heat retention at night so that the Mesozoic mammals would be able to capitalize on nocturnality more effectively and more extensively than other contemporary tetrapods (Bakker 1971). Vision, which had been exploited with a modest degree of success by the reptiles, had to give way to hearing and smell (Jerison 1973) as the primary means of guiding the early mammals around their environment in search of food or avoidance of predators. Later when the large reptiles, the archosaurs, became extinct, the then extant mammals radiated into the open diurnal niches with further consequences for both sight and hearing. The primates provide a particularly good example of mammals who moved into these diurnal adaptive zones. Related to effective homoiothermy in the earliest mammals, and perhaps to a lesser extent in their immediate reptilian forebears, was the development of a high, sustained activity level, accompanied by a stable rate of metabolism (Simpson 1967). In turn, such changes required that food be obtained more regularly and ingested more efficiently. The continuing changes in dental morphology and jaw structure and articulation were, conceivably, an outcome of these pressures.

It is likely that other significant developments that were occurring during this period also played a role in the evolution of mammalian hearing. Changes in the location and articulation of the limbs so that they were positioned directly under the body rather than to the side (Romer 1966) must have added to the speed and dexterity with which the early mammals moved about their environment. However, if the earlier reptiles depended on their position close to or on the ground for effective detection of substrate vibration (e.g., low frequency sound) by a conducting pathway through the forelimbs as Tumarkin (1968) has suggested, elevation of the body, by considerably extending this pathway, might have reduced their acoustic sensitivity. Such developments in progress may have further increased the pressure on the evolution of the mammalian middle ear and drum from the jawbones as an efficient detector of aerial sound and as a replacement for the relatively cumbersome system for detecting substrate sound (Allin 1975).

The fossil evidence indicates that many of the ancestral reptiles had a massive stapes with a relatively low areal ratio (tympanic membrane to oval window) (Hopson 1966).

Although the biomechanics are far from clear, Manley (1973) has suggested that the conduction pathway may have been the reverse of what we now know. Sound from the substrate would have been conducted through the bone and tissues to the large stapes and columella with the ear drum serving as a release for these movements. Even in some reptiles with lighter ear bones, the areal ratio was small and, if sensitive to aerial sound at all, these ears probably responded only at very low frequencies. However, there is reason to believe, at least in the therapsids, that the conducting route was centripetal—from mandible (lower jaw), to quadrate (homologous to the mammalian incus), to stapes footplate (Allin 1975). Manley (1973) has also suggested that further changes in both middle and inner ear may have been in response to pressures for more elaborate vocal communication as the mammals evolved from the therapsids. As he has indicated, in modern reptiles there is an apparent differentiation in inner ear structure between those animals that vocalize (alligators, geckos, etc.) and those that probably do not.

The changes in jaw structure, which had such important consequences for hearing in the mammals, were taking place in the carnivorous reptiles—the therapsids. Yet these animals became extinct in the face of competition with the archosaurs. The mammals descended from much smaller forms and were themselves diminutive, even shrew-sized. Is it possible, as Tumarkin (1968) has suggested, that the reduction in size of all bodily structures would have rendered the bony conducting pathway more sensitive and thus permitted the more effective use of the ossicles as a sound-conducting device at the threshold of the mammalian radiation? This raises the possibility that reduction in body size was at least partly in response to increased demand for better acoustic sensitivity. It is likely that selection favored diminution in these animals for a variety of reasons related to the development of mammalian reproductive strategy (Hopson 1973), to efficient heat loss (Bakker 1971), and to the ability of these animals to fit into diurnal shelters (trees or underground burrows) in order to avoid the large archosaurs (Bakker 1971). Allin (1975) has proposed that there was strong selection in the therapsids for improved auditory acuity and even sensitivity to high frequency sound. He has argued that, in addition to reduction in size of the key structures (the ossicles), there was a loosening of their attachments and a change in the locus of the jaw muscle insertion. Such changes would be expected to increase auditory sensitivity by decreasing the impedance of middle ear structures.

Paleontological evidence indicates quite clearly that the external ear canal was present in the mammal-like reptiles and evolved gradually with the changes that were occurring in jaw and skull in these animals (Olson 1971). It is not readily apparent what was happening to the tympanic membrane during this period. The pinna developed much later and in many mammals serves both an auditory and a thermoregulatory function (Webster 1966).

It is clear from the evidence that the evolution of such an efficient outer and middle ear conduction system was a fortuitous outcome of certain primary adaptations of a carnivore's lifestyle directly related to obtaining food and eating it. Structural examination of certain living mammals as well as embryological data are supported by fossil material; together these three lines of evidence make a convincing case that few would question. The columellar single bone conducting system in modern reptiles and birds has evolved independently and approaches in sensitivity the mammalian ossicular chain, although it is apparently much less effective in the transmission of high frequencies (Manley 1973). It is considerably less clear what was taking place over the same period in the mammalian inner ear and the more central portions of the auditory system. With the exception of the kind of information provided by endocranial casts (McKenna 1969), soft tissue rarely leaves any record, and thus there is no fossil support for speculation based entirely on material from living species.

A coiled cochlea (one turn or more) is characteristic of all living placental mammals and marsupials (Gray 1955), yet not present in monotremes (Fernandez and Schmidt 1963). It is therefore unlikely that a coiled cochlea developed before the monotremes had diverged from the rest of the mammals early in mammalian evolution (Manley 1973). There is no evidence that any of the therapsid ancestors had anything resembling a coiled cochlea (Hopson 1969, Kermack, Kermack, and Mussett 1968). The cochlear duct is attached to the saccule in living reptiles (Baird 1974) and forms a relatively short extension of the inferior part of the labyrinth (also see Miller, Chapter 6). In some lizards, crocodiles, and birds it becomes longer, is fairly straight, and contains the sensory cells for hearing. Evidence of curvature is seen more clearly in the monotremes (Griffiths 1968), but only in the marsupials and placentals does the coiling become complete. As an aside, it is unlikely that the monotremes could serve as representatives of the ancestral mammals. They appear far too specialized, for the most part, although their auditory structures, possessing some reptilian and some mammalian features, suggest a possible reptilian-mammalian transition species (Griffiths 1968, Aitken and Johnstone 1972, Aitken, Gates, and Kenyon 1979).

Prior to coiling of the cochlea in the early mammals, it is reasonable to assume that there were earlier, less spectacular changes occurring in the length of the duct and in the size and shape of the basilar membrane, and that these changes may have had important implications for frequency range, discrimination, and analysis. These structural variations are evident in the ears of modern reptiles (Wever 1978). Short elliptical membranes found in living turtles and snakes, for example, may represent the more primitive condition (Schmidt 1964) and may have given way to the more extended and often tapering membranes seen in the crocodiles and some lizards (Manley 1973, Wever 1978, Miller, Chapter 6).

It is beyond the scope of this chapter to discuss what little is known about the evolution of the auditory portion of the central nervous system beyond the cochlea. There is, of course, no fossil record, so that the data are based on morphological and embryological evidence from living species. It has been suggested that the rates of evolution (Simpson 1967) have been higher in the more peripheral parts of the system (e.g., external and middle ear), perhaps because of their greater plasticity and adaptability to environmental pressures (Webster 1966). Jerison (1973), however, has argued for substantial enlargement of the auditory part of the brain in the earliest mammals in response to the intense pressures for guidance in a nocturnal adaptive zone. Distance sensing, involving accurate localization at night, depends on binaural interaction, which in turn involves collicular and other neural centers (Masterton 1974). Further elaboration of central structures and their connections in the inner ear may have occurred in the later mammals as they moved into diurnal adaptive zones and adopted more complex forms of social organization and communication. Those additional modifications in structure and innervation may have had consequences for improved differential acuity for dimensions of acoustic stimuli such as frequency and intensity.

In sum, then, the gross morphologic features in the auditory system that character-ize the mammals are the pinna, the external ear canal, and the recessed tympanum, the three-bone ossicular chain, the coiled cochlea with considerable increase in length over that of the ancestral reptiles, and those as yet not clearly specified changes related to binaural hearing and differential resolution of acoustic signals that took place in the central nervous system. Obviously these were not the only changes of significance in the auditory system, but they were clearly important and signalled other changes that occurred at a more microscopic level, such as two different populations of receptor cells, separately and unequally innervated. One primary question to which this chapter is directed concerns the functional changes accompanying these alterations in structure that took place over many millions of years as the mammals were evolving. The ques-tion, though easily stated, is somewhat less easily answered.

4 Evolution of Function

It may be well to heed Simpson's (1958) suggestions regarding the nature of the objec-tive data ideally required for historical study of a function or process. The necessary information is of four classes:

1. descriptive and, when possible, functional morphology
2. behavior
3. the environmental conditions or selective pressures
4. the temporal sequence

The paleontologist has access to certain of the material in class 1 (structure) and class 4 (sequence). Observations of class 3 (conditions) are mainly inferential and direct ob-servations of class 2 (behavior) are completely lacking. Inferences regarding behavior may be based on fossil morphology and direct analogy with living creatures. Eating habits and locomotion have been considered, but behaviors such as hearing or seeing, where the format is considerably more complex and the relevant fossil record less en-during, present a significantly greater challenge.

Data of the first three classes (morphology, behavior, and environmental conditions) are available to the behaviorist, but such information about existing forms fails to re-veal the temporal sequence through history (class 4). Those phyletic sequences that have been adopted by behaviorists, such as goldfish, laboratory white rat, rhesus mon-key, and man, are unjustified and have no basis in historical fact. This is not to say that the study of the evolution of function (e.g., hearing) is unapproachable by the be-haviorist, but only that there are numerous pitfalls to be avoided and cautions to be observed (see Hodos and Campbell 1969).

Since historical information regarding behavior is unavailable, the behaviorist must infer largely from living species. Such inference, to be legitimate, must be based on the following (see Simpson 1958):

1. Related taxa frequently evolve in parallel, but some evolve much more rapidly than others—thus, living representatives may be placed in a series that approximates the historical sequence (e.g., marsupials and placental mammals).
2. Certain trends such as increase in body size and in behavioral complexity are sufficiently frequent that they may be supported in a given example (e.g., prosimian and Old World or possibly even New and Old World monkey).
3. Characteristics shared by living taxa can be assumed to have been present in their common ancestry (e.g., superior part of the labyrinth in primate and teleost).

Unfortunately none of the above is without exception, and the unwary are easily trapped. Homology is confused with analogy, some trends go in either direction, and one characteristic of a group may evolve at a very different rate from another characteristic. In spite of the difficulties, the wary and informed behaviorist proceeds gingerly with the help of the comparative anatomist, the embryologist and the paleontologist.

5 Hearing of the Reptilian Ancestors and Early Mammals

A direct and unequivocal answer to the question of the relation between structural and functional evolution in the auditory system cannot be realized. On the basis of fossil evidence, it is possible to speculate on the hearing of early mammals and the ancestral reptiles (McCandless, Madsen, and Parkin 1978), but such evidence is thin and usually based only on the bony structure of the middle ear. Alternatively we may rely on data obtained from living animals with assumptions regarding phyletic ordering. Here, too, there are limitations on the available data. While the behavioral findings on hearing in the mammals are, if not substantial, at least suggestive, such material on reptiles is lacking. In examining these animals, it is necessary to depend almost entirely on electrophysiological data obtained by recording from their inner ear. Such evidence provides an inadequate substitute for behavioral measures of acoustic sensitivity (Raslear 1974).

These approaches are utilized, but within the context of a somewhat broader perspective. On the basis of the available evidence, what are some of the adaptations for hearing that appear unique to the later mammals? Some of these have been considered in detail (Masterton et al. 1969, Manley 1971, 1973). These adaptations have occurred in response to continuing and diverse pressures, different probably from those to which the early mammals and mammal-like reptiles were subjected. The concern here is with the more generalized mammalian pattern. Some groups such as the bats and cetacea or desert rodents have, in carrying some of these adaptations to extreme lengths, become highly specialized. Others, such as the primates, appear to have retained the more primitive or generalized mammalian pattern but with certain significant modifications.

Hearing, for the purposes of this chapter, is considered a behavioral function. It is a response by the entire, intact animal to some form of acoustic input. In some instances, since there are no behavioral data, hearing must be inferred on the basis of electrophysiological, or even morphological, evidence. Further, hearing is not a simple unitary function but a variety of functions, some of which may have evolved for a different reason and at different rates than others. Although most of what is known concerns absolute sensitivity and frequency range (see Masterton et al. 1969), it is essential to treat other aspects as well, such as localization, frequency analysis, intensity discrimination and so on.

On the basis of the fossil evidence, it is a best guess that the mammal-like reptiles were sensitive to little more than the low frequency substrate vibrations produced by their larger, predatory relatives, the archosaurs (Henson 1974). In comparison with middle ear structures, insofar as they are known (i.e., mass of the stapes; see Henson 1974), and areal ratios (Manley 1973), with modern reptiles, the mammal-like reptiles must have been less sensitive to acoustic stimulation than most living species. Given the most liberal estimates based on an inner ear approaching that of most modern reptiles, they would, in all likelihood, have been restricted to high sound levels at frequencies well below 1000 Hz. It is, of course, possible that they did not hear at all, but this seems unlikely in view of the presence of structures suited at least for the detection of substrate vibrations or even low frequency aerial hearing.

Modern reptiles and birds descended from the archosaurs, which were separated from the premammalian reptiles early in the reptilian radiation. Thus, any sequence of living reptiles to living mammals is out of the question. Yet in spite of the antiquity of their common ancestry with the modern mammals, it is important to consider the hearing of living reptiles, for their more recent ancestors must have been subject to at least some of the same pressures as the mammal-like reptiles. Living reptiles vary considerably in inner ear morphology (Baird 1974, Wever 1978, Miller, Chapter 6) and perhaps somewhat less in middle ear structure (Henson 1974, Wever 1978). From physiological recordings of cochlear microphonics from the inner ear of reptiles (Wever 1978), sensitivity and frequency range (the two measurable parameters) appear somewhat less variable than in the mammals (also see Turner, Chapter 7).

From the numerous reptiles that have been examined for cochlear microphonics in the laboratory, two are presented in Fig. 15-1. One is from a small burrowing snake (*Rhinophis drummondhayi*) and the other is from a crocodilian (*Caiman crocodilus*) (Wever 1978). In addition to considerable differences in frequency range, there are also marked sensitivity differences in the recorded cochlear microphonics between these two animals, but their relation to actual differences in hearing threshold sensitivity is uncertain (see Raslear 1974). Behavioral threshold data obtained from one species of turtle (*Pseudemys scripta*) are shown in Fig. 15-2 from Patterson (1966). In Fig. 15-3 the threshold function for a lizard, the Tokay Gecko (*Gecko gecko*), is based on recordings from single auditory nerve fibers and provides a closer approximation to behavioral hearing threshold than the cochlear microphonic response (Manley 1972). These data illustrate extremes in frequency range and sensitivity among the reptiles and, thus, provide an indication of the variation in these parameters in the living reptiles. At one end of the continuum there are modern reptiles whose hearing is probably similar to the hearing of the very early reptiles, with sensitivity limited to fairly intense vibration of the substrate in the immediate vicinity. Wever (1978) has suggested that these ears serve as band-limited, low frequency detectors with little, if any, discriminative acuity. On the other hand, the crocodilian ear bears a striking morphological resemblance to that of most birds (see Miller, Chapter 6, Saito, Chapter 8). Together

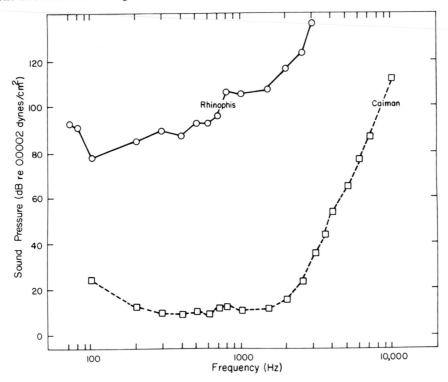

Figure 15-1. Cochlear potential sensitivity to sound in a burrowing snake (*Rhinophis drummondhayi*) and a crocodile (*Caiman crocodilus*) (from Wever 1978). Figs. 20-35 and 24-36 from Ernest Glen Wever, *The Reptile Ear: Its Structure and Function* (copyright © 1978 by Princeton University Press), pp. 739 and 962. Reprinted by permission of Princeton University Press.

with the cochlear microphonic function, these anatomical data suggest behavioral sensitivity and discriminative acuity equal to that of many birds (see Dooling, Chapter 9) and even approaching that of some mammals. Whereas many reptiles appear relatively silent and probably rely on sound only for protection from other species, others, notably the crocodiles and some species of lizards, are highly vocal and use sound in inter- and intraspecific encounters such as territorial defense and reproduction (Marcellini 1978, Gans and Maderson 1973).

6 Hearing of the Mammals

It may be argued with some plausibility that these findings from living reptiles give us a best estimate of the upper limits of hearing (sensitivity and frequency range) of which the reptiles ancestral to the mammals were capable. The morphological evidence that exists for those early reptiles would support this assumption. As the mammalian radiation developed, the selective pressures that had played a role in the evolution of

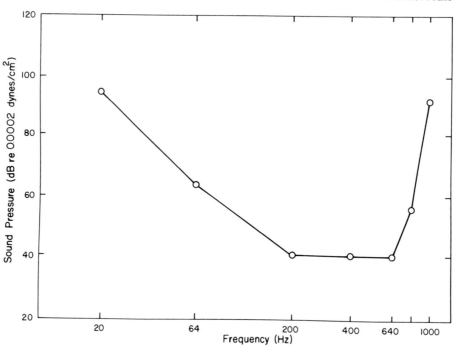

Figure 15-2. Behavioral threshold function for a turtle (*Chrysemys scripta*) (from Patterson 1966).

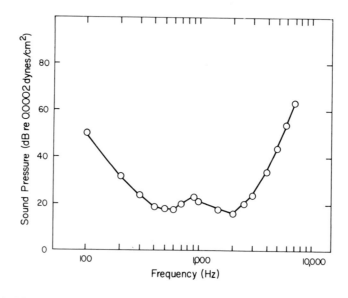

Figure 15-3. Threshold function for the Tokay Gecko (*Gecko gecko*) derived from recordings from single auditory nerve fibers (from Manley 1972).

hearing in the earlier vertebrates were probably intensified. Predator evasion, as suggested earlier, must have played a significant role in the life of the early mammals during the period when the large carnivorous reptiles were abundant. The marked enhancement in auditory sensitivity was probably a reflection of this pressure. Then, too, as Masterton et al. (1969) have suggested, the considerable expansion of the audible frequency range from their reptilian ancestors permitted the mammals to accurately localize the source of a sound in space. Coupled with the overall sensitivity changes, this increase in range would have bestowed no mean advantage on the mammals and given them a significant head start in more ways than one in their competition with the archosaurs.

Representative threshold functions for several mammalian taxa are shown in Fig. 15-4. The functions are smoothed simply to indicate form and placement on the frequency and sound pressure axes. Although behavioral threshold data from amphibians and reptiles is virtually nonexistent, it is assumed, partly on the basis of electrophysiological findings, that the expansion in frequency range and in sensitivity is chiefly characteristic of the mammals, although there is overlap with some avian species (see Dooling, Chapter 9). However, diversity is also apparent within the class, and this may be related to the specific pressures on the different groups. Such traits

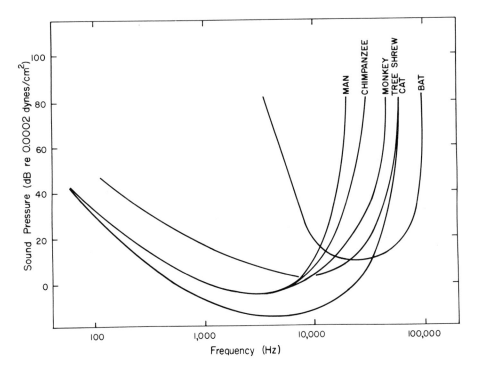

Figure 15-4. Audibility functions for several mammalian species: bat (*Eptesicus fuscus*), Dalland 1970; man (*Homo sapiens*), Sivian and White 1933; chimpanzee (*Pan troglydytes*), Elder 1934; monkey (*Macaca*), Stebbins 1970; tree shrew (*Tupaia glis*), Stebbins 1975; and cat (*Felis catus*), Miller et al. 1963.

as ultra high frequency sensitivity in the bats and cetacea, low frequency sensitivity in the felids and desert rodents, and a contraction of the frequency range in the primates, are examples.

It has been argued (Stebbins 1975, 1978) that there must have been additional pressures on the mammals related to social organization and, hence, to the development of effective intraspecific communication systems. Such pressures exerted on the auditory system and on auditory processing might be expected to result in improved differential acuity along the dimensions of intensity and frequency. While this is probably true, there is no comparative evidence from present or former reptiles. In fact, Fay (1974) has shown that at least one modern species of teleost and one avian species barely overlap the mammals in their frequency resolving capabilities.

If the primates are considered separately, the picture becomes somewhat clearer. The primates were one of the mammalian taxa that, probably after extinction of the ruling archosaurs, radiated into diurnal adaptive zones. Color vision, better visual acuity, stereopsis, and improved depth perception provided support for sound localization. Particularly among the terrestrial primates, social organization became more elaborate and intraspecific communication, accordingly, more complex. Interestingly, if Fig. 15-4 is reexamined, a gradual collapse of the high-frequency boundary of the threshold function can be seen. The sequence (which is assumed to be historical) starts with a proto-typical mammal or even a marginal primate, the tree shrew (*Tupaia glis*), at 60000 Hz and progresses through Old World monkey and chimpanzee to man at about 20000 Hz. There is no obvious reason for this retrogressive shift in the upper frequency limit of hearing and no clear structural correlate. In fact, the basilar membrane in man is somewhat longer than in other primate species.

A possible answer is seen in Fig. 15-5. Frequency difference thresholds are presented for the tree shrew, bush baby, macaque, and man. Once again, if an approximation to phyletic ordering may be suggested (insectivore–prosimian–Old World monkey–man), then we can argue for a significant enhancement in frequency resolution within the primates, related perhaps to pressures for improved intraspecific signaling systems. It is unclear what related changes occurred in an inner ear and central nervous system that altered function from a wide-band energy detector to a very narrow-band discriminator. There is some additional evidence to indicate improvements in the primates in other forms of differential acuity such as intensity discrimination (Stebbins 1978).

7 New Research Directions

Until now the mammalian auditory system has been characterized with regard to its capabilities in responding to simple, synthetic, and perhaps even artificial stimuli. Such an approach is an important first step before complex biologically relevant stimulus events can be examined. Researchers are now beginning to treat the issue of biological relevance in the laboratory and in the context of the evolution of hearing. Although there is little of substance to add to the evolutionary framework of mammalian hearing, the approach is of sufficient importance that two experimental examples may be

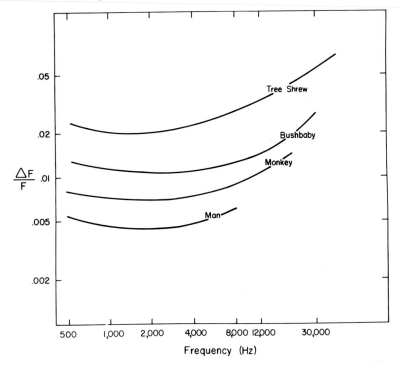

Figure 15-5. Frequency discrimination functions for several primate species: tree shrew (*Tupaia glis*), Heffner et al. 1969a; bush baby (*Galago senegalensis*), Heffner et al. 1969b; macaque (*Macaca*), Stebbins 1978; and man (*Homo sapiens*), Filling 1958 (from Stebbins 1978). Copyright by Academic Press Inc. (London) Ltd.

mentioned here. In the first laboratory experiment, subjects of one primate species learned more rapidly than did subjects of a related species to discriminate between many instances of two calls given by members of their own species (Zoloth, Petersen, Beecher, Green, Marler, Moody, and Stebbins 1979). Further, these same animals in this experiment showed evidence of a significant right ear advantage in learning and, thus, of possible left hemisphere dominance for species calls (Petersen, Beecher, Zoloth, Moody, and Stebbins 1978). In a second laboratory experiment, Old World monkeys showed clearly that their ability to accurately localize primate calls depended on the acoustic structure of the calls. The accuracy with which a call could be located in space was directly related to the extent of the frequency modulation occurring in the call (Brown, Beecher, Moody, and Stebbins 1978).

Such experiments exemplify an approach to comparative hearing that goes beyond detection and discrimination to the significance of biological stimuli in the life of the animal. It is also an approach that enlists the support of the field and evolutionary biologist in order to correctly identify those issues appropriate to the life history and phylogenesis of animal hearing and communication.

8 Summary

I have attempted to identify some of the selective pressures that have occurred in the course of mammalian evolution and have led to the structural and behavioral changes in the auditory system of modern mammals from that of reptilian and early mammalian forms. A certain amount of speculation has been unavoidable because the data are insufficient, and frequently we have had to assume phyletic ordering in living species. The purpose has not been to write the final chapter, but rather an introductory one in which questions are raised and issues highlighted. At this stage the structure-behavior relations are relatively gross. An ear canal with certain resonance properties and a three-bone ossicular chain appear to have produced a considerable enhancement in acoustic sensitivity in modern mammals relative to their reptilian progenitors. A cochlea that substantially increased its length by coiling, together with some as yet unspecified changes in receptor cell and neural innervation patterns, may have further improved sensitivity but almost certainly increased the frequency range of hearing and the ability to localize sound. It is argued that in later mammals, particularly the anthropoid primates, further morphological changes in the central nervous system resulted in greatly improved differential acuity. Selective pressures that were originally related to sound localization in nocturnal environments may have shifted somewhat as some of the mammals moved into diurnal adaptive zones, and social organization and hence intraspecific communication became more elaborate and complex. Sensitivity in the low frequency region in which these biosignals were emitted was important, and perhaps high frequency sensitivity yielded, in some of these diurnal species, to better differential acuity, particularly in this same frequency region.

Acknowledgments. The preparation of the chapter and some of the research reported therein was supported by grants from the National Science Foundation (BNS77-19254) and the National Institutes of Health (NS05785). I am grateful to J. E. Hawkins, Jr., G. A. Manley, and the editors for their critical comments on the first draft of the manuscript.

References

Aitkin, L. M., Gates, G. R., Kenyon, C. E.: Some peripheral auditory characteristics of the marsupial brush-tailed possum, *Trichosurus vulpecula.* J. Exp. Zool. 209, 317-322 (1979).

Aitkin, L. M., Johnstone, B. M.: Middle-ear function in a monotreme: the Echnida (*Tachyglossus aculeatus*). J. Exp. Zool. 180, 245-250 (1972).

Allin, E. F.: Evolution of the mammalian middle ear. J. Morph. 147, 403-438 (1975).

Baird, I. L.: Anatomical features of the inner ear in submammalian vertebrates. In: Handbook of Sensory Physiology: Vol. V/1, Auditory System. Keidel, W. D., Neff, W. D. (eds.). New York: Springer-Verlag, 1974, pp. 159-212.

Bakker, R. T.: Dinosaur physiology and the origin of mammals. Evol. 25, 636-658 (1971).

Barghusen, H. R.: The lower jaw of cynodonts (*Reptilia, Therapsida*) and the evolutionary origin of mammal-like adductor jaw musculature. Postilla (Peabody

Mus. Nat. Hist. Yale Univ.). 116, 1-49 (1968).

Brown, C. H., Beecher, M. D., Moody, D. B., Stebbins, W. C.: Localization of primate calls by Old World monkeys. Science. 201, 753-754 (1978).

Dalland, J. I.: The measurement of ultrasonic hearing. In: Animal Psychophysics: The Design and Conduct of Sensory Experiments. Stebbins, W. C. (ed.). New York: Plenum Press, 1970, pp. 21-40.

Elder, J. H.: Auditory acuity of the chimpanzee. J. Comp. Psych. 17, 157-183 (1934).

Fay, R. R.: Auditory frequency discrimination in vertebrates. J. Acoust. Soc. Am. 56, 206-209 (1974).

Fernandez, C., Schmidt, R. S.: The opossum ear and evolution of the coiled cochlea. J. Comp. Neurol. 121, 151-159 (1963).

Filling, S.: Difference Limen for Frequency. Denmark: Andelsbogtrykkeriet, 1 Odense, 1958.

Fleischer, G.: Evolutionary principles of the mammalian middle ear. Adv. Anat. Embryol. Cell Biol. 55, 1-70 (1978).

Gans, C., Maderson, P. F. A.: Sound producing mechanisms in recent reptiles: review and comment. Amer. Zool. 13, 1195-1203 (1973).

Gray, O.: A brief survey of the phylogenesis of the labyrinth. J. Laryngol. Oto. 69, 151-179 (1955).

Griffiths, M.: Echnidas. Exeter: Pergamon Press, 1968.

Heffner, H. E., Ravizza, R., Masterton, R. B.: Hearing in primitive mammals, III: Tree shrew (*Tupaia glis*). J. Aud. Res. 9, 12-18 (1969a).

Heffner, H. E., Ravizza, R., Masterton, R. B.: Hearing in primitive mammals, IV: Bushbaby (*Galago senegalensis*). J. Aud. Res. 9, 19-23 (1969b).

Henson, O. W.: Comparative anatomy of the middle ear. In: Handbook of Sensory Physiology: Vol. V/1, Auditory System. Keidel, W. D., Neff, W. D. (eds.). New York: Springer-Verlag, 1974, pp. 40-110.

Hodos, W., Campbell, C. B. G.: *Scala Naturae*: Why there is no theory in comparative psychology. Psychol. Review. 76, 337-350 (1969).

Hopson, J. A.: The origin of the mammalian middle ear. Am. Zool. 6, 437-450 (1966).

Hopson, J. A.: The origin and adaptive radiation of mammal-like reptiles and nontherian mammals. In: Comparative and Evolutionary Aspects of the Vertebrate and Central Nervous System. Petras, J. M., Noback, C. R. (eds.). Ann. N. Y. Acad. Sci. 167, 199-216 (1969).

Hopson, J. A.: Endothermy, small size, and the origin of mammalian reproduction. Am. Naturalist. 107, 446-452 (1973).

Jerison, H. J.: Evolution of the Brain and Intelligence. New York: Academic Press, 1973.

Kermack, D. M., Kermack, K. A., Mussett, F.: The Welsh pantothere *Kuehnotherium praecursoris*. J. Linn. Soc. (Zool.) 47, 407-423 (1968).

Lombard, E. R., Bolt, J. R.: Evolution of the tetrapod ear: an analysis and reinterpretation. J. Linn. Soc. (Biol.) 11, 19-76 (1979).

Manley, G. A.: Some aspects of the evolution of hearing in vertebrates. Nature. 230, 506-509 (1971).

Manley, G. A.: Frequency response of the ear of the Tokay Gecko. J. Exp. Zool. 181, 159-168 (1972).

Manley, G. A.: A review of some current concepts of the functional evolution of the ear in terrestrial vertebrates. Evolution. 26, 608-621 (1973).

Marcellini, D. L.: The acoustic behavior of lizards. In: Behavior and Neurology of Lizards. Greenberg, N., MacLean, P. D. (eds.). NIMH, 1978, pp. 287-300.

Masterton, R. B.: Adaptation for sound localization in the ear and brainstem of mammals. Fed. Proc. 33, 1904-1910 (1974).

Masterton, R. B., Heffner, H. E., Ravizza, R.: The evolution of human hearing. J. Acoust. Soc. Am. 45, 966-985 (1969).

McCandless, G. A., Madsen, J. H., Parkin, J. L.: The auditory system of an allosaurus, a late Jurassic therapod dinosaur. J. Acoust. Soc. Am. 64, S84 (1978).

McKenna, M. G.: The origin and early differentiation of therian mammals. Ann. N. Y. Acad. Sci. 167, 217-240 (1969).

Miller, J. D., Watson, C. S., Covell, W. P.: Deafening effects of noise on the cat. Acta Otolaryng. Supp. 176, 1-91 (1963).

Olson, E. C.: Vertebrate Paleozoology. New York: Wiley-Interscience, 1971.

Parrington, F. R.: The evolution of the mammalian middle and outer ears: a personal review. Biol. Rev. 54, 369-387 (1979).

Patterson, W. C.: Hearing in the turtle. J. Aud. Res. 6, 453-464 (1966).

Petersen, M. R., Beecher, M. D., Zoloth, S. R., Moody, D. B., Stebbins, W. C.: Neural lateralization of species-specific vocalizations by Japanese macaques (*Macaca fuscata*). Science. 202, 324-327 (1978).

Raslear, T. G.: The use of the cochlear microphonic response as an indicant of auditory sensitivity: review and evaluation. Psychol. Bull. 81, 791-803 (1974).

Romer, A. S.: The Vertebrate Story. Chicago: University of Chicago Press, 1959.

Romer, A. S.: Vertebrate Paleontology. Chicago: University of Chicago Press, 1966.

Schmidt, R. S.: Phylogenetic significance of lizard cochlea. Copeia. 3, 542-549 (1964).

Sivian, L. J., White, S. D.: On minimum audible sound fields. J. Acoust. Soc. Am. 4, 288-321 (1933).

Simpson, G. G.: The study of evolution: methods and present status of theory. In: Behavior and Evolution. Roe, A., Simpson, G. G. (eds.). New Haven: Yale University Press, 1958, pp. 7-26.

Simpson, G. G.: The Meaning of Evolution. New Haven: Yale University Press, 1967.

Stebbins, W. C.: Studies of hearing and hearing loss in the monkey. In: Animal Psychophysics: The Design and Conduct of Sensory Experiments. Stebbins, W. C. (ed.). New York: Plenum Press, 1970, pp. 41-66.

Stebbins, W. C.: On the evolution of hearing. (Paper presented in the symposium "Comparative Hearing in the Vertebrates" at the meeting of the Acoustical Society of America, San Francisco, 1975. Mimeographed.)

Stebbins, W. C.: Hearing of the primates. In: Recent Advances in Primatology, Vol. 1, Primate Behavior. Chivers, D., Herbert, J. (eds.). London: Academic Press, 1978, pp. 705-720.

Tumarkin, A.: Evolution of the auditory conducting apparatus in terrestrial vertebrates. In: Hearing Mechanisms in Vertebrates. De Reuck, A. V. S., Knight, J. (eds.). Boston: Little, Brown Co.. 1968, pp. 18-36.

Van Bergeijk, W. A.: The evolution of vertebrate hearing. In: Contributions to Sensory Physiology. Neff, W. D. (ed.). New York: Academic Press, 1967, pp. 1-49.

Webster, D. B.: Ear structure and function in modern mammals. Am. Zoologist. 6, 454-466 (1966).

Wever, E. G.: The Reptile Ear: Its Structure and Function. Princeton: Princeton University Press, 1978.

Zoloth, S. R., Petersen, M. R., Beecher, M. D., Green, S., Marler, P., Moody, D. B., Stebbins, W. C.: Species-specific perceptual processing of vocal sounds by monkeys. Science. 204, 870-873 (1979).

Future View

In this final chapter, Bullock presents to us a personal view of the issues and problems in the study of hearing to which multidisciplinary comparative approaches will most likely contribute significantly in the years to come.

Chapter 16

Comparative Audition: Where Do We Go from Here?

THEODORE H. BULLOCK*

1 Introduction

My assignment is to give a personal view of some of the questions in comparative audition that are likely to produce interesting new developments in the near future. My main qualification for this task is a high respect for the goals of this volume and these authors. I am thoroughly fascinated by the subject taken up, and by the progress reported. It represents an outstanding entry into the general problems of how the brain works in processing information and recognizing natural stimuli and of the ways neuroethology might tell us how brains are different in species that behave differently.

I have been an ardent admirer of auditory physiologists and anatomists for a long time. Occasionally my own laboratory has gone into decibels, as with the work on dolphins (Yanagisawa, et al. 1966, Bullock et al. 1968, Bullock and Ridgway 1971a, 1971b, 1972a, 1972b, Bullock and Gurevich 1979); sea lions (Bullock, Ridgway, and Suga 1971); insects, edentates, and bats (Suga 1966, 1967, 1968a, 1968b, 1968c, 1968d, 1969a, 1969b, 1969c, 1977, Grinnell 1970); lizards (Suga and Campbell 1967); snakes (Hartline 1971a, 1971b, Hartline and Campbell 1969); birds (Biederman-Thorson 1967, 1970a, 1970b); fish (Piddington 1971a, 1971b, 1972); sharks (Bullock and Corwin 1977, 1979, Corwin 1977b, 1978); and manatees (Bullock, Domning, and Best 1980). Still, this is the voice of a real amateur in acoustics, a general neurophysiologist, and a zoologist; so what follows has the advantage of naiveté!

What follows is a parade of problems that strike me as being probably answerable through present techniques, peculiarly in need of the comparative approach, and especially pregnant in their potential significance for understanding the human organism.

*Neurobiology Unit, Scripps Institution of Oceanography and Department of Neurosciences, School of Medicine, University of California San Diego, La Jolla, California 92093.

2 The Roles of VIIIth Nerve Acoustic Reception versus Lateral Line Mechanoreception in Lower Vertebrates

This problem is a hoary chestnut in zoology, but it looks as though it is taking a drastic new turn. Now that displacement (= velocity) wave sensitivity in the ear is being more often advocated for some fish instead of only pressure component sensitivity, the issue must be reopened: how are VIIIth nerve acoustic reception and lateral line mechanoreception different? I refer to the evidence for directional localization of sound sources at hundreds of meters distance in sharks (Nelson and Johnson 1976, Myrberg, Ha, Walewski, and Banbury 1972) and in teleosts (Schuijf 1975, 1976, Schuijf and Buwalda 1975, Schuijf and Hawkins 1976) as well as the evidence from microphonic potentials and single unit physiological preparations (Sand 1976, Fay and Olsho 1977, 1979), all indicating that besides whatever the pressure component of the acoustic wave can do, the displacement component can also be effective at presumed naturally occurring levels of intensity. A bit more comparative work is needed to allow us to tell a coherent story to our students.

I'm sure we each see it differently, but here is one view. Whereas in the days of Parker (1909) the distinction was thought to be that low frequencies stimulate the lateral line and higher frequencies stimulate the ear, van Bergeijk (1977), Dijkgraaf (1963), and others changed the distinction to emphasize the contrast between near and far sound sources. Van Bergeijk thought of the near-by stimulus as a frequency, Dijkgraaf thought of it as a bulk movement of water, both implicated the lateral line. The pressure component from distant sources has been thought to stimulate the VIIIth nerve via the pressure sensitive swim bladder or a bulla system (Blaxter, Denton, and Gray 1979, Gray and Denton 1979, Denton, Gray, and Blaxter 1979). Since that meant that the ears could not tell direction of a sound source (van Bergeijk 1964), Piddington (1971b) introduced the idea that the acoustic detectors in the labyrinth may usefully tell only whether a source is approaching or receding by the slight excess over 50% of the cycle spent in compression or in rarefaction. He showed that fish can tell the difference behaviorally. Now we know that at least some teleosts can tell the azimuth of a source and utilize the displacement component of sound even from distant sources. But that left a real mystery as to the physiology of acoustic maculae, if there are any, in fish without swim bladders like elasmobranchs. It was doubted that these animals have any acoustic reception other than a lateral line sense. Now we are told that sharks as well as teleosts not only hear sound from hundreds of meters, but can use the displacement component, and get immediate direction cues from such distant sources, in frequency ranges overlapping with that of lateral line receptors. Is the distinction between lateral line and ear function back where Parker had it, mainly in sensitivity and frequency range? Perhaps there is also a difference in the sensitivity as a function of sound source azimuth and elevation. The answer, or at least substantial new evidence, is technically accessible, but we won't have a good idea of the right emphasis unless we have information from several major taxa. Such information will add greatly to our knowledge of neuroethology and of vertebrate evolution.

3 The Roles of Centrifugal Nerve Fibers to the Ear in Different Taxa

This nut is not quite so old but it may be even harder to crack; seemingly more people have worried more about it. My main wager is that it will yield sooner and better, i.e., a more adequate proposition will emerge, if we look around for favorable taxa and compare both diverse species and diverse parts of the acoustico-lateralis, or better, the octavo-lateralis system. We are at a stage where we need proposals in order to argue that this is bad or unlikely and that is better. Ideas are more likely to spring to mind if we have more bits of the jigsaw puzzle, such as the apparent absence of centrifugals to various octavo-lateralis organs. Among these organs are several independently evolved types of electroreceptors (Szabo 1974), outer hair cells in the cochlea of *Rhinolophus ferrumequinum* (horseshoe bat; see Bruns 1976, 1979), the basilar papilla of frogs, and the lateral line free neuromasts in lampreys. Similarly there should be clues in the disparity in abundance of efferents to vestibular and cochlear hair cells and between inner and outer hair cells of the cochlea of *Felis cattus* (cat; see Spoendlin 1968). I expect that at least some ideas would come from comparing several types of neuromast organs and statolith organs such as the sacculus and utriculus. If the latter are different, then comparison of ordinary fish and flatfish, where tilt nystagmus is controlled by the sacculus instead of the utriculus (Platt 1973), should be suggestive. Anatomy and physiology of efferents and, if possible, behavioral changes with efferents stimulated or lesioned should be studied. The same is true for species of sharks with a large macula neglecta as compared to sharks with small maculae and perhaps poorer hearing. The big maculae, Corwin (1978) tells us, have 300000 hair cells and only 5000 nerve fibers, including any efferents there may be. Miller (Chapter 6) says some snakes have a high afferent-to-efferent ratio in the basilar papilla. Where the physiology is best known, as in shark semicircular canal organs, cat cochleas, and frog vestibules, I believe it will pay to look at contrasting species; that is, species with a different lifestyle—maybe *Torpedo* and *Chimaera* (ratfish); *Geomys* (gopher) and *Aplodontia* (mountain beaver), *Necturus* (mud puppy) and *Gymnophiona* (caecilian). To give zest to the search, I remind you that we are in a race with the retina people who have at present about as meager an idea of the normal functional role of centrifugals to the eye as we have to the ear.

4 The Role of Number of Hair Cells

This innocent sounding problem could have wide ramifications. What does it mean that each inner hair cell of our cochlea has many nerve fibers, each of which supplies only one hair cell, whereas some sharks must have an average of well over 60 hair cells for each nerve fiber in the macula neglecta? High ratios are likewise known in the utricle of *Lota* (Flock 1964) and all the otolithic organs of *Amia* (bowfin) (Popper and Northcutt unpublished) as well as in the outer hair cells of the mammalian cochlea. If sensitivity is enhanced by adding hair cells, especially those oriented in a common direction, why are other species of sharks so relatively deficient, and why do we have so few, at least in the inner hair cells, where most of our afferent nerve fibers go? Is there a high price for hair cells or a countervailing selection pressure for something else? Is the hair cell to nerve fiber ratio of 1:20 in our inner hair cells significant for smoothing out noise of a different kind, such as synaptic noise, without the kind of gain in sensitivity that the reverse ratio might give the outer hair cells?

I have cast these as possible examples of the principle of overcoming system noise by averaging parallel channels to detect weak signals. What is a shark doing that requires 300000 hair cells in one macula? Corwin (1977a) suspects it is detecting extremely small displacements, comparing the intensities or phases to get azimuth, integrating over time and summing receptor potentials from many hair cells to enhance sensitivity to sound from the best azimuth, hence improving directional selectivity. Possibly other maculae and lateralis organs also have diversity in number of hair cells and number of nerve fibers. It may well be that more than numbers and ratios are involved in the differences of roles among maculae or between the inner and the outer hair cell roles in our cochlea. Other corollary aspects may be at least as important but the problem will be better understood, I am sure, if it is examined in diverse taxa. Subdividing the issue into answerable formulations may be the next step; then, comparing diverse species is likely to suggest new clues.

5 The Roles of Granule Cells and of Stratification in Cochlear Nuclei

Moving into the central nervous system, I choose first an issue that is difficult but could have a broad interest beyond audition. Even a plausible but unproven idea would be a contribution. Put in the context of comparative audition: what could be the role of layers and even masses of granule cells as found in the dorsal cochlear nuclei of some species and not others in the same mammalian order? Among the rodents, *Geomys* (pocket gopher) and *Aplodontia* (mountain beaver) (Merzenich, Kitzes, and Aitkin 1973) show a granule cell layer or mass much more developed than in other members of the order; the primates *Lemur* and other prosimians likewise exhibit a much better granule cell layer than simians or apes (Moore 1979). What is the functional meaning of lamination in general? In what respects are humans and apes, with poor lamination, different functionally from prosimians? Although the human dorsal cochlear nucleus looks like that of prosimians with arrested ontogenetic development, we are not considered to be inferior in hearing in any of a number of tests that I have heard of.

Let us ask ourselves what might be the functional consequences of lamination over and beyond any ontogenetic explanation of how it comes about. I believe that, even if a developmental hypothesis adequately accounts for it, we should ask what advantage, if any, is conferred on the functioning of the more laminated cortex. In a recent symposium devoted to laminated structures (Creutzfeldt 1976), the question of the functional meaning does not appear to have been discussed. Surely it is not so self-evident a question that an explicit proposal is trivial even if very speculative and unfinished. Can we even be sure, as we are so inclined to say, that better lamination goes with more highly developed or complex function? I don't believe our knowledge of auditory performance in comparable species is good enough in respect to dorsal cochlear nucleus function. Is the optic tectum of reptiles really more complex in function than that in mammals? The torus semicircularis of gymnotoid electric fish is much more laminated than its homolog, the inferior colliculus, in mammals, or the torus of most fishes, including that of mormyrid electric fish. Elsewhere I have developed this theme and a proposition about physiological consequences of stratification (Bullock 1980). Briefly, it is suggested that stratification permits more dependence on local interactions. As evolution develops more confinement of the target of each type of axon to limited parts of the postjunctional neurons, there may be more differentiation of axonal terminals, dendrites, and output messages. With the evolution of these forms of increased specification, a tendency to bring together into one stratum the corresponding parts of each cell type would provide several advantages. One is to shorten the aggregate length of the fibers in the pre- and postjunctional arbors in comparison to the hypothetical, disordered situation where the target portions of the set of cells to which each fiber sends projections would be scattered through the depth of the cortex. Stratification would permit far more connections per unit of volume. Another consequence would be to permit short axon cells to become more important and to reach more targets with their limited arbors. Perhaps more significantly, it would permit relatively more dependence on local interactions, especially those that involve more than two neurons. Local circuitry means not only connectivity mediated by graded, local events via defined junctions, as in spikeless transmission, but also the influence of local fields of current and the action of transmitters that diffuse through some volume of tissue.

It is further proposed that an even more highly developed system might not appear as well stratified as a less evolved system, by any particular method of preparation. With sufficient complexity, a cortex could develop not more or sharper strata, as seen by any given method, but more kinds of stratification, each requiring a different kind of criterion; for example, histochemistry for various enzymes, transmitters, or other substances, current source density of evoked potentials, electron microscopy, silver impregnation or discontinuities of cell, myelin, or dendrite distribution.

On such bases, it could be that the human dorsal cochlear nucleus is more advanced than that of the prosimian. It has been noted repeatedly in this volume that human hearing is at least as good as that of other species. Yet it may be that our performance tests involve the ventral cochlear nucleus more than the dorsal or do not stress the capabilities of the latter. We need to know in what ways the acoustic analysis of a lemur is different from our own, or that of the rat from that of the gopher, in behavioral performance and in physiology attributable to this structure. Towards opening up new clues and fresh material, I wonder what other groups could be found to show a range of development of granule cells and of lamination in this nucleus, among the orders of mammals, reptiles, and birds. In teleosts, Northcutt (Chapter 3) tells us that the products of differentiation of the medial nucleus (called the posterior lateral line lobe by Maler 1979, Maler, Finger, and Karten 1974) can be cortical in structure and have a good layer of granule cells in some fishes. But in most taxa for which I have seen illustrations, there is little sign of such differentiation.

Physiologically, it is not at all hopeless to record from granule cell layers, as the elegant work of Shames, Gibson, and Welker (1978) shows. They found columnar organization and "fractured mosaic" mapping of the skin by recording with a spacing of 50 μm between electrode tracks; that means 200 tracks/mm^2! The opportunities for gaining insight into the physiology of strata, of small cells, and of local circuits, include many parts of the auditory system and all the diversity of the vertebrate classes. I think there are clues to be checked into and unknowns sufficient to keep motivated detectives busy and happy for some time.

6 The Role of the Cerebellum and Monaural Clues in Localization of Sound Sources

Here, as in the preceding item, I mix together, for the sake of flavor, two ingredients of quite distinct composition. My task, remember, is to give a personal view of what might be exciting just ahead. Brain mechanisms involved in localizing sources of sound are still in an early stage of understanding, and I expect new developments of several kinds, including the answers to what, where, and how. The "what" seems likely to include finding neurons that prefer sources at a certain distance, making the acoustic map in the brain a three dimensional model. I expect this finding will apply mainly to sound sources within a few meters and will require, instead of earphones, sources in normal surroundings with echoing objects and natural sounds and that the animal be unanesthetized with indwelling electrodes. Possibly the first place to look for this finding is in the hippocampus since we know it has an absolute or geometric map of the familiar positions of things (O'Keefe and Nadel 1978). Some units in the hippocampus become active whenever the animal goes to the corner with the water bottle, others when it goes to that corner and finds the bottle missing.

I would like also to call attention to the possibility that the cerebellum may be a system that computes or maps acoustic space, perhaps relative or egocentric space instead of absolute or geocentric space. A good many hints suggest this but there is a dearth of hard evidence. The cerebellum, as I see it now, was invented to compute and display something about the nearby extra-personal space, the position of the body and appendages in it, and the expected reafference from self-generated movements. Since systems homologous to the auditory, such as the electroreceptive system in electric fish, extensively project to the cerebellum (Bastian 1974, 1975, 1976, Behrend 1977, Russell and Bell 1978) and units in the cerebellum have small receptive fields—often with excitatory centers and inhibitory surrounds, commonly with directional motion preference, sometimes also a preferred distance—it does not seem farfetched to look for localizing signs of nearby acoustic disturbances. I can't help speculating that one reason for the huge cerebellum of cetaceans—as much as twice its relative size in the human brain—is a significant role in localizing sound sources and sources of echos.

Incidentally, as further indication of the relevance of electrosensory processing to auditory physiology, I would like to point out that the kind of computed map that Knudsen and Konishi (1978; Knudsen, Chapter 10) elegantly reported in the acoustic midbrain of the owl, was already shown by Knudsen (1978) in the electrosensory torus semicircularis of catfish with respect to both the azimuth of a small, near-field, dipole source, and the orientation of a dipole.

The other question in the heading of this section is perhaps closely related. It appears to me that there has been a relative neglect of the search for neurons that repond to sound source position when only monaural clues are available. Granted that interaural clues are usually available and influential and by themselves can give good illusions of sound source localization; it remains true, nevertheless, that we have a surprising degree of monaural localization, especially in front, and that the binaural clues are quite different in most vertebrates from those available in the human, with a relatively large head. The brain can use other clues that are much more complex than simple time difference between ears (see Knudsen, Chapter 10; Schuijf and Buwalda, Chapter 2; Suga and O'Neill 1978). It seems likely that the cerebellum contributes to this analysis.

There is an issue in the literature concerning complex analysis and interaction of modalities that seems likely to be cleared up at any time and will be extremely interesting whichever way it turns out. This is the question raised by Pöppel (1973) about Morrell's (1972) bimodal units in the cat visual cortex, area 18, that have acoustic responses with a narrow excitatory field perfectly congruent in the anesthetized preparation with the azimuth vector of their visual field. Pöppel asked what would happen if the cat deviated its eyes voluntarily without moving its head. If the visual field shifts in space accordingly but the acoustic field does not, we would be faced with a curious disparity. If the acoustic field shifts with the visual, although the head is fixed, we would be faced with a remarkable example of compensatory computation based on expectation. In an old terminology that needs reformulating today, it would be an outstanding case of efference copy, specified in both sign and magnitude.

7 Parallel Subsystems in Different Taxa

Another question of broad neurobiological interest can be phrased this way: can the acoustic processing system be divided into more or less distinct, parallel subsystems interested in different functions of audition or biological types of sounds?

In the visual system, we are used to thinking of two parallel subsystems, the retinotectal and retinogeniculostriate, plus at least four other destinations of optic nerve, fibers: the pretectal area, the accessory optic nucleus, the suprachiasmatic nucleus, and the ventral nucleus of the lateral geniculate body. Each of these subsystems, and probably some subdivisions of them, is concerned with a different sphere of visual events, gradually being sorted out by neuroethologists (e.g., Ingle 1976, 1977). Little of this sort of discourse is common in auditory physiology, although the idea has been around and some evidence is at hand. For instance, we found in the dolphin years ago (Yanagisawa et al. 1966, Bullock et al. 1968, Bullock and Ridgway 1972) that the kind of ultrasonic clicks used in echolocation and the sonic frequency, longer duration, slow onset so-called social communication sounds, are handled quite differently, in pathways that are largely separate anatomically. Could it be that there are several parallel auditory subsystems in familiar species, waiting to be untangled by ethologically designed stimuli? Cortical areas in bats, one concerned more with frequency modulation and another concerned with amplitude, suggest parallel subsystems more than sequential stages of processing (Suga 1977, Suga, O'Neill, and Manabe 1978). How about cortical auditory areas A1, A2, A3, and others in cats and monkeys; are these merely hierarchical stages in a unitary system? This seems unlikely. I know that a great deal of thought and experiment have gone into this question, and I expect major new findings and interpretations in the near future to change our picture of "the" auditory system. That day will be sooner, if specially favorable taxa are compared with ethologically chosen stimuli. I should underline that the progress of the search will depend greatly on the way in which the question is formulated, and that in vision research, they are still groping for the most appropriate formulations. I believe that whereas discovery of separable "parallel" subsystems is a major step forward, it is going to turn out that we need new concepts and terms for the functional distinctions, in part because the anatomically distinguishable pathways interact in complex ways.

That brings me to the next and most difficult issue to pose.

8 Requirements for Neuronal Interaction in Recognition of Natural Stimuli

A major area of future excitement, I am confident, will be the gradual unveiling, or rather the pick and shovel unearthing, of details here and there of the general form and principles of the recognition apparatus that mediates appropriate responses to species characteristic sounds and learned sounds—grunting, croaking, warbling, barking, chattering, individual voices, human speech, dialects, and musical styles. This domain of inquiry is singularly divisive, but I take that as a sign of ferment and promise. I take for granted disagreements over the meaning of experimental findings and over the completion of the sentence "The real question is . . . ." I regret, but expect in each conversation, a period at crossed purposes due to differences in usage of terms such as "feature detectors." We heard yesterday that "None of us in the c.n.s. is looking for feature detectors anymore." Few people ever were, if you mean, as I do, that feature detectors are units belonging to a set most responsive to a fairly complex stimulus—yet they keep cropping up. The puzzle of how the brain "recognizes" or responds appropriately to significant, complex stimuli will not come together neatly or in a logical sequence in the normal course of scientific advance. The general picture will remain clouded and ambiguous for some time, but I predict we will see tantalizing glimpses and exciting findings as we slowly and falteringly distinguish the several solutions. Surely there will be not a single universal recognition mechanism nor twenty basically different mechanisms!

There are remarkably few people working directly on the brain processing of complex natural sounds in relation to the magnitude of the problem, the range and variety of sounds, and the several types of basically distinct neuronal mechanism that will be found. There have been some findings, but this is not the place to review those findings or the variety of research strategies. I will only underline the large relevance beyond audition of any principles we may find in auditory recognition of complex natural sounds. They would bear not only on central analysis of other modalities of input but on general issues of how the brain works in respect to coding, decoding, and reencoding, to reliability and redundancy, to statistical configurations versus defined circuits and similar issues.

I know well that for some people these terms are turn-offs; they sound soft or messy. Let me try another statement of a domain of investigation that I claim will pay large dividends—not only in the remote future but as fast as we explore it. This is another way of looking at the same can of worms. The domain can be regarded as the search for, and characterization of neurons, at various levels of the central auditory system, that require the interaction of several aspects of the input. Scheich, Langner, and Koch (1977) have higher order units in a guinea fowl that "prefer" two specific tones simultaneously, plus the absence of others. Suga and Capranica, and Moffat (see Chapter 5) and others also have relevant findings. Complexity like this probably begins already in the acoustic nucleus and builds up from there on. We should expect at some level that various permissive interactions will be added such as a requirement of time of day, or hormonal state, of absence of certain visceral afferents, and of some specified recent history with respect to habituating or depressing stimuli, besides the specific recognition criteria for the natural stimulus in question.

I call even moderately complex cells "recognition cells" because they strongly suggest that they are well along in the process of appropriate response to natural stimuli. I do not imply that they are the final stage or that they necessarily act singly as decision units (Bullock 1961). They may in some cases, as Mauthner's cells do in "deciding" when to trigger a startle response (Faber and Korn 1978), but it is more reasonable to assume that usually a limited set with overlapping receptive fields (that is, overlapping labels) operates together.

The discovery of such units and the process of working out their labels are so much a matter of luck and of unsystematic approximation that we can have little idea even now, with 20 years of experience, how widespread, complex, and fixed or labile they are. The few people who are working directly on recognition cells are now very cautious and although they find units with complex requirements, they do not expect units to be found easily, if at all, that are highly specific to a single natural call and unresponsive to any other calls of the same and other species. But such units are not excluded; they may be few and hard to find. What is more likely now is some large degree of convergence, filtering and abstraction, funneling down to a modest array of complex, partially overlapping units that together signal a particular natural call. We have no idea how close we have been with electrodes to the narrowest part of the funnel for any given characteristic response. By the way, it is curious that in spite of some effort, we are still ignorant of units with temporal sequence requirements of sufficient specification to help substantially in explaining behavioral specificity in cricket chirp patterns or bird calls with important temporal structure. I wager that such units will be found.

I have dwelt on this topic because it occupies an important place between the lower level quantification of afferent unit properties and the study of "higher nervous functions." Let us turn now to the last topic, which impinges on these functions.

9 Acoustic Roles of Higher Centers in Lower Vertebrates

Tangible progress in this neglected area is beginning to occur when we hear from Capranica and Moffat (Chapter 5; Mudry, Constantine-Paton, and Capranica 1977) that they have found units, of a higher level of integration in the frog auditory thalamus than any so far found in the frog midbrain or lower levels, that fire impulses in response to combination tones like those in the guinea fowl thalamus (Scheich, Langner, and Koch 1977). The same is true when we learn that sharks have a definite acoustic projection in the telencephalon (Bullock and Corwin 1979). These are not earth-shaking surprises, but they serve to remind us of two things. The first is, our information about acoustic representation in the telencephalon and diencephalon of lower groups is really meager. This means we can surely expect proportionately large increment to that information; a few new findings will double our knowledge! The other is, we stand to learn much about c.n.s. organization in general, hierarchical processing, and higher functions when we use the leverage of auditory physiology and biobehavioral study or comparative psychophysics. Can we find unit or compound potentials signaling a missing click in a series? If so, is it more likely in the forebrain than in the midbrain? Can we find, in units of complex requirements such as Scheich's or Capranica's combination tone units, a plasticity with hormonal level, or threatening visual stimuli, or alarm substances like the Schreckstoff of teleosts?

Units and spikes are relevant and informative but they cannot tell us everything. Both graded, local processes and emergent behavior of large populations call for the use of other avenues to what is going on. We are seriously short of an adequate variety of avenues and should not neglect the ones we have. The simplest is the field potential of a small population, and in the context of comparative study of audition, the simplest level of exploitation of such field potentials is the study of evoked potentials.

Evoked potentials have been extensively and profitably exploited in laboratory mammals and humans but relatively little has been done with lower vertebrates and invertebrates. I look forward to clarification in the near future of such questions as these: In what respects are evoked potentials similar and in what respects different, in animals without a cortex? Are there late, slow waves and, if so, do they share any characteristics with those ("P-300") associated with cognitive processes in humans (Hillyard, Picton, and Regan 1978)? Is evoked potential shape a useful tool for revealing discrimination ability, difference limens and biologically meaningful stimuli? From the limited evidence available, citing only examples from teleosts (Piddington 1972, Bullock 1979) and elasmobranchs (Platt, Bullock, Czeh, Kovaćević, Konjević, and Gojkovic 1974, Bullock and Corwin 1979, Bullock 1979), I think we can answer, yes, this tool looks promising for uncovering new answers and new questions in comparative hearing.

Under eight headings I have suggested topics on which I expect something in the near future—from an interesting new finding to a substantial change in our view of nature. Whereas all of them will contribute to our understanding of acoustic reception and processing, some will also influence our general understanding of how the nervous system works.

Acknowledgment. This work was aided by grants to Dr. Theodore H. Bullock from the National Institutes of Health and the National Science Foundation.

References

Bastian, J.: Electrosensory input to the corpus cerebelli of the high frequency electric fish *Eigenmannia virescens*. J. Comp. Physiol. 90, 1-24 (1974).

Bastian, J.: Receptive fields of cerebellar cells receiving exteroceptive input in a gymnotid fish. J. Neurophysiol. 38, 285-300 (1975).

Bastian, J.: The range of electrolocation: A comparison of electroreceptor responses and the responses of cerebellar neurons in a gymnotid fish. J. Comp. Physiol. 108, 193-210 (1976).

Behrend, K.: Processing information carried in a high frequency wave: Properties of cerebellar units in the high frequency electric fish. J. Comp. Physiol. 118, 357-371 (1977).

Biederman-Thorson, M. A.: Auditory responses of neurons in the lateral mesencephalic nucleus (inferior colliculus) of the Barbary dove. J. Physiol. 193, 695-705 (1967).

Biederman-Thorson, J.: Auditory evoked responses in the cerebrum (field L) and ovoid nucleus of the ring dove. Brain Res. 24, 235-245 (1970a).

Biederman-Thorson, M.: Auditory responses of units in the ovoid nucleus and cerebrum (field L) of the ring dove. Brain Res. 24, 247-256 (2970b).

Blaxter, J. H. S., Denton, E. J., Gray, J. A. B.: The herring swimbladder as a gas reservoir for the acousticolateralis system. J. Mar. Biol. Ass. U. K. 59, 1-10 (1979).

Bruns, V.: Peripheral auditory tuning for fine frequency analysis by the CF-FM bat, *Rhinolophus ferrumequinum*. II. Frequency mapping in the cochlea. J. Comp. Physiol. 106, 87-97 (1976).

Bruns, V.: Demonstration at Animal Sonar Systems, Jersey Conference, April 1-8 (1979).

Bullock, T. H.: The problem of recognition in an analyzer made of neurons. In: Sensory Communication. Rosenblith, W. A. (ed.). Cambridge: Technology Press (1961).

Bullock, T. H.: Processing of ampullary input in the brain: Comparisons of sensitivity and evoked responses among siluroids and elasmobranchs. J. Physiol., Paris, 75, 397-407 (1979).

Bullock, T. H.: Spikeless neurons: where do we go from here? In: Neurons Without Impulses. Bush, B., Roberts, A. (eds.). Cambridge: Cambridge Univ. Press (in press).

Bullock, T. H., Corwin, J. T.: Central auditory physiology of sharks. Proc. XXVII Int. Un. Physiol. Sci., Paris. 13, 107 (1977).

Bullock, T. H., Corwin, J. T.: Acoustic evoked activity in the brain of sharks. J. Comp. Physiol. 129, 223-234 (1979).

Bullock, T. H., Gurevich, V. S.: Soviet literature on the nervous system and psychobiology of Cetacea. Internat. Rev. Neurobiol. 21, 47-127 (1979).

Bullock, T. H., Ridgway, S. H.: Neurophysiological findings relevant to echolation in marine mammals. In: Animal Orientation and Navigation. Galler, S. R., Schmidt-Koenig, K., Jacobs, G. J., Belleville, R. E. (eds.). Washington, D. C.: NASA. U. S. Government Printing Office, 1971a, pp. 373-395.

Bullock, T. H., Ridgway, S. H.: Evoked potentials in the auditory system of alert porpoises (*Cetacea*) and sea lions (*Pinnipedia*) to their own and to artificial sounds. Proc. Int. Un. Physiol. Sci. 9, 89 (1971b).

Bullock, T. H., Ridgway, S. H.: Evoked potentials in the central auditory system of alert porpoises to their own and artificial sounds. J. Neurobiol. 3, 79-99 (1972).

Bullock, T. H., Domning, D., Best, R.: Hearing in a manatee (Sirenia: *Trichechus inunguis*). J. Mammalogy 61, 130-133 (1980).

Bullock, T. H., Ridgway, S. H., Suga, N.: Acoustically evoked potentials in midbrain auditory structures in sea lions (*Pinnipedia*). Z. vergl. Physiol. 74, 372-387 (1971).

Bullock, T. H., Grinnell, A. D., Ikezono, E., Kameda, K., Katsuki, Y., Nomoto, M., Sato, O., Suga, N., Yanagisawa, K.: Electrophysiological studies of central auditory mechanisms in cetaceans. Z. vergl. Physiol. 59, 117-156 (1968).

Corwin, J. T.: Morphology of the macula neglecta in sharks of the genus *Carcharhinus*. J. Morphol. 152, 341-361 (1977a).

Corwin, J. T.: Ongoing hair cell production, maturation, and degeneration in the shark ear. Neuroscience Abstracts. 3, 4 (1977b).

Corwin, J. T.: The relation of inner ear structure to feeding behavior in sharks and rays. Scanning Electron Microscopy/1978. 2, 1105-1112 (1978).

Creutzfeldt, O. D.: Afferent and intrinsic organization of laminated structures in the brain. Exp. Brain Res., Suppl. 1: XXIII + 1-579 (1976).

Denton, E. J., Gray, J. A. B., Blaxter, J. H. S.: The mechanics of the clupeid acoustico-lateralis system: Frequency response. J. Mar. Biol. Ass. U. K. 59, 27-47 (1979).

Dijkgraaf, S.: The functioning and significance of the lateral line organs. Biol. Rev. 38, 51-105 (1963).

Faber, D., Korn, H.: Neurobiology of the Mauthner Cell. New York: Raven Press, 1978.

Fay, R. R., Olsho, L.: Response patterns of neurons in the lagenar branch of the goldfish auditory nerve. J. Acoust. Soc. Am. 61, 559 (1977).

Fay, R. R., Olsho, L.: Discharge patterns of lagenar and saccular neurones of the goldfish eighth nerve: Displacement sensitivity and directional characteristics. Comp. Biochem. Physiol. 62A, 377-386 (1979).

Flock, Å.: Structure of the macula utriculi with special reference to directional interplay of sensory responses as revealed by morphological polarization. J. Cell Biol. 22, 413-431 (1964).

Gray, J. A. B., Denton, E. J.: The mechanics of the clupeid acoustico-lateralis system: Low frequency measurements. J. Mar. Biol. Ass. U. K. 59, 11-26 (1979).

Grinnell, A. D.: Comparative auditory neurophysiology of neotropical bats employing different echolocation signals. Z. vergl. Physiol. 68, 117-153 (1970).

Hartline, P. H.: Physiological basis for detection of sound and vibration in snakes. J. Exp. Biol. 54, 349-371 (1971a).

Hartline, P. H.: Midbrain responses of the auditory and somatic vibration systems in snakes. J. Exp. Biol., 54, 373-390 (1971b).

Hartline, P. H., Campbell, H. W.: Auditory and vibratory responses in the midbrains of snakes. Science. 163, 1221-1223 (1969).

Hillyard, S. A., Picton, T., Regan, G.: Sensation, perception and attention: analysis using ERPs. In: Event Related Brain Potentials, in Man. Callaway, E., Tueting, R., Koslow, S. (eds.). New York: Academic Press, 1978.

Ingle, D.: Behavioral correlates of central visual function in anurans. In: Frog Neurobiology. Llinas, R., Precht, W. (eds.). New York: Springer-Verlag, 1976.

Ingle, D.: Detection of stationary objects by frogs (*Rana pipiens*) after ablation of optic tectum. J. Comp. Physiol. Psych. 91, 1359-1364 (1977).

Knudsen, E. I.: Functional organization in the electroreceptive midbrain of the catfish. J. Neurophysiol. 41, 350-364 (1978).

Knudsen, E. I., Konishi, M.: Space and frequency are represented separately in auditory midbrain of the owl. J. Neurophysiol. 41, 870-884 (1978).

Maler, L.: The posterior lateral line lobe of certain gymnotoid fish: quantitative light microscopy. J. Comp. Neurol. 183, 323-364 (1979).

Maler, L., Finger, T., Karten, H. J.: Differential projections of ordinary lateral line receptors and electroreceptors in the gymnotid fish, *Apteronotus (Sternarchus) albifrons*. J. Comp. Neurol. 158, 363-382 (1974).

Merzenich, M. M., Kitzes, L., Aitkin, L.: Anatomical and physiological evidence for auditory specialization in the mountain beaver (*Aplodontia rufa*). Brain Res. 58, 331-344 (1973).

Moore, J. K.: The primate cochlear nuclei: Loss of lamination as a phylogenetic process (submitted).

Morrell, F.: Visual system's view of acoustic space. Nature. 238, 44-46 (1972).

Mudry, K. M., Constantine-Paton, M., Capranica, R. R.: Auditory sensitivity of the diencephalon of the leopard frog *Rana p. pipiens*. J. Comp. Physiol. 114, 1-13 (1977).

Myrberg, A. A., Ha, S. J., Walewski, S., Banbury, J. C.: Effectiveness of acoustic signals in attracting epipelagic sharks to an underwater sound source. Bull. Marine Sci. 22, 926-949 (1972).

Nelson, D. R., Johnson, R. H.: Some recent observations on acoustic attraction of Pacific reef sharks. In: Sound Reception in Fish. Schuijf, A., Hawkins, A. D. (eds.). Amsterdam: Elsevier, 1976.

O'Keefe, J., Nadel, L.: The Hippocampus as a Cognitive Map. Oxford: Clarendon Press, 1978.

Parker, G. H.: The sense of hearing in the dogfish. Science. 29, 428 (1909).

Piddington, R. W.: Central control of auditory input in the goldfish. I. Effect of shocks to the midbrain. J. Exp. Biol. 55, 569-584 (1971a).

Piddington, R. W.: Central control of auditory input in the goldfish. II. Evidence of action in the free-swimming animal. J. Exp. Biol. 55, 585-610 (1971b).

Piddington, R. W.: Auditory discrimination between compressions and rarefactions by goldfish. J. Exp. Biol. 56, 403-420 (1972).

Platt, C. J.: Central control of postural orientation in flatfish. I. Postural change dependence on central neural changes. J. Exp. Biol. 59, 523-541 (1973).

Platt, C. J., Bullock, T. H., Czeh, G., Kovaćević, N., Konjević, Dj., Gojković, M.: Comparison of electroreceptor, mechanoreceptor and optic evoked potentials in the brain of some rays and sharks. J. Comp. Physiol. 95, 323-355 (1974).

Pöppel, E.: Comment on "Visual system's view of acoustic space". Nature. 243, 231 (1973).

Russell, C. J., Bell, C. C.: Neuronal responses to electrosensory input in mormyrid valvula cerebelli. J. Neurophysiol. 41, 1495-1510 (1978).

Sand, O.: Microphonic potentials as a tool for auditory research in fish. In: Sound Reception in Fish. Schuijf, A., Hawkins, A. D. (eds.). Amsterdam: Elsevier, 1976.

Scheich, H., Langner, G., Koch, R.: Coding of narrow-band and wide-band vocalizations in the auditory midbrain nucleus (MLD) of the guinea fowl (*Numida meleagris*). J. Comp. Physiol. 117, 245-265 (1977).

Schuijf, A.: Directional hearing of cod (*Gadus morhua*) under approximate free field conditions. J. Comp. Physiol. 98, 307-332 (1975).

Schuijf, A.: The phase model of directional hearing in fish. In: Sound Reception in Fish. Schuijf, A., Hawkins, A. D. (ed.). Amsterdam: Elsevier, 1976.

Schuijf, A., Buwalda, R. J. A.: On the mechanism of directional hearing in cod (*Gadus morhua* L.). J. Comp. Physiol. 98, 333-344 (1975).

Schuijf, A., Hawkins, A. D.: Sound Reception in Fish. Amsterdam: Elsevier, 1976.

Shames, G. M., Gibson, J. M., Welker, W. I.: Fractured somatotopy in granule cell tactile areas of rat cerebellar hemispheres revealed by micromapping. Brain Behav. Evol. 15, 94-140 (1978).

Spoendlin, H.: Ultrastructure and peripheral innervation pattern of the receptor in relation to the first coding of the acoustic message. In: Hearing Mechanism in Vertebrates. De Reuck, A. V. S., Knight, J. (eds.). Boston: Little, Brown and Co., 1968.

Suga, N.: Ultrasonic production and its reception in some neotropical Tettigoniidae. J. Insect Physiol. 12, 1039-1050 (1966).

Suga, N.: Echo-detection by single neurons in the inferior colliculus of echo-locating bats. In: Animal Sonar Systems, Biology and Bionics. Busnel, R. G. (ed.). Jouy-en-Josas, France; 1967.

Suga, N.: Hearing in some arboreal edentates in terms of cochlear microphonics. J. Aud. Res. 7, 267-270 (1968a).

Suga, N.: Analysis of frequency-modulated and complex sounds by single auditory neurones of bats. J. Physiol. 198, 51-80 (1968b).

Suga, N.: Analysis of complex sounds by single auditory neurons in relation to human speech. Proc. Int. Un. Physiol. Sci. 7, 421 (1968c).

Suga, N.: Neural responses to sound in a Brazilian mole cricket. J. Aud. Res. 8, 129-134 (1968d).

Suga, N.: Classification of inferior collicular neurones of bats in terms of responses to pure tones, FM sounds and noise bursts. J. Physiol. 200, 555-574 (1969a).

Suga, N.: Echo-location and evoked potentials of bats after ablation of inferior colliculus. J. Physiol. 203, 707-728 (1969b).

Suga, N.: Echo-location of bats after ablation of auditory cortex. J. Physiol. 203, 729-740 (1969c).

Suga, N.: Amplitude spectrum representation in the doppler-shifted-CF processing area of the auditory cortex of the mustache bat. Science. 196, 64-67 (1977).

Suga, N., Campbell, H. W.: Frequency sensitivity of single auditory neurons in the gecko, *Caleonyx variegatus*. Science. 157, 88-90 (1967).

Suga, N., O'Neill, W. E.: Mechanisms of echolocation in bats. Comments on the neuroethology of the biosonar system of 'CF-FM' bats. Trends in Neurosc. 1, 35-38 (1978).

Suga, N., O'Neill, W. E., Manabe T.: Cortical neurons sensitive to combinations of information-bearing elements of biosonar signals in the mustache bat. Science. 200, 778-781 (1978).

Szabo, T.: Anatomy of the specialized lateral line organs of electroreception. In: Handbook of Sensory Physiology, III/3. Electroreceptors and Other Specialized Receptors in Lower Vertebrates. Fessard, A. (ed.). New York: Springer-Verlag, 1974.

van Bergeijk, W. A.: Directional and non-directional hearing in fish. In: Marine Bio-Acoustics. Tavolga, W. N. (ed.). Oxford: Pergamon Press, 1964.

van Bergeijk, W. A.: The evolution of vertebrate hearing. Contrib. Sens. Physiol. 2, 1-49 (1967).

Yanagisawa, K., Sato, O., Nomoto, M., Katsuki, Y., Ikezono, E., Grinnell, A. D., Bullock, T. H.: Auditory evoked potentials from brain stem in cetaceans. Fed. Proc. 25, 464 (1966).

Index

Acoustico lateralis system, *see* Octavolateralis
  system
Ambient noise   17, 18, 268, 273−275
Amphibian papilla   143, 144, 149, 156,
  157
  innervation   104
Amplitude modulated sounds
  fishes   26−30
  man   409−414
Association cortex   391, 392
Audiogram, *see* Thresholds
Auditory central nervous system
  amphibian   101−108
  anamniotes   79−118
  birds   299, 300, 314−317, 334
  comparative   79−118, 445, 446
  elasmobranchs   80, 440
  evolution of   80−82, 110−112,
    121−124, 425, 426
  ontogeny of   81, 82
  reptiles   230−233
Auditory cortex (*see also*
    Forebrain)   375−398, 445
Auditory cortex, history of study   376−380
Auditory (eighth) nerve
  amphibians   104−106, 144, 145
  birds   324−335, 337
  efferent fibers   252−256, 441
  elasmobranchs   442
  fishes   10, 11, 22, 32, 33, 59, 60, 61, 64,
    81−85, 440, 441
  mammals   140−143, 325−335
  reptiles   216, 217

Auditory space   444
  coding   314−318

Basilar membrane mechanics   142, 143, 212,
    213, 245−247
Basilar membrane tuning   229
Basilar papilla
  amphibians   104, 130, 131, 143
  birds   244−257, 294
  reptiles   169−204, 211, 212, 425
Behavioral auditory thresholds, *see*
    Thresholds
Best frequency, *see* Tuning curves
Binaural beat   415

Centrifigal nerve fibers   107, 252, 441
Cerebral asymmetries   392, 393, 433
Cerebellum   444, 445
CF (characteristic frequency), *see* Tuning
    curves
Central Nervous System, *see* Auditory CNS
Chorda tympani   132−135
Cilia, *see* Hair cells
Cochlea
  birds   241−260
  evolution of   425, 426
  mammals   425
  microphonics, *see* Microphonic potentials
  nonlinearities   139, 330, 331
  phylogeny   170−172
  reptiles   169−204, 425
Cochlear mechanics   142, 143, 212, 213,
    245

Cochlear nucleus (*see also* Octavolateralis
    area)
    birds   334–344
    comparative   442–444
    mammals   442–444
    reptiles   230–232
Cochleotopic cortical organization   384, 385,
    389
Cocktail party effect   23, 32
Columella (stapes)
    amphibians   125, 143
    birds   243
    reptiles   206–209, 422, 424
Combination tones   145–152
Cortical cytoarchitecture   377
Critical band
    birds   272–275
    fishes   23–26
    mammals   399, 402, 405
Critical ratio
    birds   268, 272–275
    comparative   275
    fishes   23–26
    mammals   274
Culmen   211, 212, 230
Cubic difference tone   139–142

Difference tone   139, 146, 149–152, 158
Directional hearing, (*see also* Minimum
    audible angle)
    bandwidth effects   307, 312, 365
    birds   289–322
    central nervous system   314–317
    cues (*see also* Interaural cues)   290, 291,
        301, 302, 358, 359
    discrimination   54, 55
    elasmobranchs   48, 49, 57, 58, 70, 71
    elevation   298, 307–309, 311, 313, 358
    evolution of   431, 432
    external ear asymmetry   242, 303–305,
        359–362
    fishes   23, 24, 31–34, 43–78, 440
    head effects   242, 291–293, 295, 296,
        297, 303, 359–362, 363, 417
    inner ear   56–63
    mammals   50, 357–374
    man   291, 368, 413–417
    monaural cues   291, 301, 302, 304, 305,
        444–445
    multi-dimensional   301, 302

precedence effect   293, 368
primates   365, 366, 433
theories of   64–71, 308–313, 359, 369,
    413, 416, 417
Displacement sensitivity   51, 52, 57–59
Distance perception   52
Duration discrimination   270

Eighth nerve, *see* Auditory nerve
Efferent auditory fibers, *see* Auditory nerve
Electroreception   22, 31, 94
Endolymphatic potential
    birds   248–250
    reptiles   213, 214
Evoked potentials   377, 378, 448
Evolution
    amphibian ear   134–136
    aquatic vertebrate ear   79–81
    auditory CNS   81, 82, 110–112,
        121–124
    bandwidth of hearing   428–432
    cortex   391, 392
    hearing in mammals   421–434
    hearing in reptiles   427–429
    mammals   422, 423
    reptile ear   197, 198
    reptiles   176, 177
External ear
    asymmetry   242, 302–304, 359–362
    birds   242, 291, 292, 296, 302–304, 313
    directionality   304, 305
    reptiles   206
    mammals   424
Extracolumella   206, 208, 209, 243

Far-field   43–78
Feature detectors   105, 350, 351, 446, 447
Forebrain   448
    amphibians   108
    anamiotes   90, 98, 108
    birds   350
    directional hearing   315–317
Forward masking, *see* Masking
Frequency analysis
    amphibians   143, 160, 161
    birds   245–247, 271–277
    fishes   21–26, 28
    man   399–407
Frequency discrimination
    birds   271, 272, 332, 333

comparative   271, 272
fishes   19–21
mammals   432–433
Frequency resolution of ear   245

Gap detection   270
Gap junction   213
Granule cells   442

Hair cell
   anatomy   9, 10, 179, 181, 188, 189, 193,
      194, 211–213, 250–252, 253, 254, 258
   birds   250–252, 253, 254, 258
   density   180, 181, 198, 441, 442
   fishes   9, 10
   physiology   213
   reptiles   179, 180, 181, 188, 189, 198,
      213
Hair cell innervation
   birds   252–257
   fishes   10, 441, 442
   reptiles   202
Hair cell orientation patterns
   fishes   8, 11–13, 31, 58, 59, 61
   reptiles   173, 174, 211
Hearing sensitivity, see Thresholds
Helicotrema   245, 266, 267
Hindbrain, see Cochlear nucleus
Homeothermy   423

Impedance matching   208–210
Infrasound   267
Inner ear anatomy
   amphibians   121–138
   birds   243, 244, 248, 266, 267
   fishes   4, 6, 7, 11, 12, 59, 60
   reptiles   169–204, 210–212
Inner ear innervation
   amphibians   104, 105
   birds   252–257
   fishes   10, 160
   reptiles   169–204, 210–212
Intensity discrimination
   birds   269, 270
   fishes   28
   primates   432
Interaural cues
   distance   310
   envelope   359, 415, 416

intensity   291, 295–298, 300, 310, 311,
   359–362
phase   242, 291, 292, 295, 310, 311, 358,
   359, 413
spectrum   291, 298–301, 309, 359
time   290, 291, 293, 294, 297, 309–311,
   314, 358, 359, 368, 369, 414
Intermodulation distortion   139–142, 144

Lagena
   birds   244, 266
   fishes   4, 6, 7, 11, 12, 59, 60
   reptile   171, 176
Lateral inhibition   400
Lateralization   366–370, 399, 414
Lateral line   44, 45, 52, 56, 81, 83, 87, 89,
   94, 99, 100, 440, 441
Limbus   176, 179, 181

Mach bands   400
Macula neglecta   57, 58, 86, 87, 441
Masking
   birds   268, 272–277
   fishes   17, 21–26, 28, 30–32, 47, 51, 52,
      55, 56
   man   400–407
   non-simultaneous   30, 400–404, 406, 412
   psychophysical tuning curves   24–26, 28,
      276–277, 403, 404
Masking level differences   23, 369, 416, 417
Medial geniculate   380
Microphonic potentials
   amphibians   142
   birds   250, 296, 297, 305
   comparative   428
   fishes   32, 59–63
   reptiles   214, 428, 429
Midbrain
   amphibians   106–108
   birds   299, 300, 314–317, 344–351
   directional hearing   314–315
   elasmobranchs   90
   fishes   86, 90, 98, 99, 443
   reptiles   231, 232
Middle ear
   amphibians   125–128, 132, 134, 135, 143
   birds   242, 243, 266, 291, 292, 297
   evolution of   122, 123, 422–424
   fishes, see Weberian ossicles and
      Swimbladder

Middle ear [cont.]
  impedance matching 208–210
  mammals 423, 424
  reptiles 206–210
  transfer function 207, 208, 243, 244
Minimum audible angle
  birds 290, 294, 305–307, 313
  fishes 31, 49–51
  mammals 362–366, 368

Near-field 17, 18, 43–78, 440
Neural sharpening 400
Noise exposure 275
Nucleus angularis 334–344
Nucleus magnocellularis 334–344
Nucleus mesencephali lateralis pars dorsalis
  (MLD), see Midbrain

Octavolateralis area 79–118
  amphibians 103, 109
  fishes 81, 83–99, 109, 112, 441
Octavus nerve, see Auditory nerve
Octavus nuclei 87, 89, 91, 97, 98
Otolith 4, 6, 7, 16, 17, 59
Otolithic membrane 4, 6

Phase analysis (behavioral) 47, 53–56, 63,
  64, 67–71, 440
Phase locking 21, 22, 27, 63, 64, 220, 222,
  228, 331–335, 412
Pinna 424
Pitch 404, 405
Place principle
  amphibians 121–138
  birds 275, 333
  fishes 19, 21, 22
  reptiles 212, 229
Pressure gradient 45
Psychophysical tuning curves, see Masking
Pulsation threshold 402–403

"Q", (Quality factor), see Tuning curves
Quadratic difference tone 140–142, 145,
  147, 152

Receptive fields 315, 444, 446, 447
Ripple noise 404–406

Sacculus 4–17, 21, 56–64, 179, 425
Sallet 189, 191, 193, 211, 212, 230
Second filter 161

Sound conduction pathway 16, 17, 55–63,
  65, 229, 230, 423
Song learning 278, 279
Sound localization, see Directional hearing
Spontaneous neural activity
  amphibians 155
  birds 324–327, 336, 337
  fishes 21
  mammals 325, 326
  reptiles 217
Species-specific vocalization 277–279,
  433, 446, 447
  directional hearing 365
  feature detector 446, 447
  neural responses to 323, 344–351
Stapes (see also Columella) 125, 127, 128,
  243, 423, 424
Summating potential
  birds 250
  reptiles 214
Swimbladder 4, 12–15, 18, 19, 46–49, 52,
  53, 58, 59, 63, 65

Tectorial membrane
  amphibians 143
  birds 246, 247, 251
  fishes, see Otolithic membrane
  reptile 171, 179, 181, 188, 192, 193, 211,
  212, 229
Telencephalon, see Forebrain
Temporal integration (and summation)
  birds 269, 271
  comparative 270, 271
  fishes 27, 28, 271
  man 399, 407–413
Temporal modulation transfer function
  fishes 27, 28
  man 409–412
Temporal resolution
  birds 269–271
  fishes 26–30
  mammals 270, 271
  man 407–412
Thresholds (auditory sensitivity)
  birds 262–267, 292, 293
  body size 424
  comparative 262, 263
  evolution 428
  fishes 16–19
  mammals 263, 428–432
  reptiles 216, 233, 428–430

Tonotopic organization
  amphibians   107
  birds   276, 335−339
  fishes   21
  mammals   375−398
  reptiles   218, 226, 227, 229
Torus semicircularis, *see* Midbrain
Traveling wave   222, 228
Tuning curve (neural)
  amphibians   142, 143, 145, 148, 150
  birds   323−354
  fishes   21, 22, 24−26
  inhibitory area   142, 153−159, 337, 343,
    345−351

  mammals   327, 328, 383−386
  reptiles   217−220, 227, 229−231
Two-tone suppression
  behavioral   401−404
  neural   139, 142, 143, 153−161, 220,
    222, 228, 326, 400
Tympanic membrane, *see* Middle ear

Vocal ontogeny   278−279

Weber fraction for intensity   270, 273−275
Weberian ossicles   4, 13, 15, 16, 59
Whole nerve action potential   214, 215, 223